Springer Proceedings in Advanced Robotics 24

The Springer Proceedings in Advanced Robotics (SPAR) publishes new developments and advances in the fields of robotics research, rapidly and informally but with a high quality.

The intent is to cover all the technical contents, applications, and multidisciplinary aspects of robotics, embedded in the fields of Mechanical Engineering, Computer Science, Electrical Engineering, Mechatronics, Control, and Life Sciences, as well as the methodologies behind them.

The publications within the "Springer Proceedings in Advanced Robotics" are primarily proceedings and post-proceedings of important conferences, symposia and congresses. They cover significant recent developments in the field, both of a foundational and applicable character. Also considered for publication are edited monographs, contributed volumes and lecture notes of exceptionally high quality and interest.

An important characteristic feature of the series is the short publication time and world-wide distribution. This permits a rapid and broad dissemination of research results.

Indexed by SCOPUS, SCIMAGO, WTI Frankfurt eG, zbMATH.

All books published in the series are submitted for consideration in Web of Science.

More information about this series at https://link.springer.com/bookseries/15556

Oscar Altuzarra · Andrés Kecskeméthy
Editors

Advances in Robot Kinematics 2022

Springer

Editors
Oscar Altuzarra
Faculty of Engineering in Bilbao
University of the Basque Country
Bilbao, Spain

Andrés Kecskeméthy
Inst. für Mechatronik und Systemdynamik
University of Duisburg-Essen
Duisburg, Germany

ISSN 2511-1256 ISSN 2511-1264 (electronic)
Springer Proceedings in Advanced Robotics
ISBN 978-3-031-08139-2 ISBN 978-3-031-08140-8 (eBook)
https://doi.org/10.1007/978-3-031-08140-8

This Springer imprint is published by the registered company Springer Nature Switzerland AG
The registered company address is: Gewerbestrasse 11, 6330 Cham, Switzerland

Preface

This book is the compilation of proceedings presented at the 18th International Symposium on Advances in Robot Kinematics held in Bilbao (Spain) in 2022. The series of international conferences in Advances in Robot Kinematics (ARK) was organized for the first time in 1988 in Ljubljana (Slovenia), and since then, they are regularly organized, every two years, in various European cities. Since 1992, ARK is held under the patronage of the International Federation for the Promotion of Mechanism and Machine Sciences (IFToMM). Springer published the first edited book in 1991, and since 1994, a new issue has been published every two years. Each edited book is linked to the corresponding symposium, and since 2016, the proceedings books are included in the Springer Proceedings in Advanced Robotics (SPAR) series. The current book is the 16th in the series of Springer. It contains 53 contributions as result of a peer-review process that selected the newest and most original achievements in the field. A large team of reviewers contributed with their critical and insightful recommendations and helped to improve the original submissions of participants. For this, the editors of this book convey their sincere thanks to the reviewers, also in the name of the authors.

Kinematics—as the branch of theoretical mechanics dealing with the geometry of motion—remains at the forefront of robotics, not only providing the fundamental basis of robot design, analysis, and control, but irradiating also in many other scientific domains, such as biomechanics, molecular biology, rehabilitation, space exploration, bio-inspired and soft robotics, and many more. Hereby, theoretical achievements of the past offer themselves to be revisited from time to time to tackle new applications, or completely new solutions need to be sought for still existing well-known unsolved or partially solved problems, as well for settings of growing complexity, as, for example, new mechanism designs involving, e.g., cable-driven systems, tensegrity, or compliant and continuum elements that have come into play into the robotics field. All this makes kinematics a fascinating and ebullient topic, which requires new, thriving younger minds that communicate effectively with senior academics at the boundary of handing-over experience. This spirit is the essence of this series that pursues to foster the meeting of an increasing number of kinematician friends.

In closing, we would like to express some words of gratitude. First and foremost, we are grateful to the authors that contributed their work to this project with a renewed enthusiasm and strong commitment. The reader will find senior contributors to this series along with young researchers in the area, an example of the vitality of this topic, and a combination that maintains the high standards of quality of this symposium. We are indebted to all members of the *Compmech* research group in Bilbao for their dedicated service to organize this symposium. Special thanks are also due to the University of the Basque Country UPV/EHU for its support providing the conference venue, and to the generous sponsorship from the Spanish Association of Mechanical Engineers (AEIM). Finally, we acknowledge the support from Springer staff to make this publication possible.

We edited this new ARK book with the hope that, again, it will reach many hands of scholars, researchers, and students who are attracted to the unique field of robot kinematics, and that it will be helpful to set ground and reach further achievements.

April 2022

Oscar Altuzarra
Andrés Kecskeméthy

Organization

Conference Chairman

O. Altuzarra	University of the Basque Country (UPV/EHU), Spain

Conference Co-chairman

A. Kecskemethy	University of Duisburg, Germany

Honorary Chairmen

B. Roth	Stanford University, USA
J. Lenarčič	J. Stefan Institute, Ljubljana, Slovenia

Scientific Committee

J. Angeles	McGill University, Montreal, Canada
B. Siciliano	University of Naples Federico II, Italy
M. Carricato	University of Bologna, Italy
M. Husty	University of Innsbruck, Austria
O. Khatib	Stanford University, USA
J. M. McCarthy	University of California at Irvine, USA
V. Parenti-Castelli	University of Bologna, Italy
F. Park	Seoul National University, Korea
A. Perez	Universidad Politécnica de Cataluña, Spain
J. Selig	London South Bank University, UK
M. M. Stanisic	University of Notre Dame, USA
F. Thomas	Institute of Industrial Robotics, Spain
P. Wenger	Ecole Centrale de Nantes, France

Organizing Committee

Alfonso Hernández	University of the Basque Country (UPV/EHU), Spain
Enrique Amezua	University of the Basque Country (UPV/EHU), Spain
Charles Pinto	University of the Basque Country (UPV/EHU), Spain
Victor Petuya	University of the Basque Country (UPV/EHU), Spain
Erik Macho	University of the Basque Country (UPV/EHU), Spain
Francisco J. Campa	University of the Basque Country (UPV/EHU), Spain
Javier Corral	University of the Basque Country (UPV/EHU), Spain
Mikel Díez	University of the Basque Country (UPV/EHU), Spain
Mónica Urizar	University of the Basque Country (UPV/EHU), Spain
Saioa Herrero	University of the Basque Country (UPV/EHU), Spain
Paul Diego	University of the Basque Country (UPV/EHU), Spain
Rubén Valenzuela	University of the Basque Country (UPV/EHU), Spain
Massimo Cichella	University of Bologna, Italy

Contents

Cable Driven Manipulators

Computational Kinematics

Theoretical Kinematics

Möbius Linkages

Manfred L. Husty(✉)

Unit of Geometry and Surveying, University Innsbruck,
Joanneum Research Robotics, Klagenfurt, Austria
manfred.husty@uibk.ac.at

Abstract. Möbius linkages are n-axes closed kinematic chains in which the axes form a discrete Möbius strip. It was conjectured by previous papers [1,2] that in some limit cases these linkages have less degrees of freedom than one would expect by their number of joints. In this paper the limit angles for 7R, 8R and 9R Möbius linkages are derived, the pathological degree of freedom is proven and some interesting kinematic features of these linkages are presented.

1 Introduction

Closed kinematic chains have been in the main interest of researchers since many years. Especially "pathological" (overconstrained) closed 4R, 5R and 6R linkages have been discussed a lot because of their unexpected degrees of freedom (dof). Whereas 4R and 5R linkages are classified (se e.g. [5]), a complete classification of overconstrained 6-R linkages is still open. In this paper a different class of paradoxical linkages is introduced and discussed, namely linkages which have less dofs as one would expect by their number of joints. These linkages will be called Möbius linkages. Möbius linkages are n-axes closed kinematic chains where the oriented axes form a discrete Möbius strip (Fig. 1).

These types of linkages have been the topic of two disputed papers [1,2]. In both papers the authors start from the well known kaleidocycle linkages [3], which are n-axes closed kinematic chains with special kinematic properties. These properties comprise a constant twist angle of the axes, zero offset and equal distances between the axes. In [1,2] the authors discuss closed n-axes kaleidocycles having the "Möbius" property, but do not use the term Möbius linkage although in [2] Möbius is mentioned in the title. In [3] kaleidocycles with the "Möbius" property are called "twisted" kaleidocycles. As opposed to the claim of the papers [1,2], namely having invented these types of linkages, linkages having the reversing property of the first and the last coordinates system were already mentioned by M. Engel on a lost web page. Parts of the web page were recovered by the French mathematician M. Starck and included in his web page [4]. M. Engel had provided a mathematical derivation of the motion of the kaleidocycles, including the twisted ones. But actually M. Engel had only shown that n-axes ($n > 7$) twisted or untwisted kleidocycles allow an everting motion similar to the well known everting motion of the Bricard linkage. Nothing is said

O. Altuzarra and A. Kecskémthy (Eds.): ARK 2022, SPAR 24, pp. 3–12, 2022.
https://doi.org/10.1007/978-3-031-08140-8_1

Fig. 1. Axes on a Möbius strip

that this 1 dof motion is just among the higher dofs of these linkages. Therefore, an unexperienced reader could get the impression that the everting motion is the only motion that is possible for the kaleidocycles. The papers [1,2] on the other hand come up with the new conjecture that there is a "limit" angle for every n-axes Möbius linkage which reduces the motion of the Möbius kaleidocycles to a one dof (everting) motion of the linkage. Unfortunately for all of the conjectured properties only numerical evidences are discussed in detail. In [1] equations for the limit angles in the 7R and 9R case are mentioned but without stringent mathematical derivation and clear statement of the assumptions. In this paper the 7R, 8R and 9R Möbius linkages are discussed. Especially, the equations of the limit angles are derived and some kinematic background of the unexpected reduction of the motion capability is explained. In Sect. 2 the mathematical methods are presented and in the following sections the three Möbius linkages are treated. In the last section the results are summarized and some open questions are addressed.

2 Mathematical Basics

For the mathematical description of Möbius linkages and the derivation of their kinematic properties the Study parametrization of the Eucclidean displacement group $SE(3)$ is used. A Euclidean displacement is a mapping

$$\gamma\colon \mathbb{R}^3 \to \mathbb{R}^3, \quad \mathbf{x} \mapsto \mathbf{A}\mathbf{x} + \mathbf{a} \tag{1}$$

where $\mathbf{A} \in \mathrm{SO}(3)$ is a proper orthogonal three by three matrix and $\mathbf{a} \in \mathbb{R}^3$ is a vector. The entries of $\mathbf{A}$ fulfill the orthogonality condition $\mathbf{A}^T \cdot \mathbf{A} = \mathbf{I}_3$, where $\mathbf{I}_3$ is the three by three identity matrix. As usual, this mapping is written as a linear transformation with a four by four transformation matrix. Note that in this paper the homogenizing coordinate is written as first coordinate.

$$\begin{bmatrix} 1 \\ \mathbf{x} \end{bmatrix} \mapsto \begin{bmatrix} 1 & \mathbf{o}^T \\ \mathbf{a} & \mathbf{A} \end{bmatrix} \cdot \begin{bmatrix} 1 \\ \mathbf{x} \end{bmatrix}. \tag{2}$$

Study's kinematic mapping $\varkappa$ maps an element α of SE(3) to a point $X \in \mathbb{P}^7$. If the homogeneous coordinate vector of X is $[x_0 : x_1 : x_2 : x_3 : y_0 : y_1 :$

$y_2 : y_3]^T$, the kinematic pre-image of X is the displacement α described by the transformation matrix

$$\mathbf{T} = \frac{1}{\Delta}\begin{bmatrix} \Delta & 0 & 0 & 0 \\ p & x_0^2 + x_1^2 - x_2^2 - x_3^2 & 2(x_1x_2 - x_0x_3) & 2(x_1x_3 + x_0x_2) \\ q & 2(x_1x_2 + x_0x_3) & x_0^2 - x_1^2 + x_2^2 - x_3^2 & 2(x_2x_3 - x_0x_1) \\ r & 2(x_1x_3 - x_0x_2) & 2(x_2x_3 + x_0x_1) & x_0^2 - x_1^2 - x_2^2 + x_3^2 \end{bmatrix} \tag{3}$$

where

$$\begin{aligned} p &= 2(-x_0y_1 + x_1y_0 - x_2y_3 + x_3y_2), \\ q &= 2(-x_0y_2 + x_1y_3 + x_2y_0 - x_3y_1), \\ r &= 2(-x_0y_3 - x_1y_2 + x_2y_1 + x_3y_0), \end{aligned} \tag{4}$$

and $\Delta = x_0^2 + x_1^2 + x_2^2 + x_3^2$. The lower three by three sub-matrix is a proper orthogonal matrix if and only if

$$x_0y_0 + x_1y_1 + x_2y_2 + x_3y_3 = 0 \tag{5}$$

and not all x_i are zero. When these conditions are fulfilled we call $[x_0 : \cdots : y_3]^T$ the *Study parameters* of the displacement α (for detailed information on this approach see [6]).

The Möbius linkage is described by a sequence of coordinate transformations which are closed by the reverting matrix $\mathbf{I}^*$

$$\mathbf{M}_1 \cdot \mathbf{G}_1 \cdot \mathbf{M}_2 \cdot \mathbf{G}_2 \cdot \cdots \cdot \mathbf{M}_n \cdot \mathbf{G}_n = \mathbf{I}^*, \tag{6}$$

where

$$\mathbf{M}_i = \begin{bmatrix} 1 & 0 & 0 & 0 \\ 0 & \cos(v_i) & -\sin(v_i) & 0 \\ 0 & \sin(v_i) & \cos(v_i) & 0 \\ 0 & 0 & 0 & 1 \end{bmatrix}, \quad \mathbf{G}_i = \begin{bmatrix} 1 & 0 & 0 & 0 \\ a_i & 1 & 0 & 0 \\ 0 & 0 & \cos(u_i) & -\sin(u_i) \\ d_i & 0 & \sin(u_i) & \cos(u_i) \end{bmatrix} \quad \text{and} \tag{7}$$

$$\mathbf{I}^* = \begin{bmatrix} 1 & 0 & 0 & 0 \\ 0 & 1 & 0 & 0 \\ 0 & 0 & -1 & 0 \\ 0 & 0 & 0 & -1 \end{bmatrix}. \tag{8}$$

a_i, d_i and u_i are the usual Denavit Hartenberg (DH) parameters. In case of a Möbius linkage the DH-parameter simplify a lot: $d_i = 0$ and all distances between the rotation axes a_i are equal and can be set without loss of generality to $a_i = 1$. All angles between the rotation axes are the same $u_i = u = const.$ For the kinematic analysis of the Möbius linkage the kinematic chain Eq. 6 is split into two parts:

$$\mathbf{M}_1 \cdot \mathbf{G}_1 \cdot \mathbf{M}_2 \cdot \mathbf{G}_2 \cdot \cdots \cdot \mathbf{M}_m\mathbf{G}_m = \mathbf{I}^* \cdot \mathbf{G}_n^{-1} \cdot \cdot \mathbf{M}_n^{-1} \ldots \mathbf{G}_{m+1}^{-1}\mathbf{M}_{m+1}^{-1}. \tag{9}$$

The left part is called the forward chain and the right part is called backward chain. The coordinate system which performs the coordinate transformation between the m and the $m+1$ axis is the coupler system of the chain. The goal for the following analysis is first of all to obtain the limit angles u_i which just allow an assembly of the chain and then to prove the decreased dof for the 8R and the 9R Möbius linkage in this limit case. The pursued algorithm will be discussed in detail for the 7R Möbius linkage which according to the well known Grübler-Kutzbach-Chebyshev formula anyway has one dof. But as it will turn out the limit case also for this linkage unveils some interesting kinematic and geometric properties which enlighten already the strange mobility of the other two linkages.

3 7R-Möbius Linkage

In case of the 7R linkage the forward chain contains 4 revolute axes and the backward chain has three axes. At first the forward kinematics of the left chain and the backward kinematics of the right chain are computed in Study coordinates. This yields two eight vectors $\mathbf{x}_f$ and $\mathbf{x}_b$ of the parametric representations of both chains

$$\mathbf{x}_f = \mathbf{x}(v_1, v_2, v_3, v_4) \qquad \mathbf{x}_b = \mathbf{x}(v_5, v_6, v_7). \tag{10}$$

Now we aim to obtain a single equation which contains the input angle v_1, the output angle v_7 and the design parameter α, which is the algebraic value of the angle u obtained by half tangent substitution. Therefore, in a next step implicit constraint equations for both chains are derived using LIA (Linear Implicitization Algorithm) (see [7] or [8]). In case of the forward chain the motion parameters v_2, v_3 and v_4 are eliminated and LIA comes up with two linear polynomials and two quadratic polynomials (P_1^2, P_2^2). The linear polynomials are displayed in the list F (Eq. 11), the quadratic polynomials are very big and are left out due to limitations of space.

$$\begin{aligned} F = [&\alpha^6 v_1^2 x_2 - 2\alpha^5 v_1^2 y_2 + \alpha^6 x_2 + \alpha^4 v_1^2 x_2 - 4\alpha^4 v_1^2 y_3 - 2\alpha^5 y_2 - 4\alpha^4 v_1 y_0 + 4 \\ &\alpha^3 v_1^2 y_2 + \alpha^4 x_2 + 8\alpha^3 v_1 y_1 - \alpha^2 v_1^2 x_2 + 4\alpha^2 v_1^2 y3 - 4\alpha^3 y_2 + 4\alpha^2 v_1 y_0 \\ &- 2\alpha v_1^2 y_2 - \alpha^2 x_2 - v_1^2 x_2 - 2\alpha y_2 - x_2, \alpha^6 v_1^2 x_1 - 2\alpha^5 v_1^2 y_1 + \alpha^6 x_1 + \alpha^4 v_1^2 x_1 \\ &- 2\alpha^5 y_1 - 4\alpha^4 v_1 y3 - 4\alpha^3 v_1^2 y_1 + \alpha^4 x_1 - 4\alpha^4 y_0 + 8\alpha^3 v_1 y_2 - \alpha^2 v_1^2 x_1 + 4\alpha^3 y_1 \\ &+ 4\alpha^2 v_1 y3 - 2\alpha v_1^2 y_1 - \alpha^2 x_1 + 4\alpha^2 y_0 - v_1^2 x_1 - 2\alpha y_1 - x_1, P_1^2, P_2^2] \end{aligned} \tag{11}$$

The backward chain contains three revolute joints and it is well known (see. e.g. [8]) that the implicit equations corresponding to a 3R chain, after eliminating two of the three motion parameters, are all linear in Study parameters. Here v_5 and v_6 are eliminated and one obtains the list B containing four linear polynomials

$$
\begin{aligned}
B = [&-2\alpha^3 v_7 y_0 - 2\alpha^3 y_3 - \alpha^2 v_7 x_0 + 3\alpha^2 x_3 + 8\alpha^2 y_2 - 2\alpha v_7 y_0 + 6\alpha y_3 - v_7 x_0 + 3x_3,\\
&- 2\alpha^3 v_7 y_3 + 2\alpha^3 y_0 - \alpha^2 v_7 x_3 - 3\alpha^2 x_0 + 8\alpha^2 y_1 - 2\alpha v_7 y_3 - 6\alpha y_0 - v_7 x_3 - 3x_0,\\
&- 2\alpha^3 v_7 y_0 + 4\alpha^3 x_2 - 2\alpha^3 y_3 - \alpha^2 v_7 x_0 + 3\alpha^2 x_3 - 2\alpha v_7 y_0 - 2\alpha y_3 - v_7 x_0 - x_3,\\
&- 2\alpha^3 v_7 y_3 + 4\alpha^3 x_1 + 2\alpha^3 y_0 - \alpha^2 v_7 x_3 - 3\alpha^2 x_0 - 2\alpha v_7 y_3 + 2\alpha y_0 - v_7 x_3 + x_0]
\end{aligned}
\tag{12}
$$

The eight equations in the lists B and F describe the 7R chain completely. Actually, there are now two possibilities how one could proceed. The equations contain the eight (homogeneous) Study parameters and two (v_1 and v_7) motion parameters. A first possibility would be to eliminate all Study parameters, then one could expect to obtain a single equation in v_1, v_7. This would be the input output (IO) equation of the 7R linkage. The other possibility would be to eliminate all but two Study parameters and v_1 and v_7 then one equation in the remaining two Study parameters would remain, which could be used to parameterize the coupler motion of the 7R chain. Note that in both algorithms the design parameter α would be still in the equations! It turns out that the first possibility is easier to handle and therefore the IO-equation will be computed. Before doing this the dehomogenizing equation $x_0 = 1$ is substituted into both lists. This can be done without loss of generality because $x_0 = 0$ comprises the 180° turns and one can assume that not the full motion of the chain will consist of 180° turns. The list B is then used to compute all y_i linearly. The results are substituted into the list F. After substitution it turns out that two of the four equations are still linear in x_1 and x_2. They are solved for those two unknowns and the results are substituted into the two remaining equations. Surprisingly, the two remaining equations are still quadratic in x_3, therefore it can be eliminated using the resultant method. The resultant R of the two equations, after eliminating x_3, factors into eight terms.

$$
R = \alpha^4 (v_1^2 + 1)^2 (\alpha^2 + 1)^2 R_1 R_2 R_3 R_4, \tag{13}
$$

where R_i are polynomials in α, v_1, v_7. It is evident that the vanishing of the first three terms in R has no meaning, therefore one has to discuss the remaining four terms. R_4 is the denominator of the linear solving of the Study parameters, therefore it should not vanish and R_1 and R_2 are complex. The only remaining equation is R_3:

$$
R_3 = 25\alpha^8 v_1^4 v_7^4 + \ldots [62\ \textit{terms}] \cdots + 250 v_7^2 + 225 = 0 \tag{14}
$$

In view of the fact that α is the algebraic value of the constant angle u between the axes of the linkage, this equation is the IO-equation of all possible Möbius 7R linkages. One can visualise this one parameter set of IO-equations as surface in the three unknowns v_1, v_7 and α (Fig. 2). Each horizontal section ($\alpha = \textit{const}$) corresponds to the IO equation of a Möbius 7R linkage. It turns out that each section consists of two curves of degree four and it is obvious that there must be a limit angle because there is a gap between the two parts of the surface. As one can see on the right side of Fig. 2 the two curves of each horizontal slice are symmetric with respect to the line $v_1 = -v_7$ and they have double points on this

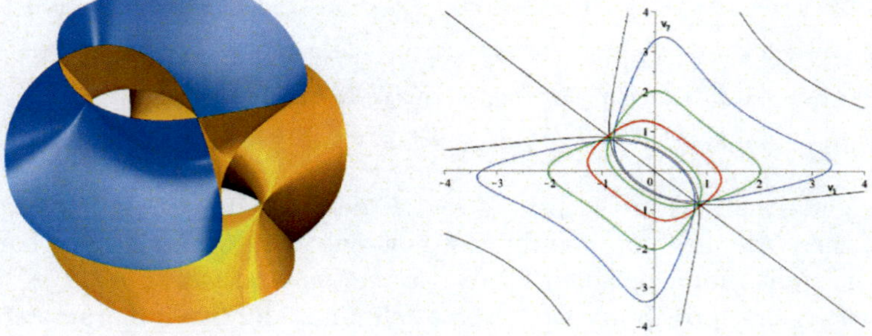

(a) (b)

Fig. 2. (a) IO-surface. (b) Different IO-curves of 7R Möbius linkages

line. The kinematic interpretation of this fact is, that the 7R Möbius linkages have two assembly modes and there are two poses where they can branch from one mode into the other. The limit case occurs when the horizontal section is tangent to the surface. As a tangent plane normally touches a surface in a point, one should expect that this tangency point corresponds to a rigid assembly of the linkage. This expectation could be motivated by the rigid limit assembly of a four bar when the sum of the coupler and both arms equals the base (see [1,2]).

To obtain the horizontal ($\alpha = const.$) tangent planes of R_3 one has to look for solutions of the set of equations

$$W = \left[R_3, \frac{\partial R_3}{\partial v_1}, \frac{\partial R_3}{\partial v_7} \right]. \tag{15}$$

W is a set of three equations in three unknowns and therefore one should expect a discrete set of solutions. Using resultants to eliminate v_7 from all three combinations of the elements of W yields a surprising result: all three resultants have a squared polynomial of degree six in α with only even powers in common:

$$(140\alpha^6 - 56\alpha^4 + 35\alpha^2 - 25)^2 = 0. \tag{16}$$

Equation 16 has only one pair of real roots:

$$\alpha = \pm \frac{\sqrt{210}\sqrt{\sqrt[3]{694477 + 183750\sqrt{15}}\left(\left(694477 + 183750\sqrt{15}\right)^{2/3} + 28\sqrt[3]{694477 + 183750\sqrt{15}} - 2891\right)}}{210\sqrt[3]{694477 + 183750\sqrt{15}}} \tag{17}$$

The geometric interpretation of this fact is that the lowest horizontal tangent plane touches the surface in a curve instead of a point! Figure 3 shows on the left side R_3, the horizontal tangent plane and the tangency curve and on the right side the curve itself.

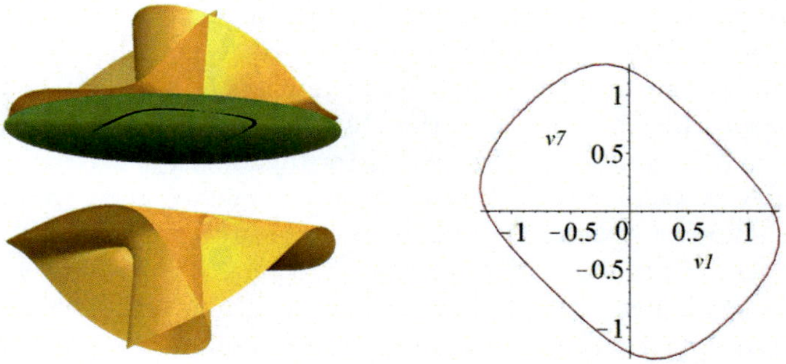

Fig. 3. R_3 with horizontal tangent plane and tangency curve

The equation of the tangency curve, which is the IO-equation of the limit Möbius 7R, is obtained by substituting one of the real roots of Eq. 17 into Eq. 14:

$$\begin{aligned}
&1496000 v_1 v_7 \left(694477+183750\sqrt{15}\right)^{2/3}\sqrt{15}+2619939840 v_1{}^2 v_7{}^2+18172000\, v_1 v_7 \sqrt[3]{694477+183750\sqrt{15}}\sqrt{15}\\
&-5368176\, v_1 v_7 \left(694477+183750\sqrt{15}\right)^{2/3}-136726128\, v_1 v_7 \sqrt[3]{694477+183750\sqrt{15}}\\
&-43500 \left(694477+183750\sqrt{15}\right)^{2/3}\sqrt{15}+2619939840\, v_1{}^2+4445710416\, v_1 v_7+2619939840\, v_7{}^2\\
&-23850750 \sqrt[3]{694477+183750\sqrt{15}}\sqrt{15}-210845 \left(694477+183750\sqrt{15}\right)^{2/3}\\
&+92121715 \sqrt[3]{694477+183750\sqrt{15}}+889517335=0
\end{aligned} \tag{18}$$

The curve is of degree four and is elliptic. Therefore it cannot be rationally parameterized. But it is biquadratic and can be used to parameterize the coupler motion of the Möbius 7R in closed form with square roots. The limit linkage has several interesting kinematic and geometric properties which cannot be explained in detail due to lack of space.

4 8R Möbius Linkage

In case of the 8R Möbius linkage the algorithm will be roughly the same as in the 7R case but of course it will be more complicated. The 8R chain is split into a 4R forward and a 4R backward chain. As a general 8R chain should have 2 dofs, we will try to obtain a single equation in three of the motion parameters v_i and the design parameter α. This single equation then can be interpreted as a 3-dim hypersurface in a 4-dim ambient space. As in the 7R case three motion parameters v_2, v_3, v_4 are eliminated using LIA from the forward chain. The result is the same as in the 7R case and the set F consists of two linear and two quadratic polynomials (Eq. 11). In the backward chain only two motion parameters v_5, v_7 are eliminated and LIA comes up with four linear equations. These equations are solved for $y_i, i = 0 \dots 3$ and the obtained resultes are substituted into the set F. The resulting polynomials in F contain the Study parameters x_0, x_1, x_2, x_3 and the motion parameters v_1, v_7, v_8 and are denoted K_1, K_2, K_3, K_4. Again, without loss of generality x_0 is set to 1. It turns out that K_1 and K_2 are still linear in the remaining Study parameters x_1, x_2 and are therefore solved for these two Study parameters. The results are substituted into K_3 and K_4. After substitution one observes that the remaining polynomials are still quadratic in x_3. This is of course essentially the same as in the 7R case! The resultant, eliminating x_3, consists of 8 factors from which only one factor (denoted F_3) is either non trivial or non complex. This factor has 575 terms and is therefore to big to be displayed here. F_3 is the IO-equation for the one parameter set of 2 dof Möbius 8R chains. Four IO-surfaces out of the one parameter set are displayed in Fig. 4. As in the 7R case and conjectured by [1,2] there should be a limit angle where the chain can just be assembled and which should yield a single dof motion. Therefore, one has to look for the lowest horizontal tangent 3-space. The horizontal tangent spaces are obtained from the solutions of the set of equations V:

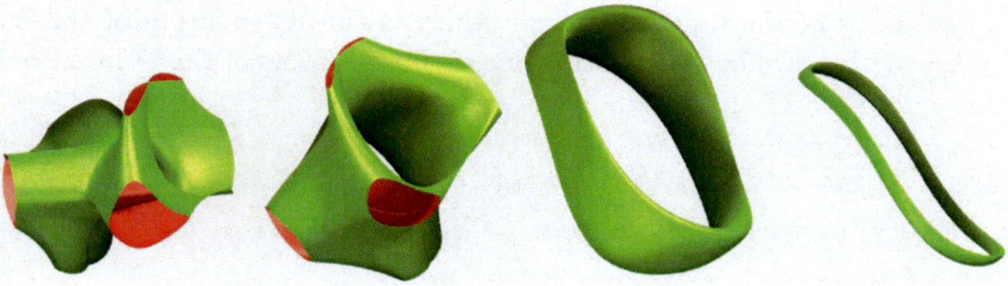

Fig. 4. Horizontal slices of F_3 at angles $\alpha = 0.8, 0.5, 0.37, 0.361$

$$V = \left[F_3, \frac{\partial F_3}{\partial v_1}, \frac{\partial F_3}{\partial v_7}, \frac{\partial F_3}{\partial v_8}\right]. \tag{19}$$

It turns out, after some lengthy and tricky computations, that the "lowest" tangent space is obtained from the equation

$$\begin{aligned} &6125\alpha^{24} - 21700\alpha^{22} + 59610\alpha^{20} - 191156\alpha^{18} + 241731\alpha^{16} - 314504\alpha^{14} \\ &+427500\alpha^{12} - 314504\alpha^{10} + 241731\alpha^{8} - 191156\alpha^{6} + 59610\alpha^{4} - 21700\alpha^{2} + 6125 = 0. \end{aligned} \tag{20}$$

This equation has a remarkable structure. First of all it has only even powers and secondly it is palindromic, i.e. the coefficients are completely symmetric! Palindromic monomials are special because up to degree 9 there exists a solution formula. Equation 4 can only be reduced to degree 12 and an additional check with Galois theory shows that this equation can only be solved numerically. Two real solutions exist which essentially yield mirrored solutions.

$$\alpha = .60048499935171120259, \qquad \alpha* = 1.6653205343674007619 \tag{21}$$

which correspond to the angles

$$\varphi = 61.968369637850872817°, \qquad \varphi* = 118.03163036214912718° \tag{22}$$

Figure 5 shows the limit curve in the "lowest" tangent space, a CAD model and a 3D printed version of the limit Möbius linkage, which allows also practically only a 1 dof motion.

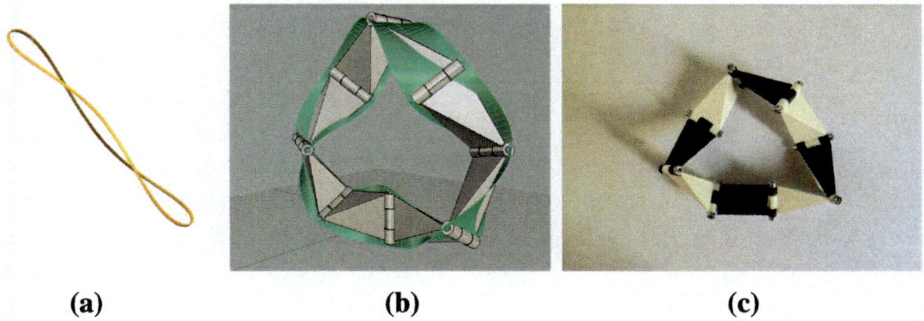

Fig. 5. (a) Limit curve, (b) CAD model, (c) 3d-print

5 9R Möbius Linkage

Because of symmetry (every third joint angle is the same!) this case is easier to handle than the 8R case. After elimination of five joint parameters and the Study parameters a single equation in four joint parameters and the design parameter α is obtained $P(\alpha, v_i) = 0, i = 1\ldots, 4$. Using the same algorithm as before, namely differentiating with respect to the remaining joint parameters 4 equations P_1, P_2, P_3, P_4 are obtained. All of them are big, but using some sophisticated cuts of the surfaces one can obtain the equation for the lowest tangent space:

$$
\begin{aligned}
&12\alpha^6 + 3\alpha^2 - 1 = 0,\\
&\alpha = \frac{\sqrt{6}\sqrt{\sqrt[3]{9+6\sqrt{3}}}\left(\left(9+6\sqrt{3}\right)^{2/3} - 3\right)}{6\sqrt[3]{9+6\sqrt{3}}} = .51153470499520369571,
\end{aligned}
\tag{23}
$$

The numerical values reported in [1] were .5115056365 and in [2] .5115373414. Having obtained the equation for the limit angle it is also possible to deduce an IO-equation for the linkage in the joint parameters v_4 and v_5. From this equation all the other joint parameters can be computed:

$$
\begin{aligned}
&4\,v_4\,v_5\,\left(9+6\sqrt{3}\right)^{2/3}\sqrt{3} + 2\,v_4\,v_5\,\sqrt[3]{9+6\sqrt{3}}\sqrt{3} - 5\,v_4\,v_5\,\left(9+6\sqrt{3}\right)^{2/3} - 4\,{v_4}^2{v_5}^2\\
&+ 4\sqrt{3}\left(9+6\sqrt{3}\right)^{2/3} - 9\,v_4\,v_5\,\sqrt[3]{9+6\sqrt{3}} + 2\,\sqrt[3]{9+6\sqrt{3}}\sqrt{3} - 5\,\left(9+6\sqrt{3}\right)^{2/3} - 4\,{v_4}^2\\
&+ 7\,v_4\,v_5 - 4\,{v_5}^2 - 9\,\sqrt[3]{9+6\sqrt{3}} + 3 = 0,
\end{aligned}
\tag{24}
$$

The corresponding curve to this equation is visualized in Fig. 6. Equation 24 could be used to parameterize the motion in closed form because it is biquadratic.

6 Conclusion

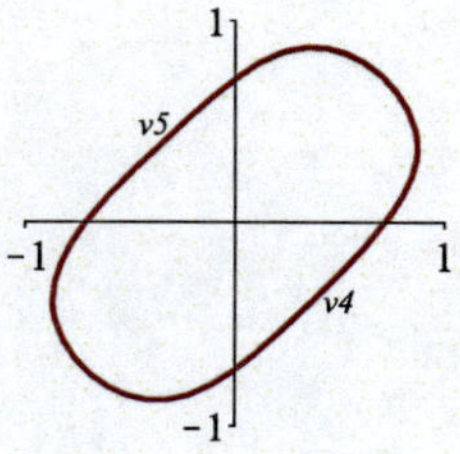

Fig. 6. IO-curve of the limit 9R Möbius linkage

In this paper Möbius linkages were introduced and analyzed. Möbius linkages are closed kinematic chains where the closing is done with a 180° rotation of the last coordinate system. It was proven that each of the three discussed linkages possesses a limit twist angle which causes a special motion behaviour of the linkage. First kinematic properties of these linkages were derived. But the discussion showed that many open questions remain unanswered: is there a chance to prove the conjecture that for all n-axes Möbius linkages exists a limit angle which causes a highly reduced single dof? It is conjectured that a better coordinate system might simplify the computations a lot and that the use of line geometry might enlighten the kinematic properties better than the methods presented. Nevertheless it is hoped that the presented results might open a new research topic within linkage kinematics.

References

1. Schönke, J., Fried, E.: Single degree of freedom everting ring linkages with nonorientable topology. PNAS **116**(1), 90–95 (2019)
2. Kaji, S.: A Closed Linkage Mechanism Having the Shape of a Discrete Möbius Strip, Semitsu Conference Paper, 6p (unpublished)
3. Schattschneider, D., Walker, W.: M .C. Escher Kaleidocycles. Ballentine Books, New York (1977)
4. http://www.polyhedra-world.nc/index3_.htm. Accessed 07 Dec 2021
5. Karger, A.: Classification of 5R closed kinematic chains with self mobility. Mech. Mach. Theory **33**(1), 213–222 (1998)
6. Husty, M.L., Schröcker H.-P.: Kinematics and algebraic geometry. In: McCarthy, M. (ed.) 21st Century Kinematics, pp. 85–123. Springer, London (2012). https://doi.org/10.1007/978-1-4471-4510-3_4
7. Walter, D.R., Husty, M.L.: On implicitization of kinematic constraint equations. Mach. Des. Res. **26**, 218–226 (2010)
8. Husty, M.L., Walter, D.R.: Mechanism constraints and singularities-the algebraic formulation. In: Müller, A., Zlatanov, D. (eds.), Singular Configurations of Mechanisms and Manipulators, CISM International Centre for Mechanical Sciences vol. 589, pp. 102–180 (2019)

On Origami-Like Quasi-mechanisms with an Antiprismatic Skeleton

G. Nawratil(✉)

Institute of Discrete Mathematics and Geometry & Center for Geometry and Computational Design, TU Wien, Vienna, Austria
nawratil@geometrie.tuwien.ac.at

Abstract. We study snapping and shaky polyhedra which consist of antiprismatic skeletons covered by polyhedral belts composed of triangular faces only. In detail, we generalize Wunderlich's trisymmetric sandglass polyhedron in analogy to the generalization of the Jessen orthogonal icosahedron to Milka's extreme birosette structures, with the additional feature that the belt is developable into the plane as the Kresling pattern. Within the resulting 2-dimensional family of origami-like sandglasses we study the 1-parametric sets of quasi-mechanisms which are either *shaky* or have an *extremal snap*, i.e. one realization is on the boundary of self-intersection. Moreover, we evaluate the capability of these snapping/shaky quasi-mechanisms to flex on base of the *snappability* index and the novel *shakeability* index, respectively.

Keywords: Model flexors · Quasi-mechanisms · Snapping · Shakiness · Snappability · Shakeability · Origami

1 Introduction

We consider polyhedral structures, where the bottom face α and the parallel top face β are regular convex n-gons $A_0, \ldots, A_{n-1}$ (with center A) and $B_0, \ldots, B_{n-1}$ (with center B), respectively, for $n \geq 3$ and a side length of 1. Moreover, these two faces are twisted against each other by a rotation about the orthogonal axis AB (cf. Fig. 1a). Then this antiprismatic skeleton is covered by a polyhedral belt composed of triangular faces only. Moreover, all resulting polyhedra discussed in the paper do not possess continuous isometric deformations; i.e. they are rigid from the mathematical point of view. But the physical models of theses polyhedra can flex due to non-destructive elastic deformations of material (or backlash/tolerances in hinges), i.e. small changes in the intrinsic metric (given by the edge lengths) have significant effects on the spatial shape. Therefore these structures are also known as *model flexors* or *quasi-mechanisms*.

If the inner geometry (intrinsic metric together with the combinatorial structure) of the polyhedron is fixed, then the embedding of the polyhedron into the Euclidean 3-space E^3 is in general not uniquely determined; i.e. the polyhedron

O. Altuzarra and A. Kecskémthy (Eds.): ARK 2022, SPAR 24, pp. 13–21, 2022.
https://doi.org/10.1007/978-3-031-08140-8_2

Fig. 1. The antiprismatic skeleton for $n = 3$ (a), its belt given by the Kresling pattern (b), the resulting octahedral structure (c) and its configuration after the snap (d).

has different incongruent realizations[1]. Based on these basic notations we can distinguish the following two kinds of quasi-mechanisms:

(i) Snapping quasi-mechanism: The shape variation results from the snap (caused by deformation) of a given realization into another one. An example for this is the Siamese dipyramid, which even snaps between three realizations [1,2].
(ii) Shaky quasi-mechanism: Now the deformed states originate from a given *shaky* (also known as *singular* or *infinitesimal flexible*) realization. The best known example for this kind of model flexion is the Jessen orthogonal icosahedron [1,3,4] (cf. Fig. 2a). Note that shakiness can be seen as the limit of snapping when the related realizations converge to coincidence [2,5].

There are also examples, like the four-horn [2,6], which use both functional principles (i & ii) at the same time.

Review. We proceed with a review on model flexors with an antiprismatic skeleton:

(1) Let us start with the Kresling pattern [7], a flat strip of congruent triangles (cf. Fig. 1b) which can be folded up and closed to a belt for the skeleton. The resulting antiprismatic structure has a bi-stable behavior, which was already known to Wunderlich (cf. [8] and for $n = 3$ the more detailed study [9]). During the snap, the relative motion of α and β is composed of a rotation about AB and a change in height (see Fig. 1c,b). A detailed literature review on these structures (as well as a study of related ones) is given by the author in [10].
(2) Next, we consider so-called extreme birosettes [11], which can be seen as generalizations of the Jessen orthogonal icosahedron (case $n = 3$). In this case the belt consists of $2n$ equilateral triangles with side length of 1 and $2n$ *petals*, which are skew rhombi of side length 1 broken along one of its diagonals[2] of length p. The resulting structure is symmetric with respect to (a) rotations of $\frac{2\pi}{n}$ about AB and (b) a reflection at the birosette center

[1] In this paper the word realization always refers to an undeformed embedding into E^3.

[2] Note that the birosette degenerates into an antiprism if p converges towards zero.

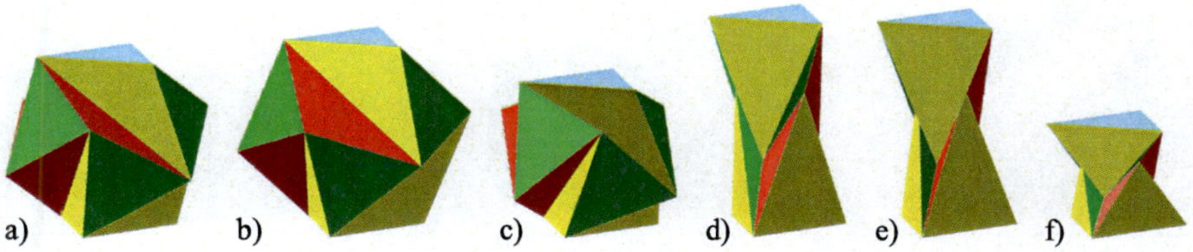

Fig. 2. (a) Jessen orthogonal icosahedron [3], (b,c) two snapping realizations of a non-extremal birosette for $n = 3$ with $p = 1.6 < p_+ = 4/\sqrt{6}$, (d,e) the two realizations of the snapping sandglass icosahedron of Wunderlich [13, Fig. 6], (f) shaky sandglass icosahedron [13, Fig. 4].

(midpoint of AB) which has to be complemented by a rotation of $\frac{\pi}{n}$ about AB for even values of n (see Figs. 2a and 3). The relative instantaneous motion of α and β is translative in direction of AB. Moreover, the maximal value p_+ of p such that the birosette can be assembled under consideration of the mentioned symmetry, yields the *extreme* birosette. Note that birosettes for $n = 3$ are also known as Douady shaddocks [4,12].

(3) Finally, we recall Wunderlich's trisymmetric sandglass polyhedron [13], which is a snapping icosahedron possessing the same symmetry and combinatorial structure as the birosette for $n = 3$. In this case the belt consists of six congruent isosceles triangles, whose bases of length 1 are hinged to the equilateral triangles α and β, respectively, and the gaps between them are filled by further 12 congruent isosceles triangles (see Fig. 2d,e). During the snap, the relative motion of α and β is a translation along AB.

Remark 1. By considering the limit mentioned in (ii) one can also produce shaky quasi-mechanisms from the snapping structures given in (1) studied in [8,9] and (3) illustrated in Fig. 2f. Also snapping birosettes can be generated by choosing $p < p_+$ as displayed in Fig. 2b,c. ◇

Outline. Based on this review we generalize Wunderlich's sandglass polyhedron to arbitrary n in analogy to the birosette construction, with the additional feature that the belt is developable into the plane as the Kresling pattern (cf. Sect. 2), which allows an efficient production. Within the resulting 2-dimensional family of origami-like sandglasses (for arbitrary $n \geq 3$) we study the 1-parametric sets of quasi-mechanisms which are either *shaky* (cf. Sect. 4) or have an *extremal snap* (cf. Sect. 3), i.e. one realization is on the boundary of self-intersection. Moreover, in these two sections we also evaluate the capability of these snapping/shaky quasi-mechanisms to flex on base of the *snappability* index (cf. [2,14]) and the novel *shakeability* index, respectively. The paper is concluded in Sect. 5.

Note that due to the page limit, we have to refer to the corresponding arXiv version [15] for some details or additional information on the study at hand.

Fig. 3. Extreme birosettes for $n = 4, 5, 6$ (a,b,c) with yellow/red pedals, where the unit-cells are framed in black.

2 Preliminary Considerations

Assume that we have given a polyhedron with the combinatorial structure of a birosette for arbitrary $n \geq 3$. In order to clarify its degree of freedom (dof), we use the notation of Fig. 3a, where the vertices of a unit-cell[3] of the belt are labeled. The relative position of β with respect to α has 6 dofs and each vertex C_i and D_i has further 3 dofs, which yields in total $6+6n$ dofs. Moreover, we have $8n$ distance constraints implied by the edges of the structure, with exception of those belonging to the skeleton. Therefore the dof of the structure is computed by $F := 6(n+1) - 8n$ yielding for $n \geq 3$ a value of $F \leq 0$, which shows that these structures are in general rigid[4].

Remark 2. It is an open question whether flexible polyhedrons exist having the same combinatorial structure as a birosette with $n \geq 3$. ◇

The assumptions on the edge lengths (inside a unit-cell) of a birosette can be weaken as follows without destroying its symmetry:

$$L_1 := \overline{B_0D_0} = \overline{A_1C_1},\ L_2 := \overline{B_0C_1} = \overline{A_0D_0},\ L_3 := \overline{D_0C_1} = \overline{C_1D_1},\ L_4 := \overline{B_0D_1} = \overline{A_0C_1} \tag{1}$$

beside the unit length of $\overline{A_0A_1}$ and $\overline{B_0B_1}$. Moreover, in the remainder of the paper we assume that these skeleton edges are undeformable under the model flexibility in contrast to the other edge lengths $L_1, \ldots, L_4 > 0$, which can vary[5].

One can compute (cf. [15, App. 5.1] for details) the condition for the developability of the obtained generalized birosette belt in terms of $L_1, \ldots, L_4$. Under consideration of the sandglass condition $L_1 = L_4$ this origami condition reads as:

$$Q_3 = Q_1 + Q_2 - \sqrt{Q_2(4Q_1 - 1)} \quad \text{with} \quad Q_i := L_i^2 \tag{2}$$

Remark 3. Wunderlich further assumed in the study of the trisymmetric sandglass [13] that $L_2 = L_3$ holds, which simplifies the related equations considerably and allows a very compact treatment. We do not make this assumption here. ◇

3 A repetitive rotation of the unit-cell around the axis AB about the angle $\frac{\pi}{n}$ generates the polyhedral belt and increases the indices of the vertices by one (mod n).

4 Note that taking the symmetry of the birosette into account yields $F = 0$ for all $n \geq 3$.

5 Note that the study [11] on birosettes is more restrictive as only the length p of the diagonal is allowed to change.

3 Snapping Quasi-mechanisms

As the generalized Wunderlich sandglasses have the same symmetry as the birosettes, the essential vertices of the unit-cell can be coordinatized similarly to [11] by

$$A_0 = (R, 0, H), \quad B_0 = (Rc, Rs, -H), \quad D_0 = (r, 0, -h), \quad C_1 = (rc, rs, h),$$

using the abbreviations $c := \cos\frac{\pi}{n}$ and $s := \sin\frac{\pi}{n}$, respectively. Moreover, R equals $\frac{1}{2s}$ due to the unit-length of the skeleton edges. With this parametrization we only remain with the following three equations:

$$q_1 : Q_1 - \overline{B_0D_0}^2 = 0, \quad q_2 : Q_2 - \overline{B_0C_1}^2 = 0, \quad q_3 : Q_3 - \overline{D_0C_1}^2 = 0, \tag{3}$$

as the fourth one, namely $Q_4 - \overline{B_0D_1}^2 = 0$, is identical with q_1. Thus for a given set of squard edge lengths Q_1, Q_2, Q_3 the corresponding realizations can be computed by solving q_1, q_2, q_3 for H, h, r, which is straight forward.

As already mentioned, we want to restrict to structures possessing an *extremal snap*, i.e. one realization is on the boundary of self-intersection. Such configurations are illustrated in Fig. 5 (right) and have the advantage that the self-blocking of the faces increases the structure's load carrying capacity. Moreover, the self-covered areas can be provided with holes. In this way the configuration of Fig. 5 (right) is still tight, thus it is called *closed*, in contrast to the *open* one of Fig. 5 (left). Therefore, such structures can for example be used as *pressure relief valves*.

At the closed state the points A_0, B_0, D_0, C_1 are coplanar. This is the case for $(2rs - 1)(2Hrs + h) = 0$ where the factors imply a dihedral angle of π and 0, respectively, along the edge D_0C_1. Therefore we set $r = -\frac{h}{2Hs}$ and plug this expression into q_1, q_2, q_3. By eliminating the unknowns H, r, s from these three equations together with $c^2 + s^2 - 1 = 0$ by means of resultants, we end up with

$$\begin{aligned} &4cQ_2Q_1 - 2cQ_2^2 - 2Q_1^2 - 28Q_2Q_1 - 2Q_2^2 + Q_1 + 5Q_2 - 2cQ_1^2 \\ &+ W^{3/2}\sqrt{Q_2} + 8Q_2^{3/2}\sqrt{W} + 4Q_1\sqrt{Q_2}\sqrt{W} = 0 \end{aligned} \tag{4}$$

by taking into account Eq. (2) and $W := 4Q_1 - 1$. This equation is linear[6] in c and plotted as gray surface in Fig. 4a. Moreover, it can be solved explicitly for Q_2 (e.g. with MAPLE), which yields four branches. We let Q_1 run within the interval $]0.25; 5]$ in steps of 0.01 for $n = 3, \ldots, 6$ and check if this value implies a pair of snapping sandglass realizations. To do so, we compute the realizations and check in a first step if they are free of self-intersections, and in a second step if they can snap into each other by the methods presented by the author in [2,14]. It turns out that only the branch of Eq. (4), which yields the lowest value for Q_2, carries solutions of our problem. Animations of these families of snapping sandglasses (for $n = 3, \ldots, 6$) can be downloaded from the author's homepage[7].

[6] For $Q_1 = Q_2$ it is independent of c but then it can only vanish for $Q_1 = Q_2 = 0$, a contradiction.

[7] https://www.dmg.tuwien.ac.at/nawratil/publications.html.

Snappability. According to [2,14] the snap between two sandglass realizations has to pass a shaky configuration (with squared edge lengths S_1, S_2, S_3) at the maximum state of deformation (see Fig. 5 (center)), which is used for the evaluation of the snapping capability in terms of the *snappability*. By considering the belt as a joint-bar structure[8] this index σ, which is based on the total elastic strain energy density of the framework, can be computed as follows (cf. [2,14]):

$$\sigma := \left(4n\frac{(Q_1-S_1)^2}{8L_1^3} + 2n\frac{(Q_2-S_2)^2}{8L_2^3} + 2n\frac{(Q_3-S_3)^2}{8L_3^3}\right)/(4nL_1 + 2nL_2 + 2nL_3) \quad (5)$$

In Fig. 4c the snappability of the computed snapping sandglass realizations, which are free of self-intersections, are displayed. The structures causing the maximal snappability are visualized in Fig. 5. Additional graphs concerning the change in volume, height or waist radius during the snap are given in [15, App. 5.4] as well as the crease pattern of the structure displayed in Fig. 5a.

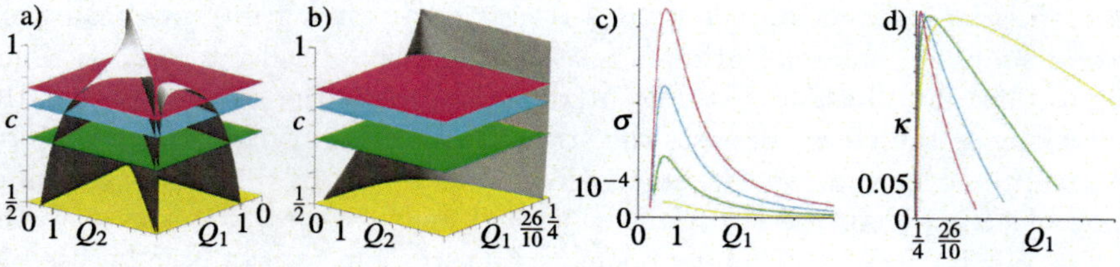

Fig. 4. Visualization of (a) Eq. (4) and (b) the shakiness condition, which only yields feasible values for Q_2 within a narrow domain of Q_1, where $Q_1 = \frac{1}{4}$ is an asymptote for the Q_2 values. The horizontal c-planes for $n = 3, \ldots, 6$ are colored in yellow, green, cyan and magenta, respectively. The same color-coding is used for the graphs of the snappability (c) and the shakeability (d).

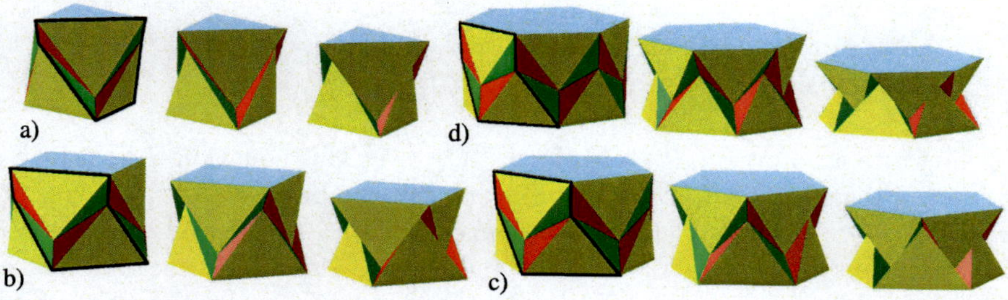

Fig. 5. Illustration of the snap from the open state (left) over the passed shaky configuration (center) to the closed state (right) for the structures causing the maximal snappability for $n = 3$ (a), $n = 4$ (b), $n = 5$ (c) and $n = 6$ (d). Moreover, the unit-cells of the open state are framed in black.

[8] It can also be interpreted as a panel-hinge framework but the formula is more complicated [2,14].

4 Shaky Quasi-mechanisms

For the computation of shaky origami-like sandglasses we proceed as follows. We eliminate from q_1, q_2, q_3 of Eq. (3) the unknowns H, h by means of resultant, such that we end up with a polynomial in r. In order that this polynomial has a solution of higher order, its discriminant with respect to r has to vanish. From the resulting expression we eliminate s by applying the resultant method with respect to $c^2 + s^2 - 1 = 0$. Under consideration of Eq. (2) we end up with the shakiness condition of the form $w_4c^4 + w_3c^3 + w_2c^2 + w_1c + w_0 = 0$ whose coefficients are given in [15, App. 5.2] due to their length. This quartic equation in c, which is plotted as gray surface in Fig. 4b, cannot be solved explicitly for Q_2 but we can evaluate it numerically.

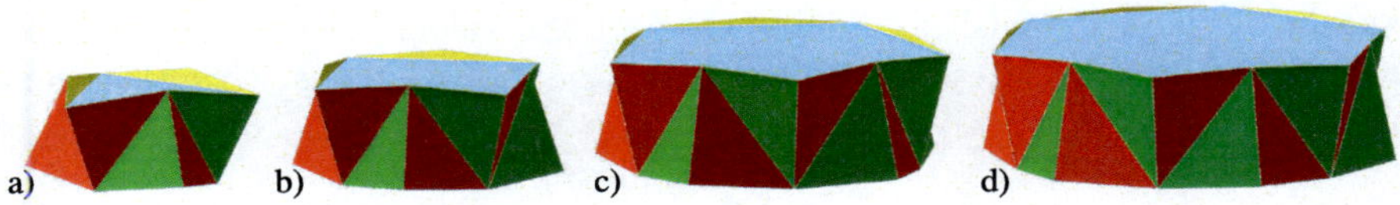

Fig. 6. Shaky structures causing the maximal shakeability for $n = 3, 4, 5, 6$ (a,b,c,d).

The shakiness of the structure is in general related with a non-trivial infinitesimal isometric deformation of dimension 1. The corresponding velocity vectors $\mathbf{v}(X_i)$ of the vertices X_i $(i = 0, \dots, n-1)$ with $X \in \{A, B, C, D\}$ are determined[9] up to a non-zero factor (fixed by Eq. (6) given later on) and can be computed by applying the *projection theorem*. For details the interested reader is referred to [15, App. 5.3].

Shakeability. In order to evaluate the structure's capability to shake, we introduce the so-called *shakeability* κ, which is defined as curvature of the snappability function over the space of squared edge lengths in direction associated with the infinitesimal mobility. As the snappability function is already dimensionless we also have to normalize the velocity vectors in such a way. This can e.g. be achieved by the condition[10], that the mean of the relative instantaneous changes of the squared edge lengths is equal to 1. For our sandglass structure this normalization reads as follows:

$$\left(4n\frac{\|\mathbf{v}(B_0)-\mathbf{v}(D_0)\|^2}{Q_1} + 2n\frac{\|\mathbf{v}(B_0)-\mathbf{v}(C_1)\|^2}{Q_2} + 2n\frac{\|\mathbf{v}(D_0)-\mathbf{v}(C_1)\|^2}{Q_3}\right)/(8n) = 1 \qquad (6)$$

Assumed that this condition holds, we can set $S_1 = Q_1 + t\|\mathbf{v}(B_0) - \mathbf{v}(D_0)\|^2$, $S_2 = Q_2 + t\|\mathbf{v}(B_0) - \mathbf{v}(C_1)\|^2$ and $S_3 = Q_3 + t\|\mathbf{v}(D_0) - \mathbf{v}(C_1)\|^2$ and plug

[9] Note that the faces α and β are assumed to translate instantaneously along the z-axis with the same speed but in opposite direction; i.e. $\mathbf{v}(A_i) = -\mathbf{v}(B_i)$. Therefore the symmetry of the structure remains intact by attaching the velocity vectors $\mathbf{v}(X_i)$ to the corresponding vertices X_i.

[10] One can also think of other normalizations, which will effect the resulting value for κ.

these expressions into Eq. (5), which now depends quadratically on t; i.e. $\sigma(t)$. According to the well-known curvature formula the shakeability κ can then be computed as

$$\kappa := \left.\frac{\sigma''}{(1+\sigma'^2)^{3/2}}\right|_{t=0} = \sigma''|_{t=0} \tag{7}$$

We let Q_1 run within the interval $]0.25; 0.31]$ in steps of 0.001 for $n = 3, \ldots, 6$ and compute for the associated shaky realization the shakeability, which is displayed in Fig. 4d. The structures causing the maximal shakeability are visualized in Fig. 6.

Animations of the families of shaky origami-like sandglasses for $n = 3, \ldots, 6$ can be downloaded from the author's homepage (cf. see footnote 7) and the crease pattern of the structure displayed in Fig. 6a is given in [15, App. 5.4].

Remark 4. Shaky realizations with smaller κ are more shaky, but note that the shakeability cannot vanish as in this case all the velocity vectors have to be the same; i.e. the infinitesimal isometric deformation is trivial (instantaneous translation). ◇

5 Conclusion and Future Work

We generalized Wunderlich's sandglass polyhedron in analogy to the birosette construction, with the additional feature that the belt of the antiprismatic skeleton is developable. One can think of two further generalizations of the obtained snapping or shaky quasi-mechanisms; namely (a) to omit the sandglass condition and/or the origami condition for the study of the generalized birosette structures (cf. Eq. (1)) and (b) to use an antifrustum as skeleton (i.e. α and β have different radii). For the latter case one can proceed similar to [10,16]. Moreover, we introduced the *shakeability* to evaluate the capability of a shaky quasi-mechanism to flex as counterpart to the snappability [2,14]. As this index is dimensionless, it enables a comparison of shaky structures differing in the inner geometry, which is also subject to future research as well as the influence of the normalization condition (cf. see footnote 10).

Acknowledgments. The author is supported by grant P 30855-N32 of the Austrian Science Fund FWF and by FWF project F77 (SFB "Advanced Computational Design", subproject SP7).

References

1. Goldberg, M.: Unstable polyhedral structures. Math. Mag. **51**(3), 165–170 (1978)
2. Nawratil, G.: Snappability and singularity-distance of pin-jointed body-bar frameworks. Mech. Mach. Theory **167**, 104510 (2022)
3. Jessen, B.: Orthogonal icosahedron. Nordisk Mat. Tidskr. **15**(2), 90–96 (1967)
4. Gorkavyy, V., Kalinin, D.: On model flexibility of the Jessen orthogonal icosahedron. Beiträge Algebra Geomet. Contrib. Algebra Geomet. **57**(3), 607–622 (2016). https://doi.org/10.1007/s13366-016-0287-5

5. Stachel, H.: What lies between rigidity and flexibility of structures. Serb. Arch. J. **3**(2), 102–115 (2011)
6. Wunderlich, W., Schwabe, C.: Eine Familie von geschlossenen gleichflächigen Polyedern, die fast beweglich sind. Elem. Math. **41**(4), 88–93 (1986)
7. Kresling, B.: Natural twist buckling in shells: from the Hawkmoth's bellows to the deployable Kresling-pattern and cylindrical Miura-ori. In: Abel, J.F., Cooke, J.R. (eds.) Proceedings of the 6th International Conference on Computation of Shell and Spatial Structures, Ithaca, NY (2008)
8. Wunderlich, W.: Snapping and shaky antiprisms. Math. Mag. **52**(4), 235–236 (1979)
9. Wunderlich, W.: Starre, kippende, wackelige und bewegliche Achtflache. Elem. Math. **20**(2), 25–32 (1965)
10. Nawratil, G.: Multi-stable design of triangulated origami structures on cones of revolution. arXiv preprint arXiv:2110.10986 (2021)
11. Gor'kavyi, V.A., Milka, A.D.: Birosettes are model flexors. Ukr. Math. J. **70**(7), 1022–1041 (2018)
12. Douady, A.: Le shaddock à six becs. Bull. A.P.M.E.P. **281**, 699–701 (1971)
13. Wunderlich, W.: Kipp-Ikosaeder II. Elem. Math. **37**(3), 84–89 (1982)
14. Nawratil, G.: On the snappability and singularity-distance of frameworks with bars and triangular plates. In: Holderbaum, W., Selig, J.M. (eds.) IMA 2020. SPAR, vol. 21, pp. 144–152. Springer, Cham (2022). https://doi.org/10.1007/978-3-030-91352-6_15
15. Nawratil, G.: On origami-like quasi-mechanisms with an antiprismatic skeleton. arXiv preprint arXiv:2108.01950 (2021)
16. Ishida, S., Nojima, T., Hagiwara, I.: Mathematical approach to model foldable conical structures using conformal mapping. J. Mech. Des. **136**, 091007 (2014)

Line-Point Constraints and Robot Surgery

J. M. Selig(✉)

School of Engineering, London South Bank University, London SE1 0AA, UK
seligjm@lsbu.ac.uk

Abstract. The space of rigid-body displacements that move a line so that its remains in contact with a fixed point is studied. This constraint variety is related to robot surgery where a straight, rigid cannula is inserted into the patient through a trocar. A surgical robot manipulates the cannula so the insertion point is fixed. The space of displacements determined by a pair of these constraints is also studied briefly. This correspond to a pair cannulas with their ends rigidly connected.

1 Introduction

This work is inspired by robot surgery, however the subject of this article is the geometry of robots. In robot surgery, tools are introduced into the patient's body through a trocar or "port". This port is a point on the patient's body through which the robot inserts a long thin cannula, the robot can also rotate the cannula about the point defined by the port. The rigid-body motion of a cannula is thus subject to a geometric constraint. The line defined by the cannula must pass through the point specified by the port. See [2] for a brief review of the history of this subject from the point of view of kinematics.

We begin with some mathematical background. The use of dual quaternions to represent rigid-body displacements dates back to E. Study at the beginning of the 20th century and is well known, see for example [3]. A rigid-body displacement can be represented in two ways by dual quaternions. A dual quaternion has the form,

$$g = (a_0 + a_1 i + a_2 j + a_3 k) + \varepsilon(c_0 + c_1 i + c_2 j + c_3 k).$$

Here, i, j and k form a base for the quaternions satisfying the usual rules for quaternions $i^2 = j^2 = k^2 = -1$, $ij = -ji = k$ and so forth. The element ε is the dual unit which commutes with all quaterions but squares to zero, $\varepsilon^2 = 0$. The Study parameters, $a_0, a_1, \ldots, c_3$, are real numbers. Not all dual quaterions represent rigid-body displacements. Taking only the dual quaternions satisfying the equations,

$$a_0^2 + a_1^2 + a_2^2 + a_3^2 = 1 \tag{1}$$

and

$$a_0 c_0 + a_1 c_1 + a_2 c_2 + a_3 c_3 = 0 \tag{2}$$

O. Altuzarra and A. Kecskémthy (Eds.): ARK 2022, SPAR 24, pp. 22–29, 2022.
https://doi.org/10.1007/978-3-031-08140-8_3

gives elements that comprise the double cover of the group of rigid-body displacements. This is the group Spin(3) $\rtimes$ $\mathbb{R}^3$. Both, g and $-g$ give the same rigid-body displacement. The elements satisfying these two equations form an affine algebraic variety.

Alternatively, we can think of the Study parameters as homogeneous coordinates in a 7-dimensional projective space $\mathbb{P}^7$. This has the effect of identifying the elements g and $-g$ since they only differ by multiplication by -1, a non-zero constant. Equation (1) cannot apply to these elements as it is not homogeneous. Equation (2) does apply, and dual quaternions satisfying this homogeneous equations form a 6-dimensional projective quadric variety usually known as the Study quadric. (A quadric is simply a variety of degree 2). Every rigid-body displacement corresponds to a unique dual quaternion in the Study quadric. Some elements of the Study quadric however, are not rigid-body displacements. These elements lie on the 3-dimensional plane lying entirely in the Study quadric and are defined by the equations $a_0 = a_1 = a_2 = a_3 = 0$. This 3-plane will be referred to as A_∞.

Away from A_∞, dual quaternions in the Study quadric can be written as,

$$g = (a_0 + a_1 i + a_2 j + a_3 k) + \varepsilon(c_0 + c_1 i + c_2 j + c_3 k) = r + \varepsilon\frac{1}{2}tr$$

where r is an ordinary quaternion representing a pure rotation and $t = t_x i + t_y j + t_z k$ is a pure quaternion representing a translation with translation vector $\boldsymbol{t} = (t_x,\, t_y,\, t_z)^T$. The action of such a rigid-body displacement on a point $(x,\, y,\, z)$ is given by the product,

$$\left(r+\varepsilon\frac{1}{2}tr\right)\left(1+\varepsilon(xi+yj+zk)\right)\left(r^-+\varepsilon\frac{1}{2}r^-t\right) = (rr^-)\left(1+\varepsilon\left(rxi+yj+zk)r^-+t\right)\right)$$

where $^-$ denotes the quaternion conjugate so that rr^- is a real scalar. Further, details can be found in several standard texts including [3].

In algebraic geometry the concepts of Veronese and Segre embeddings are fundamental. The Veronese embedding maps a projective space $\mathbb{P}^n$ into a projective space of larger dimension. Consider the n-dimensional projective space $\mathbb{P}^n$ with homogeneous $n+1$ coordinates, $x_0, x_1, \ldots, x_n$. The image of the degree d-Veronese embedding is parameterised by all $\binom{n+d-1}{d}$ degree d monomials in the x_i coordinates. That is, the image lies in $\mathbb{P}^m$ with coordinates $y_0, y_1, \ldots, y_m$ where $m = \binom{n+d-1}{d} - 1$ and $y_0 = x_0^d$, $y_1 = x_0^{n-1}x_1, \ldots, y_m = x_n^d$. The image of the this map is known as a Veronese variety, it can be shown to lie on a number of quadric hypersurfaces in $\mathbb{P}^m$. The simplest example is the degree 2 embedding of $\mathbb{P}^1$ in $\mathbb{P}^2$. This Veronese variety is a conic curve parameterised as $(x_0^2 : x_0x_1 : x_1^2)$ which is clearly the conic determined by the homogeneous equation $y_0y_2 - y_1^2 = 0$. Veronese varieties are generally not complete intersections, so Bézout's theorem cannot give their degree. However, it is possible to find the degree of a general Veronese variety, in [1, p.231] it is shown that the degree of the degree d embedding of $\mathbb{P}^n$ is d^n.

The Segre embedding maps the Cartesian product of projective spaces into another projective space. If the original projective spaces are $\mathbb{P}^n \times \mathbb{P}^m$ with coordinates x_i and y_j then the image of the Segre embedding, is parameterised by the possible products of pairs of coordinates. If z_k are the homogeneous coordinates of a $\mathbb{P}^q$ where $q = (n+1)(m+1) - 1 = nm + n + m$, then the image of the Segre embedding, the Segre variety, is parameterised by $z_0 = x_0 y_0$, $z_1 = x_0 y_1, \ldots, z_q = x_n y_m$. The simplest example here is the embedding of $\mathbb{P}^1 \times \mathbb{P}^1$ in $\mathbb{P}^3$, given by, $z_0 = x_0 y_0$, $z_1 = x_0 y_1$, $z_2 = x_1 y_0$ and $z_3 = x_1 y_1$. This Segre variety is given by the quadric surface $z_0 z_3 - z_1 z_2 = 0$. In general, Segre varieties are not complete intersections but lie on several quadric hypersurfaces. The degree of the Segre variety $\mathbb{P}^n \times \mathbb{P}^m$ is $\binom{n+m}{n}$, see [1, p.233].

2 The Displacement Variety

In [6] the problem of finding the subvariety of displacements that move a point so that it remains in contact with a fixed plane was studied. Here the inverse problem of how to move a plane so that is remains in contact with a fixed point is addressed first. In [7], it was shown that reversing the order of an open loop kinemetic chain inverts the displacements that the end-effector can perform. If the variety determined by the possible displacements is given implicitly by equations in the Study parameters, then the equations satisfied by the inverses can be found by changing the signs of the coefficients which multiply odd numbers of the parameters a_1, a_2, a_3 and c_1, c_2, c_3.

Here we begin by looking at the varieties defined in terms other representations of the group of rigid-body displacements. Suppose we represent the fixed point by the extended vector,

$$\tilde{p} = \begin{pmatrix} p_x \\ p_y \\ p_z \\ 1 \end{pmatrix} \quad \text{and the plane by the 4-vector,} \quad \pi = \begin{pmatrix} n_x \\ x_y \\ n_z \\ -d \end{pmatrix},$$

where n_x, n_y and n_z are the components of the unit normal vector to the plane and d is the perpendicular distance from the plane to the origin. If the plane contains the point we have the following relation,

$$\tilde{p}^T \pi = p_x n_x + p_y n_y + p_z n_z - d = 0.$$

The action of the group of rigid-body displacements on the plane is given by the inverse-transpose of the standard 4×4 representation of $SE(3)$,

$$\pi' = \begin{pmatrix} R & 0 \\ -\boldsymbol{t}^T R & 1 \end{pmatrix} \pi = M\pi,$$

where, as usual, R is a 3×3 rotation matrix and $\boldsymbol{t}$ the translation vector of the displacement. So, the displacements of the plane which preserve the contact between the point $\tilde{p}$ and the plane π must satisfy the equation,

$$\tilde{p}^T M \pi = \tilde{p}^T \begin{pmatrix} R & 0 \\ -\boldsymbol{t}^T R & 1 \end{pmatrix} \pi = 0. \tag{3}$$

Substituting the Study parameters $a_0, \ldots c_0, \ldots c_3$, for the components of the 4×4 matrix M gives,

$$M = \begin{pmatrix} a_0^2 + a_1^2 - a_2^2 - a_3^2 & 2(a_1a_2 - a_0a_3) & 2(a_1a_3 + a_0a_2) & 0 \\ 2(a_1a_2 + a_0a_3) & a_0^2 - a_1^2 + a_2^2 - a_3^2 & 2(a_2a_3 - a_0a_1) & 0 \\ 2(a_1a_3 - a_0a_2) & 2(a_2a_3 + a_0a_1) & a_0^2 - a_1^2 - a_2^2 + a_3^2 & 0 \\ \tau_x & \tau_y & \tau_z & \Delta \end{pmatrix}, \quad (4)$$

where

$$\begin{aligned} \tau_x &= 2(a_1c_0 - a_0c_1 - a_3c_2 + a_2c_3), \\ \tau_y &= 2(a_2c_0 + a_3c_1 - a_0c_2 - a_1c_3), \\ \tau_z &= 2(a_3c_0 - a_2c_1 + a_1c_2 - a_0c_3), \end{aligned} \quad (5)$$

and $\Delta = a_0^2 + a_1^2 + a_2^2 + a_3^2$, which has been included to make the equations homogeneous. Equation (3) can then be written as,

$$\mathbf{g}^T Q(p, \pi)\mathbf{g} = 0,$$

with $\mathbf{g}^T = (a_0, a_1, a_2, a_3, c_0, c_1, c_2, c_3)$. These quadrics can be written as 8×8 symmetric matrices, see [6] for details on how this can be done.

As examples, suppose the fixed point is the origin, $\tilde{p}_0^T = (0, 0, 0, 1)$ and the plane is initially the xy-plane, $\pi_{xy}^T = (0, 0, 1, 0)$ or initially the xz-plane $\pi_{xz}^T = (0, 1, 0, 0)$, the quadrics for the possible displacements of these planes will be,

$$\mathbf{g}^T Q(\tilde{p}_0, \pi_{xy})\mathbf{g} = 2(a_3c_0 - a_2c_1 + a_1c_2 - a_0c_3)$$

and

$$\mathbf{g}^T Q(\tilde{p}_0, \pi_{xz})\mathbf{g} = 2(a_2c_0 + a_3c_1 - a_0c_2 - a_1c_3).$$

Now, it is tempting to think that the conditions for a single point to lie on a line can be thought of as a pair of point-plane constraints where the same point is constrained to a pair of planes intersecting along the given line. For example, the displacements which move the x-axis so that it remains in contact with the origin, lie in the intersection of the Study quadric with both $Q(\tilde{p}_0, \pi_{xy})$ and $Q(\tilde{p}_0, \pi_{xz})$. However, this is not the whole story.

Consider how this subspace of displacements could be parameterised. The dual quaternions which preserve the incidence of a point with a line form a $\mathbb{P}^3 \times \mathbb{P}^1$ Segre variety in the Study quadric. Suppose the point is initially located at the origin and is constrained to remain on the x-axis. Clearly any translation along the line will be such an element as will any rotation about the origin. Combining these gives a 4-dimensional set of dual quaternions that can be parameterised as,

$$g = (\mu_0 + \mu_1 i + \mu_2 j + \mu_3 k)(\lambda_0 + \varepsilon\lambda_1 i),$$

where μ_i and λ_j are arbitrary parameters. The parameters $(\lambda_0 : \lambda_1)$ can be thought of as homogeneous coordinates for a $\mathbb{P}^1$ and $(\mu_0 : \mu_1 : \mu_2 : \mu_3)$ for $\mathbb{P}^3$. The image in $\mathbb{P}^7$ is given in coordinates by $g = a + \varepsilon c$ with,

$$\begin{aligned} a &= \lambda_0\mu_0 + \lambda_0\mu_1 i + \lambda_0\mu_2 j + \lambda_0\mu_3 k, \\ c &= -\lambda_1\mu_1 + \lambda_1\mu_0 i + \lambda_1\mu_3 j - \lambda_1\mu_2 k. \end{aligned} \quad (6)$$

It is clear that any point-on-a-line can be transformed to this configuration using a suitable conjugation in the $SE(3)$. This Segre variety lies in several quadrics in $\mathbb{P}^7$, the equations of these quadrics can be found by asserting that the rank of the following matrix is 1 or less:

$$\begin{pmatrix} a_0 & a_1 & a_2 & a_3 \\ c_1 & -c_0 & -c_3 & c_2 \end{pmatrix}.$$

This gives 6 quadrics,

$$\begin{aligned} Q_1 &= a_0c_0 + a_1c_1 = 0, & Q_4 &= a_1c_3 - a_2c_0 = 0, \\ Q_2 &= a_0c_3 + a_2c_1 = 0, & Q_5 &= a_1c_2 + a_3c_0 = 0, \\ Q_3 &= a_0c_2 - a_3c_1 = 0, & Q_6 &= a_2c_2 + a_3c_3 = 0. \end{aligned}$$

The variety will lie on any linear combination of these six. In particular, it is straightforward to check that the Study quadric and the quadrics $\mathbf{g}^T Q(\tilde{p}_0, \pi_{xy})\mathbf{g}$ and $\mathbf{g}^T Q(\tilde{p}_0, \pi_{xz})\mathbf{g}$, all lie in the linear system formed by these quadrics, the Study quadric is, $Q_1 + Q_6 = a_0c_0 + a_1c_1 + a_2c_2 + a_3c_3 = 0$, see (2). The other two are, $\mathbf{g}^T Q(\tilde{p}_0, \pi_{xy})\mathbf{g} = 2(Q_5 - Q_2) = 0$ and $\mathbf{g}^T Q(\tilde{p}_0, \pi_{xz})\mathbf{g} = -2(Q_3 + Q_4) = 0$.

The degree of such a Segre variety is $\binom{n+m}{n} = \binom{3+1}{1} = 4$. This suggests that there is another component to the intersection of the 3 quadrics. The intersection of the quadrics $\mathbf{g}^T Q(\tilde{p}_0, \pi_{xy})\mathbf{g}$ and $\mathbf{g}^T Q(\tilde{p}_0, \pi_{xz})\mathbf{g}$ is a five dimensional variety, that can be parameterised as,

$$\begin{aligned} a_0 &= \lambda_0\mu_0, & c_0 &= -\lambda_1\mu_1 + \lambda_2\mu_0, \\ a_1 &= \lambda_0\mu_1, & c_1 &= \lambda_1\mu_0 + \lambda_2\mu_1, \\ a_2 &= \lambda_0\mu_2, & c_2 &= \lambda_1\mu_3 + \lambda_2\mu_2, \\ a_3 &= \lambda_0\mu_3, & c_3 &= -\lambda_1\mu_2 + \lambda_2\mu_3. \end{aligned}$$

This is a linear projection of the Segre variety, $\mathbb{P}^2 \times \mathbb{P}^3$. Now, substituting this parameterisation into the Study quadric gives,

$$a_0c_0 + a_1c_1 + a_2c_2 + a_3c_3 = \lambda_0\lambda_2(\mu_0^2 + \mu_1^2 + \mu_2^2 + \mu_3^2)$$

Hence the intersection of all three quadrics consists of two 4-dimensional varieties. For the first $\lambda_2 = 0$ and the Segre variety discussed above is recovered. The second component arises when $(\mu_0^2 + \mu_1^2 + \mu_2^2 + \mu_3^2) = 0$. All solutions to this condition will be complex and are thus not physically valid displacements. When $\lambda_0 = 0$ the result is the 3-plane A_∞, which lies in the both the Segre variety $\mathbb{P}^1 \times \mathbb{P}^3$ and the complex residual variety.

The displacement variety described above can be realised with an SP dyad. That is, the end effector of a linkage consisting of a spherical and a prismatic joint will be able to adopt all displacements that maintain contact between a line parallel to the prismatic joint passing through the centre of the spherical joint. If the final link of this mechanism is rigidly attached to, say an RR linkage then, in general, the resulting closed-loop mechanism will have 4 assembly configurations. This is because the displacement variety of the RR mechanism is known to be the intersection of the Study quadric with a 3-plane. The general number of assembly configurations is the intersection of this variety with the $\mathbb{P}^3 \times \mathbb{P}^1$ Segre variety traced by the SP linkage. The Segre variety lies in the Study quadric so intersecting with the 3-plane gives 4 points; the degree of the Segre variety.

3 Two Ports

In robot surgery, several cannulas are used. Suppose the micro grippers at the ends of two such cannulas hold a rigid body, a needle perhaps. What rigid displacements can the body be subjected to now? In terms of mechanisms, this question is equivalent to asking for the coupler variety of a single loop mechanism consisting of two SP legs. The solution should be the intersection of two $\mathbb{P}^3 \times \mathbb{P}^1$ Segre varieties, but any such Segre variety will contain the 3-plane A_∞. Since we expect the variety to be 2-dimensional, the intersection cannot be a complete intersection. To get around this difficulty a birational map can be applied to the varieties. In [4] a birational map between the Study quadric and a variety defined by the standard homogeneous representation of $SE(3)$ was studied. Here, a map to the variety defined by the inverse-transpose to the standard representation will be used. This map is very similar to the one given in [4], in fact only few signs are changed. The map from the Study quadric is essentially given by Eqs. (4) and (5), the non-zero elements of the matrix M are taken as homogeneous coordinates in a $\mathbb{P}^{12}$. The point of doing this is that the exceptional set of the map, the set on which the map is not defined, is just A_∞.

The inverse map is given by,

$$\begin{aligned}
a_0 &= -2(\Delta + m_{11} + m_{22} + m_{33})\Delta,\\
a_1 &= 2(m_{23} - m_{32})\Delta,\\
a_2 &= 2(m_{31} - m_{13})\Delta,\\
a_3 &= 2(m_{12} - m_{21})\Delta,\\
c_0 &= -((m_{23} - m_{32})\tau_x + (m_{31} - m_{13})\tau_y + (m_{12} - m_{21})\tau_z),\\
c_1 &= (-(\Delta + m_{11} + m_{22} + m_{33})\tau_x - (m_{12} - m_{21})\tau_y + (m_{31} - m_{13})\tau_z),\\
c_2 &= ((m_{12} - m_{21})\tau_x - (\Delta + m_{11} + m_{22} + m_{33})\tau_y - (m_{23} - m_{32})\tau_z),\\
c_3 &= (-(m_{31} - m_{13})\tau_x + (m_{23} - m_{32})\tau_y - (\Delta + m_{11} + m_{22} + \rho_{33})\tau_z).
\end{aligned}$$

As with the homogeneous representation, the image of the Study quadric in the $\mathbb{P}^{12}$ with coordinates m_{ij}, τ_k and Δ, can be shown to be the join of the 3-dimensional Veronese variety with a disjoint 2-plane. Call this variety Y, the degree of Y is thus 8, see [4]. The image of the two quadrics, $\mathbf{g}^T Q(\tilde{p}_0, \pi_{xy})\mathbf{g} = 0$ and $\mathbf{g}^T Q(\tilde{p}_0, \pi_{xz})\mathbf{g} = 0$ are simply the hyperplanes $\tau_y = 0$ and $\tau_z = 0$. On the other hand the image of the Segre variety can be parameterised as,

$$M_p = \begin{pmatrix} \lambda_0(\mu_0^2 + \mu_1^2 - \mu_2^2 - \mu_3^2) & 2\lambda_0(\mu_1\mu_2 - \mu_0\mu_3) & 2\lambda_0(\mu_1\mu_3 + \mu_0\mu_2) & 0 \\ 2\lambda_0(\mu_1\mu_2 + \mu_0\mu_3) & \lambda_0(\mu_0^2 - \mu_1^2 + \mu_2^2 - \mu_3^2) & 2\lambda_0(\mu_2\mu_3 - \mu_0\mu_1) & 0 \\ 2\lambda_0(\mu_1\mu_3 - \mu_0\mu_2) & 2\lambda_0(\mu_2\mu_3 + \mu_0\mu_1) & \lambda_0(\mu_0^2 - \mu_1^2 - \mu_2^2 + \mu_3^2) & 0 \\ -\lambda_1\Delta & 0 & 0 & \lambda_0\Delta \end{pmatrix} \tag{7}$$

with $\Delta = \mu_0^2 + \mu_1^2 + \mu_2^2 + \mu_3^2$. This can be found either by multiplying an arbitrary translation in the x-direction by a general rotation about the origin or by substituting the parameterisation given in (6) into the map for the inverse-transpose representation (Eqs. (4) and (5)) and cancelling the common factor λ_0. This can be thought of as the join of the Veronese variety with a single point, $\Delta = m_{ij} = \tau_y = \tau_z = 0$. Hence the image of the Segre variety has degree 8 and can thus be identified with the intersection of Y with the two hyperplanes.

Returning to the original problem, suppose we choose coordinates so that one point is located at the origin and the other at the point $\tilde{p}_a$, the x-axis will be aligned with the initial position of the first line and the second point will initially lie on the planes π_1 and π_2. Now, substitute the parameterisation for the first Segre variety (7), into the two equations for the second line-plane constraint,

$$\tilde{p}_a^T M_p \pi_1 = 0, \quad \text{and} \quad \tilde{p}_a^T M_p \pi_2 = 0.$$

The result will be a pair of equations linear in the "λ" variables and quadratic in the "μ" parameters. In fact it is clear that the equations will have the form, $-n_{ix}\Delta\lambda_1 + q_i\lambda_0 = 0$, where $i = 1, 2$; q_1 and q_2 are quadratic functions of the μ_js and n_{ix} is the x component of the normal vector to π_i. To get non-trivial solutions for the λs we must have that $q_1 n_{2x} - q_2 n_{1x} = 0$. This is a quadratic equation in the four homogeneous parameters $\mu_0, \ldots, \mu_3$, hence determines a 2-dimensional quadric in a $\mathbb{P}^3$. This is itself a Segre variety, $\mathbb{P}^1 \times \mathbb{P}^1$. So the μ_i can be written as linear functions of homogeneous parameters $\alpha_j\beta_k$ where j, $k = 0, 1$. That is, the μs are separately linear in the α and β variables. The solutions for the λs however, are quadratic in these variables, since, $\lambda_0 = n_{1x}\Delta$ and $\lambda_1 = q_1$. Substituting these results back into the parameterisation given in (6) will result in a parameterisation of a 2-dimensional variety in the Study quadric that is the image of $\mathbb{P}^1 \times \mathbb{P}^1$ by a map that is separately cubic in the coordinates of each $\mathbb{P}^1$. Notice that, at any configuration the two lines can rotate about the axis joining the two fixed points $\tilde{p}_0$ and $\tilde{p}_a$.

Alternatively, if we use Eq. (7) to map this 2-dimensional variety into Y, the inverse-transpose representation of $SE(3)$, the image will have a parameterisation that is degree 4 in both the α and β parameters. Since the image variety is the intersection of Y with a pair of hyperplanes, it will have degree 8.

Finally, a parallel mechanism consisting of 3 SP linkages, that is three cannulas holding a rigid object, we would expect a finite number of assembly configurations. This is the intersection of the group variety Y with six hyperplanes. In general, the number of assemblies is then just the degree of Y, that is 8. However, if for example the lines are mutually parallel then a one degree-of-freedom motion is possible. This corresponds to the a linear dependency between the 6 hyperplanes.

4 Conclusions

The study of displacement varieties seems to be fundamental to the theory of mechanisms and linkages. Although the inspiration for this work comes from problems in surgical robotics it is hoped that there are wider applications of the results and techniques outlined. There are several other interesting problems relevant to robot surgery and other application. For example, how can general 6R robots be controlled to respect a geometrical constraint such as the line-on-point constraint studied here?

References

1. Harris, J.: Algebraic Geometry: A First Course. Springer, New York (1992). https://doi.org/10.1007/978-1-4757-2189-8
2. Kuo, C.H., Dai, J.S.: Robotics for minimally invasive surgery: a historical review from the perspective of kinematics. In: Yan, H.S., Ceccarelli, M. (eds.) International Symposium on History of Machines and Mechanisms. Springer, Dordrecht (2009). https://doi.org/10.1007/978-1-4020-9485-9_24
3. Selig, J.M.: Geometrical Fundamentals of Robotics, 2nd edn. Springer, New York (2005). https://doi.org/10.1007/b138859
4. Selig, J.M.: On the geometry of the homogeneous representation for the group of proper rigid-body displacements. Romanian J. Tech. Sci. Appl. Mech. (Special issue on "New Trends in Advanced Robotics") **58**(1–2), 27–50 (2013)
5. Selig, J.M.: Some rational vehicle motions. In: Thomas, F., Pérez Gracia, A. (eds.) Computational Kinematics. MMS, vol. 15, pp. 21–29. Springer, Dordrecht (2014). https://doi.org/10.1007/978-94-007-7214-4_3
6. Selig, J.M.: Quadratic constraints on rigid-body displacements. ASME J. Mech. Robot. **2**(4), 041009 (2010)
7. Stigger, T., Husty, M.L.: Constraint equations of inverted kinematic chains. In: Zeghloul, S., Romdhane, L., Laribi, M.A. (eds.) Computational Kinematics. MMS, vol. 50, pp. 491–498. Springer, Cham (2018). https://doi.org/10.1007/978-3-319-60867-9_56

On the Use of Ternary Products to Characterize the Dexterity of Spatial Kinematic Chains

Bertold Bongardt[1(✉)] and Andreas Müller[2]

[1] Technische Universität Braunschweig, Institut für Robotik und Prozessinformatik, 38106 Braunschweig, Germany
b.bongardt@tu-bs.de

[2] Johannes Kepler Universität Linz, Institut für Robotik, 4040 Linz, Austria
a.mueller@jku.at

1 Introduction

Due to high dimensions and nonlinear behavior, the characterization of the entire workspace of a spatial robotic system represents an ambitious computation task. The *singular configurations* of a kinematic chain are the workspace's subsets in which its tangent space degenerates exceptionally in certain senses [10]. The 'transmission behavior' of a manipulator between its input and output space is quantified by the so-called *manipulability* for each configuration of its workspace in the domains of generalized velocities and efforts [15]. The concepts of singularity and manipulability are typically computed via the manipulator's Jacobian matrix: The 'singularity degree' is expressed by the Jacobian's determinant and the 'manipulability degree' by the proportion of the Jacobian's smallest to the largest singular value [7,8].

The approach for computing singular configurations of general regional 3R chains was supplemented by introducing three Gramian matrices, obtained from certain blocks of the geometric Jacobian matrix, and their determinants [18]. Due to their geometry, the Gramians were identified as of 'orientation', 'position', and 'attitude' type. The novel 'attitude' singularity type unifies the 'orientation' and of 'position' singularities, it matches geometrically to joint axes becoming 'coplanar' in the particular 'singular' joint configuration, and it has been identified via the transference principle and singularities of non-orthogonal spherical 3R chains [13].

This manuscript proposes the novel idea of ternary products for dual vectors. By generalizing the concept of the ternary scalar product for a pair of oriented lines to a sequence of oriented lines, the three Gramian matrices defined previously [18] are obtained in a uniform manner as three distinct faces of the 'same coin': a ternary Gramian matrix. In a similar fashion, the novel idea of the ternary dyadic product for dual vectors is proposed and generalized to the ternary dyadic matrix. Thus, the ternary Gramian matrix, that only holds for well-actuated and under-actuated manipulators, is complemented by a suitable concept that holds for well-actuated and over-actuated kinematic chains. By

O. Altuzarra and A. Kecskémthy (Eds.): ARK 2022, SPAR 24, pp. 30–38, 2022.
https://doi.org/10.1007/978-3-031-08140-8_4

these means, the analysis – limited to singularities of regional 3R chains formerly [18] – is extended to quantify singularity and manipulability aspects of arbitrary spatial kinematic chains in a systematic and uniform manner.

2 Geometry of Spears

A directed line in space is briefly called a spear and is represented in homogeneous coordinates by a Plücker vector $\boldsymbol{\Lambda} = \left(\begin{smallmatrix} \boldsymbol{n} \\ \boldsymbol{m} \end{smallmatrix}\right) \in \mathbb{R}^6$ with direction $\boldsymbol{n} \in \mathbb{R}^3$ and moment $\boldsymbol{m} = \boldsymbol{a} \times \boldsymbol{n} \in \mathbb{R}^3$ about the origin with the Plücker condition $\boldsymbol{n} * \boldsymbol{m} = 0$, a particular Grassmann condition [3]. Alternately, a spear is often identified as a dual vector $\boldsymbol{\Lambda} \cong \boldsymbol{n} + \epsilon \cdot \boldsymbol{m} \in \tilde{\mathbb{R}}^3$ [3]. The adjoint representation of a spear is given by the matrix $\mathrm{ad}(\boldsymbol{\Lambda}) = \left(\begin{smallmatrix} \mathrm{ad}(\boldsymbol{n}) & \boldsymbol{0} \\ \mathrm{ad}(\boldsymbol{m}) & \mathrm{ad}(\boldsymbol{n}) \end{smallmatrix}\right) \in \mathbb{R}^{6\times 6}$ with $\mathrm{ad}(\boldsymbol{x}) = \left(\begin{smallmatrix} 0 & -x_3 & x_2 \\ x_3 & 0 & -x_1 \\ -x_2 & x_1 & 0 \end{smallmatrix}\right) \in \mathbb{R}^{3\times 3}$ [16]. In the following, distinct concepts are reflected that have been taken into account for defining scalar products of oriented lines in space. In the sequel, the novel concept of the ternary scalar product of dual quantities is then introduced.

3 Real Products and Dual Products

Real products. The so-called 'reciprocal product' [2] of two spears $\boldsymbol{\Lambda}_\mathsf{A}$ and $\boldsymbol{\Lambda}_\mathsf{B}$ is defined as the sum of 'blended' products of their directions and moments

$$\boldsymbol{\Lambda}_\mathsf{A} \diamond \boldsymbol{\Lambda}_\mathsf{B} := \boldsymbol{n}_\mathsf{A} * \boldsymbol{m}_\mathsf{B} + \boldsymbol{m}_\mathsf{A} * \boldsymbol{n}_\mathsf{B} \ , \tag{1}$$

where $\boldsymbol{x} * \boldsymbol{y} := \boldsymbol{x}^T \cdot \boldsymbol{y}$ denotes the scalar product of vectors and $\boldsymbol{X} * \boldsymbol{Y} := \boldsymbol{X}^T \cdot \boldsymbol{Y}$ subsumes the scalar products of all vector pairs (matrix columns) in the remainder of the document. By means of the swap operator $\triangle := \left(\begin{smallmatrix} \boldsymbol{O} & \boldsymbol{I} \\ \boldsymbol{I} & \boldsymbol{0} \end{smallmatrix}\right)$ according to [9,12], transforming screws in axis coordinates in ray coordinates [17] and vice versa, an alternate formulation of the reciprocal product reads

$$\boldsymbol{\Lambda}_\mathsf{A} \diamond \boldsymbol{\Lambda}_\mathsf{B} \equiv \boldsymbol{\Lambda}_\mathsf{A} * \triangle \cdot \boldsymbol{\Lambda}_\mathsf{B} \ . \tag{2}$$

The geometric meaning of a vanishing reciprocal product $\boldsymbol{\Lambda}_\mathsf{A} \diamond \boldsymbol{\Lambda}_\mathsf{B} = 0$ is that $\boldsymbol{\Lambda}_\mathsf{A}$ and $\boldsymbol{\Lambda}_\mathsf{B}$ are coplanar[1]: either coincident, intersecting, or parallel.

An alternate real-number scalar product is called the 'orthogonal product' [5] of two spears $\boldsymbol{\Lambda}_\mathsf{A}$ and $\boldsymbol{\Lambda}_\mathsf{B}$ and defined as sum of direction and moment products

$$\boldsymbol{\Lambda}_\mathsf{A} * \boldsymbol{\Lambda}_\mathsf{B} := \boldsymbol{n}_\mathsf{A} * \boldsymbol{n}_\mathsf{B} + \boldsymbol{m}_\mathsf{A} * \boldsymbol{m}_\mathsf{B} \ . \tag{3}$$

The orthogonal product of (3) is "inconsistent in regard both to dimensions and units of measurements" and depends on the "origin for the rectangular Cartesian frame in which the screws are represented" [9].

---

[1] Also denoted by the $\boldsymbol{\Lambda}_\mathsf{A} \diamond \boldsymbol{\Lambda}_\mathsf{B} = 0 \Leftrightarrow \tilde{\vartheta}_\mathsf{AB} \asymp 0 \Leftrightarrow \sin(\vartheta_\mathsf{AB}) = 0 \vee d_\mathsf{AB} = 0$ notation in [18].

Dual products. For introducing dual products for oriented lines in space, the concept of a dual number is recalled. A dual number $\tilde{x} = x + \epsilon \cdot \mathring{x}$ is the sum of one number $x \in \mathbb{R}$ and a second number $\mathring{x} \in \mathbb{R}$ multiplied with the dual unit ϵ featuring $\epsilon^2 = 0$ [4]. The set of dual numbers $\tilde{\mathbb{R}}$ forms the dual number plane $\tilde{\mathbb{R}} = \mathbb{R}{+}\epsilon{\cdot}\mathbb{R}$, whose unit circle is given by the two lines with elements of the form $\hat{\tilde{x}} \in \{\pm 1{+}\epsilon{\cdot}\mathring{x} \,|\, \mathring{x} \in \mathbb{R}\}$, with the norm $\|\tilde{x}\|^2 = \langle \tilde{x}, \tilde{x} \rangle := (x{+}\epsilon{\cdot}\mathring{x}){\cdot}(x{-}\epsilon{\cdot}\mathring{x}) = \|x\|^2$.

The 'dual scalar product' [1,2,11] of two spears $\boldsymbol{\Lambda}_{\mathsf{A}}$ and $\boldsymbol{\Lambda}_{\mathsf{B}}$ is given – by identifying $\boldsymbol{\Lambda}_{\mathsf{A}} \cong \boldsymbol{n}_{\mathsf{A}} + \epsilon \cdot \boldsymbol{m}_{\mathsf{A}}$ and $\boldsymbol{\Lambda}_{\mathsf{B}} \cong \boldsymbol{n}_{\mathsf{B}} + \epsilon \cdot \boldsymbol{m}_{\mathsf{B}}$ and by 'transferring' [6] the standard scalar product – with the expression

$$\boldsymbol{\Lambda}_{\mathsf{A}} \circledast \boldsymbol{\Lambda}_{\mathsf{B}} := \boldsymbol{n}_{\mathsf{A}} * \boldsymbol{n}_{\mathsf{B}} + \epsilon \cdot (\boldsymbol{\Lambda}_{\mathsf{A}} \diamond \boldsymbol{\Lambda}_{\mathsf{B}}) \,, \tag{4}$$

and incorporates the reciprocal scalar product from (1) as its dual part. For sake of completeness, the 'dual motor product' or 'motor product' [1,2,11] of two spears $\boldsymbol{\Lambda}_{\mathsf{A}}$ and $\boldsymbol{\Lambda}_{\mathsf{B}}$ is introduced with

$$\boldsymbol{\Lambda}_{\mathsf{A}} \times \boldsymbol{\Lambda}_{\mathsf{B}} := \left(\boldsymbol{n}_{\mathsf{A}} \times \boldsymbol{n}_{\mathsf{B}},\, \boldsymbol{n}_{\mathsf{A}} \times \boldsymbol{m}_{\mathsf{B}} + \boldsymbol{m}_{\mathsf{A}} \times \boldsymbol{n}_{\mathsf{B}}\right)^T , \tag{5}$$

obtained by 'transferring' [6] the standard cross product to dual vectors.

4 Ternary Products

The ternary scalar product of a pair of spears is introduced as an extension of the dual scalar product according to the scheme

reciprocal scalar product(1) → dual scalar product(4) → ternary scalar product(9)

neglecting to evaluate the characteristic identity $\epsilon^2 = 0$ of the dual unit. Similarly, the ternary dyadic product for dual vectors is introduced according to

(vector)dyadicproduct → (dual vector) ternary dyadic product (7)

skipping the second stage of the 'dual dyadic product for dual vectors' that has not been introduced in the literature yet to the best of the authors' knowledge.

Ternary scalar product. The ternary scalar product of two dual vectors $\tilde{\boldsymbol{x}}_{\mathsf{A}} = \boldsymbol{x}_{\mathsf{A}} + \epsilon \cdot \mathring{\boldsymbol{x}}_{\mathsf{A}}$ and $\tilde{\boldsymbol{x}}_{\mathsf{B}} = \boldsymbol{x}_{\mathsf{B}} + \epsilon \cdot \mathring{\boldsymbol{x}}_{\mathsf{B}}$ is defined via the expansion

$$\left\langle \tilde{\boldsymbol{x}}_{\mathsf{A}}, \tilde{\boldsymbol{x}}_{\mathsf{B}} \right\rangle_{\blacksquare} := \boldsymbol{x}_{\mathsf{A}} * \boldsymbol{x}_{\mathsf{B}} + \epsilon \cdot (\boldsymbol{x}_{\mathsf{A}} * \mathring{\boldsymbol{x}}_{\mathsf{B}} + \mathring{\boldsymbol{x}}_{\mathsf{A}} * \boldsymbol{x}_{\mathsf{B}}) + \epsilon^2 \cdot (\mathring{\boldsymbol{x}}_{\mathsf{A}} * \mathring{\boldsymbol{x}}_{\mathsf{B}}) \,, \tag{6}$$

matching the dual scalar product of $\tilde{\boldsymbol{x}}_{\mathsf{A}}$ and $\tilde{\boldsymbol{x}}_{\mathsf{B}}$ if the characteristic constraint $\epsilon^2 = 0$ was evaluated. The ternary scalar product does not generalize the multiplicative *default* operation of dual vectors: the (dual) vector-generalization of dual-number multiplication is given by generalized cross product in (5). The ternary product of (6) is a (ternary) inner product, that is a (ternary) generalization of a *similarity* [14]. Formally, the ternary scalar product of (6) is a collection of three reals or one ternary number $\tilde{\tilde{x}} = x + \epsilon \cdot \mathring{x} + \epsilon^2 \cdot \breve{x}$ as an element of ternary numbers $\tilde{\tilde{\mathbb{R}}} = \mathbb{R} + \epsilon \cdot \mathbb{R} + \epsilon^2 \cdot \mathbb{R}$, just introduced as closure of dual numbers for the scalar product $\langle .,. \rangle_{\scriptscriptstyle\blacksquare} : \tilde{\mathbb{R}}^n \times \tilde{\mathbb{R}}^n \to \tilde{\tilde{\mathbb{R}}}$ defined in (6).

Ternary dyadic product. The ternary dyadic product of two dual vectors $\tilde{\boldsymbol{x}}_{\mathsf{A}} = \boldsymbol{x}_{\mathsf{A}} + \epsilon \cdot \mathring{\boldsymbol{x}}_{\mathsf{A}}$ and $\tilde{\boldsymbol{x}}_{\mathsf{B}} = \boldsymbol{x}_{\mathsf{B}} + \epsilon \cdot \mathring{\boldsymbol{x}}_{\mathsf{B}}$ is defined via the expansion

$$\tilde{\boldsymbol{x}}_{\mathsf{A}} \otimes_{\scriptscriptstyle\blacksquare} \tilde{\boldsymbol{x}}_{\mathsf{B}} := \boldsymbol{x}_{\mathsf{A}} \otimes \boldsymbol{x}_{\mathsf{B}} + \epsilon \cdot (\boldsymbol{x}_{\mathsf{A}} \otimes \mathring{\boldsymbol{x}}_{\mathsf{B}} + \mathring{\boldsymbol{x}}_{\mathsf{A}} \otimes \boldsymbol{x}_{\mathsf{B}}) + \epsilon^2 \cdot (\mathring{\boldsymbol{x}}_{\mathsf{A}} \otimes \mathring{\boldsymbol{x}}_{\mathsf{B}}) \tag{7}$$

where $\boldsymbol{x} \otimes \boldsymbol{y} := \boldsymbol{x} \cdot \boldsymbol{y}^T$ denotes the dyadic product (dyad) of two vectors and the dyadic $\boldsymbol{X} \otimes \boldsymbol{Y} := \boldsymbol{X} \cdot \boldsymbol{Y}^T$ for $\boldsymbol{X}, \boldsymbol{Y} \in \mathbb{R}^{n \times k}$ is the k-sum of dyads of vector pairs (matrix columns) $\boldsymbol{X} \otimes \boldsymbol{Y} = \sum_i (\boldsymbol{x}_i \otimes \boldsymbol{y}_i)$ [2]. The ternary dyadic product of (7) complements the ternary scalar product of (6). Formally, it is a collection of three (3×3)-matrices with real entries or one (3×3)-matrix with ternary entries. The product $\otimes_{\scriptscriptstyle\blacksquare} : \tilde{\mathbb{R}}^n \times \tilde{\mathbb{R}}^n \to \tilde{\tilde{\mathbb{R}}}^{n \times n}$ is a tensor product and does not interfere with the constitutive default multiplication of dual quantities.

Ternary spear products. For a unit spear $\hat{\boldsymbol{\Lambda}} = \binom{\hat{\boldsymbol{n}}}{\boldsymbol{m}}$, with unit direction $\hat{\boldsymbol{n}}$ and moment $\boldsymbol{m}$, any anchor $\boldsymbol{a} \in \boldsymbol{\Lambda}$ fulfills $\boldsymbol{a} \times \hat{\boldsymbol{n}} = \boldsymbol{m}$. The particular 'zero anchor'

$$\underline{\boldsymbol{a}} := \hat{\boldsymbol{n}} \times \boldsymbol{a} \times \hat{\boldsymbol{n}} \tag{8}$$

is the point on $\boldsymbol{\Lambda}$ with $\underline{\boldsymbol{a}} = \operatorname{argmin}_{\boldsymbol{a} \in \boldsymbol{\Lambda}}(\|\boldsymbol{a}\|)$ and $\underline{\boldsymbol{a}} * \hat{\boldsymbol{n}} = 0$. Also, the equality $\boldsymbol{m} * \boldsymbol{m} = \|\underline{\boldsymbol{a}}\|^2$ holds for the zero anchor. The ternary inner and dyadic products of two spears $\hat{\boldsymbol{\Lambda}}_{\mathsf{A}}$ and $\hat{\boldsymbol{\Lambda}}_{\mathsf{B}}$ are reformulations of (6) and (7)

$$\langle \hat{\boldsymbol{\Lambda}}_{\mathsf{A}}, \hat{\boldsymbol{\Lambda}}_{\mathsf{B}} \rangle_{\scriptscriptstyle\blacksquare} := \hat{\boldsymbol{n}}_{\mathsf{A}} * \hat{\boldsymbol{n}}_{\mathsf{B}} + \epsilon \cdot (\hat{\boldsymbol{n}}_{\mathsf{A}} * \boldsymbol{m}_{\mathsf{B}} + \boldsymbol{m}_{\mathsf{A}} * \hat{\boldsymbol{n}}_{\mathsf{B}}) + \epsilon^2 \cdot (\boldsymbol{m}_{\mathsf{A}} * \boldsymbol{m}_{\mathsf{B}}) \tag{9}$$

$$\hat{\boldsymbol{\Lambda}}_{\mathsf{A}} \otimes_{\scriptscriptstyle\blacksquare} \hat{\boldsymbol{\Lambda}}_{\mathsf{B}} := \hat{\boldsymbol{n}}_{\mathsf{A}} \otimes \hat{\boldsymbol{n}}_{\mathsf{B}} + \epsilon \cdot (\hat{\boldsymbol{n}}_{\mathsf{A}} \otimes \boldsymbol{m}_{\mathsf{B}} + \boldsymbol{m}_{\mathsf{A}} \otimes \hat{\boldsymbol{n}}_{\mathsf{B}}) + \epsilon^2 \cdot (\boldsymbol{m}_{\mathsf{A}} \otimes \boldsymbol{m}_{\mathsf{B}}) \,. \tag{10}$$

In comparison to the dual scalar product of (4), the additional summand $\boldsymbol{m}_{\mathsf{A}} * \boldsymbol{m}_{\mathsf{B}}$ in (9) of the ternary product is interpreted geometrically as

$$\boldsymbol{m}_{\mathsf{A}} * \boldsymbol{m}_{\mathsf{B}} = \cos(\vartheta_{\mathsf{AB}}) \cdot \|\underline{\boldsymbol{a}}_{\mathsf{A}}\| \cdot \|\underline{\boldsymbol{a}}_{\mathsf{B}}\|$$

with $\vartheta_{\mathsf{AB}} := \operatorname{ang}(\boldsymbol{m}_{\mathsf{A}}, \boldsymbol{m}_{\mathsf{B}})$. Thus, the ternary product of two spears from (9) is reformulated in geometric terms as

$$\langle \hat{\boldsymbol{\Lambda}}_{\mathsf{A}}, \hat{\boldsymbol{\Lambda}}_{\mathsf{B}} \rangle_{\scriptscriptstyle\blacksquare} = \cos(\vartheta_{\mathsf{AB}}) + \epsilon \cdot (-d_{\mathsf{AB}} \cdot \sin(\vartheta_{\mathsf{AB}})) + \epsilon^2 \cdot (\underline{\eta}_{\mathsf{AB}} \cdot \cos(\vartheta_{\mathsf{AB}})) \,,$$

where $\underline{\eta}_{\mathsf{AB}} := \|\boldsymbol{m}_{\mathsf{A}}\| \cdot \|\boldsymbol{m}_{\mathsf{B}}\| = \|\underline{\boldsymbol{a}}_{\mathsf{A}}\| \cdot \|\underline{\boldsymbol{a}}_{\mathsf{B}}\|$. Evaluating (9) for two identical arguments, the ternary squared norm ('normance') of a unit screw $\$_{\mathsf{A}}$ is obtained with

$$\langle \hat{\$}_{\mathsf{A}}, \hat{\$}_{\mathsf{A}} \rangle_{\scriptscriptstyle\blacksquare} = 1 + \epsilon \cdot 2 \cdot h + \epsilon^2 \cdot \|\underline{\boldsymbol{a}}_{\mathsf{A}}\|^2 \,,$$

indicating the squared unit length of the unit direction $\hat{\boldsymbol{n}}$ in primal order (with an empty measurement unit), the pitch $h = \hat{\boldsymbol{n}} * \boldsymbol{m}$ in dual order (with 'meter' unit), and the squared length of the lever with respect to the origin in ternal order (with 'squared-meter' unit). In Fig. 1, a sketch of the geometry of two example spears in space in relation to the origin as the evaluation point is provided.

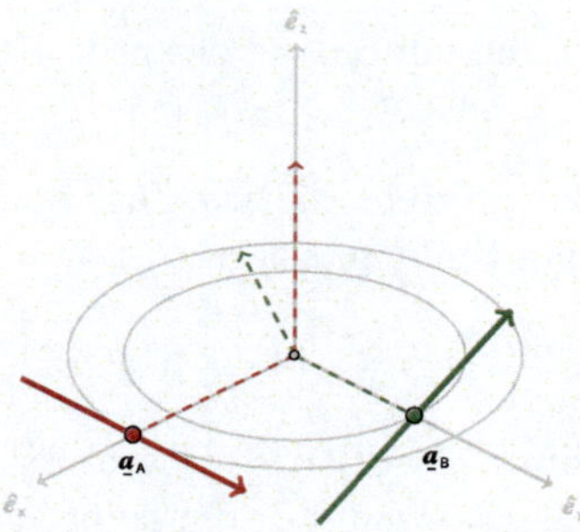

Fig. 1. Spears $\hat{\mathbf{\Lambda}}_{\mathsf{A}}$ and $\hat{\mathbf{\Lambda}}_{\mathsf{B}}$ in skew setup with $\boldsymbol{n}_{\mathsf{A}} = (0,1,0)^T$, $\boldsymbol{n}_{\mathsf{B}} = (-2,0,1)^T$ and zero anchors $\underline{\boldsymbol{a}}_{\mathsf{A}} = (1,0,0)^T$ and $\underline{\boldsymbol{a}}_{\mathsf{B}} = (0,3/4,0)^T$ with lever length $\|\underline{\boldsymbol{a}}\| = \|\boldsymbol{m}\|$.

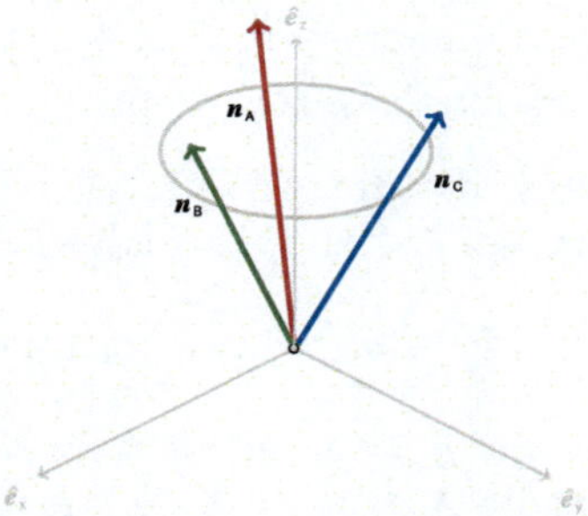

Fig. 2. Isonorm directions $\boldsymbol{n}_{\mathsf{A}}, \boldsymbol{n}_{\mathsf{B}}$, And $\boldsymbol{n}_{\mathsf{C}}$ from cone $x^2 + y^2 = \kappa \cdot z^2$, with maximal uniformity: the directions have identical scalar products (Gramian off-diagonals).

5 Ternary Product Matrices

Ternary Gramian. The three Gramian matrices of orientation, attitude, and position type have been defined [18] by

$$\begin{aligned} \boldsymbol{G}_{\mathsf{R}} &= \mathrm{Gram}_{\,\mathsf{R}}\big((\hat{\mathbf{\Lambda}}_{\mathsf{ij}})_{\mathsf{ij}}\big) := \hat{\boldsymbol{N}} * \hat{\boldsymbol{N}} \\ \boldsymbol{G}_{\mathsf{A}} &= \mathrm{Gram}_{\,\mathsf{A}}\big((\hat{\mathbf{\Lambda}}_{\mathsf{ij}})_{\mathsf{ij}}\big) := \hat{\boldsymbol{J}} * \boldsymbol{K} \\ \boldsymbol{G}_{\mathsf{P}} &= \mathrm{Gram}_{\,\mathsf{P}}\big((\hat{\mathbf{\Lambda}}_{\mathsf{ij}})_{\mathsf{ij}}\big) := \boldsymbol{M} * \boldsymbol{M} \end{aligned} \tag{11}$$

where the 'reciprocal Jacobian' [18] is defined as $\boldsymbol{K} := \triangle \cdot \hat{\boldsymbol{J}}$. The attitude Gramian $\boldsymbol{G}_{\mathsf{A}}$ is restated by $\boldsymbol{G}_{\mathsf{A}} = \hat{\boldsymbol{J}} \diamond \hat{\boldsymbol{J}}$, indicating the reciprocal product of (2) explicitly. The three Gramians appear as distinct 'faces' of the ternary Gramian $\boldsymbol{G}_{\blacksquare} = \boldsymbol{G}_{\mathsf{R}} + \epsilon \cdot \boldsymbol{G}_{\mathsf{A}} + \epsilon^2 \cdot \boldsymbol{G}_{\mathsf{P}}$, defined by the ternary spear product $\langle\, .,. \,\rangle_{\blacksquare}$ of (9) as

$$\boldsymbol{G}_{\blacksquare} = \mathrm{Gram}_{\blacksquare}\big((\hat{\mathbf{\Lambda}}_{\mathsf{ij}})_{\mathsf{ij}}\big) := \big(\langle\, \hat{\mathbf{\Lambda}}_{\mathsf{ij}}, \hat{\mathbf{\Lambda}}_{\mathsf{kl}} \,\rangle_{\blacksquare}\big)_{\mathsf{ij,kl}} \tag{12}$$

as a ternary generalization of a Gramian of a vector sequence. The determinant $\det(\boldsymbol{G})$ of a product matrix $\boldsymbol{G}$ matches the product of all its eigenvalues

$$V = \det(\boldsymbol{G}) = \prod_i \lambda_i(\boldsymbol{G}) \tag{13}$$

and provides an index for the squared volume ('volumance'). For the ternary Gramian matrix $\boldsymbol{G}_{\blacksquare}$, the determinant reads $\det(\boldsymbol{G}_{\blacksquare}) = \det(\boldsymbol{G}_{\mathsf{R}}) + \epsilon \cdot \det(\boldsymbol{G}_{\mathsf{A}}) + \epsilon^2 \cdot \det(\boldsymbol{G}_{\mathsf{P}})$. The minimal fraction 'reg($\boldsymbol{G}$)' of the eigenvalues of a Gramian matrix $\boldsymbol{G}$

$$U = \mathrm{reg}(\boldsymbol{G}) := \mathrm{cond}^{-1}(\boldsymbol{G}) = \frac{\lambda_{\mathsf{min}}(\boldsymbol{G})}{\lambda_{\mathsf{max}}(\boldsymbol{G})} = \min_{i,j}\left(\frac{\lambda_i(\boldsymbol{G})}{\lambda_j(\boldsymbol{G})}\right) \leq 1 \tag{14}$$

indicates the uniformity of the eigenvalues. In Fig. 2, an example of directions on a cone with maximal uniformity U is indicated. For the ternary Gramian matrix $\boldsymbol{G}_{\blacksquare}$, the uniformity reads $\mathrm{reg}(\boldsymbol{G}_{\blacksquare}) = \mathrm{reg}(\boldsymbol{G}_{\mathsf{R}}) + \epsilon \cdot \mathrm{reg}(\boldsymbol{G}_{\mathsf{A}}) + \epsilon^2 \cdot \mathrm{reg}(\boldsymbol{G}_{\mathsf{P}})$.

Table 1. Ternary Gramian matrix and ternary dyadic matrix for spear sequences.

Gramian matrix $\boldsymbol{G}_{\blacksquare}$	Dyadic matrix $\boldsymbol{H}_{\blacksquare}$
$\boldsymbol{G}_{\blacksquare}$ of a k-sequence $\hat{\boldsymbol{J}}$ of spears contains the ternary scalar products for all spear pairs in $\hat{\boldsymbol{J}}$ and $\boldsymbol{K}$	$\boldsymbol{H}_{\blacksquare}$ of a k-sequence $\hat{\boldsymbol{J}}$ of spears consists of the sum of the ternary dyadic products of k spear pairs in $\hat{\boldsymbol{J}}$ and $\boldsymbol{K}$
The ternary scalars 'det($\boldsymbol{G}_{\blacksquare}$)' and 'reg($\boldsymbol{G}_{\blacksquare}$)' yield criteria for sequences of independent directions, spears, moments	The ternary scalars 'det($\boldsymbol{H}_{\blacksquare}$)' and 'reg($\boldsymbol{H}_{\blacksquare}$)' yield criteria for sequences of redundant directions, spears, moments

Ternary Dyadic. In formal correspondence to the three Gramian matrices of (11), the three dyadic matrices are introduced for spears with

$$\begin{aligned} \boldsymbol{H}_{\mathsf{R}} &= \mathrm{Dyad}_{\mathsf{R}}\big((\hat{\boldsymbol{\Lambda}}_{\mathsf{ij}})_{\mathsf{ij}}\big) := \hat{\boldsymbol{N}} \otimes \hat{\boldsymbol{N}} \\ \boldsymbol{H}_{\mathsf{A}} &= \mathrm{Dyad}_{\mathsf{A}}\big((\hat{\boldsymbol{\Lambda}}_{\mathsf{ij}})_{\mathsf{ij}}\big) := \hat{\boldsymbol{J}} \otimes \boldsymbol{K} \\ \boldsymbol{H}_{\mathsf{P}} &= \mathrm{Dyad}_{\mathsf{P}}\big((\hat{\boldsymbol{\Lambda}}_{\mathsf{ij}})_{\mathsf{ij}}\big) := \boldsymbol{M} \otimes \boldsymbol{M} \;, \end{aligned} \tag{15}$$

where the dyadic $\boldsymbol{A} \otimes \boldsymbol{B}$ for matrices $\boldsymbol{A}, \boldsymbol{B} \in \mathbb{R}^{n \times k}$ corresponds to the k-sum of dyads $\boldsymbol{A} \otimes \boldsymbol{B} = \sum_i (\boldsymbol{a}_i \otimes \boldsymbol{b}_i) \in \mathbb{R}^{n \times n}$. The ternary dyadic matrix $\boldsymbol{H}_{\blacksquare} = \boldsymbol{H}_{\mathsf{R}} + \epsilon \cdot \boldsymbol{H}_{\mathsf{A}} + \epsilon^2 \cdot \boldsymbol{H}_{\mathsf{P}}$ subsumes the three matrices of (15) and is defined, resembling the form of the ternary Gramian matrix in (12), as

$$\boldsymbol{H}_{\blacksquare} = \mathrm{Dyad}_{\blacksquare}\big((\hat{\boldsymbol{\Lambda}}_{\mathsf{ij}})_{\mathsf{ij}}\big) := \sum_{\mathsf{ij}} \big(\hat{\boldsymbol{\Lambda}}_{\mathsf{ij}} \otimes_{\blacksquare} \hat{\boldsymbol{\Lambda}}_{\mathsf{ij}}\big) \;, \tag{16}$$

by means of the ternary dyadic product of (10). The dyadic matrix features the ternary determinant $V_{\blacksquare} = \det(\boldsymbol{H}_{\blacksquare}) = \det(\boldsymbol{H}_{\mathsf{R}}) + \epsilon \cdot \det(\boldsymbol{H}_{\mathsf{A}}) + \epsilon^2 \cdot \det(\boldsymbol{H}_{\mathsf{P}})$. For a summand $\boldsymbol{H} = \boldsymbol{B} \otimes \boldsymbol{B}$ with $\boldsymbol{B} \in \mathbb{R}^{n \times k}$, the determinant equals $\det(\boldsymbol{H}) = \sum_{i_1 < \ldots < i_n} (\det(\boldsymbol{b}_{i_1} \,|\, \boldsymbol{b}_{i_2} \,|\, \ldots \,|\, \boldsymbol{b}_{i_n}))^2$ indicating a $\binom{k}{n}$-fold sum of squared n-volumes. The dyadic matrix $\boldsymbol{H}_{\blacksquare}$ features the ternary uniformity $U_{\blacksquare} = \mathrm{reg}(\boldsymbol{H}_{\blacksquare}) = \mathrm{reg}(\boldsymbol{H}_{\mathsf{R}}) + \epsilon \cdot \mathrm{reg}(\boldsymbol{H}_{\mathsf{A}}) + \epsilon^2 \cdot \mathrm{reg}(\boldsymbol{H}_{\mathsf{P}})$. A comparison of the geometric characteristics of the ternary Gramian matrix and the ternary dyadic matrix and their associated indices, the determinant and the uniformity, is outlined in Table 1.

6 Generalized Product Matrix and Indices

For rectangular matrices $\boldsymbol{A}, \boldsymbol{B} \in \mathbb{R}^{k \times n}$, the generalized product is defined[2] by

$$\boldsymbol{Q}_{AB} = \mathrm{Prod}\big(\boldsymbol{A}, \boldsymbol{B}\big) := \begin{cases} \boldsymbol{A} * \boldsymbol{B} & \text{for } k \leq n \\ \boldsymbol{A} \otimes \boldsymbol{B} & \text{for } k > n \end{cases} , \tag{17}$$

as a tool to introduce three particular products $\boldsymbol{Q}_{\mathsf{R}} = \mathrm{Prod}_{\mathsf{R}}\big((\hat{\boldsymbol{\Lambda}}_{\mathsf{ij}})_{\mathsf{ij}}\big) := \mathrm{Prod}\big(\hat{\boldsymbol{N}}, \hat{\boldsymbol{N}}\big)$, $\boldsymbol{Q}_{\mathsf{A}} = \mathrm{Prod}_{\mathsf{A}}\big((\hat{\boldsymbol{\Lambda}}_{\mathsf{ij}})_{\mathsf{ij}}\big) := \mathrm{Prod}\big(\hat{\boldsymbol{J}}, \boldsymbol{K}\big)$, and $\boldsymbol{Q}_{\mathsf{P}} = \mathrm{Prod}_{\mathsf{P}}\big((\hat{\boldsymbol{\Lambda}}_{\mathsf{ij}})_{\mathsf{ij}}\big) :=$

[2] For identical arguments, the generalized product of (17) matches the 'denominators' employed in computing the ('left' and 'right') generalized Moore–Penrose inverses.

Table 2. Comparison of concepts for the three geometric aspects of spear systems.

Type	Orientation	Attitude	Position
Unit	ϵ^0	ϵ	ϵ^2
Order	Primal	Dual	Ternal
Components	Directions $\boldsymbol{n} \leftrightarrow \boldsymbol{n}$	Blends $\boldsymbol{n} \leftrightarrow \boldsymbol{m}$	Moments $\boldsymbol{m} \leftrightarrow \boldsymbol{m}$
Ternary volume criterion	$V_{\mathsf{R}} = \det(\boldsymbol{Q}_{\mathsf{R}})$ $\propto$ direction volumance	$V_{\mathsf{A}} = \det(\boldsymbol{Q}_{\mathsf{A}})$ $\propto$ spear volumance	$V_{\mathsf{P}} = \det(\boldsymbol{Q}_{\mathsf{P}})$ $\propto$ moment volumance
Ternary shape criterion	$U_{\mathsf{R}} = \mathrm{reg}(\boldsymbol{Q}_{\mathsf{R}})$ $\propto$ direction uniformity	$U_{\mathsf{A}} = \mathrm{reg}(\boldsymbol{Q}_{\mathsf{A}})$ $\propto$ spear uniformity	$U_{\mathsf{P}} = \mathrm{reg}(\boldsymbol{Q}_{\mathsf{P}})$ $\propto$ moment uniformity
Validity	'global'	'intrinsic'	'local'

$\mathrm{Prod}(\boldsymbol{M}, \boldsymbol{M})$ to unify the Gramian products of (11) and dyadic products of (15) in a common expression. The matrices $\boldsymbol{Q}_{\mathsf{R}}$, $\boldsymbol{Q}_{\mathsf{A}}$, $\boldsymbol{Q}_{\mathsf{P}}$ are elements of $\boldsymbol{Q}_{\blacksquare}$, the generalized ternary product matrix

$$\boldsymbol{Q}_{\blacksquare} = \mathrm{Prod}_{\blacksquare}((\hat{\boldsymbol{\Lambda}}_{\mathsf{ij}})_{\mathsf{ij}}) := \boldsymbol{Q}_{\mathsf{R}} + \epsilon \cdot \boldsymbol{Q}_{\mathsf{A}} + \epsilon^2 \cdot \boldsymbol{Q}_{\mathsf{P}} \,. \tag{18}$$

The indices $\det(\boldsymbol{Q}_{\blacksquare})$ and $\mathrm{reg}(\boldsymbol{Q}_{\blacksquare})$ of the generalized ternary matrix $\boldsymbol{Q}_{\blacksquare}$ provide three-fold measures for the squared volume ('volumance'), as in (13), and for the uniformity of shape by means of the ratio of extreme eigenvalues (14):

$$V_{\blacksquare} = \det(\boldsymbol{Q}_{\blacksquare}) = \det(\boldsymbol{Q}_{\mathsf{R}}) + \epsilon \cdot \det(\boldsymbol{Q}_{\mathsf{A}}) + \epsilon^2 \cdot \det(\boldsymbol{Q}_{\mathsf{P}}) \tag{19}$$

$$U_{\blacksquare} = \mathrm{reg}(\boldsymbol{Q}_{\blacksquare}) = \mathrm{reg}(\boldsymbol{Q}_{\mathsf{R}}) + \epsilon \cdot \mathrm{reg}(\boldsymbol{Q}_{\mathsf{A}}) + \epsilon^2 \cdot \mathrm{reg}(\boldsymbol{Q}_{\mathsf{P}}) \tag{20}$$

The ternary concepts of the volumance $V_{\blacksquare}$ and of the uniformity $U_{\blacksquare}$ to characterize spear systems, and thus singularities and manipulability of general spatial kinematic chains, in three distinct geometric aspects are summarized in Table 2.

7 Conclusions

In this paper, three distinct Gramian matrices, that previously appeared as unrelated concepts associated to a 3-sequence of oriented lines in space, are consolidated as aspects of the unifying concept of the ternary Gramian matrix. For spatial kinematic chains of arbitrary length, the ternary dyadic matrix is introduced as an analogue concept that complements the ternary Gramian. The two ternary matrix products are deduced from ternary generalizations of the scalar product and the dyadic product of dual vectors. The Gramian and dyadic product matrices are united by a generalized ternary product which permits to analyze the dexterity of a spatial kinematic chain in a common form. The relevance of the ternary determinant and uniformity of the generalized product for specifying volume and shape aspects of a sequence of spatial spears is indicated.

In the future, the concepts, proposed in this manuscript in a compact form, should be specified and illustrated in greater detail: The geometric significance of the different product types can be motivated by means of vector sequences and the transference principle; the physical meaning of the ternary performance indices can be exemplified with concrete instances of spatial kinematic chains. Finally, the idea of 'ternary numbers', which was associated to dual-number algebras in a pragmatic manner, needs to be considered systematically from a mathematical point of view.

Acknowledgements. The second author acknowledges that this work has been supported by the "LCM-K2 Center for Symbiotic Mechatronics" within the framework of the Austrian COMET-K2 program.

References

1. von Mises, R.: Motorrechnung, ein neues Hilfsmittel der Mechanik. Z. Angew. Math. Mech. **4**, 2 (1924)
2. Brand, L.: Vector and Tensor Analysis. Wiley (1947)
3. Woo, L., Freudenstein, F.: Application of Line geometry to theoretical kinematics and the kinematic analysis of mechanical systems. J. Mech. **5**, 3 (1970)
4. Veldkamp, G.R.: On the use of dual numbers, vectors and matrices in instantaneous, spatial kinematics. Mech. Mach. Theory **11**, 2 (1976)
5. Sugimoto, K., Duffy, J.: Application of linear algebra to screw systems. Mech. Mach. Theory **17**, 1 (1982)
6. Selig, J.M.: A note on the principle of transference. In: 19th Mechanisms Conference (ASME) (1986)
7. Murray, R.M., Li, Z., Sastry, S.S.: A Mathematical Introduction to Robotic Manipulation. CRC Press (1994)
8. Staffetti, E., Bruyninckx, H., De Schutter, J.: On the invariance of manipulability indices. In: Advances in Robot Kinematics. Springer, Dordrecht (2002). https://doi.org/10.1007/978-94-017-0657-5_7
9. Davidson, J.K., Hunt, K.H.: Robots and Screw Theory: Applications of Kinematics and Statics to Robotics. Oxford University Press (2004)
10. Donelan, P.S.: Singularity-theoretic methods in robot kinematics. Robotica **25**, 06 (2007)
11. Pennestrì, E., Stefanelli, R.: Linear algebra and numerical algorithms using dual numbers. Multibody Syst. Dyn. **18** (2007)
12. McCarthy, J.M., Soh, G.S.: Geometric Design of Linkages. Springer, New York (2010). https://doi.org/10.1007/978-1-4419-7892-9
13. Bongardt, B.: Geometric Characterization of the Workspace of Non-orthogonal Rotation Axes. J. Geomet. Mech. (2014)
14. Deza, M.M., Deza, E.: Encyclopedia of Distances. Springer, Heidelberg (2014). https://doi.org/10.1007/978-3-662-44342-2
15. Lynch, K.M., Park, F.C.: Modern Robotics: Mechanics, Planning, and Control. Cambridge University Press (2017)
16. Bongardt, B.: The adjoint trigonometric representation of displacements and a closed- form solution to the IKP of general 3C chains. J. Appl. Math. Mech. **100**, 6 (2020)

17. Dai, J.S., Sun, J.: Geometrical revelation of correlated characteristics of the ray and axis order of the Plücker coordinates in line geometry. Mech. Mach. Theory **153** (2020)
18. Bongardt, B., Müller, A.: On orientation, position, and attitude singularities of general 3R chains. Math. Robot. (2022)

Extended Rotation Matrix for Kinematics of Pointing Mechanisms

Shaoping Bai(✉)

Department of Materials and Production, Aalborg University, Aalborg, Denmark
shb@mp.aau.dk

Abstract. This paper introduces a new formation of rotation, which is developed with two non-parallel vectors. A transformation matrix, called extended rotation matrix (ERM), is thus formulated. In particular, the matrix contains two known vectors from one body and their cross product, with which all other vectors or points in the same body can be uniquely described by a set of values associated with the matrix, namely, their alternative coordinates. Using ERM and alternative coordinates, kinematic equations of pointing mechanisms can be formulated uniquely and conveniently, without any redundancy of parametrization. A case study of pointing mechanisms is included to demonstrate the advantage of the new formulation.

1 Introduction

Rotation matrix is a classic mathematic tool to transform coordinates within two coordinate frames and to describe rigid-body rotation. According to Euler's Theorem, a rigid-body rotation, or simply a rotation, is represented by a 3×3 proper orthogonal matrix. Many types of parameters can be used to express rotation matrix, such as Euler angles, Euler parameters, natural invariants, etc. [1–3]. For all representations, at least three parameters are needed.

While rotation matrix is widely used in kinematics and robotics, a limitation to represent space rotation in less than 3-DOF, such as 1-DOF spherical linkages, or 2-DOF spherical mechanisms, is noticed. Ideally, the rotation matrix for these type of mechanisms should correspondingly have 1 or 2 parameters to match the number of DOF. Unfortunately, any aforementioned representation of rotation matrix contains parameters more than needed. Parameter redundancy exists for these mechanisms. Thus in formulating the kinematic equations, we have to eliminate redundancy in parametrization.

The interest of this study is the kinematics of pointing mechanisms. This type of mechanisms has two DOFs and provides orientations with respect to desired longitude and latitude on the unit sphere [4]. Pointing mechanisms have a wide range of applications, covering five-axis machining tools [5], solar panel tracker [6], micro assembly cell [7], etc. They can be constructed by adopting either serial or parallel kinematic chains [8]. For serial pointing mechanisms, two motors (actuators) drive directly to rotate for azimuth and elevation angles. On

O. Altuzarra and A. Kecskéméthy (Eds.): ARK 2022, SPAR 24, pp. 39–46, 2022.
https://doi.org/10.1007/978-3-031-08140-8_5

the other hand, parallel pointing mechanisms require to calculate the drive angle of each actuators from their inverse kinematics. Following foregoing introduction, parameter redundancy of rotation matrix is desirably to be eliminated.

In this work, we introduce a new formulation of rotation free of the rotation parameters introduced. The idea is to use three non-coplanar vectors to establish a frame of coordinates to describe the orientation of a rigid body. Within the new frame, any point or vector attached to the body is uniquely expressed by a new set of coordinates, with which constraint equations of pointing mechanisms are readily defined and solved for inverse and forward kinematic problems.

2 Extended Rotation Matrix and Alternative Coordinates

Prior to formulating kinematic equations of pointing mechanisms, we introduce first the concept of extended rotation matrix.

By definition, a rotation matrix is a matrix that is used to perform a rotation in Euclidean space. Algebraically, a rotation matrix in n-dimensions is a $n \times n$ special orthogonal matrix, i.e. an orthogonal matrix whose determinant is 1. Geometrically, each column in a rotation matrix represents the unit vector of an axis of the body attached coordinate frame. Take 3-D rotation as example, a rotation matrix $\mathbf{Q}$ can be written as

$$\mathbf{Q} = [\mathbf{e}_x, \mathbf{e}_y, \mathbf{e}_z] \tag{1}$$

where $\mathbf{e}_x$, $\mathbf{e}_y$, $\mathbf{e}_z$ and the unit vectors of $x-$, $y-$ and $z-$axis of the body-attached frame, respectively.

Now we extend the rotation matrix from special orthogonal matrix to a more general one. Instead of orthogonal unit vectors $\mathbf{e}_x$, $\mathbf{e}_y$, and $\mathbf{e}_z$, we use three non-coplanar vectors to establish the body attached frame. They consist a basis of $\mathbb{R}^3$. Let $\mathbf{e}_1, \mathbf{e}_2, \mathbf{e}_3$ are three non-coplanar 3-dimensional vectors, then an arbitrary vector $\mathbf{v}$ of the same dimension can be written as a linear combination of three vectors,

$$\mathbf{v} = c_1\mathbf{e}_1 + c_2\mathbf{e}_2 + c_3\mathbf{e}_3 \tag{2}$$

where $c_i, i = 1, 2, 3$ are real coefficients.

Apparently, there are infinitive sets of $\mathbf{e}_i, i = 1, 2, 3$ that can be selected to form a basis, but they are generally not helpful for the problem at hand, as they do not contribute to reduce redundancy of parametrization.

In this formulation, we assume two vectors on the body are known, then a basis is constructed from the two vectors, along with their cross product. By this way, an orientation representation with minimum parameters is obtained.

We take an arbitrary rigid body in the form of spherical triangle ABC, shown in Fig. 1, as an example to describe the new formation. Assume unit vectors $\mathbf{n}_1$ and $\mathbf{n}_2$ parallel to OA and OB are known. A frame can hence be established with both vectors and their cross product, namely, $\mathbf{n}_1$, $\mathbf{n}_2$ and $\mathbf{n}_d = \mathbf{n}_1 \times \mathbf{n}_2$. They are linearly independent and form a basis of $\mathbb{R}^3$.

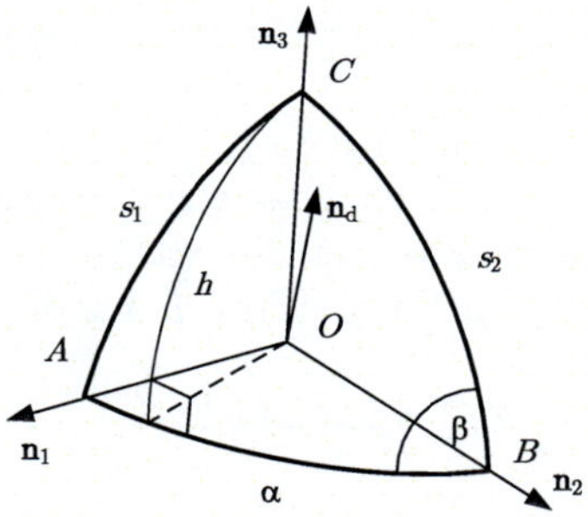

Fig. 1. Illustration of parameterized coordinates

With reference to Fig. 1, the unit vector parallel to OC, noted as $\mathbf{n}_3$, can be expressed as

$$\mathbf{n}_3 = \mu\mathbf{n}_1 + \nu\mathbf{n}_2 + \lambda\mathbf{n}_d = \mathbf{M}\mathbf{n}_3^\star, \quad \mathbf{n}_3^\star = [\mu, \nu, \lambda]^T \tag{3}$$

where μ, ν and λ are dimensionless parameters. In this formulation, $\mathbf{M}$ is referred as *extended rotation matrix* (ERM), while $\mathbf{n}_3^\star$ is a vector containing associated coordinates of point C, which are the set of *alternative coordinates.* Following a previous work [9], we refer them as *parameterized coordinates.*

The parameterized coordinates can be uniquely determined from either their geometry or known values of three vectors. They can be uniquely determined from Eq. (3)

$$\mathbf{n}_3^\star = \mathbf{M}^{-1}\mathbf{n}_3 \tag{4}$$

When the dimensions of the rigid body are known, the parameterized coordinates are readily determined as well. We premultiply vectors $\mathbf{n}_1^T$, $\mathbf{n}_2^T$, and $\mathbf{n}_d^T$ separately on both sides of Eq. (3), which yields

$$\cos s_1 = \mu + \nu\cos\alpha \tag{5a}$$

$$\cos s_2 = \mu\cos\alpha + \nu \tag{5b}$$

$$p = \lambda\sin^2\alpha \tag{5c}$$

from which we obtain

$$\mu = \frac{\cos s_1 - \cos s_2\cos\alpha}{\sin^2\alpha}, \; \nu = \frac{\cos s_2 - \cos s_1\cos\alpha}{\sin^2\alpha}, \; \lambda = \frac{p}{\sin^2\alpha} = \frac{\sin h}{\sin\alpha} \tag{6}$$

where $p = (\mathbf{n}_1 \times \mathbf{n}_2)\cdot\mathbf{n}_3$. From the spherical trigonometry, it can be shown that $p = \sin\alpha\sin h$, where $h = \sin^{-1}(\sin\beta\sin s_1)$.

As we can see, the parameterized coordinates are determined purely from geometric relations of three vectors. They are invariants and hence frame independent.

3 Geometric and Algebraic Properties of ERMs

Geometrically, rotation is a transformation that moves points without changing the distances between them. This geometric property is preserved with the ERMs, as the alternative coordinates remain unchanged with the new formation.

Algebraically, a rotation matrix satisfies orthogonal condition and unitary determinant, namely,

$$\mathbf{Q}^T\mathbf{Q} = \mathbf{1}; \quad \det(\mathbf{Q}) = 1 \tag{7}$$

where $\mathbf{1}$ stands for the 3×3 identity matrix.

For ERMs, it has to be noted that both conditions are not satisfied. Instead, the new matrix leads to

$$\mathbf{M}^T\mathbf{M} = \begin{bmatrix} 1 & \cos\alpha & 0 \\ \cos\alpha & 1 & 0 \\ 0 & 0 & \sin^2\alpha \end{bmatrix} \tag{8}$$

and

$$\det(\mathbf{M}) = \sin^2\alpha \tag{9}$$

where α is the angle made by $\mathbf{n}_1$ and $\mathbf{n}_2$. From the above identities, the inverse matrix to the RHS of eq. (8) is found as

$$\mathbf{N} = \frac{1}{\sin^2\alpha}\begin{bmatrix} 1 & -\cos\alpha & 0 \\ -\cos\alpha & 1 & 0 \\ 0 & 0 & 1 \end{bmatrix} \tag{10}$$

Thus the inverse matrix of $\mathbf{M}$ is

$$\mathbf{M}^{-1} = \mathbf{N}\mathbf{M}^T \tag{11}$$

Matrix $\mathbf{M}$ can be expressed in another way

$$\mathbf{M} = [\mathbf{m}_1, \mathbf{m}_2, \mathbf{m}_3] \tag{12}$$

where vectors $\mathbf{m}_i, i = 1, 2, 3$ are non coplanar vectors, which can be written as

$$\mathbf{m}_i = c_{i1}\mathbf{e}_x + c_{i2}\mathbf{e}_y + c_{i3}\mathbf{e}_z = \mathbf{Q}\mathbf{c}_i, \quad i = 1, 2, 3 \tag{13}$$

where $c_{ij}, j = 1, 2, 3$ are real coefficients. We thus have

$$\mathbf{M} = \mathbf{Q}[\mathbf{c}_1, \mathbf{c}_2, \mathbf{c}_3] = \mathbf{Q}\mathbf{C} \tag{14}$$

where matrix $\mathbf{C}$ is a mapping matrix for the conventional rotation matrix and the extended rotation matrix. Obviously, $\det(\mathbf{C}) = \det(\mathbf{M}) = \sin^2\alpha$.

The relations between the local coordinates and the alternative coordinates can now be derived. Let $\mathbf{Q}$ be the rotation matrix rotating a rigid body from frame $\mathcal{B}$ to frame $\mathcal{A}$. For an arbitrary vector $\mathbf{s}$ in frame $\mathcal{B}$, the following relation holds:

$$\mathbf{s} = \mathbf{Q}\mathbf{s}' \tag{15}$$

where $\mathbf{s}'$ is the local vector expressed in frame $\mathcal{A}$.

On the other hand, the same vector can be expressed in terms of alternative coordinates as

$$\mathbf{s} = \mathbf{M}\mathbf{s}^{\star} = \mathbf{Q}\mathbf{C}\mathbf{s}^{\star} \tag{16}$$

from which we find a useful identity

$$\mathbf{s}' = \mathbf{C}\mathbf{s}^{\star} \tag{17}$$

4 A 2-DOF Pointing Mechanism

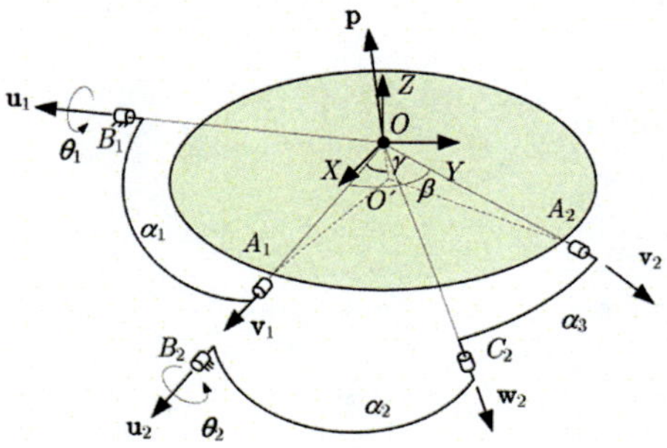

Fig. 2. A spherical pointing mechanism centered at point O

Now we can derive explicitly the kinematics equation of pointing mechanisms with the new formulation. A spherical five-bar spherical linkage is shown in Fig. 2. This is a 2-DOF parallel spherical manipulator constructed with two limbs. One RR limb of single curved link B_1A_1 has a dimension of α_1, the other RRR limb of two curved links B_2C_2 and B_2A_2 having dimensions of α_2 and α_3, respectively. The two revolute joints B_1 and B_2 on the fixed base are described by unit vectors $\mathbf{u}_1$ and $\mathbf{u}_2$, while the two revolute joints A_1 and A_2 on the moving platform are described by unit vectors $\mathbf{v}_1$ and $\mathbf{v}_2$, respectively. The unit vector parallel to the axis of the revolute joint C_2 is denoted by $\mathbf{w}_2$. The pointing direction is noted by a unit vector $\mathbf{p} = [x, y, z]^T$, its magnitude unity implying

$$g(x, y, z) = x^2 + y^2 + z^2 - 1 = 0 \tag{18}$$

Without loss of generality, we assume the mobile platform is a cone with its half angle specified by γ. We start from the first limb B_1A_1. The link rotates about a fixed axis for an angle θ_1,

$$\mathbf{v}_1 = \mathbf{R}_1\mathbf{v}_{10} \tag{19}$$

where $\mathbf{v}_{10}$ is the unit vector parallel to the initial position of OA_1. Moreover, matrix $\mathbf{R}_1$ describes the rotation of link B_1A_1 about OB_1. Using *natural invariants* of the rigid-body rotation, matrix $\mathbf{R}_1$ takes the form [2]

$$\mathbf{R}_1 = \mathbf{u}_1\mathbf{u}_1^T + \sin\theta_1\tilde{\mathbf{u}}_1 + \cos\theta_1(\mathbf{1} - \mathbf{u}_1\mathbf{u}_1^T) \tag{20}$$

where the sign '~' on top of a vector denotes the *anti-skew symmetric matrix* derived from the vector.

The first constraint equation is hence formulated upon vector $\mathbf{v}_1$ expressed above,

$$\mathbf{p}^T\mathbf{v}_1 = \cos(\pi - \gamma) = -\cos\gamma \tag{21}$$

which implies the angle between OA_1 and the pointing direction remains constant.

For the second limb, curved link B_2C_2 rotates about the axis OB_2, which is described, similar to Eq. (20), by

$$\mathbf{R}_2 = \mathbf{u}_2\mathbf{u}_2^T + \sin\theta_2\tilde{\mathbf{u}}_2 + \cos\theta_2(\mathbf{1} - \mathbf{u}_2\mathbf{u}_2^T) \tag{22}$$

upon which the unit vector $\mathbf{w}_2$ is expressed as

$$\mathbf{w}_2 = \mathbf{R}_2\mathbf{w}_{20} \tag{23}$$

where $\mathbf{w}_{20}$ is the unit vector parallel to OC_2 at the reference position.

The geometric constraint equation of the second limb is

$$\mathbf{w}_2^T\mathbf{v}_2 = \cos\alpha_3 \tag{24}$$

Using Eq. (3), we can express unit vector $\mathbf{v}_2$ in terms of vectors $\mathbf{p}$ and $\mathbf{v}_1$ as

$$\mathbf{v}_2 = \mathbf{M}\mathbf{v}_2^\star, \quad \mathbf{M} = [\mathbf{p}, \mathbf{v}_1, \mathbf{p}\times\mathbf{v}_1] \tag{25}$$

The alternative coordinates are

$$\mathbf{v}_2^\star = [\mu_2, \nu_3, \lambda_2]^T = \mathbf{M}_0^{-1}\mathbf{v}_{20} = [\mathbf{p}_0, \mathbf{v}_{10}, \mathbf{p}_0\times\mathbf{v}_{10}]^{-1}\mathbf{v}_{20} \tag{26}$$

where $\mathbf{v}_{20}$ is the unit vector parallel to the initial position of OC_2.

Equations (21) and (24) stand for the system of constraint equations of the 2-DOF pointing mechanism. The system contains two inputs angles and the 3-dimensional vector $\mathbf{p}$, which is subject to unity length constraints. The system enables to solve directly the forward position problem for any specified angles of θ_1 and θ_2, or the inverse position problem with a specified pointing vector.

Equations (21) and (24) can adopt *spherical coordinates*, namely, *longitude* and *latitude*. This will then yield another system of constraint equation of with two inputs, in terms of either driving angles or spherical coordinates of pointing vector, and two unknowns only.

5 Example

We include an example to demonstrate the new approach. In this example, the geometry of a 2-DOF pointing mechanism is given by $\mathbf{u}_1 = [1, 0, 0]^T$, $\mathbf{u}_2 = [0, 1, 0]^T$, and $\{\alpha_1, \alpha_2, \alpha_3, \beta, \gamma\} = \{90, 60, 60, 65, 75\}$ degrees. Let $\mathbf{p} = [0.3551, 0.0719, 0.9320]^T$. Using (21) and (24), four inverse kinematic solutions are immediately found, as listed in Table 1. Figure 3 displays all four configurations of the pointing mechanism for the four solutions.

Table 1. Inverse kinematic solutions

No.	1	2	3	4
$[\theta_1, \theta_2]$	[1.7742, 1.2102]	[1.7742, −3.0900]	[−1.9284, −0.4538]	[−1.9284, −2.5035]

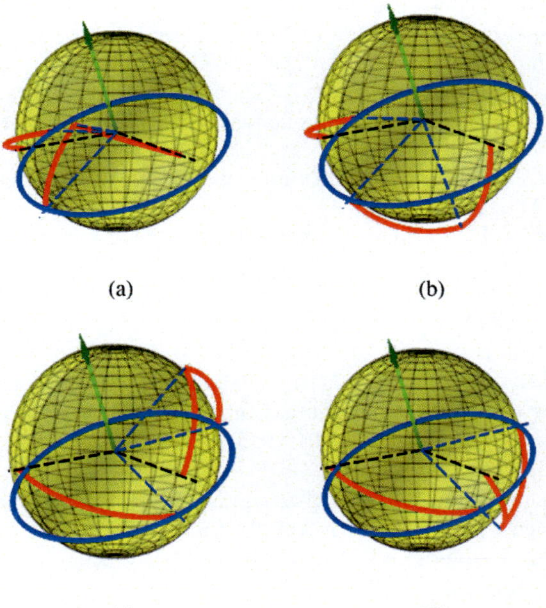

(a) (b) (c) (d)

Fig. 3. Graphic display of the inverse kinematic solutions, (a) solution #1, (b) solution #2, (c) solution #3, (d), solution #4. In the figure, red curves are moving links, while the blue circle stands for the mobile platform, with a pointing vector in green normal to it.

6 Conclusions

In this work, a new method of alternative rotation representation, namely, ERM, is proposed to simplify the kinematic constraint equations of parallel pointing mechanisms. With the ERM, rotation equation is established by taking advantage of two known vectors attached to the mobile platform, which avoids to use the classic rotation matrix. By this way, the redundancy of rotation parametrization for the pointing mechanisms is eliminated. Both the inverse and forward position problem can be readily solved.

The developed method has advantages of simple formulation for mechanisms of fewer degrees of freedom. In this paper, an example of position analysis of a 2-DOF parallel pointing mechanism is included, which demonstrates the effectiveness of the new method. Cases of using ERM in single-DOF linkage kinematic analysis can be found in [9, 10]. Application of the method to other parallel mechanisms will be considered in future study.

References

1. Stuelpnagel, J.: On the parametrization of the three-dimensional rotation group. SIAM Rev. **6**(4), 422–430 (1964)
2. Angeles, J.: Fundamentals of Robotic Mechanical Systems. MES, vol. 124. Springer, Cham (2014). https://doi.org/10.1007/978-3-319-01851-5
3. Bai, S., Li, X., Angeles, J.: A review of spherical motion generation using either spherical parallel manipulators or spherical motors. Mech. Mach. Theory **140**, 377–388 (2019)
4. Li, W., Angeles, J., Valášek, M.: Contributions to the kinematics of pointing. Mech. Mach. Theory **108**, 97–109 (2017)
5. Weck, M., Staimer, D.: Parallel kinematic machine tools - current state and future potentials. CIRP Annals **51**(2), 671–683 (2002)
6. Nagaraja, T.K., et al.: Design and development of solar panel tracking mechanism. In: International Conference on Sustainable Engineering and Technology (ICONSET 2018), p. 020014 (2018)
7. Palmieri, G., Callegari, M., Carbonari, L., Palpacelli, M.C.: Mechanical design of a mini pointing device for a robotic assembly cell. Meccanica **50**(7), 1895–1908 (2015). https://doi.org/10.1007/s11012-015-0132-1
8. Nie, L., Ding, H., Kecskeméthy, A., Gan, J., Wang, J., Ting, K.: Singularity and branch identification of a 2 degree-of-freedom (DOF) seven-bar spherical parallel manipulator. Mech. Sci. **11**(2), 381–393 (2020)
9. Bai, S.: Algebraic coupler curve of spherical four-bar linkages and its applications. Mech. Mach. Theory **158**, 104218 (2021)
10. Bai, S., Li, Z., Angeles, J.: Exact path synthesis of RCCC linkages for a maximum of nine prescribed positions. ASME J. Mech. Robot. **14**(2), 021011 (2022)

A Linearization Method Based on Lie Algebra for Pose Estimation in a Time Horizon

Nicolas Torres Alberto[1(✉)], Lucas Joseph[2], Vincent Padois[3], and David Daney[4]

[1] Stellantis, Centre Technique Vélizy, Vélizy-Villacoublay, France
nicolas.torres@stellantis.com
[2] CNRS, Solvay, Univ. Bordeaux, Bordeaux, France
lucas.joseph-ext@solvay.com
[3] Auctus, Inria Bordeaux, Talence, France
vincent.padois@inria.fr
[4] Auctus, Inria Bordeaux, Talence, France
david.daney@inria.fr

Abstract. In this paper, a strategy for linear pose estimation over a time horizon is presented. This linearization is crucial for the computationally efficient formulation of predictive and optimization-based control problems in robotics. The proposed approach is based on a truncation of the Magnus expansion for the approximation of the exponential map derivative and employs Lie algebra to represent position and orientation, allowing for a unified vectorial representation in vector form that can be integrated linearly over time, offering a convenient formulation for optimization solvers. The method shows promising results for precision and computation times.

1 Introduction

Collaborative robotics has fueled a growing demand for more complex control schemes [1]. Indeed, during human-robot interactions, robots evolve in a changing environment. Consequently, they are required to adapt reactively to unpredictable events while satisfying a set of constraints in order to provide safety guarantees.

In this context of complex robotic applications, optimization-based controllers offer a convenient way to formulate tasks to be achieved under constraints. These problems can be formulated over one control time step, i.e. reactively [5], but can also lead to much more robust behaviours when considering control over receding time horizons. Most humanoid control approaches rely on such a Model Predictive Control (MPC) paradigm [6] but these type of approaches remain to be proven useful in other complex robotics contexts.

One of the key issues when considering a receding horizon formulation is the ability to locally predict the evolution of the robot state given a control horizon while maintaining the computation cost low enough to allow for frequent optimization updates. While recent works have shown the ability to perform such

O. Altuzarra and A. Kecskémthy (Eds.): ARK 2022, SPAR 24, pp. 47–56, 2022.
https://doi.org/10.1007/978-3-031-08140-8_6

locally optimal computation in a non linear way [7], linear formulations remain advantageous both from a computational and control problem formulation point of view. Yet, to the best of our knowledge, an efficient linearization method for both 3D position + orientation[1] while offering a convenient formulation for optimization algorithms remains to be proposed.

The objective of this paper is to present a way to estimate, at each time step, the robot pose trajectory in a receding horizon by reconstructing it from the initial pose and a known input twist horizon.

The proposed method relies on the Lie algebra $\mathfrak{se}^3$ associated to the tangential space of a pose described in $\mathbb{SE}^3$. This method (based on the Magnus Expansion (ME) [4]) leads to a linear expression of the transformation between two poses, using the logarithmic map. In its truncated form, it leads to a linear propagation/integration formula which has an explicit algebraic form and takes a vector form adapted to the writing of recursive state propagation general formulas.

In Sect. 2, some fundamentals about Lie groups are presented. Section 3 presents the linearization strategy. Finally, Sect. 4 presents a simple kinematic simulation to showcase the results.

2 Fundamentals

This Section introduces the fundamental modeling aspects underlying the general objectives of this article. Lie groups and their algebras are first presented in vector and matricial forms. Afterwards, the receding horizon linearized system form pursued in this paper is presented.

2.1 Lie Groups and Algebras

A pose is composed of a position $\boldsymbol{p} \in \mathbb{R}^3$ and an orientation. For this paper, orientations are represented as a rotation matrix $\boldsymbol{R}$ that belongs to the Special Orthogonal Group ($\mathbb{SO}^3$), a Lie group. The robot pose $\boldsymbol{\mathcal{X}} = (\boldsymbol{R}, \boldsymbol{p})$ is composed of two elements, belonging to the Special Euclidean Group ($\mathbb{SE}^3$), another Lie group.

Given a Lie group $\mathcal{G}$, there exists a Lie algebra $\mathfrak{g}$ which is the logarithm map of the Lie group $\mathcal{G}$ such that $\log : \mathcal{G} \rightarrow \mathfrak{g}$, with an inverse operation $\exp : \mathfrak{g} \rightarrow \mathcal{G}$. This algebra corresponds to the tangent space of an infinitesimal increments.

In the case of pose, a infinitesimal increment is called a twist $\boldsymbol{T} \in \mathfrak{se}^3$, composed of an infinitesimal orientation increment (or rotational speed) $\boldsymbol{\omega} \in \mathfrak{so}^3$ (that in term corresponds to the Lie algebra of $\mathbb{SO}^3$) and linear component (linear velocity) $\boldsymbol{v} \in \mathbb{R}^3$.

Given $\boldsymbol{\xi} = \log \boldsymbol{\mathcal{X}}$, $\boldsymbol{\xi}$ represents the equivalent pose increment that will produce the same linear and rotational displacement as the homogenous transformation $\boldsymbol{\mathcal{X}}$(also referred as a pose here) when integrated for 1s. By abuse of

[1] Typically of the robot end-effector.

language, $\boldsymbol{\xi}$ might also be referred as a pose throughout this paper (as it can be used to retrieve the pose with $\boldsymbol{\mathcal{X}} = e^{\boldsymbol{\xi}}$). Even though both $\boldsymbol{T}$ and $\boldsymbol{\xi}$ both belong in $\mathfrak{se}^3$, only $\boldsymbol{T}$ will be used to represent infinitesimal increments throughout this paper. The same can be said for $\boldsymbol{\phi} = \log \boldsymbol{R}$, with $\boldsymbol{\omega}$ and $\boldsymbol{\phi}$ both belonging in $\mathfrak{so}^3$, but only $\boldsymbol{\omega}$ representing an infinitesimal increment.

It is important to remark that $\boldsymbol{\phi} \in \mathfrak{so}^3$ yields a one to one correspondence to $\boldsymbol{R} \in \mathbb{SO}^3$ through the log map as long as $||\boldsymbol{\phi}|| \leq \pi$. This in turn extends to $\boldsymbol{\xi}$ and $\boldsymbol{\mathcal{X}}$ with $\mathfrak{se}^3$ and $\mathbb{SE}^3$.

Due to the skew-symmetric nature of $\mathfrak{so}^3$, $\boldsymbol{\omega}$ admits a vector parametrization in $\mathbb{R}^3$ and a matricial form in $\mathbb{R}^{3\times3}$. One can go from one to the other through the reciprocal operations: the hat map $\cdot^\wedge : \mathbb{R}^3 \rightarrow \mathbb{R}^{3\times3}$ and the vee map $\cdot^\vee : \mathbb{R}^{3\times3} \rightarrow \mathbb{R}^3$. This extends to $\mathfrak{se}^3$ between $\mathbb{R}^6$ and $\mathbb{R}^{4\times4}$ as:

$$\boldsymbol{T}^\wedge = \begin{bmatrix} \boldsymbol{\omega}^\wedge & \boldsymbol{v} \\ 0 & 0 \end{bmatrix} \text{ with } \boldsymbol{\omega}^\wedge = \begin{bmatrix} 0 & -\omega_3 & \omega_2 \\ \omega_3 & 0 & -\omega_1 \\ -\omega_2 & \omega_1 & 0 \end{bmatrix} \tag{1}$$

$$\text{with: } \boldsymbol{\omega} = [\omega_1, \omega_2, \omega_3]^T, \boldsymbol{v} = [v_1, v_2, v_3]^T, \boldsymbol{T} = [\boldsymbol{v}^T\ \boldsymbol{\omega}^T]^T \tag{2}$$

Even though the two forms are equivalent, the vector form is employed in this article,[2] unless specified.

The Lie algebra is equipped with the commutation operator $[\boldsymbol{X}, \boldsymbol{Y}] = \boldsymbol{XY} - \boldsymbol{YX} = \mathrm{ad}_{\boldsymbol{X}}\boldsymbol{Y}$ for $\boldsymbol{X}, \boldsymbol{Y} \in \mathfrak{g}$. This is also referred as the *adjoint action* endomorphism, as it is a linear map onto itself such that $\mathrm{ad}_{\mathfrak{g}}\cdot : [\mathfrak{g}, \mathfrak{g}] \rightarrow \mathfrak{g}$.

For the case of: $\boldsymbol{\phi} \in \mathfrak{so}^3$ and $\boldsymbol{\xi} = [\boldsymbol{\rho}^T, \boldsymbol{\phi}^T]^T \in \mathfrak{se}^3$ (with $\boldsymbol{\rho} \in \mathbb{R}^3$, the linear component of $\boldsymbol{\xi}$), the ad operation is:

$$\text{for } \mathfrak{so}^3 : \mathrm{ad}_{\boldsymbol{\phi}_1}\boldsymbol{\phi}_2 = [\boldsymbol{\phi}_1, \boldsymbol{\phi}_2] = \boldsymbol{\phi}_1 \times \boldsymbol{\phi}_2 = {\boldsymbol{\phi}_1}^\wedge . \boldsymbol{\phi}_2 \qquad \text{with: } \mathrm{ad}_{\boldsymbol{\phi}_1} = {\boldsymbol{\phi}_1}^\wedge \tag{3}$$

$$\text{for } \mathfrak{se}^3 : \mathrm{ad}_{\boldsymbol{\xi}_1}\boldsymbol{\xi}_2 = [\boldsymbol{\xi}_1, \boldsymbol{\xi}_2] = \begin{bmatrix} \boldsymbol{\phi}_1 \times \boldsymbol{\rho}_2 + \boldsymbol{\rho}_1 \times \boldsymbol{\phi}_2 \\ \boldsymbol{\phi}_1 \times \boldsymbol{\phi}_2 \end{bmatrix} \quad \text{with: } \mathrm{ad}_{\boldsymbol{\xi}_1} = \begin{bmatrix} {\boldsymbol{\phi}_1}^\wedge & {\boldsymbol{\rho}_1}^\wedge \\ \boldsymbol{0} & {\boldsymbol{\phi}_1}^\wedge \end{bmatrix} \tag{4}$$

For a more comprehensive introduction into Lie groups, refer to [11].

2.2 Receding Horizon

Let us define a dynamic system with vector state $\boldsymbol{x}$ and input $\boldsymbol{u}$. Discretizing a time horizon in H time steps of duration Δt, ones obtains $\boldsymbol{x}(t_k) = \boldsymbol{x}_k$, $\boldsymbol{u}(t_k) = \boldsymbol{u}_k$ for $t_k = k\Delta t$ where $k = 0, 1, \dots, H$. Notice t_0 (the starting time of the horizon) can be chosen arbitrarily without loss of generality.

[2] The log and exp operations are defined for matrices. By abuse of notation, throughout this article they will also operate on the lie algebra vector form (given the fact that it is easy to retrieve the matrix form through the hat map). For the same reason, assume the results of the log and exp operations are expressed in vector form.

One can formulate a linear receding horizon expression as:

$$\boldsymbol{x}_{k+1} = \boldsymbol{A}\boldsymbol{x}_k + \boldsymbol{B}\boldsymbol{u}_k \text{ for } k = 0, 1, \ldots, H-1 \tag{5}$$

where $\boldsymbol{A}$ and $\boldsymbol{B}$ are respectively the state system and input matrix. This linear discrete time dynamic model expression can straightforwardly be written in a recursive matricial form over the whole horizon. This form is central to the formulation of MPC as a constrained Quadratic Program (QP).

3 Linear Pose Estimation

In this Section, the propagation of the pose through a manifold is proposed.

3.1 Pose Estimation in a Horizon

For a robot pose $\boldsymbol{\mathcal{X}} \in \mathbb{SE}^3$, let us reconstruct a pose trajectory $\boldsymbol{\mathcal{X}}(t)$ in time given some initial horizon state $\boldsymbol{\mathcal{X}}(t_0) = \boldsymbol{\mathcal{X}}_0$ and some input twist trajectory $\boldsymbol{T}(t) \in \mathfrak{se}^3$, consisting of infinitesimal increments of pose in time.

As described in Sect. 2.1, the robot pose belongs to a Lie group and its twist, to its algebra (its tangential space).

Discretizing a time horizon in H time steps of duration Δt, we obtain $\boldsymbol{\mathcal{X}}(t_k) = \boldsymbol{\mathcal{X}}_k$, $\boldsymbol{T}(t_k) = \boldsymbol{T}_k$ for $t_k = k\Delta t$ where $k = 0, 1, \ldots, H$. An infinitesimal twist can be integrated to obtain the next pose as:

$$\boldsymbol{\mathcal{X}}_{k+1} = \boldsymbol{\mathcal{X}}_k e^{\boldsymbol{T}_k \Delta t} \tag{6}$$

This enables expressing any state in the horizon $\boldsymbol{\mathcal{X}}_{k+1}$ as a function of $\boldsymbol{\mathcal{X}}_0$ and the successive $\boldsymbol{T}_k$ applied:

$$\boldsymbol{\mathcal{X}}_{k+1} = \boldsymbol{\mathcal{X}}_0 \underbrace{\prod_{i=0}^{k} e^{\boldsymbol{T}_i \Delta t}}_{f(\boldsymbol{T}_0, \boldsymbol{T}_1, \ldots, \boldsymbol{T}_k)} \tag{7}$$

The focus of this paper is to showcase a linear approximation of $\boldsymbol{\xi}(t)$ (the logarithmic map of $\boldsymbol{\mathcal{X}}(t)$), using Eq. 7.

3.2 From the Magnus Expansion to an Estimate of the Logarithm Map

The objective of this Section is to find an approximation of $\boldsymbol{\xi}(t)$ in the form of Eq. 5, with system state $\boldsymbol{\xi}$ and input $\boldsymbol{T}$, by exploiting Eq. 7.

Consider the following differential equation:

$$\boldsymbol{Y}'(t) = \boldsymbol{Y}(t)\boldsymbol{U}(t), \text{ with } \boldsymbol{Y}(t_0) = \boldsymbol{Y}_0 \tag{8}$$

with $\boldsymbol{Y}(t) \in \mathcal{G}$ and $\boldsymbol{U}(t) \in \mathfrak{g}$ (as proposed in [12]). It is possible to find a solution in the form:

$$\boldsymbol{Y}(t) = \boldsymbol{Y}_0 e^{\boldsymbol{\Omega}(t)} \tag{9}$$

Consequently, $\boldsymbol{Y}'(t) = \boldsymbol{Y}_0 e^{\boldsymbol{\Omega}(t)} \mathrm{dexp}_{\boldsymbol{\Omega}(t)} \boldsymbol{\Omega}'(t)$ from which, based on Eq. 8, one can identify $\boldsymbol{U}(t)$ as

$$\boldsymbol{U}(t) = \mathrm{dexp}_{\boldsymbol{\Omega}(t)} \boldsymbol{\Omega}'(t) \tag{10}$$

where dexp_x denotes the derivative of the exponential map[3] at x.

Actually, the main interest lies in finding $\boldsymbol{\Omega}(t)$ that satisfies Eq. 8. We can retrieve $\boldsymbol{\Omega}'(t)$ from Eq. 10 by inverting dexp:

$$\boldsymbol{\Omega}' = \mathrm{dexp}_{\boldsymbol{\Omega}}^{-1} \boldsymbol{U}(t) = \sum_{n=0}^{\infty} \frac{B_n}{n!} \mathrm{ad}_{\boldsymbol{\Omega}}^n \boldsymbol{U}(t) \tag{11}$$

$$\mathrm{ad}_{\boldsymbol{\Omega}}^n = \underbrace{[\boldsymbol{\Omega}(t), [\ldots, [\boldsymbol{\Omega}(t)}_{n-times}, \boldsymbol{U}(t)]]] \tag{12}$$

where B_n are the Bernoulli numbers. The relation between Eq. 11 and dexp^{-1} is developed in [10] and [9]. Note that $\boldsymbol{\Omega}(t) = \int_{t_o}^{t} \boldsymbol{U}(x)dx$ iif $\boldsymbol{U}$ is constant or commutes with its primitive (i.e. $[\boldsymbol{U}(t), \int \boldsymbol{U}(t)dt] = 0$). Otherwise, $\boldsymbol{\Omega}(t)$ is an infinite series known as the *Magnus expansion* [3,8].

The current paper uses a truncated version of Eq. 11 up to $n = 2$, referred as ME2 from now on. Remembering that $B_n = 0$ for all odd $n > 1$ (which implies that ME3 $\equiv$ ME2), yields:

$$\boldsymbol{\Omega}'(t) \approx \boldsymbol{U}(t) + \frac{[\boldsymbol{\Omega}(t), \boldsymbol{U}(t)]}{2} + \frac{[\boldsymbol{\Omega}(t), [\boldsymbol{\Omega}(t), \boldsymbol{U}(t)]]}{12} \tag{13}$$

Given Eq. 13, we can now see how Eq. 9 is in fact the time continuous version of Eq. 7. Furthermore, $\boldsymbol{U}(t)$ can be interpreted as the time continuous version of the trajectory twist $\boldsymbol{T}(t)$ while $\boldsymbol{\Omega}(t)$ incorporates its cumulative effects in time over the initial pose $\boldsymbol{\mathcal{X}}_0$. Note that $\boldsymbol{Y}(t)$ and $\boldsymbol{U}(t)$ do not necessarily commute: this means that $\boldsymbol{Y}_0\boldsymbol{U}_0\boldsymbol{U}_1 \neq \boldsymbol{Y}_0\boldsymbol{U}_1\boldsymbol{U}_0$ (or for any other order for that matter). This is exactly why $\boldsymbol{\Omega}(t)$ cannot be found via a trivial integration and why the Magnus expansion is required.

By discretizing $\boldsymbol{\Omega}_k$, we can construct an equivalent for Eq. 7:

$$\boldsymbol{\mathcal{X}}_{k+1} = \boldsymbol{\mathcal{X}}_0 e^{\boldsymbol{\Omega}_k(\boldsymbol{T}_0, \boldsymbol{T}_1, \ldots, \boldsymbol{T}_k)} \tag{14}$$

and computing the logarithm map of the pose as $\boldsymbol{\xi}_{k+1} = \log(\boldsymbol{\mathcal{X}}_{k+1})$, we arrive at:

$$\boldsymbol{\xi}_{k+1} = \boldsymbol{K} + \boldsymbol{\Omega}_k(\boldsymbol{T}_0, \boldsymbol{T}_1, \ldots, \boldsymbol{T}_k) \tag{15}$$

[3] The derivative of the exponential map is often presented as $\frac{d}{dt}e^{x(t)} = e^x \frac{(1-e^{\mathrm{ad}_x})}{\mathrm{ad}_x}\frac{dx}{dt}$ or, with a different notation but equivalently: $\frac{d}{dt}e^{x(t)} = \mathrm{dexp}_{x(t)} x'(t)$.

where $\boldsymbol{K} \in \mathfrak{se}^3$ corresponds to some initial conditions that will satisfy for $\boldsymbol{\mathcal{X}}_0$.

Finally, linearizing Eq. 15 at the start of the horizon $\boldsymbol{\xi}_0$ yields:

$$\boldsymbol{\xi}_{k+1} = \boldsymbol{\xi}_0 + \Delta t.\text{dexp}_{\boldsymbol{\xi}_0}^{-1}.(\boldsymbol{T}_0 + \boldsymbol{T}_1 + \ldots + \boldsymbol{T}_k) \tag{16}$$

Two remarkable properties about this equation are:

- $\boldsymbol{\xi}$ is a 6-vector in $\mathfrak{se}^3$, a more space-efficient (equivalent) representation of pose, compared to a 4×4 matrix in $\mathbb{SE}^3$, allowing to be more easily embedded in optimization problems.
- because the commutator is an $\mathfrak{se}^3$ operation (refer to Sect. 2.1), the result of Eq. 16 also lies in $\mathfrak{se}^3$ (for small enough Δt).

This linear approximation implies the assumption that $\text{dexp}_{\boldsymbol{\xi}_0}$ does not change much during the horizon, which is an accepted compromise for any linearization approach and is also why the approximation gets worse with longer horizons.

It can be seen that equation Eq. 16 resembles Eq. 5, enabling its use in linear MPC. This equation constitutes the main contribution of this paper.

4 Results

The goal is to estimate, at each time step, the robot pose trajectory in a receding horizon $\boldsymbol{\mathcal{X}}_k$ by reconstructing it from the initial horizon pose $\boldsymbol{\mathcal{X}}_0$ and a perfectly known input twist $\boldsymbol{T}_k$.

The objective is not to estimate $\boldsymbol{\mathcal{X}}_k$ directly in a matricial form. Instead, one can estimate $\boldsymbol{\xi}_k$ using Eq. 16. The pose can be retrieved from $\boldsymbol{\mathcal{X}}_k = e^{\boldsymbol{\xi}_k}$.

Let the starting and ending poses of a robot end-effector path be $\boldsymbol{\mathcal{X}}_0, \boldsymbol{\mathcal{X}}_f \mathbb{SE}^3$ as depicted in Fig. 1. In this experiment, to represent a typical movement of a Franka Emika Panda in its workspace, two paths with different positional and rotational displacement were generated. For trajectory 1: $||\boldsymbol{p}_f - \boldsymbol{p}_0|| = 0.71\text{m}$ and $||ang(\delta)|| = 0.94\text{rad}$; while for trajectory 2: $||\boldsymbol{p}_f - \boldsymbol{p}_0|| = 0.32\text{m}$ and $||ang(\delta)|| = 0.90\text{rad}$. The rotational displacement (the norm of the angular component $ang(\delta)$) is computed from $\delta = \log(\boldsymbol{\mathcal{X}}_0^{-1}\boldsymbol{\mathcal{X}}_f)$.

Defining $\alpha \in [0, 1]$ to discretize the path between both poses as: $\boldsymbol{\mathcal{X}}_k^* = \boldsymbol{\mathcal{X}}_0 e^{\alpha_k \delta}$ yields a ground truth reference velocity twist $\boldsymbol{T}_k^* = \frac{\log(\boldsymbol{\mathcal{X}}_k^{*-1}\boldsymbol{\mathcal{X}}_{k+1}^*)}{\Delta t}$.

This implies that the reference trajectory can be reconstructed with:

$$\boldsymbol{\mathcal{X}}_{k+1} = \boldsymbol{\mathcal{X}}_k e^{\Delta t \boldsymbol{T}_k^*} \tag{17}$$

$$\text{where: } \log(\boldsymbol{\mathcal{X}}_k) = \boldsymbol{\xi}_k \tag{18}$$

The Ruckig library [2] is used[4] to compute a normalized $\alpha(t)$ in time to find the time-optimal trajectory, following a trapezoidal trajectory (in velocity), as shown in Fig. 2.

[4] The code used for experiments in this paper is publicly available at: https://gitlab.inria.fr/auctus-team/publications/shared-paper-code/ark2022-lin.

We showcase the precision of linear integration of a robot's end-effector pose by using Eq. 16 throughout an H-step horizon and comparing it against Eq. 17, the ground truth. Figure 3 depicts the obtained errors with respect to the perfect propagation.

Assuming a receding horizon approach where only horizon estimation matters, the pose is reset back to ground truth at the end of each horizon (every 300 ms) to showcase the maximum error obtained with the method and to exemplify how the precision worsens towards the end of the horizon.

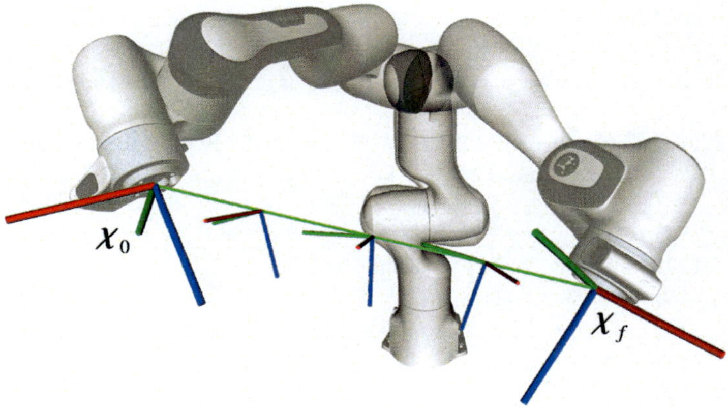

Fig. 1. Depiction of the path taken taking the end-effector of a Franka Emika Panda from a starting to the ending poses $\mathcal{X}_0, \mathcal{X}_f \in \mathbb{SE}3$.

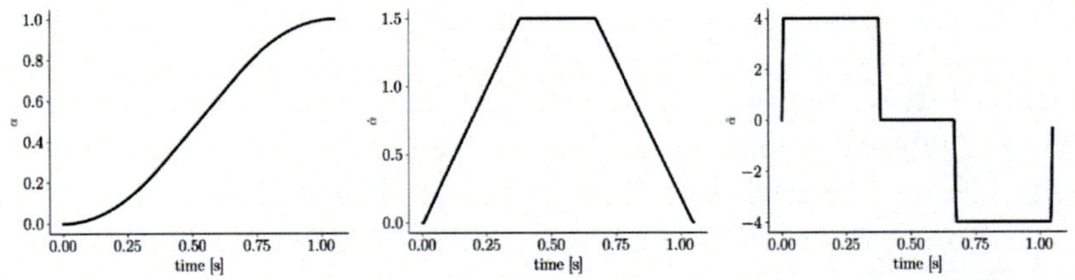

(a) $\alpha(t)$ (b) $\dot{\alpha}(t)$ (c) $\ddot{\alpha}(t)$

Fig. 2. Normalized $\alpha(t)$ for a time-optimal trajectory generated with a trapezoidal profile (in velocity).

In Table 1, the computation time and error averages obtained for the ME2 approximation are presented. The method achieves a linear and angular error of 1.09±6.1 mm and 0.048±0.27° (for the longest trajectory), respectively. This is done in about two thirds of the time it takes to do it with the exponential (exact but non linear). ME2 offers not only a precision level practical for many purposes but it could offer a speed improvement when both methods of integration are

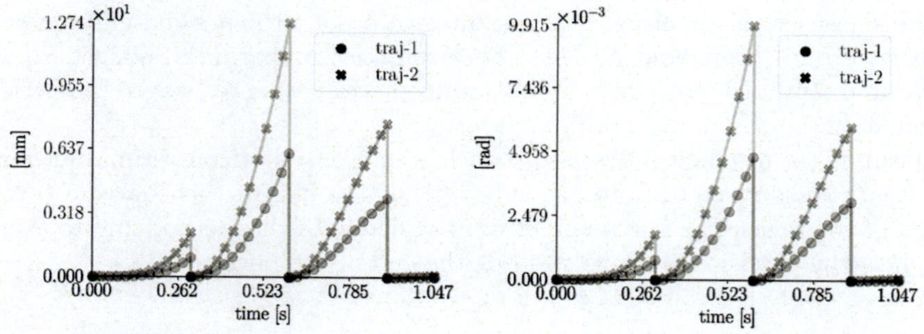

(a) Norm of linear error (b) Norm of angular error

Fig. 3. Variable speed integration using ME2 for a 300 ms horizon.

Table 1. Shows the average compute time and error for the horizons shown in Fig. 3.

Trajectory	Compute time			Linear error			Angular error		
	Avg	Rel	Max	Avg	Max	Std	Avg	Max	Std
ref-1	0.23 µs	100.0%	1.57 µs	1.09 mm	6.15 mm	±1.47 mm	0.05°	0.27°	±0.06°
traj-1	0.15 µs	62.4%	3.74 µs						
ref-2	0.18 µs	100.0%	0.82 µs	2.26 mm	12.74 mm	±2.99 mm	0.10°	0.57°	±0.13°
traj-2	0.13 µs	72.3%	0.23 µs						

possible (the exponential integration method cannot be embedded in a QP, which is the focus of this paper). This computational time improvement could also be exploited on robots with limited resources.

To further illustrate the effects of enlarging the horizon duration, we show in Fig. 4 the same experiment run multiple times for different values of H (number of time steps). It can be seen that the longer the horizon, the greater the error is. Thus, a compromise must be made in order to define a horizon that allows to predict further enough in time without compromising precision.

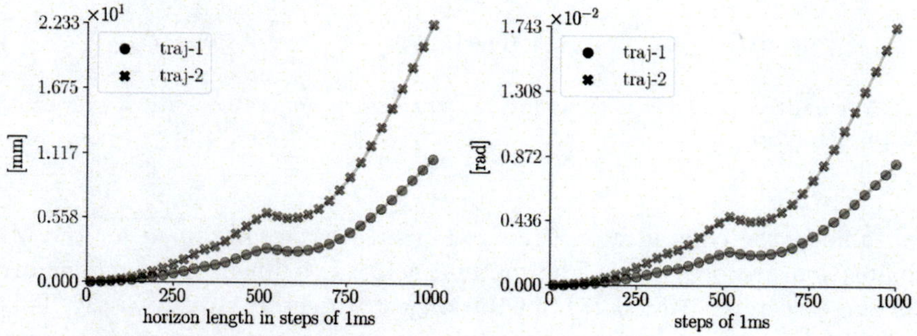

(a) Norm of linear error (b) Norm of angular error

Fig. 4. Average error for different horizon lengths.

5 Conclusions

Lie algebras offer a convenient framework for pose description and system linearization on manifolds, allowing for vector expressions that can be embedded in a linear MPC problem, unlike their matricial forms.

The variety of strategies available offers an opportunity for the community to develop the tools required for more intelligent robotics applications that can optimize through predicted horizons to solve complex problems like safety in human-robot interactions, energy efficiency and trajectory feasibility of dynamically constrained systems.

In future research, we would like to employ this algorithm for the dynamic replanification (based on acceleration increments instead of velocity) of end-effector trajectories subject to kino-dynamic constraints.

The estimation of a bound on the maximum error induced by the proposed linear approximation is also part of the future work.

Acknowledgements. This work was funded by the ANRT and Stellantis Group and performed within the framework of the OpenLab AI.

References

1. Ajoudani, A., Zanchettin, A.M., Ivaldi, S., Albu-Schäffer, A., Kosuge, K., Khatib, O.: Progress and prospects of the human-robot collaboration. Auton. Robot. **42**(5), 957–975 (2018)
2. Berscheid, L., Kröger, T.: Jerk-limited real-time trajectory generation with arbitrary target states. In: Robotics: Science and Systems XVII (2021)
3. Blanes, S., Casas, F., Oteo, J., Ros, J.: The Magnus expansion and some of its applications. Phys. Rep. **470**(5–6), 151–238 (2008). https://doi.org/10.1016/j.physrep.2008.11.001
4. Casas, F., Iserles, A.: Explicit Magnus expansions for nonlinear equations. J. Phys. A: Math. Gen. **39**, 5445–5461 (2006). https://doi.org/10.1088/0305-4470/39/19/S07
5. Escande, A., Mansard, N., Wieber, P.B.: Hierarchical quadratic programming: fast online humanoid-robot motion generation. Int. J. Robot. Res. **33**(7), 1006–1028 (2014)
6. Ibanez, A., Bidaud, P., Padois, V.: Optimization-based control approaches to humanoid balancing. In: Goswami, A., Vadakkepat, P. (eds.) Humanoid Robotics: A Reference, pp. 1541–1567. Springer, Dordrecht (2019). https://doi.org/10.1007/978-94-007-6046-2_71
7. Kleff, S., Meduri, A., Budhiraja, R., Mansard, N., Righetti, L.: High-frequency nonlinear model predictive control of a manipulator. In: ICRA 2021 (2021)
8. Magnus, W.: On the exponential solution of differential equations for a linear operator (1954). https://doi.org/10.1002/CPA.3160070404
9. Rossmann, W.: Lie Groups: An Introduction Through Linear Groups. Oxford University Press, Oxford (2006)
10. Schur, F.: Zur Theorie der endlichen Transformationsgruppen. Mathematische Annalen **38**, 263–286 (1891). https://eudml.org/doc/157545

11. Solà, J., Deray, J., Atchuthan, D.: A micro Lie theory for state estimation in robotics. ArXiv (2018)
12. Zanna, A.: Collocation and relaxed collocation for the Fer and the Magnus expansions. SIAM J. Numer. Anal. **36**, 1145–1182 (1999). https://doi.org/10.1137/S0036142997326616

Hyper-multidual Algebra and Higher-Order Kinematics

Daniel Condurache(✉)

Technical University of Iasi, D. Mangeron Street no. 59, 700050 Iasi, Romania
daniel.condurache@tuiasi.ro

Abstract. This paper proposes a novel computing method for higher-order accelerations vector field in the case of rigid body motion and multibody systems, using the property of the hyper-multidual algebra. The equations that allow the determination of higher-order accelerations for the spatial serial kinematic chains are the main subject of this paper. The results are closed-form and coordinate-free. Furthermore, the velocities, accelerations, jerks, and jounces fields are offered for an mC general manipulator.

Keywords: Hyper-multidual algebra · Higher-order kinematics · Lie group $SE(3)$ · Lie algebra $se(3)$

1 Introduction

The development of high-precision robotic systems, artificial vision systems, and molecular dynamics requires new procedures to calculate higher-order accelerations [1–7]. The key to the proposed approach starts with the property of rigid body displacements group of forming a Lie group, accompanied by its Lie algebra. A previous result offers an isomorphic representation of the Lie group $SE(3)$ with the group of the orthogonal dual tensors and Lie algebra se(3) of the Lie algebra of dual vectors. The results obtained using dual algebras completely solve the problem of finding the field of higher-order accelerations, using a set of results obtained by the previous papers [1–3]. Moreover, it extended for the multidual commutative algebra [4]. This paper proposes a novel computing method for studying the higher-order accelerations fields for rigid body motion and serial kinematic lover-pair spatial chains, using the calculus with the dual and the hyper-multidual algebras. The "automatic differentiation" feature of the hyper-multidual functions is used.

2 Higher-Order Acceleration Field in Rigid Body Motion

Let be a rigid motion given by a curve in Lie group of the rigid displacements $SE(3)$ given by the homogenous matrix $\mathbf{g} = \begin{bmatrix} \mathbf{R} & \boldsymbol{\rho} \\ 0 & 1 \end{bmatrix}$ where $\mathbf{R} \in SO(3)$ is a proper orthogonal tensor [2, 3, 5], $\mathbf{R} = \mathbf{R}(t)$, and $\boldsymbol{\rho} = \boldsymbol{\rho}(t)$ a functions of a time variable n-times differentiable. As

O. Altuzarra and A. Kecskémthy (Eds.): ARK 2022, SPAR 24, pp. 57–64, 2022.
https://doi.org/10.1007/978-3-031-08140-8_7

shown in [1–3], the vector field of the higher-order accelerations is given by the below matrix:

$$\mathbf{\Psi}_n = \begin{bmatrix} \mathbf{\Phi}_n & \mathbf{a}_n \\ 0 & 0 \end{bmatrix}, \tag{1}$$

where the second order tensor $\mathbf{\Phi}_n$ and the vector $\mathbf{a}_n$ is given by the below equations:

$$\mathbf{\Phi}_n = \mathbf{R}^{(n)}\mathbf{R}^T, \tag{2}$$

$$\mathbf{a}_n = \boldsymbol{\rho}^{(n)} - \mathbf{\Phi}_n\boldsymbol{\rho}. \tag{3}$$

The n-th order acceleration of a point of the rigid body given by the position vector $\mathbf{r}$, denoted by $\mathbf{a}_{\mathbf{r}}^{[n]}$, can be computed with the following relation:

$$\mathbf{a}_{\mathbf{r}}^{[n]} = \mathbf{a}_n + \mathbf{\Phi}_n\mathbf{r};\, n \in \mathbb{N}. \tag{4}$$

In [1–3], an iterative procedure is described to determine the vector invariant $\mathbf{a}_n$ and instantaneous tensor invariant $\mathbf{\Phi}_n$ using the time derivative of spatial twist or the dual angular velocity (dual twist) of rigid body motion. For $n = 1, 2, 3, 4$ obtain, respectively, the velocity, acceleration, jerk, and jounce fields. Next, a new non-iterative procedure that permits the determination of the higher-order acceleration field using the specific differential transformation will be presented. The "automatic differentiation" feature of the hyper-multidual functions is used to obtain the higher-order derivative of a rigid body pose. No need for further differentiation of the body poses with respect to time is used.

3 Dual and Hyper-multidual Algebra

Let be $\underline{\mathbb{R}} = \mathbb{R} + \varepsilon_0\mathbb{R};\, \varepsilon_0 \neq 0,\, {\varepsilon_0}^2 = 0$, the set of dual numbers. We will introduce the set of hyper-multidual (HMD) numbers by: $\widehat{\underline{\mathbb{R}}} = \underline{\mathbb{R}} + \varepsilon\underline{\mathbb{R}} + \cdots \varepsilon^n\underline{\mathbb{R}};\, \varepsilon \neq 0,\, \varepsilon^{n+1} = 0$, $n \in \mathbb{N}$ a natural number. For n = 1, on obtain hyper-dual numbers [5].

Two generic elements $\hat{\underline{x}}, \hat{\underline{y}} \in \widehat{\underline{\mathbb{R}}}$ will be written as following:

$$\hat{\underline{x}} = \underline{x}_0 + \underline{x}_1\varepsilon + \cdots + \underline{x}_n\varepsilon^n;\, \underline{x}_k \in \underline{\mathbb{R}},\, k = \overline{0, n}, \tag{5}$$

$$\hat{\underline{y}} = \underline{y}_0 + \underline{y}_1\varepsilon + \cdots + \underline{y}_n\varepsilon^n;\, \underline{y}_k \in \underline{\mathbb{R}},\, k = \overline{0, n}. \tag{6}$$

We will denote by $\underline{x}_0 = Du\hat{\underline{x}}$, $Re\hat{\underline{x}} = Re\underline{x}_0$ and $Mu\hat{\underline{x}} = \sum_{k=1}^{n} \underline{x}_k\varepsilon^k$ the dual part, real part, and respectively, multidual part of the HMD number $\hat{\underline{x}}$.

We will define the operations of additions and, respectively, multiplication of two HMD numbers by:

$$\hat{\underline{x}} + \hat{\underline{y}} = \sum\nolimits_{k=0}^{n} \left(\underline{x}_k + \underline{y}_k\right)\varepsilon^k, \tag{7}$$

$$\hat{\underline{x}}\hat{\underline{y}} = \sum\nolimits_{k=0}^{n} \left(\sum\nolimits_{p=0}^{k} \underline{x}_p\underline{y}_{p-k}\right)\varepsilon^k. \tag{8}$$

It can be easily proved that the set $\widehat{\mathbb{R}}$ with the addition operation (7) and multiplication operation (8) is a commutative ring with unit. An element $\hat{\underline{x}} \in \widehat{\mathbb{R}}$ is invertible if and only if $Re\hat{\underline{x}} \neq 0$. The zero divisor from the ring $\widehat{\mathbb{R}}$ is characterized by $Re\hat{\underline{x}} = 0$. An element from $\widehat{\mathbb{R}}$ is either invertible or zero divisor [6].

Let be $\underline{f} : \underline{\mathbb{I}} \subseteq \underline{\mathbb{R}} \to \underline{\mathbb{R}}, \underline{f} = \underline{f}(\underline{x})$ a n-times differentiable dual function with dual variable [6]. We will define the HMD function $\underline{f}(\hat{\underline{x}})$ of HMD variable $\hat{x}$ by equation:

$$\underline{f}(\hat{\underline{x}}) = \underline{f}(\underline{x}_0) + \sum_{k=1}^{n} \frac{\Delta^k}{k!} \underline{f}^{(k)}(\underline{x}_0) \tag{9}$$

where we denoted by Δ:$\Delta = \hat{\underline{x}} - \underline{x}_0 = \sum_{k=1}^{n} \underline{x}_k \varepsilon^k$.

Using the Eq. (9), we will define the following HMD functions:

$$\sin\hat{\underline{x}} = \sin\underline{x}_0 + \sum_{k=1}^{n} \frac{\Delta^k}{k!} \sin\left(\underline{x}_0 + k\frac{\pi}{2}\right), \tag{10}$$

$$\cos\hat{\underline{x}} = \cos\underline{x}_0 + \sum_{k=1}^{n} \frac{\Delta^k}{k!} \cos\left(\underline{x}_0 + k\frac{\pi}{2}\right), \tag{11}$$

$$\hat{\underline{x}}^{-1} = \frac{1}{\underline{x}_0} + \sum_{k=1}^{n} (-1)^k \Delta^k \frac{1}{\underline{x}_0{}^{k+1}}; Re\hat{\underline{x}} \neq 0. \tag{12}$$

Hyper-multidual Vectors

Let be $\underline{\mathbf{V}}_3 = \mathbf{V}_3 + \varepsilon_0 \mathbf{V}_3$ the set of dual vectors, where $\mathbf{V}_3$ is the vector set on three-dimensional Euclidean space. We will introduce the set of HMD vectors by:

$$\widehat{\underline{\mathbf{V}}}_3 = \underline{\mathbf{V}}_3 + \varepsilon \underline{\mathbf{V}}_3 + \cdots + \varepsilon^n \underline{\mathbf{V}}_3; \varepsilon \neq 0, \varepsilon^{n+1} = 0. \tag{13}$$

It can be easily proven that the set of vectors $\widehat{\underline{\mathbf{V}}}_3$ is a free module of rank 3 over the commutative ring $\widehat{\mathbb{R}}$. We will define the scalar product respectively the vector product of two vectors from $\widehat{\mathbb{V}}_3$ by:

$$\hat{\underline{\mathbf{a}}} \cdot \hat{\underline{\mathbf{b}}} = \sum_{k=0}^{n} \sum_{p=0}^{k} \left(\underline{\mathbf{a}}_p \cdot \underline{\mathbf{b}}_{k-p}\right) \varepsilon^k, \tag{14}$$

$$\hat{\underline{\mathbf{a}}} \times \hat{\underline{\mathbf{b}}} = \sum_{k=0}^{n} \sum_{p=0}^{k} \left(\underline{\mathbf{a}}_p \times \underline{\mathbf{b}}_{k-p}\right) \varepsilon^k. \tag{15}$$

Hyper-multidual Second-Order Tensors

An $\widehat{\mathbb{R}}$-linear application of $\widehat{\underline{\mathbf{V}}}_3$ into $\widehat{\underline{\mathbf{V}}}_3$ is called a second order Euclidean HMD tensor. Denote $\mathbf{L}\left(\widehat{\underline{\mathbf{V}}}_3, \widehat{\underline{\mathbf{V}}}_3\right)$ be the set of HMD tensors. Any $\widehat{\underline{\mathbf{T}}} \in \mathbf{L}\left(\widehat{\underline{\mathbf{V}}}_3, \widehat{\underline{\mathbf{V}}}_3\right)$, the transposed tensor denoted by $\widehat{\underline{\mathbf{T}}}^T$ is defined by: $\widehat{\underline{\mathbf{V}}}_1 \cdot \left(\widehat{\underline{\mathbf{T}}}\widehat{\underline{\mathbf{V}}}_1\right) = \widehat{\underline{\mathbf{V}}}_2 \cdot \left(\widehat{\underline{\mathbf{T}}}^T \widehat{\underline{\mathbf{V}}}_1\right); \forall \widehat{\underline{\mathbf{V}}}_1, \widehat{\underline{\mathbf{V}}}_2 \in \widehat{\underline{\mathbf{V}}}_3$.

The determinant of HMD tensor $\widehat{\underline{\mathbf{T}}}$ is defined by:

$$\langle \widehat{\underline{\mathbf{T}}}\hat{\mathbf{v}}_1, \widehat{\underline{\mathbf{T}}}\hat{\mathbf{v}}_2, \widehat{\underline{\mathbf{T}}}\hat{\mathbf{v}}_3 \rangle = \det\widehat{\underline{\mathbf{T}}} \langle \widehat{\underline{\mathbf{V}}}_1, \widehat{\underline{\mathbf{V}}}_2, \widehat{\underline{\mathbf{V}}}_3 \rangle. \tag{16}$$

$\forall \underline{\widehat{\mathbf{V}}}_1, \underline{\widehat{\mathbf{V}}}_2, \underline{\widehat{\mathbf{V}}}_3 \in \underline{\widehat{\mathbf{V}}}_3, Re\left(\langle \underline{\widehat{\mathbf{V}}}_1, \underline{\widehat{\mathbf{V}}}_2, \underline{\widehat{\mathbf{V}}}_3 \rangle\right) \neq 0$, will denote by $\langle \underline{\widehat{\mathbf{V}}}_1, \underline{\widehat{\mathbf{V}}}_2, \underline{\widehat{\mathbf{V}}}_3 \rangle = \underline{\widehat{\mathbf{V}}}_1 \cdot \left(\underline{\widehat{\mathbf{V}}}_2 \times \underline{\widehat{\mathbf{V}}}_3\right)$ the triple product of HDN vectors. For any vector $\underline{\widehat{\mathbf{a}}} \in \underline{\widehat{\mathbf{V}}}_3$, the associated skew-symmetric tensor will be denoted by $\underline{\widehat{\mathbf{a}}}\times$ and will be defined by equation $(\underline{\widehat{\mathbf{a}}}\times)\underline{\widehat{\mathbf{b}}} = \underline{\widehat{\mathbf{a}}} \times \underline{\widehat{\mathbf{b}}}, \forall \underline{\widehat{\mathbf{b}}} \in \underline{\widehat{\mathbf{V}}}_3$. The set of skew-symmetric HMD tensors is structured as a free $\widehat{\underline{\mathbb{R}}}$- free module of rank 3 and is isomorphic with $\underline{\widehat{\mathbf{V}}}_3$.

4 Differential HMD Transform of Dual Time Functions

This section will introduce a differential transformation that associates a dual function of a real variable an HMD function of a real variable. This transformation permits a simultaneous determination of the higher-order acceleration vector fields for a given rigid body motion.

So, being, $\underline{f} : \mathbb{I} \subseteq \mathbb{R} \to \underline{\mathbb{R}}, \underline{f} = \underline{f}(t)$, a dual function of a real-time variable, n-times differentiable, $n \in \mathbb{N}$. To this function, it will be associated the HMD function of the real variable given by the following equation:

$$\widehat{\underline{f}} = \underline{f} + \varepsilon \dot{\underline{f}} + \cdots + \frac{\varepsilon^n}{n!} \underline{f}^{(n)}. \tag{17}$$

The properties of this transformation are given by the following theorem:

Theorem 1. *Being $\underline{f}$ and $\underline{g}$ two dual functions by time variable of class $C^n(\mathbb{I})$. The following properties take place*:

$$\widehat{\underline{f} + \underline{g}} = \widehat{\underline{f}} + \widehat{\underline{g}}, \tag{18}$$

$$\widehat{\underline{f}\underline{g}} = \widehat{\underline{f}}\widehat{\underline{g}}, \tag{19}$$

$$\widehat{\lambda \underline{f}} = \lambda \widehat{\underline{f}}, \forall \lambda \in \mathbb{R}, \tag{20}$$

$$\widehat{\underline{f}(\underline{\alpha})} = \underline{f}(\widehat{\underline{\alpha}}), \underline{\alpha} \in C^n(\mathbb{I}), \tag{21}$$

$$\widehat{\dot{\underline{f}}} = \dot{\widehat{\underline{f}}}. \tag{22}$$

Proof: The properties of differential transform direct result from the definition (17). We have proof only of the property (19). By definition (17) we have:

$$\widehat{\underline{f}\underline{g}} = \sum_{k=0}^{n} \frac{\varepsilon^k}{k!} \left(\underline{f}\underline{g}\right)^{(k)} = \sum_{k=0}^{n} \frac{\varepsilon^k}{k!} \sum_{p=0}^{k} C_k^p \underline{f}^{(p)} \underline{g}^{(k-p)}, \tag{23}$$

$$= \sum_{k=0}^{n} \frac{\varepsilon^k}{k!} \sum_{p=0}^{k} \frac{k!}{p!(k-p)!} \underline{f}^{(p)} \underline{g}^{(k-p)} = \sum_{k=0}^{n} \sum_{p=0}^{k} \frac{\underline{f}^{(p)}}{p!} \frac{\underline{g}^{(k-p)}}{(k-p)!} \varepsilon^k = \widehat{\underline{f}}\widehat{\underline{g}} \tag{24}$$

5 Higher-Order Kinematics with Differential Transform

Let a rigid body motion parameterization by the homogenous matrix in Sect. 2. The equation gives the orthogonal dual tensor $\underline{\mathbf{R}} = \underline{\mathbf{R}}(t) \in \underline{SO}(3), \forall t \in \mathbb{I} \subseteq \mathbb{R}$ [1]:

$$\underline{\mathbf{R}} = \left(\underline{\mathbf{I}} + \varepsilon_0 \boldsymbol{\rho} \times\right) \mathbf{R} \tag{25}$$

parameterized the same general rigid body motion. An exponential form of orthogonal dual tensor $\underline{\mathbf{R}}$ is [2]:

$$\underline{\mathbf{R}} = \exp(\underline{\alpha \mathbf{u}}) = \underline{\mathbf{I}} + (\sin\underline{\alpha})\underline{\mathbf{u}} \times + \left(1 - \cos\underline{\alpha}\right)\left(\underline{\mathbf{u}}\times\right)^2. \tag{26}$$

Applying the HMD differential transform defined by Sect. 4 for orthogonal dual tensor $\underline{\mathbf{R}} = \underline{\mathbf{R}}(t)$, it results:

$$\widehat{\underline{\mathbf{R}}} = \underline{\mathbf{R}} + \dot{\underline{\mathbf{R}}}\varepsilon + \cdots + \frac{\underline{\mathbf{R}}^{(n)}}{n!}\varepsilon^n \tag{27}$$

Theorem 2. *For any $\underline{\mathbf{R}} = \underline{\mathbf{R}}(t) \in \underline{SO}(3)$, a n times order differentiable function, the below relation takes place:*

$$\widehat{\underline{\mathbf{R}}}\widehat{\underline{\mathbf{R}}}^T = \widehat{\underline{\mathbf{R}}}^T \widehat{\underline{\mathbf{R}}} = \hat{\underline{\mathbf{I}}}, \tag{28}$$

$$\det\widehat{\underline{\mathbf{R}}} = 1. \tag{29}$$

Proof. The proof of this theorem is obtain applying Theorem 1 the equations: $\underline{\mathbf{R}}\,\underline{\mathbf{R}}^T = \underline{\mathbf{R}}^T\underline{\mathbf{R}} = \underline{\mathbf{I}}$, $\det\underline{\mathbf{R}} = 1$, relationships that define the set of dual orthogonal tensors $\underline{SO}(3)$. We will call HMD tensors respecting Eqs. (28), (29), HMD orthogonal tensors.

If orthogonal dual tensor is expressed by Eq. (26), a Rodrigues-like formula:

$$\widehat{\underline{\mathbf{R}}} = \hat{\underline{\mathbf{I}}} + (\sin\widehat{\underline{\alpha}})\hat{\underline{\mathbf{u}}} \times + \left(1 - \cos\widehat{\underline{\alpha}}\right)\left(\hat{\underline{\mathbf{u}}}\times\right)^2 = \exp(\widehat{\underline{\alpha}}\hat{\underline{\mathbf{u}}}\times) \tag{30}$$

is obtain, where $\hat{\underline{\mathbf{I}}}$ denotes the unit HMD tensor, and:

$$\widehat{\underline{\alpha}} = \underline{\alpha} + \dot{\underline{\alpha}}\varepsilon + \cdots + \frac{\underline{\alpha}^{(n)}}{n!}\varepsilon^n, \tag{31}$$

$$\hat{\underline{\mathbf{u}}} = \underline{\mathbf{u}} + \dot{\underline{\mathbf{u}}}\varepsilon + \cdots + \frac{\underline{\mathbf{u}}^{(n)}}{n!}\varepsilon^n. \tag{32}$$

Using the previous Theorem 1 and Eq. (25), the unique decomposition results:

$$\widehat{\underline{\mathbf{R}}} = \left(\hat{\underline{\mathbf{I}}} + \varepsilon_0 \hat{\boldsymbol{\rho}} \times\right)\widehat{\mathbf{R}} : \varepsilon_0 \neq 0, \varepsilon_0{}^2 = 0, \tag{33}$$

where:

$$\widehat{\mathbf{R}} = \mathbf{R} + \dot{\mathbf{R}}\varepsilon + \cdots + \frac{\mathbf{R}^{(n)}}{n!}\varepsilon^n, \tag{34}$$

$$\hat{\boldsymbol{\rho}} = \boldsymbol{\rho} + \dot{\boldsymbol{\rho}}\varepsilon + \cdots + \frac{\boldsymbol{\rho}^{(n)}}{n!}\varepsilon^n. \tag{35}$$

Let the HMD orthogonal tensor:

$$\underline{\widehat{\boldsymbol{\Psi}}} = \underline{\widehat{\mathbf{R}}\mathbf{R}^T}. \tag{36}$$

By Eqs. (34), (35), considering $\mathbf{a}_n = \boldsymbol{\rho}^{(n)} - \boldsymbol{\Phi}_n\boldsymbol{\rho}$, a unique decomposition result:

$$\underline{\widehat{\boldsymbol{\Psi}}} = \left(\mathbf{I} + \varepsilon_0\hat{\mathbf{a}}\times\right)\widehat{\boldsymbol{\Phi}};\ \varepsilon_0 \neq 0,\ {\varepsilon_0}^2 = 0. \tag{37}$$

Tensor $\widehat{\boldsymbol{\Phi}}$ and vector $\hat{\mathbf{a}}$ is expressed as:

$$\widehat{\boldsymbol{\Phi}} = \mathbf{I} + \boldsymbol{\Phi}_1\varepsilon + \cdots + \frac{\boldsymbol{\Phi}_n}{n!}\varepsilon^n, \tag{38}$$

$$\hat{\mathbf{a}} = \mathbf{a}_1\varepsilon + \frac{\mathbf{a}_2}{2}\varepsilon^2 \cdots + \frac{\mathbf{a}_n}{n!}\varepsilon^n. \tag{39}$$

In conclusion (see Sect. 2) HMD orthogonal tensor (36) captures all the information about the higher-order acceleration field of rigid body motion of order $k = \overline{1, n}$. For $n = 4$ obtain, simultaneously, the velocity, acceleration, jerk, and jounce fields.

In case $\underline{\mathbf{u}} = \text{const}$ ($\underline{\hat{\mathbf{u}}} = \underline{\mathbf{u}}$) HMD orthogonal tensor (36) are:

$$\underline{\widehat{\boldsymbol{\Psi}}} = \underline{\widehat{\mathbf{R}}\mathbf{R}^T} = \exp\left(\underline{\hat{\alpha}}\underline{\mathbf{u}}\times\right)\exp\left(-\underline{\alpha}\underline{\mathbf{u}}\times\right) = \exp\left(\Delta\underline{\hat{\alpha}}\underline{\mathbf{u}}\times\right). \tag{40}$$

With Rodrigues's like formula, we obtain:

$$\underline{\widehat{\boldsymbol{\Psi}}} = \underline{\hat{\mathbf{I}}} + \left(\sin\Delta\underline{\hat{\alpha}}\right)\underline{\mathbf{u}}\times + \left(1 - \cos\Delta\underline{\hat{\alpha}}\right)\left(\underline{\mathbf{u}}\times\right)^2 = \exp\left(\Delta\underline{\hat{\alpha}}\underline{\mathbf{u}}\times\right) \tag{41}$$

where with $\Delta\underline{\hat{\alpha}}$ was denoted the multidual part of $\underline{\hat{\alpha}}$. In previous equation the mapping's $\sin\Delta\underline{\hat{\alpha}}$ and $\cos\Delta\underline{\hat{\alpha}}$ are polynomial and not transcendent, considering that$(\Delta\underline{\hat{\alpha}})^p = 0;\ p \geq n+1$.

6 Higher-Order Kinematics of General mC Manipulator

Let denote by C_k, $k = \overline{0, m}$, a spatial kinematic chain of the m bodies. The rigid body C_0 is the base of kinematic chain. The relative motion of the rigid body C_k with respect to reference frame attached to C_{k-1} is described by the orthogonal dual tensor ${}^{k-1}\underline{\mathbf{R}}_k \in \underline{SO}(3)$, $k = \overline{1, m}$. The relative motion properties of the terminal body C_m (the end effector) with respect to reference frame attached to C_0 are described by the orthogonal dual tensor [2, 7]:

$$\underline{\mathbf{R}} = {}^0\underline{\mathbf{R}}_1\,{}^1\underline{\mathbf{R}}_2 \ldots {}^{m-1}\underline{\mathbf{R}}_m, \tag{42}$$

$$\underline{\mathbf{R}} = exp(\underline{\alpha}_1\,{}^0\underline{\mathbf{u}}_1\times)exp(\underline{\alpha}_2\,{}^1\underline{\mathbf{u}}_2\times)\ldots exp(\underline{\alpha}_m\,{}^{m-1}\underline{\mathbf{u}}_m\times). \tag{43}$$

If unit dual vectors ${}^{k-1}\underline{\mathbf{u}}_k = \text{const}, k = \overline{1,m}$, the spatial kinematic chain is named general mC manipulator. The dual angles $\underline{\alpha}_p, p = \overline{1,m}$:

$$\underline{\alpha}_p = \alpha_p(t) + \varepsilon_0 d_p(t), t \in \mathbb{I} \subseteq \mathbb{R} \tag{44}$$

characterize the relative motion of the rigid body C_p with respect to C_{p-1}. The link between rigid bodies C_p and C_{p-1} is a general cylindrical joint C. On obtain revolute R, prismatic P, or helical H joints, in special cases. The functions $\underline{\alpha}_p, p = \overline{1,m}$, are assumed to be n times differentiable, $\forall t \in \mathbb{I} \subseteq \mathbb{R}$. The following theorem can be proved:

Theorem 3. *The HMD orthogonal tensor, which then characterizes the vector fields of higher-order acceleration on the terminal body of mC manipulator given by the kinematic mapping (43), results by Brockett-like formula:*

$$\underline{\widehat{\mathbf{\Psi}}} = \exp[\Delta(\underline{\widehat{\alpha}}_1)\underline{\mathbf{u}}_1\times]\exp[\Delta(\underline{\widehat{\alpha}}_2)\underline{\mathbf{u}}_2\times]\ldots\exp[\Delta(\underline{\widehat{\alpha}}_m)\underline{\mathbf{u}}_m\times] \tag{45}$$

where $\underline{\mathbf{u}}_1 = {}^0\underline{\mathbf{u}}_1$, *and:*

$$\underline{\mathbf{u}}_k = {}^0\underline{\mathbf{R}}_1{}^1\underline{\mathbf{R}}_2\ldots{}^{k-2}\underline{\mathbf{R}}_{k-1}{}^{k-1}\underline{\mathbf{u}}_k, k = \overline{2,m} \tag{46}$$

is unit dual vector corresponding to instantaneous screw joint k, $k = \overline{1,m}$, *resolved in the reference frame of body base* C_0, *and* $\Delta(\underline{\widehat{\alpha}}_k)$ *denote the multidual part of HMD variable* $\underline{\widehat{\alpha}}_k, k = \overline{1,m}$.

Proof: By Eq. (43) and Theorem 1 results:

$$\underline{\widehat{\mathbf{R}}} = exp(\underline{\widehat{\alpha}}_1{}^0\underline{\mathbf{u}}_1\times)exp(\underline{\widehat{\alpha}}_2{}^1\underline{\mathbf{u}}_2\times)\ldots exp(\underline{\widehat{\alpha}}_m{}^{m-1}\underline{\mathbf{u}}_m\times), \tag{47}$$

$$\underline{\mathbf{R}}^T = exp(-\underline{\alpha}_1{}^0\underline{\mathbf{u}}_1\times)exp(-\underline{\alpha}_2{}^1\underline{\mathbf{u}}_2\times)\ldots exp(-\underline{\alpha}_m{}^{m-1}\underline{\mathbf{u}}_m\times). \tag{48}$$

Considering equation $\underline{\widehat{\mathbf{\Psi}}} = \underline{\widehat{\mathbf{R}}}\underline{\mathbf{R}}^T$, and the generic properties of orthogonal dual tensors: $\underline{\mathbf{R}}\exp(\underline{\widehat{\alpha}}\underline{\mathbf{u}}\times)\underline{\mathbf{R}}^T = \exp[\underline{\widehat{\alpha}}(\underline{\mathbf{R}}\underline{\mathbf{u}})\times], \forall\underline{\mathbf{R}} \in \underline{SO}(3)$ result Eq. (45) and (46).

7 Conclusion

A general method for studying the vector field of an arbitrary higher-order accelerations vector field is described. It is proved that all information regarding the properties of the distribution of higher-order accelerations is contained in the specified HMD orthogonal tensor. Furthermore, higher-order kinematics properties of lower-pair serial chains with HMD algebra are proposed. The "automatic differentiation" feature of the hyper-multidual functions is used. The results interest the theoretical kinematics, higher-order kinematics analysis in the case of a serial manipulator, control theory, and multibody kinematics.

References

1. Condurache, D.: Higher-order kinematics of rigid bodies. A tensors algebra approach. In: Kecskeméthy, A., Geu Flores, F., Carrera, E., Elias, D.A. (eds.) Interdisciplinary Applications of Kinematics. MMS, vol. 71, pp. 215–225. Springer, Cham (2019). https://doi.org/10.1007/978-3-030-16423-2_20
2. Condurache, D.: Higher-order relative kinematics of rigid body motions: a dual lie algebra approach. In: Lenarcic, J., Parenti-Castelli, V. (eds.) ARK 2018. SPAR, vol. 8, pp. 83–91. Springer, Cham (2019). https://doi.org/10.1007/978-3-319-93188-3_10
3. Condurache, D.: Higher-order acceleration centers and kinematic invariants of rigid body. In: The 5th Joint International Conference on Multibody System Dynamics, Lisbon (2018)
4. Condurache, D.: Multidual algebra and higher-order kinematics. In: Pisla, D., Corves, B., Vaida, C. (eds.) EuCoMeS 2020. MMS, vol. 89, pp. 48–55. Springer, Cham (2020). https://doi.org/10.1007/978-3-030-55061-5_7
5. Cohen, A., Shoham, M.: Application of hyper-dual numbers to multibody kinematics. ASME. J. Mech. Robot. **8**(1), 011015 (2016)
6. Messelmi, F.: Differential calculus of multidual functions. Ann. Rev. Chaos Theory Bifurcations Dyn. Syst. **10**, 1–15 (2021)
7. Müller, A.: An overview of formulae for the higher-order kinematics of lower-pair chains with applications in robotics and mechanism theory. Mech. Mach. Theory **142**, 103594 (2019)

Geometry Based Analysis of 3R Serial Robots

Durgesh Salunkhe[1(✉)], Jose Capco[2], Damien Chablat[1], and Philippe Wenger[1]

[1] LS2N, Nantes, France
{durgesh.salunkhe,damien.chablat,philippe.wenger}@ls2n.fr
[2] University of Innsbruck, Innsbruck, Austria
jose.capco@uibk.ac.at

Abstract. Cuspidal robots can travel from one inverse kinematic solution (IKS) to another without meeting a singularity. This property can be analyzed by understanding the inverse kinematic model (IKM) as well as the singularities in the joint space and in the workspace. In this article, we revisit the geometrical interpretation of the IKM with conics. The conditions of getting different conics and their implication on singularities are discussed and the observations regarding the nature of the conics are presented. Further, a sufficient condition for a 3R robot to be binary (i.e. with up to 2 IKS) as well as quaternary (i.e. with up to 4 IKS) is put forth by analyzing the geometrical interpretation of the IKM. The possibility to derive a necessary and sufficient condition is presented too.

1 Introduction

Cuspidal robots can travel from one inverse kinematic solution (IKS) to another without encountering a singularity. For these robots, posture identification is very difficult [1], which makes the task of trajectory planning more challenging [2]. It is known that a 3R robot can have at most four IKS, and it is generally preferred to choose a robot geometry that maximizes the size of regions with four IKS. This allows an end-user to choose IKS from different regions to counter the collision issues in the workspace. Robots that have 4 IKS regions in their workspace are referred to as *quaternary* robots, while robots that have at most 2 IKS are referred to as *binary* robots [3]. It is important to note that though quaternary robots have their advantages, they can be cuspidal too, while on the other hand binary robots cannot be cuspidal [1]. A particular class of 3R robots, *orthogonal* 3R robots, have been studied extensively. These robots have three mutually orthogonal joint axes. A D-H parameter based condition for an orthogonal 3R robot to be quaternary was provided by algebraic analysis in [3]. An extension of such an analysis to non-orthogonal 3R robots is more challenging, and no conditions for binary or quaternary non-orthogonal robots have been reported yet. Recently, the cuspidality of *generic* 3R robots was analyzed by using a geometric interpretation of the inverse kinematic model (IKM) [4]. This work studied the intersection of a conic with a unit circle and derived important observations regarding the existence of reduced aspects as well as the necessary condition for cuspidality in generic 3R robots.

O. Altuzarra and A. Kecskeméthy (Eds.): ARK 2022, SPAR 24, pp. 65–72, 2022.
https://doi.org/10.1007/978-3-031-08140-8_8

The presented work reports few more properties of the conic and its implication on the maximum number of IKS in the workspace. The classification presented provides a simple and clear geometric interpretation for the condition of binary and quaternary robots. This work can be extended to have a necessary and sufficient condition for a generic 3R robot to be quaternary. The following work is divided into three sections: Sect. 2 revisits the geometrical interpretation of the inverse kinematics of 3R robots and singularities. This section also discusses the classification of 3R robots based on the geometry of the IKM as well as discusses the implication of the same in joint space and workspace. Section 3 shows a case of binary and quaternary robot by analyzing the geometrical properties of the IKM. Section 4 concludes the work by discussing the implications of the contribution and addressing a few pointers to future work.

2 Inverse Kinematic Model

Let $\mathbf{x} = (x, y, z)$ be the vector of coordinates of the robot's end effector in the workspace $\mathcal{W} \subset \mathbb{R}^3$ at a configuration $\mathbf{q} = (\theta_1, \theta_2, \theta_3)$ in the joint space $\mathcal{J} = S^1 \times S^1 \times S^1$. The mapping between $\mathcal{J}$ and $\mathcal{W}$, denoted by $f : \mathcal{J} \rightarrow \mathcal{W}$, defines the direct kinematics: $\mathbf{x} = f(\mathbf{q}), \mathbf{x} \in \mathcal{W}, \mathbf{q} \in \mathcal{J}$. The elements in the preimage $f^{-1}(\mathbf{x})$ are the IKS of $\mathbf{x}$. A robot configuration associated with an IKS is called a *posture*.

Solving the inverse kinematics of 3R serial robots was first reported in [5] where it was noted that the solutions correspond to the intersection of a conic with a circle in c_3s_3-plane, where c_3 and s_3 denote $\cos\theta_3$ and $\sin\theta_3$, respectively. Using the classical D-H parameters to describe the geometry of the robot (see [4]), the solution to the IKM is revisited briefly (Fig. 1). Let, $R = \rho^2 + z^2$, where $\rho^2 = x^2 + y^2 = g(\theta_2, \theta_3)$. The terms R and z can be written as

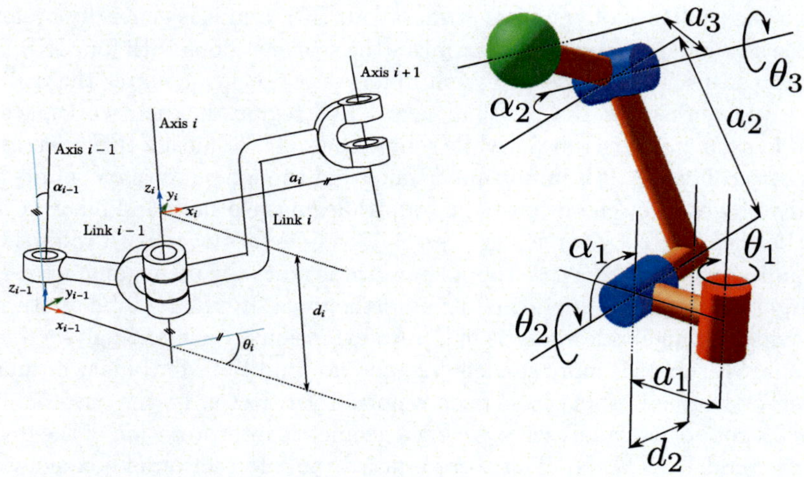

Fig. 1. D-H convention used and a schematic of a 3R serial robot

$$R = (F_1 \cos\theta_2 + F_2 \sin\theta_2)\, 2a_1 + F_3$$
$$z = (F_1 \sin\theta_2 - F_2 \cos\theta_2) \sin\alpha_1 + F_4$$

where $F_i = g_i(\theta_3)$, for $i = 1,..,4$. Upon rearrangement, we obtain the general equation of a conic in c_3s_3-plane as given in (1).

$$A_{xx}\, c_3^2 + 2A_{xy}\, c_3 s_3 + A_{yy}\, s_3^2 + 2B_x\, c_3 + 2B_y\, s_3 + C = 0 \tag{1}$$

The coefficients of the conic are skipped for brevity, but they are functions of the D-H parameters and of, (R, z) as shown in (2),

$$\begin{aligned} A_{xx}, A_{xy}, A_{yy} &= f_1(a_1, a_2, a_3, d_2, \alpha_1, \alpha_2) \\ B_x, B_y, C &= f_2(a_1, a_2, a_3, d_2, d_3, \alpha_1, \alpha_2, R, z) \end{aligned} \tag{2}$$

The inverse kinematic solutions are defined by the intersection points between the conic (1) and the unit circle $c_3^2 + s_3^2 = 1$ in c_3s_3-plane. This conic can be a hyperbola, parabola or an ellipse depending on the D-H parameters. An example of each one is shown in Fig. 2.

Performing the tangent half-angle substitution, $t = \tan\frac{\theta_3}{2}$, we get a quartic inverse kinematic polynomial $M(t) = at^4 + bt^3 + ct^2 + dt + e$ similar to the one mentioned in [6]. The coefficients of $M(t)$ are functions of the D-H parameters and of R and z. The solutions to the polynomial equation, $M(t) = 0$, are the intersection points between the conic and the circle and are labeled as $\mathbf{m}_\psi$, where $\psi \in \{i, j, k, l\}$ in the c_3s_3-plane.

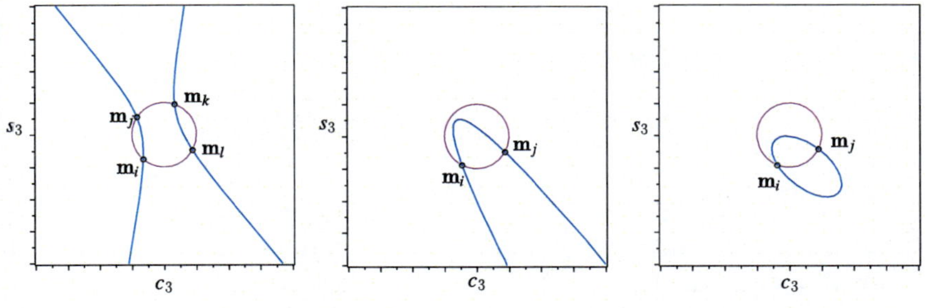

Fig. 2. The conic and unit circle in c_3s_3-plane for robots with different α parameters and at different points. Robot parameters (left): $d = [0, 1, 0], a = [1, 2, \frac{3}{2}], \alpha = [\frac{\pi}{2}, \frac{\pi}{6}, 0], (\rho, z) = (2.46, 0.15)$ (center): $d = [0, 1, 0], a = [1, 2, \frac{3}{2}], \alpha = [\frac{\pi}{3}, \frac{\pi}{2}, 0], (\rho, z) = (2.33, -0.26)$ (right): $d = [0, 1, 0], a = [1, 2, \frac{3}{2}], \alpha = [\frac{\pi}{6}, \frac{\pi}{2}, 0], (\rho, z) = (2.4, 0.6)$.

2.1 Singularities

It has been reported in [6] that the singularities in the workspace correspond to all the points where the root multiplicity of $M(t) = 0$ is greater than or equal to 2. Their geometric interpretation in the conic is two intersection points, $\mathbf{m}_i, \mathbf{m}_j$, merging together at a tangent point between the conic and the circle.

The nonsingular change of posture in cuspidal robots relates to two intersection points interchanging their position, with at least one of them not meeting any other intersection point [4]. This interpretation is also helpful to understand why binary robots are compulsorily noncuspidal. It has been shown in [4] that the orientation of the conic remains constant. Thus, if we have only two intersection points, then they cannot interchange their position without meeting each other at a tangent point.

2.2 Degenerate Conic

The nature of the conic depends on the sign of the determinant of **N**, where **N** is the Hessian of the conic. The degeneracy of a conic is given by $\det(\mathbf{D}) = 0$, where **D** is the Hessian of the quadratic form. We know that the hyperbola ($\det(\mathbf{N}) < 0$) degenerates into two intersecting lines, while the ellipse ($\det(\mathbf{N}) > 0$) degenerates to a point. The degenerate case of a parabola is of particular interest as it degenerates to two parallel lines, and they can be distinct or coincident. The presented work discusses the case when a parabola degenerates to two coincident lines, resulting in two multiple roots. This case is important to analyze as this is a special degeneracy case, and it is expected to show a special property in the workspace too. Following are the conditions for a parabola to degenerate into two coincident lines:

$$\begin{cases} \det(\mathbf{N}) = 0 \\ \det(\mathbf{D}) = 0 \\ B_x^2 + B_y^2 - (A_{xx} + A_{yy})\,C = 0 \end{cases} \tag{3}$$

Solving $\det(\mathbf{N}) = 0$ for d_2 yields:

$$d_2 = \pm\frac{\sqrt{(a_1 + a_2)(a_1 - a_2)(sa_1 - sa_2)(sa_1 + sa_2)}}{sa_1\, sa_2} \tag{4}$$

We conclude that the parameters a_3 and d_3 do not play any role to define a parabola. Upon substituting any value from (4) into $\det(\mathbf{D}) = 0$ and solving for R, we obtain the same solution. Solving the last equation in (3) for z, the solutions R and z take the following form, provided that $ca_2 \neq 0$ and $sa_1 \neq sa_2$:

$$\begin{aligned} R &= \frac{f(z)}{sa_1^2 - sa_2^2} \\ z &= \frac{ca_1}{ca_2}\,(d_3\, sa_1\, sa_2 + d_2\, ca_2) \end{aligned} \tag{5}$$

From (5), it is interesting to note that a robot such that ($ca_2 \neq 0$, $sa_1 \neq sa_2$) and corresponding to a parabola will always have a point in the workspace (R, z) such that its geometric interpretation is a parabola degenerating into a coincident line. When $ca_2 = 0$ (resp. $sa_1 = sa_2$), z (resp. R) is indeterminate. A point satisfying (5) is a tangency point between two loci of critical values, as shown in Fig. 3.

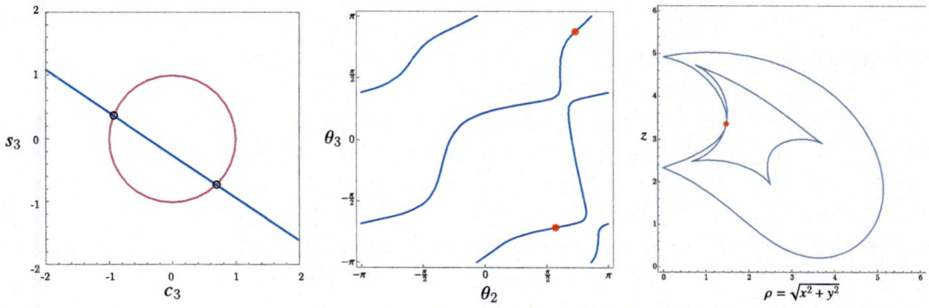

Fig. 3. Degenerate parabola case: joint space (center) and workspace (right). The point shown in red is associated with two coincident lines in the c_3s_3-plane (left). Robot parameters: $d = [0, 2.828, 0.5]$, $a = [1, 2, \frac{3}{2}]$, $\alpha = [\frac{\pi}{6}, \frac{\pi}{3}, 0]$, $(\rho, z) = (1.471, 3.315)$.

3 Special Classes of Robots

In this section, we present a sufficient condition for a 3R robot to be binary and quaternary, respectively. The motivation for the search of binary robots comes from the well-known property that two circles have at most two distinct intersections. Section 3.1 discusses the neighbourhood of such binary robots. We claim that the parameters corresponding to all the ellipses that are in a sufficiently small neighbourhood of the parameters corresponding to a circle, also result in binary robots. Section 3.2 discusses a condition for a hyperbola to compulsorily have four intersections with the unit circle in the c_3s_3-plane, thus resulting in a quaternary robot.

3.1 Binary Robots

We now consider a robot such that its associated conic is an ellipse. We prove that if the ellipse is almost a circle, then many of these robots are binary. We consider a Lemma in Geometry whose proof is straightforward and thus not presented in the text:

Lemma 1. *Consider the unit circle S^1 and a number, $e \in (0, 1)$*

(i) There is an ellipse C with eccentricity e such that $\#C \cap S^1 = 4$
(ii) As $e \to 0$ the ellipses with eccentricity e with property (i) will have centers that approach the origin (center of S^1) and have minor and major semi-axes that approach length 1.

We now make an important remark that was not emphasized in [4]: given a robot, the eccentricity of its associated conic is fixed. Indeed, the eccentricity is only dependent on the entries of $\mathbf{N}$ (see [4,5]), and these entries only depend on the D-H parameters of the robot and not on the position of the end-effector, as shown in (2). By combining this observation along with the lemma, we can prove the following theorem:

Theorem 1. *There are infinitely many binary robots whose associated conic is an ellipse (that is not a circle).*

Proof. We claim that it suffices to have one binary generic robot with this condition. The associated ellipse for such a robot will never degenerate, and so the minor and major axes must achieve their (non-zero) minimum values. From previous Lemma, these lengths must lie in an interval $I \subset \mathbb{R}$ centered at 1 (radius of an ellipse) for the ellipse to intersect the unit circle four times and for a fixed sufficiently small eccentricity e. If a generic binary robot is given with an associated ellipse of eccentricity e, then the minimum major (or minor) axes (recall the axes' length now depends on the end-effector position) is outside I. This minimum value is not in the boundary of I and is continuously dependent on the D-H parameters. One parameter that does not affect the eccentricity but does affect the minimum major/minor axes' length is d_3. So we may perturb d_3 within a small interval of d_3 of the given binary robot and still obtain a binary robot.

To conclude the proof, we give an example of a binary robot whose associated conic is an ellipse. The D-H parameters of this robot are (Fig. 4):

$$a = [-\frac{1503}{1879}, -1, -1] \quad d = [0, 0, 3] \quad \alpha = [-2.21, \frac{-\pi}{2}, 0]$$

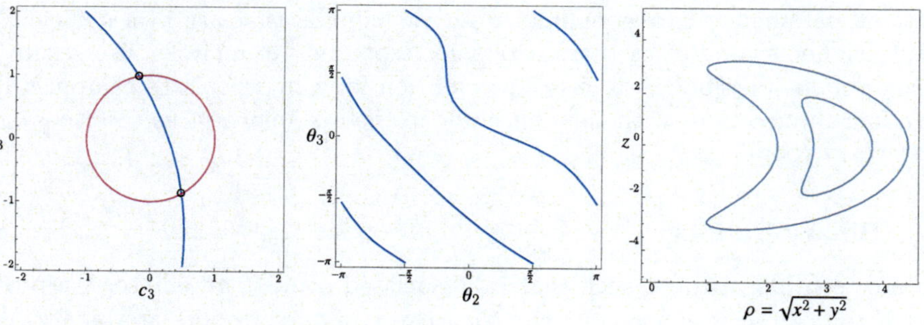

Fig. 4. An example of a binary robot's representation in c_3s_3*−plane*, the joint space and the workspace. Robot parameters: $d = [0, 0, 3]$, $a = [-\frac{1503}{1879}, -1, -1]$, $\alpha = [-2.21, \frac{-\pi}{2}, 0]$.

As the eccentricity of the associated ellipse approaches 0 (the ellipse becomes more like a circle) we will get more such binary robots. However, we will always be able to find a robot that is quaternary for such an eccentricity (as long as it is not 0). This is a conjecture that we aim to prove in the future.

3.2 Quaternary Robot

In this subsection, we discuss the case of a 3R robot such that the hyperbola degenerates and the center of the conic is inside the circle. It is straightforward to argue why such a robot is compulsorily quaternary. If the intersecting lines have their intersection point inside a circle, then each line will intersect the circle twice, thus yielding four intersection points in total. If c_x, c_y is the center of the conic in the c_3s_3-plane, then the sufficient condition for a quaternary robot is:

$$\begin{cases} \det(\mathbf{N}) < 0 \\ \det(\mathbf{Q}) = 0 \\ \sqrt{c_x^2 + c_y^2} < 1 \end{cases} \tag{6}$$

To illustrate the simplicity of the derivation of the sufficient condition, we present a case of orthogonal 3R robots (see an example in Fig. 5).

$$\begin{aligned} \det(\mathbf{N}) &= -\frac{a_3^4\, d_2^2}{a_1^2} \\ \det(\mathbf{Q}) &= \frac{a_3^4\, d_2^2\, (d_3^2 - z^2)}{a_1^2} \\ c_x^2 + c_y^2 &= \frac{aR^2 + bR + c}{4\, d_2^2\, a_3^2} \end{aligned} \tag{7}$$

In (7), a and b are functions of the D-H parameters only and are not expressed fully for brevity. It is clear from (7) that an orthogonal 3R robot always corresponds to a hyperbola in c_3s_3-plane. The condition for degeneracy depends only on d_3 and z while the condition for the center of the conic to lie inside the circle is a quadratic in R. It is important to note that the degeneracy depends only on z and not on R and thus, for $z = d_3$, the conic is always degenerate. This property also leads to some interesting observations about the hyperbolas corresponding to an orthogonal 3R robot but are not discussed here for lack of space.

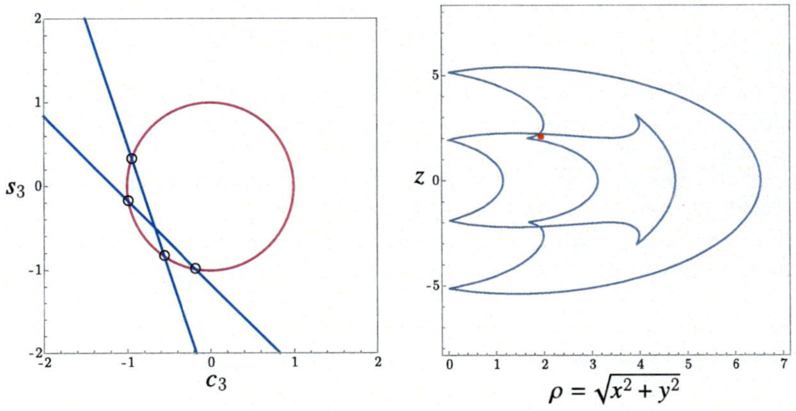

Fig. 5. An example of an orthogonal quaternary robot corresponding to an hyperbola. Robot parameters : $d = [0, 1, 2], a = [1, 2, 3], \alpha = [\frac{-\pi}{2}, \frac{\pi}{2}, 0], (\rho, z) = (2, 2)$.

4 Conclusions and Future Work

In this work, we have revisited the geometric interpretation of the inverse kinematic model of 3R robots. The interpretation of singularities as well as the nonsingular change of posture have been briefly introduced. The special case of a parabola degenerating to two coincident lines is presented along with its interpretation in the workspace. The work also presented a sufficient condition for a 3R serial chain to be binary (ellipse case) (Sect. 3.1) or quaternary (hyperbola case) (Sect. 3.2) by using geometric observations. The advantages of the geometry based analysis is that the conditions for binary or quaternary robots can be extended to more generic cases of 3R robots without resorting to complex algebraic derivations. In future work, we aim to present a necessary and sufficient condition for a generic 3R robot to be binary or quaternary. This will allow the designer to include the condition while optimizing for a workspace with 4 IKS. We will also prove a few conjectures on binary and quaternary robots whose associated conic is an ellipse.

Acknowledgements. The authors are supported by the joint French and Austrian ECARP project: ANR-19-CE48-0015, FWF I4452-N. The authors also thank Christoforos Spartalis for his contribution in the initial stages of the work.

References

1. Wenger, P.: A new general formalism for the kinematic analysis of all nonredundant manipulators. In: Proceedings of the 1992 IEEE International Conference on Robotics and Automation, Nice, France, pp. 442–447 (1992)
2. Wenger, P.: Uniqueness domains and regions of feasible paths for cuspidal manipulators. IEEE Trans. Rob. **20**(4), 745–750 (2004)
3. Wenger, P., Chablat, D., Baili, M.: A DH-parameter based condition for 3R orthogonal manipulators to have 4 distinct inverse kinematic solutions. J. Mech. Des. **127**, 150–155 (2005)
4. Salunkhe, D.H., Spartalis, C., Capco, J., Chablat, D., Wenger, P.: Necessary and sufficient condition for a generic 3R serial manipulator to be cuspidal. Mech. Mach. Theory **171**, 104729 (2022)
5. Pieper, D.L.: The kinematics of manipulators under computer control. Ph.D. thesis, Stanford University, USA, October 1968
6. Kohli, D., Spanos, J.: Workspace analysis of mechanical manipulators using polynomial discriminants. J. Mech. Transm. Autom. Des. **107**(2), 209–215 (1985)

Synthesis of Gripper Mechanisms Derived from Baranov Chains

Fernando Vinicius Morlin(✉), Vinícius Noal Artmann, Andrea Piga Carboni, and Daniel Martins

Federal University of Santa Catarina, Florianópolis, Santa Catarina, Brazil
fernando.v.morlin@posgrad.ufsc.br, daniel.martins@ufsc.br

Abstract. In this paper, an approach to generate grippers using Baranov chains is explored. First, a brief review about grippers is presented. Later, the approach of using Baranov chains for the synthesis of gripper mechanisms is discussed, which is based on concepts of group and graph theory. Then, examples are developed to illustrate the potential results. The examples are also used to investigate a set of desired characteristics of a gripper mechanism, such as number of contact points, mobility, symmetry and variety.

1 Introduction

The use of robots in production processes has been widely applied in recent decades, aiming to supplement or replace humans by doing dull, dirty, or dangerous works. Currently, the efforts are directed to expand the possibilities of application beyond industrial applications, such as the use of grippers for medical applications [16]. A common task for robots is the manipulation of objects. However, there is still a difficulty in developing grippers with adaptability and dexterity like the human hand. While technology is not enough, designers develop specific grippers for groups of tasks or objects. It is possible to find grippers with two or three fingers, with rigid or flexible fingers, employing *vacuum*, pneumatic, hydraulic or servo-electric as force-exerting element [3,12].

The grasping process demands sufficient control to warranty the equilibrium, stability, dexterity, and dynamic behavior [11]. So, developing grippers is a complex task, it requires finding a solution that satisfies aspects like object shape, object rigidity, and reliable control. A bad choice during the early stage of the design process can select a bad kinematics configuration that could hinder gripper performance [2]. In this way, the paper aims to explore how the kinematic characteristics of a chain can influence the design of a gripper.

In the last decades, several researchers worked improving the design process of grippers [1–3,6,17]. However, to the best of authors' knowledge, only a few works are focused on the application of group theory in Baranov chains as a tool to select the kinematic chain and identify the fixed links and objects. Thus, based on a previous paper [9], the work herein presented is developed using Baranov chains to generate grippers, where links are removed to change the mobility of the

O. Altuzarra and A. Kecskémethy (Eds.): ARK 2022, SPAR 24, pp. 73–80, 2022.
https://doi.org/10.1007/978-3-031-08140-8_9

system. Thus, in Sect. 2, the process of generating Baranov chains is reviewed and traditional definitions about mechanisms are discussed. In Sect. 3, the process of synthesizing grippers is investigated, showing how the same kinematic chain can create different grippers and how it can be useful for the designer.

2 Synthesis Process

In previous work, the synthesis of gripper mechanisms has been approached using the concept of Baranov chain [9]. Recall that every mechanism has a corresponding kinematic chain, which is an assemblage of links that are connected by joints, when one of the links is fixed to the ground, the kinematic chains is transformed into a mechanism [19]. Thus, a Baranov chain, also called Baranov truss, is a rigid kinematic chain that contains no rigid sub-chains, i.e., all its subchains have mobility greater than or equal to 1 [5]. In this regard, a robotic gripper is a structure that in fact has non-zero mobility. However, once the object is held by the mechanism, a rigid structure is formed, i.e., the gripper-object system is interpreted as a Baranov chain. Therefore, the synthesis process can be structured from the selection of a Baranov chain, in which one of the links of the rigid chain is the object [9].

An important aspect which needs to be considered is the type of contact between a grasped object and a finger-tip. For point contact with friction and spatial manipulation, the point on the finger-tip is constrained to a point on the grasped object, which is commonly considered to be equivalent to a spherical kinematic pair [18]. However, planar manipulation is assumed here, and the planar motion is a special case of the general spatial motion, where all points of the mechanism links describe paths located in parallel planes. Thus, in the plane, the contact between the finger-tip and the object is simplified such that a spherical kinematic pair is equivalent to a revolute pair.

Furthermore, there is a direct relationship between the number of contact points and the mobility of the gripper mechanism. For example, if the object is represented by a binary link, i.e., with two points of contact between the finger-tips and the object, then the gripper mechanism has 1 degree of freedom while moving. Similarly, for three points of contact, the gripper mechanism has 3 degrees of freedom. This relationship is verified from the mobility equation [4], which is applied to compute the number of degrees of freedom of a mechanism, written as $M = \lambda(n - j - 1) + j$, where λ is the order of the screw system to which all screws belong, n is the number of links, j is the number of joints. Thus, once the vertex representing the object has been chosen, the mobility of the mechanism is known, and vice versa. The variation of mobility ΔM due to the variation of the number of links n and joints j to the kinematic chain can be calculated as $\Delta M = \lambda\Delta n + \Delta j(1 - \lambda)$, where $\lambda = 3$ is the screw system for planar mechanisms [9].

Thus, the design of a gripper is similar to the design of a parallel manipulator. For a parallel manipulator, three levels are possible for the synthesis process [13]. In the first level, kinematic chains are enumerated. Secondly, mechanisms are enumerated by defining a fixed link. Finally, in the third level, parallel manipulators are enumerated by selecting a vertex to be the end-effector. The synthesis of robotic grippers may follow three similar steps. In the third level, however, instead of selecting an end-effector, the vertex representing the object is chosen. Furthermore, graph theory and group theory can be applied to reduce the number of isomorphic gripper mechanisms.

It is well-known that a graph univocally represents a kinematic chain, provided that vertices represent links and edges represent joints. Thus, from group theory, the symmetries of this graph can be examined, so that topologically equivalent mechanisms are discarded. In this case, the symmetric links of the chain are identified from the automorphism group of the corresponding graph. A graph is said to be symmetric when it has more than one automorphism group and symmetric links are identified by orbits of the automorphism group of the graph associated to the kinematic chain [13]. For the selection of a fixed link, the automorphism group of a non-colored graph is generated, and, then, only one representative of each orbit must be chosen. A graph in this work is said to be non-colored when no vertices have been highlighted or identified. However, on order to select a fixed link, the symmetries of a colored graph must be examined, where the colored vertex represents the fixed link of the gripper mechanism. Finally, only one representative of each orbit of the colored graph must be selected to represent the object. See [13–15] for more details on applications of group theory to the analysis of mechanisms and parallel manipulators.

As above-stated, the object is considered as a link in a Baranov chain. When the object is held, the structure is rigid. When the object is released, the chain has motion. So it is also interesting to analyse the effects of removing a link in the structure of the kinematic chain. A Baranov chain is a closed chain, this means that the corresponding graph is biconnected, i.e., each vertex is connected to at least two neighbors. However, since a link is being removed, the gripper mechanism obtained can be either closed, hybrid or open. In general, it is preferable that the mechanism obtained has a closed chain, which ensures more rigidity to the structure. Therefore, in the approach herein proposed, the structure of the mechanism is verified after removing the object, in order to identify whether the chain is closed, open, or hybrid.

An interesting characteristic for a robotic gripper is the existence of an axis of symmetry in the mechanism. This symmetry would reduce the complexity of the design and also guarantee that the object is uniformly held. Therefore, the dimensional symmetries of the gripper mechanism should also be analysed.

Finally, another point of interest refers to the positioning of the actuator. Based on the concept of variety, the definition of the position of the actuators in the chain is simplified. The variety V can be defined as the difference between the overall mobility M of the mechanism and the lowest mobility among the mobility of all closed subchains [7,18], that is, $V = M - \min : \left\{ M\left(G_k'\right) \forall G_k' \in B_s \right\}$, where M is the mobility of the chain, $M(G'_k)$ is the mobility of the k biconnected subchain contained in set of all possible biconnected subgraphs (B_s) [7]. Thus, if a mechanism has null variety, the lowest mobility of its closed subchains is equal to the mobility of the mechanism itself. Therefore, the actuator can be positioned at any joint of any subchain, because no conflict of actuation occurs. Thus, the variety of the resulting mechanism is evaluated.

3 Exploring the Synthesis Process of Grippers

As discussed in Sect. 2, it is possible to enumerate Baranov trusses to represent the kinematic chains of grippers. Three steps are executed: 1) Baranov chains are enumerated; 2) one link is selected to be the fixed base; 3) another link is selected to represent the object. In this section, examples are presented to illustrate the proposed approach and to discuss some relevant characteristics of the kinematic chains of grippers.

It is well-known that there are three different Baranov chains with three independent loops [5,8] (Fig. 1). Furthermore, as stated in Sect. 1, symmetric links (or topologically equivalent links) are identified by the orbits of the automorphism group of the corresponding graph. The kinematic chain of Fig. 1a has three orbits, each of them formed by the following vertices: $O_1 = \{0, 1, 2, 3\}$, $O_2 = \{4, 5\}$ and $O_3 = \{6\}$. Similarly, the kinematic chain of Fig. 1b has four orbits: $O_1 = \{0, 5\}$, $O_2 = \{1, 6\}$, $O_3 = \{2, 4\}$ and $O_4 = \{3\}$; while the the kinematic chain of Fig. 1c has orbits: $O_1 = \{0, 2, 4\}$, $O_2 = \{1, 3, 5\}$, and $O_3 = \{6\}$.

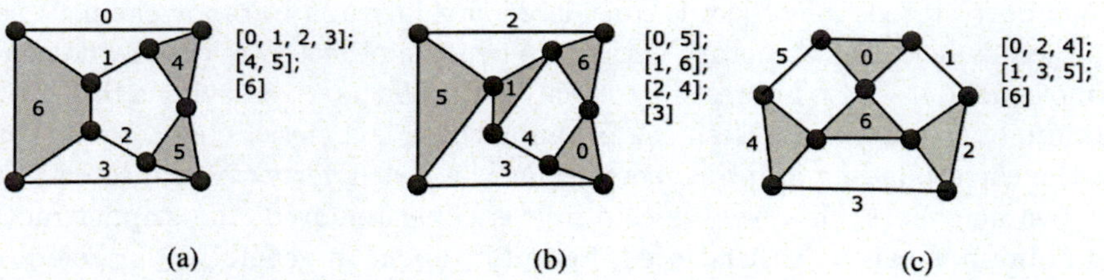

Fig. 1. Examples of Baranov trusses and their corresponding orbits

Using the kinematic chain of Fig. 1a as an example, we may set a color to link 6, considering it to be the fixed link of the gripper, this process is illustrated in Fig. 1a. Now, the orbits are reevaluated considering the coloured link; the output is: $O_1 = \{0, 1, 2, 3\}$, $O_2 = \{4, 5\}$ and $O_3 = \{6\}$. Link 6 is a single element of its orbit, but, if we consider link 4 as fixed. the reevaluation of orbits is equivalent to the results choosing the link 5 as fixed, it occurs because the link 4 and 5 are vertices of the same orbit, Figs. 2b and 2c show the equivalence. It means that the enumeration process performed considering link 5 as fixed does not need to be performed, reducing the processing cost of the algorithm.

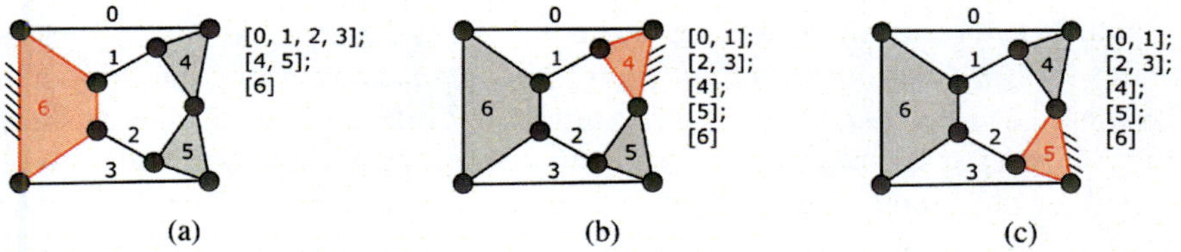

Fig. 2. Examples of different fixed links for the same Baranov chain

Now, with the fixed link defined and with the orbits of the colored graph computed, we can evaluate the different possibilities of grippers by just changing the link that represents the object. Considering the example of Fig. 2a, where the link 6 is considered as fixed, we have two possibilities of different kinematic chains for grippers. A possibility is choosing a vertex of orbit $O_1 = \{0, 1, 2, 3\}$ as object, and another possibility is choosing a vertex of orbit $O_2 = \{4, 5\}$. Figures 3a and 3b represent the kinematic chain considering the links 1 and 4 as objects, respectively.

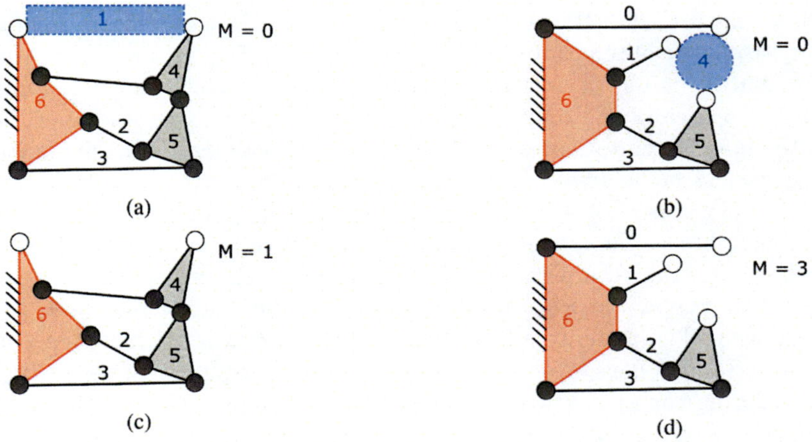

Fig. 3. Two different kinematic chains of grippers created from the same Baranov truss.

After choosing a link as an object, we can remove this link and analyse a new kinematic chain. This new kinematic chain represents the gripper while it is not holding an object. Automating these steps the authors believe it is possible to enumerate all possibilities of grippers. Also, it would be important to classify these results with respect to the base link, the object link, the finger links and the interior links, for example. In this way, characteristics of the generated grippers are now discussed.

Before removing the object link, all the kinematic chains are equivalent to a parallel manipulator, hence the kinematic chains are closed. After deleting the object link, the resultant kinematic chain can be closed, Fig. 3c, but it can also be open (Fig. 4 shows an example where the kinematic chain of the grippers is fully-open). Another possibility is obtaining a hybrid kinematic chain, as shown in Fig. 3d, where the links $6-2-5-3$ form a loop, while the link 0 and 1 do not create any closed loop. A closed kinematic chain is more rigid if compared with a hybrid or open chain, for the same amount of links. Furthermore, open chains can demand more actuators to control the mechanism.

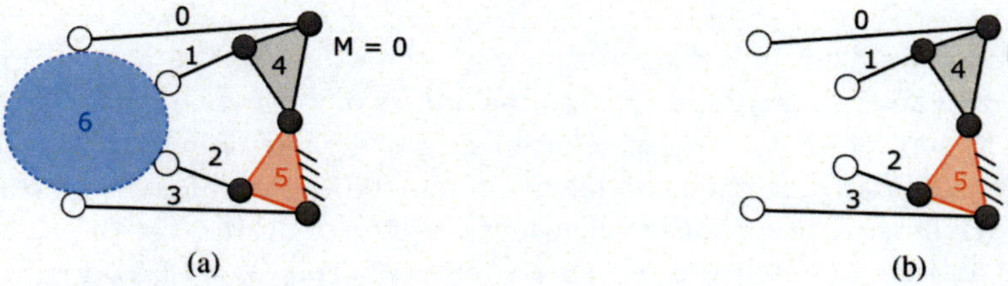

Fig. 4. An example of gripper with open chain during the open stage

The *number of contacts* must be considered in the design process. It depends on the grasping requirements which can change according to the shape, roughness, weight of the object, and the grasping strategies. The number of contacts follows the degree of the link chosen as the object. An example is the Fig. 4, where the link chosen to represent the object is a quaternary link, consequently, the object must be held with four points of contact to create a Baranov chain.

While the gripper is holding the object, the *mobility* is always equal to zero. But the mobility of the kinematic chain changes when the object is free. This is another important characteristic that must be considered during the design process. The grippers of Figs. 3c, 3d and 4b have mobility equal to 1, 3 and 5, respectively. The mobility affects the control complexity of the gripper, it is really important to have a clear view of the grasping requirements to define the mobility during the open stage. It is important to emphasize that this work is focused on the topological synthesis of grippers which is the initial stage of a gripper design. Selecting a kinematic chain with zero mobility is not sufficient to guarantee that the gripper performs the grasping activities. In a later stage of the project, it is necessary to evaluate other characteristics such as the capacity of grasping classifying the grasp as redundant, indeterminate, graspable, and defective [10].

In terms of geometrical symmetries, such as a reflection symmetry, it is possible to color the vertex related to the object link and then evaluate the *symmetry* one more time. This evaluation does not have the aim of eliminating automorphisms, but to identify symmetry lines that can facilitate the design process. For example, if a kinematic chain has a symmetry axis between the fixed link and the object, it is possible to perform a dimensional optimization of only half of the gripper, and extend the optimization to the other half. Another possibility is to actuate a chain with more than one degree of freedom with only one actuator.

To avoid problems with the positioning of the actuators of the grippers, the *variety* of the kinematic chain must be considered. When the joints belonging to the set of actuated joints are all locked, the mobility of the mechanism is zero. If the variety of a chain with j joints is $V = 0$, the actuated pairs may be defined at random; any M of the j joints may be chosen for the purpose, where M is the number of actuators. On the other hand, if the variety is $V = 1$, $M-1$ joints may be selected at random as actuated joints, but the Mth actuated joint must be chosen from a reduced set of joints, and so on [18]. Therefore, it is a simpler task to set the actuated joint combination in a gripper with zero variety, because every closed subchain has enough degrees of freedom to accommodate the actuators, so that the gripper kinematics is not affected by actuation conflicts.

4 Conclusion

This paper introduced a method to generate grippers from Baranov chains. The method was explained and characteristics of the graphs and mechanisms were explored, such as graph connectivity, symmetry, and variety of the mechanisms. A scheme based on colored graphs and symmetry analysis was organized to facilitate the understanding of the method. At the end of the method, some steps were added to analyse characteristics of the generated grippers.

Some examples were prepared to explore these relevant characteristics, where it was possible to discuss some cases where these characteristics are relevant. For the following steps, the authors aim to automate the synthesis process to create an atlas organizing the possibilities of grippers generated from Baranov chains. The atlas can raise these characteristics and others can also be added.

Acknowledgements. This study was financed in part by the Coordenação de Aperfeiçoamento de Pessoal de Nível Superior - Brasil (CAPES) and by the Conselho Nacional de Desenvolvimento Científico e Tecnológico (CNPq).

References

1. Belfiore, N.P., Pennestrì, E.: An atlas of linkage-type robotic grippers. Mech. Mach. Theory **32**(7), 811–833 (1997)
2. Erdman, A.G., Thompson, T., Riley, D.R.: Type selection of robot and gripper kinematic topology using expert systems. Int. J. Robot. Res. **5**(2), 183–189 (1986)
3. Fan Yu, C.: Gripping mechanisms for industrial robots. Mech. Mach. Theory **17**(5), 299–311 (1982)
4. Gogu, G.: Chebychev-Grübler-Kutzbach's criterion for mobility calculation of multi-loop mechanisms revisited via theory of linear transformations. Eur. J. Mech.-A/Solids **24**(3), 427–441 (2005)
5. Huang, P., Ding, H.: Structural synthesis of Baranov trusses with up to 13 links. J. Mech. Des. **141**(7), 072301 (2019)
6. Lee, J.: Kinematic synthesis of industrial robot hand/gripper-a creative design approach. Robot. Auton. Syst. **4**(3), 257–263 (1988)

7. Martins, D., Carboni, A.P.: Variety and connectivity in kinematic chains. Mech. Mach. Theory **43**(10), 1236–1252 (2008)
8. Morlin, F.V., Carboni, A.P., Martins, D.: Reconciling enumeration contradictions: complete list of Baranov chains with up to 15 links with mathematical proof. J. Mech. Des. **143**(8), 083304, 10 p. (2021)
9. Morlin, F.V., Pereira Barreto, R.L., Piga Carboni, A., Martins, D.: A method for generating gripper mechanisms using Baranov chains. In: Pucheta, M., Cardona, A., Preidikman, S., Hecker, R. (eds.) MuSMe 2021. MMS, vol. 94, pp. 3–10. Springer, Cham (2021). https://doi.org/10.1007/978-3-030-60372-4_1
10. Prattichizzo, D., Trinkle, J.C.: Grasping. In: Springer Handbook of Robotics, pp. 671–700. Springer, Cham (2016). https://doi.org/10.1007/978-3-540-30301-5_29
11. Rincon, L.M., et al.: A grasp synthesis method for a three finger gripper. In: 2018 Latin American Robotic Symposium, 2018 Brazilian Symposium on Robotics (SBR) and 2018 Workshop on Robotics in Education (WRE). IEEE (2018)
12. Samadikhoshkho, Z., Kourosh, Z., Janabi-Sharifi, F.: A brief review on robotic grippers classifications. In: 2019 IEEE Canadian Conference of Electrical and Computer Engineering (CCECE). IEEE (2019)
13. Simoni, R., Carboni, A.P., Martins, D.: Enumeration of parallel manipulators. Robotica **27**(4), 589–597 (2009)
14. Simoni, R., Doria, C.M., Martins, D.: Symmetry and invariants of kinematic chains and parallel manipulators. Robotica **31**(1), 61–70 (2013)
15. Simoni, R., Simas, H., Martins, D.: Selecting kinematic structures of parallel manipulators using symmetry and connectivity. In: International Design Engineering Technical Conferences and Computers and Information in Engineering Conference, vol. 57144. American Society of Mechanical Engineers (2015)
16. Tai, K., El-Sayed, A.-R., Shahriari, M., Biglarbegian, M., Mahmud, S.: State of the art robotic grippers and applications. Robotics **5**(2), 11 (2016). https://doi.org/10.3390/robotics5020011
17. Tamimi, A., Perez-Gracia, A., Pucheta, M.: Structural synthesis of hands for grasping and manipulation tasks. In: Lenarčič, J., Merlet, J.-P. (eds.) Advances in Robot Kinematics 2016. SPAR, vol. 4, pp. 61–70. Springer, Cham (2018). https://doi.org/10.1007/978-3-319-56802-7_7
18. Tischler, C.R., Samuel, A.E., Hunt, K.H.: Kinematic chains for robot hands-II. Kinematic constraints, classification, connectivity, and actuation. Mech. Mach. Theory **30**(8), 1217–1239 (1995)
19. Tsai, L.W.: Mechanism Design: Enumeration of Kinematic Structures According to Function. CRC Press, Boca Raton (2000)

6R Linkages with Hidden Singularities

Zijia Li[1] and Andreas Müller[2(✉)]

[1] Academy of Mathematics and Systems Science, Chinese Academy of Sciences, Beijing, China
lizijia@amss.ac.cn
[2] Johannes Kepler University, Linz, Austria
a.mueller@jku.at

Abstract. Mechanisms may possess kinematic singularities that are smooth points of the configuration space (c-space). At such points, a mechanism becomes shaky as its instantaneous mobility changes, but in contrast to c-space singularities, like bifurcations or cusps, these kinematic singularities are not reflected in the c-space. They are therefore called *hidden singularities*. Very few publications have addressed the analysis of hidden singularities. Recent research, employing methods from algebraic geometry, shows that they are often due to embedded points of the c-space variety but may also be due to singularities of the c-space variety where real and complex components intersect. Facilitating future research on the origin of hidden singularities necessitates a comprehensive list of mechanisms exhibiting this phenomenon. To this end, a constructive approach for the synthesis of 1-DOF 6R linkages with hidden singularities of prescribed rank is proposed in this paper. Various examples are shown that have hidden singularities of rank 3 or 4. Also, an example is shown that has a singularity which is a bifurcation as well as an embedded point.

Keywords: Bricard 6R linkages · Octahedral · Singularities · Hidden singularities · Configuration space · Shaky · Higher-order analysis · Kinematic tangent cone

1 Introduction

An interesting class of mechanisms, which was initially considered as a curiosity, are those that can only perform well behaved smooth finite motions, i.e., no bifurcations or cusps, but yet possess configurations where the instantaneous differential DOF changes. Kinematically, such mechanisms may exhibit kinematic singularities that are not due to bifurcations of motion branches and are thus not visible from the geometry of the configuration space (c-space) V. These are called *hidden singularities* since they are not reflected in the c-space geometry. From a practical perspective, these mechanisms deserve attention since they may become instantaneously shaky (leading to extreme stress) while still being controllable by a defined set of coordinates (locally or globally). In this

O. Altuzarra and A. Kecskémthy (Eds.): ARK 2022, SPAR 24, pp. 81–89, 2022.
https://doi.org/10.1007/978-3-031-08140-8_10

sense, it is essential to identify such points to ensure stability. From a theoretical perspective, and to better understand such mechanisms, this phenomenon is worth investigating as the origin(s) and conditions for hidden singularities are unknown. However, an indispensable prerequisite is to list different variants (or even an entire catalog) of such mechanisms. This paper addresses the synthesis of 1-DOF 6R linkages with hidden singularities. It turns out that the hidden singularities are due to 'embedded points'; the c-space V is locally the union of a single point and a smooth curve. With this observation, 6R linkages are also synthesized with c-space singularities (bifurcations), which are embedded points simultaneously. Thus, depending on the particular situation, the linkages may possess smooth finite motions only, or they may additionally possess bifurcation singularities.

One of the earliest recognitions of a mechanism that becomes shaky without passing through a c-space singularity was the 6R linkage constructed from Goldberg linkages presented in [2]. It should be mentioned that this is a special case of the double Bennett intersection as discussed in [19]. The analysis in [5,13] showed that its c-space V is a smooth manifold. Yet at two configurations, the differential DOF increases to $\delta_{\text{diff}} = 2$, while for any point in its neighborhood, it is $\delta_{\text{diff}} = \delta_{\text{loc}} = 1$. Moreover, it is a kinematic singularity. On the other hand, any Bricard octahedral linkage of the third type must have at least two such configurations (they are 'collapsible' [1], which means that the joint axes can become coplanar, and the tangents of the half of joint angles are linearly related).

2 Notation and Terminology

The vector of joint angle is denoted with $\mathbf{q} = (q_1, \dots q_6) \in \mathbb{T}^6$. The loop closure is imposed by means of geometric constraints with *constraint map* $f : \mathbb{T}^6 \rightarrow SE(3)$ defined by the product of exponentials $f(\mathbf{q}) = \exp(\mathbf{Y}_1 q_1) \exp(\mathbf{Y}_2 q_2) \cdot \ldots \cdot \exp(\mathbf{Y}_6 q_6)$, where $\mathbf{Y}_i \in \mathbb{R}^6$ is the screw coordinate vector associated to joint i, deduced from a global reference frame. The c-space is the real variety $V := \{\mathbf{q} \in \mathbb{T}^6 | f(\mathbf{q}) = \mathbf{I}\} = f^{-1}(\mathbf{I}) \cap \mathbb{T}^6$. The first-order loop closure constraints at $\mathbf{q} \in V$ are $\mathbf{J}(\mathbf{q})\mathrm{d}\mathbf{q} = \mathbf{0}$, with the 6×6 *constraint Jacobian* $\mathbf{J}(\mathbf{q}) : \mathbb{R}^6 \rightarrow \mathbb{R}^6 \cong se(3), \mathbf{J}(\mathbf{q}) = \big(\mathbf{S}_1(\mathbf{q}), \cdots, \mathbf{S}_6(\mathbf{q})\big)$, where the columns are the instantaneous joint screw coordinate vectors in configuration given by $\mathbf{S}_i(\mathbf{q}) = \mathbf{Ad}_{f_i(\mathbf{q})} \mathbf{Y}_i$, with $f_i(\mathbf{q}) := \exp(\mathbf{Y}_1 q_1) \cdot \ldots \cdot \exp(\mathbf{Y}_i q_i)$. Possible instantaneous motions at $\mathbf{q} \in V$ are $\mathbf{x} \in \ker \mathbf{J}(\mathbf{q})$. In the following, rank always refers to the rank of the Jacobian. The *differential DOF* at $\mathbf{q} \in V$ is $\delta_{\text{diff}}(\mathbf{q}) := \dim \ker \mathbf{J}(\mathbf{q})$. The local (finite) DOF at $\mathbf{q} \in V$, denoted $\delta_{\text{loc}}(\mathbf{q})$, is the local dimension of the variety V, which may change as in case of kinematotropic linkages.

A configuration $\mathbf{q} \in V$ is a *c-space singularity* iff, V is not a smooth manifold at $\mathbf{q}$. Configuration $\mathbf{q} \in V$ is a *kinematic singularity* iff the differential DOF δ_{diff} is not constant in any neighborhood of $\mathbf{q}$ in V. A mechanism is (locally) shaky iff there is a smooth point $\mathbf{q} \in V$, which is not a kinematic singularity and $\delta_{\text{diff}}(\mathbf{q}) \neq \delta_{\text{loc}}(\mathbf{q})$. That is, the mechanism as such is shaky only if its differential DOF exceeds its finite local DOF at smooth points. A configuration can be regarded shaky whenever $\delta_{\text{diff}}(\mathbf{q}) \neq \delta_{\text{loc}}(\mathbf{q})$, thus singularities are clearly shaky configurations.

3 Local Mobility Analysis

The notion of c-space singularities rests on the differential geometric properties of the real point set V. A geometric analysis of the c-space does hence not reveal a kinematic singularity due to a hidden singularity. A means to separate necessarily differential instantaneous motions from such that are tangents to finite motions, is the kinematic tangent cone [14–16]. Denote with $\mathscr{C}_{\mathbf{q}} := \{\gamma : (-\varepsilon, \varepsilon) \rightarrow V | \gamma$ is $C^{\infty}, \gamma(0) = \mathbf{q}, \varepsilon > \mathbb{R}^{+}\}$ the class of smooth arcs in V passing through $\mathbf{q}$. The *kinematic tangent cone* of V at $\mathbf{q} \in V$ is defined as $C_{\mathbf{q}}^{\mathrm{K}} V := \{\dot{\gamma} | \gamma \in \mathscr{C}_{\mathbf{q}}\} \subset \mathbb{R}^{n}$. In other words, it consists of tangents to smooth finite motions (which is a refinement of the tangent cone $C_5(V, \mathbf{q})$ introduced by Whitney [20]. It thus reveals possible finite motions and the corresponding finite mobility at an arbitrary (regular or singular) configuration $\mathbf{q}$. It is generally not a vector space but a cone. In regular configurations, $C_{\mathbf{q}}^{\mathrm{K}} V$ is identical to the tangent space $T_{\mathbf{q}} V$.

A constructive definition of the kinematic tangent cone is based on higher-order constraints. Constraints of order i can be expressed as $H^{(i)}(\mathbf{q}, \mathbf{x}_1, \mathbf{x}_2, \ldots, \mathbf{x}_i) = \mathbf{0}$. Here, $\mathbf{x}_1, \ldots, \mathbf{x}_i \in \mathbb{R}^n$ represent ith-order displacements; clearly, $H^{(1)}(\mathbf{q}, \mathbf{x}) = \mathbf{J}(\mathbf{q}) \mathbf{x}$. The higher-order constraint mapping $H^{(i)}(\mathbf{q}, \mathbf{x}_1, \mathbf{x}_2, \ldots, \mathbf{x}_i) : \mathbb{R}^n \times \cdots \times \mathbb{R}^n \rightarrow \mathbb{R}^6$ can be easily evaluated in closed form are recursively by means of simple screw products of the instantaneous screw coordinates [16] $C_{\mathbf{q}}^{\mathrm{K}} V = K_{\mathbf{q}}^{\kappa} \subset \ldots \subset K_{\mathbf{q}}^{3} \subset K_{\mathbf{q}}^{2} \subset K_{\mathbf{q}}^{1}$ where in this sequence, the cone of order i is defined as

$$K_{\mathbf{q}}^{i} := \{\mathbf{x} | \exists \mathbf{x}_1, \mathbf{x}_2, \ldots, \mathbf{x}_i \in \mathbb{R}^n : H^{(1)}(\mathbf{q}, \mathbf{x}_1) = \mathbf{0}, H^{(2)}(\mathbf{q}, \mathbf{x}_1, \mathbf{x}_2) = \mathbf{0},$$
$$H^{(3)}(\mathbf{q}, \mathbf{x}_1, \mathbf{x}_2, \mathbf{x}_3) = \mathbf{0}, \ldots, H^{(i)}(\mathbf{q}, \mathbf{x}_1, \ldots, \mathbf{x}_i) = \mathbf{0}\} \quad (1)$$

which satisfies the inclusion $K_{\mathbf{q}}^{i-1} \subseteq K_{\mathbf{q}}^{i}$, and $K_{\mathbf{q}}^{1} = \ker \mathbf{J}(\mathbf{q})$. Notice that the higher-order constraint is different from the higher-order differential, as used in [8]. This is discussed in detail in [15].

The analysis further involves stratifying V into submanifolds of smooth motions with certain rank. To this end, introduce L_k : $= \{\mathbf{q} \in \mathbb{V}^n | f(\mathbf{q}) = \mathbf{I}, \operatorname{rank} \mathbf{J}(\mathbf{q}) < k\} \subset V$. A smooth motion with rank $\mathbf{J} < k$ is a smooth curve in L_k. The set of tangents to such curves through $\mathbf{q} \in V$ forms the *kinematic tangent cone to* L_k, denoted with $C_{\mathbf{q}}^{\mathrm{K}} L_k \subseteq C_{\mathbf{q}}^{\mathrm{K}} V$. Computationally, L_k is the subset of V defined by the vanishing of all k-minors of $\mathbf{J}$, and its kinematic tangent is determined by the sequence $C_{\mathbf{q}}^{\mathrm{K}} L_k = K_{\mathbf{q}}^{k,\kappa} \subset \ldots \subset K_{\mathbf{q}}^{k,3} \subset K_{\mathbf{q}}^{k,2} \subset K_{\mathbf{q}}^{k,1}$, where $K_{\mathbf{q}}^{k,i} \subset K_{\mathbf{q}}^{k}$ is defined by the vanishing of the derivatives of all minors up to rank k. This admits symbolic or numeric evaluation by means of simply screw products [16]. Due to the lower semi-continuity of the rank, $L_r \cap U(\mathbf{q}) \equiv \emptyset, r < \operatorname{rank} \mathbf{J}(\mathbf{q})$.

Notice that the local analysis relies on a-priori knowledge of the singularity, and the corresponding instantaneous screw/Plücker coordinates. With the proposed constructive method, the singularity is known by design. Moreover, due to the symmetry of most of the linkages, there is another 'mirror' singularity. However, in general, it must be checked whether other kinematic singularities exist in c-space. To this end, algebraic methods [4] can be used, e.g., using primary ideals [5]. The local analysis is always applicable in any (regular or singular) configuration.

4 Synthesis Method for 1-DOF 6R Linkages with Hidden Singularities of Prescribed Rank

The method starts by selecting one of the possible line systems with a certain rank $r < 5$ [12]. This is the *target line/screw system* that the Plücker line coordinates of the joint axes are supposed to form in the (hidden) singularity. An exhaustive list of different types of rank r screw systems was reported in [3,9–12,18] . In the following, the notation according to [12] is used. The second step is to select a known 1-DOF 6R linkage whose geometry can be chosen (according to the type of linkage) so that it possesses a configuration where the joint axes form the target line system. This linkage is the *candidate linkage.* Its particular geometry is then fixed so that there is a configuration in which the joint screws form the desired target line system. This configuration is a singularity of the linkage (regular points have rank 5). The third and final step is to perform a *local mobility analysis* at the singularity to check whether this is a kinematic singularity only or also a c-space singularity.

The 'synthesis' part of the method thus consists in designing a linkage that possesses a kinematic singularity in which the joint screws form a specific rank r system (steps 1 and 2). The 'analysis' part (step 3) consists in checking the nature of this singularity. The procedure can be summarized as follows:

1. *Target Line System:* Select a system of lines with certain rank (either 3 or 4), and certain type (e.g., 3d, 4b, 4c, 4d, etc.) according to the classification in [12].
2. *Candidate Linkage:* Find a 1-DOF 6R linkage and its generator that possesses a configuration with joint screws forming the rank r system of step 1. Start from a known linkage type. Preferably, use a linkage with a low degree c-space curve, making it likely that the singularity is an embedded point and not a bifurcation.
3. *Mobility Analysis:* Check the finite mobility by means of the (higher-order) local analysis in terms of the instantaneous screw/Plücker coordinates.

5 Examples

The method is demonstrated for synthesizing 6R linkages with singularities of rank 3 and 4. The target line system is thus one of the type 3 and 4 systems listed in [12]. In particular, type 3d, 4b, 4c, and 4d are used. The authors are not aware of a candidate 6R linkage that possesses a configuration where the joint screws form a type 3b system. In the following, the singularity is the reference

Table 1. Type of target line system at the singularity according to [12], the candidate linkages, and the DH-parameters for the particular realizations.

Line system	Candidate linkage	DH-parameter
1.) 3d	Third type of Bricard octahedral	$\{c_i\}=\{\frac{3}{5},\frac{24}{25},\frac{6555}{6893},\frac{392196}{531325},\frac{28193}{34465},\frac{19075}{21253}\}$ $\{b_i\}=\{0,0,0,0,0,0\}$ $\{d_i\}=\{-\frac{625}{852},-\frac{275}{28},\frac{20625}{3731},\frac{236946875}{191055449},\frac{40286655475}{7105972272},-\frac{2656625}{1407504}\}$
2.) 3d	Plane-symmetric Bricard octahedral	$\{c_i\}=\{\frac{21}{29},-\frac{67\sqrt{6989}}{6989},\frac{67\sqrt{6989}}{6989},\frac{2\sqrt{29}}{29},\frac{8\sqrt{89}}{89},-\frac{9\sqrt{2581}}{2581}\}$ $\{b_i\}=\{0,0,0,0,0,0\}$ $\{d_i\}=\{\frac{\sqrt{29}}{10},\frac{\sqrt{29}}{2},\frac{\sqrt{241}}{5},-\frac{\sqrt{29}}{10},1,\frac{\sqrt{89}}{10}\}$
3.) 4b	Line-symmetric Bricard	$\{c_i\}=\{\frac{231}{425},\frac{2116}{2125},0,\frac{231}{425},\frac{2116}{2125},0\}$ $\{b_i\}=\{\frac{13005}{7954},-\frac{38675}{12723},-3,\frac{13005}{7954},-\frac{38675}{12723},-3\}$ $\{d_i\}=\{-\frac{22253}{7954},\frac{539831345}{101198742},-\frac{48512}{12723},-\frac{22253}{7954},\frac{539831345}{101198742},-\frac{48512}{12723}\}$
4.) 4c	Special double Bennett	$\{c_i\}=\{0,-\frac{4}{5},-\frac{7}{25},\frac{12}{37},-\frac{24}{25},\frac{57}{185}\},\ d_i=0,i=1,\ldots,6$ $\{b_i\}=\{\frac{1}{2},-\frac{1}{2},\frac{5}{24},\frac{5}{2},-\frac{5}{2},-\frac{493}{176}\}$ $\{d_i\}=\{0,0,0,0,0,0\}$
5.) 4d	Line-symmetric Bricard	$\{c_i\}=\{\frac{12}{25},0,\frac{16}{25},\frac{12}{25},0,\frac{16}{25}\}$ $\{b_i\}=\{\frac{375}{481},0,-\frac{125}{123},\frac{375}{481},0,-\frac{125}{123}\}$ $\{d_i\}=\{-\frac{1100}{59163},\frac{673}{481},-\frac{187}{123},-\frac{1100}{59163},\frac{673}{481},-\frac{187}{123}\}$

configuration $\mathbf{q}_0 = \mathbf{0}$. For the synthesized linkage, the geometry parameters, derived from the DH-parameters, are listed in Table 1. The parameters listed are the offsets d_i, the Bennett ratios $b_i = a_i/\sin\alpha_i$, where α_i the twist angles, and $c_i := \cos\alpha_i$.

1.) **Type 3d/Third type of Bricard octahedral:** The target line system is of type 3d (rank 3, all axes are coplanar) defined by screw coordinates $\mathbf{Y}_1 = (1,0,0,0,0,0)^T$, $\mathbf{Y}_2 = \left(\frac{3}{5},\frac{4}{5},0,0,0,1\right)^T$, $\mathbf{Y}_3 = \left(\frac{4}{5},\frac{3}{5},0,0,0,-2\right)^T$, $\mathbf{Y}_4 = \left(\frac{32616}{34465},\frac{11137}{34465},0,0,0,-\frac{23558}{6893}\right)^T$, $\mathbf{Y}_5 = \left(\frac{351990947}{732484645},\frac{642367596}{732484645},0,0,0,-\frac{92697951}{146496929}\right)^T$, $\mathbf{Y}_6 = \left(\frac{19075}{21253},\frac{9372}{21253},0,0,0,\frac{4840}{21253}\right)^T$. A 6R linkage derived from the third type Bricard octahedra is used as candidate linkage, as it has two configurations with joint axes forming a 3d system, which are the two coplanar configurations. This type of 6R linkages is also a sub-family of the third type of angle-symmetric 6R linkages [6,7]. Furthermore, a 6R linkage that has a configuration where all axes are coplanar (case 3d) must be a special linkage derived from the Bricard octahedra (three types) or the extension [17]. The particular choice of DH parameters is listed in Table 1. The constraint Jacobian has rank $\mathbf{J}(\mathbf{q}_0) = 3$, the linkage has the differential DOF $\delta_{\text{diff}}(\mathbf{q}_0) = 3$, with first-order motions $\ker \mathbf{J}(\mathbf{q}_0)$. The second-order cone $K^2_{\mathbf{q}_0}$ is the union of two vector spaces

$$K^2_{\mathbf{q}_0} = K^{2(I)}_{\mathbf{q}_0} \cup K^{2(II)}_{\mathbf{q}_0} \quad \text{with } K^{2(I)}_{\mathbf{q}_0} = \{\mathbf{x} = (u,2u,3u,-u,-2u,-3u)|u \in \mathbb{R}\}$$
$$K^{2(II)}_{\mathbf{q}_0} = \{\mathbf{0} \in \mathbb{R}^6\}$$

This is identical to all higher-order cones, i.e. $C^{\mathrm{K}}_{\mathbf{q}}V = K^2_{\mathbf{q}_0}$. This reveals that there is a single curve passing through $\mathbf{q}_0$, with tangent space $K^{2(I)}_{\mathbf{q}_0}$.

The existence of $K_{\mathbf{q}_0}^{2(II)}$ shows that the defining equations factorize indicating that $\mathbf{q}_0$ is a multiple solution, and thus certainly a kinematic singularity. In general it must be checked whether $\mathbf{q}_0$ is a kinematic singularity, since the rank may constantly be $r = 3$ in a neighborhood, and thus the linkage be shaky and $\mathbf{q}_0$ a regular point. Checking for a kinematic singularity amounts to investigate motion with certain rank. It turns out that $K_{\mathbf{q}_0}^{5,i} = \{\mathbf{0} \in \mathbb{R}^6\}, i = 1, 2, \ldots$, and hence $C_{\mathbf{q}_0}^{\mathrm{K}} L_4 = C_{\mathbf{q}_0}^{\mathrm{K}} L_5 = \{\mathbf{0} \in \mathbb{R}^6\}$, so that $\mathbf{q}_0$ is an isolated point with rank 3, while all neighboring points have rank 5. In summary, the local analysis shows that there is a single smooth curve (with rank 5) passing through $\mathbf{q}_0$, with tangent space $K_{\mathbf{q}_0}^{2(I)}$. But there is also a single point, namely the embedded point $\mathbf{q}_0$, with rank 4 and tangent space $K_{\mathbf{q}_0}^{2(II)}$.

2.) **Type 3d/Plane-symmetric Bricard octahedral:** This example shows that one may not get a hidden singularity yet an embedded point in c-space. The target line system is of type 3d again, defined by screw coordinates $\mathbf{Y}_1 = \left(\frac{1}{5}, -\frac{1}{2}, 0, 0, 0, -\frac{1}{2}\right)^T$, $\mathbf{Y}_2 = \left(-1, -\frac{5}{2}, 0, 0, 0, -\frac{5}{2}\right)^T$, $\mathbf{Y}_3 = \left(-\frac{4}{5}, 3, 0, 0, 0, -2\right)^T$, $\mathbf{Y}_4 = \left(\frac{1}{5}, \frac{1}{2}, 0, 0, 0, -\frac{1}{2}\right)^T$, $\mathbf{Y}_5 = (1, 0, 0, 0, 0, 0)^T$, $\mathbf{Y}_5 = \left(\frac{4}{5}, \frac{1}{2}, 0, 0, 0, 0\right)^T$. There are only the three types of Bricard's octahedral linkages that can attain a configuration where all axes are coplanar. Here a type of plane-symmetric is considered. The particular linkage geometry is shown in Table 1, which concludes step 1 and 2. It remains to check the mobility in the singularity. The differential DOF is $\delta_{\text{diff}}\left(\mathbf{q}_0\right) = 3$. The second-order cone $K_{\mathbf{q}_0}^2$ is the union of three vector spaces $K_{\mathbf{q}_0}^2 = K_{\mathbf{q}_0}^{2(I)} \cup K_{\mathbf{q}_0}^{2(II)} \cup K_{\mathbf{q}_0}^{2(III)}$, with

$$
\begin{aligned}
K_{\mathbf{q}_0}^{2(I)} &= \{\mathbf{x} = (u, \tfrac{\sqrt{5}-5}{25} u, -\tfrac{\sqrt{5}-5}{20} u, -u, -\tfrac{2(\sqrt{5}-1)}{5} u, \tfrac{\sqrt{5}-1}{2} u) | u \in \mathbb{R}\} \\
K_{\mathbf{q}_0}^{2(II)} &= \{\mathbf{x} = (u, -\tfrac{\sqrt{5}+5}{25} u, \tfrac{\sqrt{5}+5}{20} u, -u, \tfrac{2(\sqrt{5}+1)}{5} u, -\tfrac{\sqrt{5}+1}{2} u) | u \in \mathbb{R}\} \\
K_{\mathbf{q}_0}^{2(III)} &= \{\mathbf{0} \in \mathbb{R}^6\}.
\end{aligned}
$$

This is identical to the higher-order cones, so that $C_{\mathbf{q}_0}^{\mathrm{K}} V = K_{\mathbf{q}_0}^2$. The local analysis of the set of points of rank less than 5 and 4, respectively, yields $K_{\mathbf{q}_0}^{5,i} = \{\mathbf{0} \in \mathbb{R}^6\}, i = 1, 2, \ldots$, and hence $C_{\mathbf{q}_0}^{\mathrm{K}} L_4 = C_{\mathbf{q}_0}^{\mathrm{K}} L_5 = \{\mathbf{0} \in \mathbb{R}^6\}$. In conclusion, this shows that there are two 1-dim manifolds intersection (to which $K_{\mathbf{q}_0}^{2(I)}$ and $K_{\mathbf{q}_0}^{2(II)}$ is the respective tangent space), and that $\mathbf{q}_0$ is a bifurcation. The tangent space to the set of rank 4 singularities is the 0-dim vector space $K_{\mathbf{q}_0}^{5,i}$, which shows that all points in the neighborhood have rank 5. The 0-dim vector space $K_{\mathbf{q}_0}^{2(III)}$ is the tangent space to a 0-dim component, hence $\mathbf{q}_0$ is an embedded point, but not a hidden singularity.

3.) **Type 4b/Line-symmetric Bricard:** The target line system is of type 4b (rank 4, all lines concurrent with two skew lines) generated by $\mathbf{Y}_1 = (\frac{4}{5}, 0, -\frac{3}{5}, 0, -\frac{12}{5}, 0)^T$, $\mathbf{Y}_2 = (\frac{24}{85}, -\frac{4}{5}, -\frac{9}{17}, 0, -\frac{72}{85}, \frac{32}{25})^T$, $\mathbf{Y}_3 = (\frac{9}{25}, -\frac{4}{5}, -\frac{12}{25}, 0, -\frac{27}{25}, \frac{9}{5})^T$, $\mathbf{Y}_4 = (-\frac{4}{5}, 0, -\frac{3}{5}, 0, 12/5, 0)^T$, $\mathbf{Y}_5 = (-\frac{24}{85}, -\frac{4}{5}, -\frac{9}{17}, 0, \frac{72}{85}, \frac{32}{25})^T$, $\mathbf{Y}_6 = (-\frac{9}{25}, \frac{4}{5}, -\frac{12}{25}, 0, \frac{27}{25}, \frac{9}{5})^T$. Such a constellation exists for the line-symmetric Bricard linkage. The geometry is chosen as in Table 1. The second-order cone $K^2_{\mathbf{q}_0}$ is the union of two 1-dim vector spaces

$$K^2_{\mathbf{q}_0} = K^{2(I)}_{\mathbf{q}_0} \cup K^{2(II)}_{\mathbf{q}_0} \text{ with } K^{2(I)}_{\mathbf{q}_0} = \{\mathbf{x} = (u, -\tfrac{3825}{1199}u, \tfrac{2720}{1199}u, u, -\tfrac{3825}{1199}u, \tfrac{2720}{1199}u) | u \in \mathbb{R}\}$$
$$K^{2(II)}_{\mathbf{q}_0} = \{\mathbf{x} = (u, \tfrac{340}{33}u, -\tfrac{340}{33}u, -u, -\tfrac{340}{33}, \tfrac{340}{33}) | u \in \mathbb{R}\}. \tag{2}$$

The third-order cone is $K^3_{\mathbf{q}_0} = K^{3(I)}_{\mathbf{q}_0} \cup K^{3(II)}_{\mathbf{q}_0}$ with $K^{3(I)}_{\mathbf{q}_0} = K^{2(I)}_{\mathbf{q}_0}$ and $K^{3(II)}_{\mathbf{q}_0} = \{\mathbf{0} \in \mathbb{R}^6\}$. All higher-order cones are identical to $K^3_{\mathbf{q}_0}$. Next, the set of smooth motions with certain rank is analyzed. Clearly, for points with rank less than 6, it is $C^{\mathrm{K}}_{\mathbf{q}_0} L_6 = C^{\mathrm{K}}_{\mathbf{q}} V$. Further computation yields $K^{5,i}_{\mathbf{q}_0} = \{\mathbf{0} \in \mathbb{R}^6\}, i = 1, 2, \ldots$, and hence $C^{\mathrm{K}}_{\mathbf{q}_0} L_5 = \{\mathbf{0} \in \mathbb{R}^6\}$. Since $\operatorname{rank} \mathbf{J}(\mathbf{q}_0) = 5$, this is the kinematic tangent cone to the set of points of rank 4. Thus the c-space is locally a smooth curve with tangent space $K^{2(I)}_{\mathbf{q}_0}$, in which is embedded the 0-dim manifold $\{\mathbf{q}_0\}$, with tangent space $K^{3(II)}_{\mathbf{q}_0}$, and $\mathbf{q}_0$ a hidden singularity. Interestingly, in this singularity, the linkage possesses second-order shakiness, since the differential motions $\mathbf{x} \in K^{2(II)}_{\mathbf{q}_0} \supset K^{3(II)}_{\mathbf{q}_0}$ do not correspond to finite motions.

4.) **Type 4c/Special double Bennett:** The target line system is of type 4c (rank 4, all lines perpendicularly intersecting a common line) defined by the screws $\mathbf{Y}_1 = (1, 0, 0, 0, 0, 0)^T, \mathbf{Y}_2 = (0, 1, 0, \frac{1}{2}, 0, 0)^T, \mathbf{Y}_3 = (\frac{3}{5}, -\frac{4}{5}, 0, -\frac{16}{25}, -\frac{12}{25}, 0)^T, \mathbf{Y}_4 = (\frac{3}{5}, \frac{4}{5}, 0, \frac{4}{5}, -\frac{3}{5}, 0)^T, \mathbf{Y}_5 = (-\frac{104}{185}, \frac{153}{185}, 0, \frac{38097}{13690}, \frac{12948}{6845}, 0)^T$, $\mathbf{Y}_6 = (\frac{57}{185}, -\frac{176}{185}, 0, -\frac{86768}{34225}, -\frac{28101}{34225}, 0)^T$. A special double Bennett linkage has such a configuration where the joint screws form a linear variety of rank 4, where all lines are mutually perpendicular and intersect a common line, and is used as candidate. The constraint Jacobian has $\operatorname{rank} \mathbf{J}(\mathbf{q}_0) = 4$, and the linkage has the differential DOF $\delta_{\text{diff}}(\mathbf{q}_0) = 2$. The second-order cone $K^2_{\mathbf{q}_0}$ is the union of two vector spaces

$$K^2_{\mathbf{q}_0} = K^{2(I)}_{\mathbf{q}_0} \cup K^{2(II)}_{\mathbf{q}_0} \quad \text{with } K^{2(I)}_{\mathbf{q}_0} = \{\mathbf{x} = (u, 2u, u, -3u, -2u, -3u) | u \in \mathbb{R}\}$$
$$K^{2(II)}_{\mathbf{q}_0} = \{\mathbf{0} \in \mathbb{R}^6\}. \tag{3}$$

This is the kinematic tangent cone, i.e. $C^{\mathrm{K}}_{\mathbf{q}} V = K^2_{\mathbf{q}_0}$, since it is identical to all higher-order cones. The analysis of the set of points of rank 4 yields $C^{\mathrm{K}}_{\mathbf{q}_0} L_5 = K^{5,i}_{\mathbf{q}_0} = \{\mathbf{0} \in \mathbb{R}^6\}$. In summary, there is a single 1-dim manifold passing through $\mathbf{q}_0$, with tangent space $C^{\mathrm{K}}_{\mathbf{q}} V$, characterizing the geometry of V as a set. Yet at $\mathbf{q}_0$ the rank of $\mathbf{J}$ drops so that this is a kinematic singularity. Being an embedded point is revealed by $K^{2(II)}_{\mathbf{q}_0}$ which is the tangent space to the 0-dim manifold $\{\mathbf{q}_0\}$ of points with rank 4.

5.) **Type 4d/Line-symmetric Bricard:** A type 4d system (rank 4, where four lines are in a common plane and two lines are passing through one point in that plane) is used as target line system, where $\mathbf{Y}_1 = (\frac{3}{5}, \frac{4}{5}, 0, 0, 0, 0)^T, \mathbf{Y}_2 = (0, \frac{3}{5}, \frac{4}{5}, 1, 0, 0)^T, \mathbf{Y}_3 = (0, -\frac{4}{5}, \frac{3}{5}, -1, 0, 0)^T, \mathbf{Y}_4 = (\frac{3}{5}, -\frac{4}{5}, 0, 0, 0, 0)^T, \mathbf{Y}_5 = (0, -\frac{3}{5}, -\frac{4}{5}, 1, 0, 0)^T$, $\mathbf{Y}_6 = (0, \frac{4}{5}, -\frac{3}{5}, -1, 0, 0)^T$. A line-symmetric Bricard linkage possesses such a configuration (parameters in Table 1). The second-order cone $K^2_{\mathbf{q}_0}$ is the union of two vector spaces

$$K^2_{\mathbf{q}_0} = K^{2(I)}_{\mathbf{q}_0} \cup K^{2(II)}_{\mathbf{q}_0} \quad \text{with} \quad \begin{aligned} K^{2(I)}_{\mathbf{q}_0} &= \{\mathbf{x} = (0, u, u, 0, u, u) | u \in \mathbb{R}\} \\ K^{2(II)}_{\mathbf{q}_0} &= \{\mathbf{x} = (u, -\tfrac{12}{25}u, \tfrac{16}{25}u, -u, \tfrac{12}{25}u, -\tfrac{16}{25}u) | u \in \mathbb{R}\}. \end{aligned}$$

Higher-order analysis yields $K^3_{\mathbf{q}_0} = K^{3(I)}_{\mathbf{q}_0} \cup K^{3(II)}_{\mathbf{q}_0}$ with $K^{3(I)}_{\mathbf{q}_0} = K^{2(I)}_{\mathbf{q}_0}$ and $K^{3(II)}_{\mathbf{q}_0} = \{\mathbf{0} \in \mathbb{R}^6\}$, and this is identical to all higher-order cones, i.e. $C^{\mathrm{K}}_{\mathbf{q}} V = K^3_{\mathbf{q}_0}$. Since $K^3_{\mathbf{q}_0} \subset K^2_{\mathbf{q}_0}$, in this singularity, the linkage shows second-order shakiness. Further, $K^{5,i}_{\mathbf{q}_0} = \{\mathbf{0} \in \mathbb{R}^6\}, i = 1, 2, \ldots$, and hence $C^{\mathrm{K}}_{\mathbf{q}_0} L_4 = C^{\mathrm{K}}_{\mathbf{q}_0} L_5 = \{\mathbf{0} \in \mathbb{R}^6\}$. The configuration is a hidden singularity where the linkage is shaky of order two.

6 Conclusion

This paper proposed an approach for synthesizing 1-DOF 6R linkages exhibiting so-called hidden singularities where the joint axes form a specified line system. Hidden singularities are kinematic singularities (differential DOF changes) and at the same time smooth points of the c-space (c-space is a smooth manifold). The 'synthesis' part of the method involves specifying a desired screw system of a particular type (target line system) and identifying a 1-DOF mobile linkage (candidate linkage) that possesses a configuration where its joints form the target line system. The 'analysis' part consists of checking whether this is a hidden singularity through a local mobility analysis. The method is thus constructive. It also reveals whether the hidden singularities are embedded points in c-space. Interestingly, for all examples, hidden singularities are due to embedded points in the c-space. Future work shall address a sufficient condition that the singularities are embedded points and not c-space singularities. The origin of hidden singularities is yet to be explored. This will be approached with algebraic geometry. Preliminary results reveal that they are due to embedded points but also due to intersections of the real and the complex components of the variety, which are not apparent in V.

Acknowledgements. This research was funded by the Austrian Science Fund (FWF) [I 4452-N].

References

1. Baker, J.E.: On Bricard's doubly collapsible octahedron and its planar, spherical and skew counterparts. J. Franklin Inst. **332**(6), 657–679 (1995)
2. Chen, Y., You, Z.: Spatial overconstrained linkages - the lost jade. In: Koetsier, T., Ceccarelli, M. (eds.) Explorations in the History of Machines and Mechanisms, History of Mechanism and Machine Science, vol. 15, pp. 535–550. Springer, Cham (2012). https://doi.org/10.1007/978-94-007-4132-4_37
3. Gibson, C., Hunt, K.: Geometry of screw systems-2: classification of screw systems. Mech. Mach. Theory **25**(1), 11–27 (1990)

4. Husty, M., Pfurner, M., Schröcker, H.P., Brunnthaler, K.: Algebraic methods in mechanism analysis and synthesis. Robotica **25**(6), 661–675 (2007)
5. Li, Z., Müller, A.: Mechanism singularities revisited from an algebraic viewpoint. In: 43rd Mechanisms and Robotics Conference (MR)/ASME International Design Engineering Technical Conferences (IDETC), Anaheim, CA, USA, 18–21 August 2019
6. Li, Z., Nawratil, G., Rist, F., Hensel, M.: Invertible paradoxic loop structures for transformable design. In: Computer Graphics Forum, vol. 39, pp. 261–275 (2020)
7. Li, Z., Schicho, J.: Classification of angle-symmetric 6R linkages. Mech. Mach. Theory **70**, 372–379 (2013)
8. Liu, G., Lou, Y., Li, Z.: Singularities of parallel manipulators: a geometric treatment. IEEE Trans. Robot. Autom. **19**(4), 579–594 (2003)
9. Martinez, J.R., Duffy, J.: Classification of screw systems-I. One- and two-systems. Mech. Mach. Theory **27**(4), 459–470 (1992)
10. Martinez, J.R., Duffy, J.: Classification of screw systems-II. Three-systems. Mech. Mach. Theory **27**(4), 471–490 (1992)
11. McCarthy, J.M., Soh, G.S.: Geometric Design of Linkages, vol. 11. Springer, Cham (2010). https://doi.org/10.1007/978-1-4419-7892-9
12. Merlet, J.P.: Geometry and kinematic singularities of closed-loop manipulators. Lab. Rob. Autom. **4**, 85 (1992)
13. Müller, A.: Local kinematic analysis of closed-loop linkages-mobility, singularities, and shakiness. ASME J. Mech. Rob. **8**, 041013, 11 p. (2016)
14. Müller, A.: Higher-order analysis of kinematic singularities of lower pair linkages and serial manipulators. ASME J. Mech. Rob. **10**(1), 011008, 13 p. (2018)
15. Müller, A.: Local investigation of mobility and singularities of linkages. In: Müller, A., Zlatanov, D. (eds.) Singular Configurations of Mechanisms and Manipulators. CICMS, vol. 589, pp. 181–229. Springer, Cham (2019). https://doi.org/10.1007/978-3-030-05219-5_5
16. Müller, A.: An overview of formulae for the higher-order kinematics of lower-pair chains with applications in robotics and mechanism theory. Mech. Mach. Theory **142**, 103594 (2019)
17. Nawratil, G.: Flexible octahedra in the projective extension of the Euclidean 3-space and their application. Habilitation thesis. Inst. Discrete Math. Geom., TU Vienna (2011)
18. Pottmann, H., Wallner, J.: Computational Line Geometry. Springer, Cham (2009). https://doi.org/10.1007/978-3-642-04018-4
19. Selig, J., Li, Z.: Double bennett mechanisms with assembly modes of different dimensions. In: International Conference on Reconfigurable Mechanisms and Robots (ReMAR), pp. 1–6 (2018)
20. Whitney, H.: Local properties of analytic varieties. Differential and Combinatorial Topology, A Symposium in Honor of M. Morse. Princeton University Press (1965)

Cable Driven Manipulators

Best Operation Regions in a Planar Cable Driven System

Leonardo Mejia[1(✉)], Fabricio Frantz[1], Daniel Ponce[1], and Daniel Martins[2]

[1] Federal University of Santa Catarina, Rua João Pessoa 2514, Florianópolis CEP 89036-004, Brazil
{leonardo.mejia.rincon,daniel.alejandro}@ufsc.br

[2] Federal University of Santa Catarina, RR. Eng. Agronômico Andrei Cristian Ferreira, Florianópolis CEP 88040-900, Brazil

Abstract. This paper presents a new method to identify the best operation regions for a planar cable-driven system based on the calculation of the maximum isotropic force at all the discrete points within the manipulator's workspace. The proposed method is based on the obtention of four vectorial subspaces in which a new mathematical closed-form solution proposed by the authors is applied in order to obtain the maximum isotropic force generated for each vectorial subspace. Once obtained the maximum isotropic force for each vectorial subspace, is chosen the minimum of them as the maximum generalized isotropic force of the system. The proposed method is applied in four different study cases in which the positions of the vertices in the mobile platform are modified in order to obtain a rectangular workspace, a rectangular trapezoidal workspace, a trapezoidal workspace, and a parallelogramic workspace.

1 Introduction

The development of robots allowed several unhealthy activities, previously performed by humans, to be executed by mechanical systems, and as a result of this substitution process human beings could concentrate on intellectual activities. Among the main characteristics defining the cable-driven systems we can find, the low production cost, the high dimension for the workspace and, the fast winding of the cables stand out, allowing unique dynamic characteristics [1–4].

Despite the desirable characteristics of these mechanical systems, and the high efforts by the academic community, several research aspects are still open [3,5]. Among them, and the main motivation of this work, we can find the definition for the best operation regions in cable-driven systems based in their maximum isotropic force at the manipulator's end-effector.

2 Topology for the Studied Manipulator

Cable driven parallel robots are a special class of parallel mechanisms whose legs were replaced by cables. These devices were developed by Landsberger &

O. Altuzarra and A. Kecskémthy (Eds.): ARK 2022, SPAR 24, pp. 93–100, 2022.
https://doi.org/10.1007/978-3-031-08140-8_11

Sheridan in 1985 to surpass some of the workspace limitations of traditional parallel robots [6].

Specifically, in this work, a cable-driven system type $1R2T$ as the one presented in Fig. 1 is studied. This manipulator allows the 3 planar movements of a rigid body, being one rotation on the z axis and two translations on the $x-y$ axis [7].

The mobile platform in this system is a rectangular rigid body with dimensions M_h and M_b as shown in Fig. 1, also, the mobile platform can be rotated by an angle ϕ. The end-effector of this manipulator is located at the centroid of the mobile platform at point E.

On the other hand, the fixed platform is a polyhedron defined by the vertices A_1-A_4, and which positions will be modified in the results section of the current paper. The cables of this system are represented in Fig. 1 as blue lines connecting the joints $A_n - B_n \;\; \forall \; n = 1, ..., 4$.

3 Static Model for the Studied System

In the static analysis of mechanical systems, the objective is to determine the existing relation between the forces and moments on the joints and the efforts existing at the interface when there is contact with the medium [8]. A force or torque can be applied to the actuated joints of the mechanical system and calculate the efforts that the robot imposes on the medium, or even, external forces can be applied at the interface and the torques and forces on the actuated joints can be calculated in order to balance these external forces [9,10].

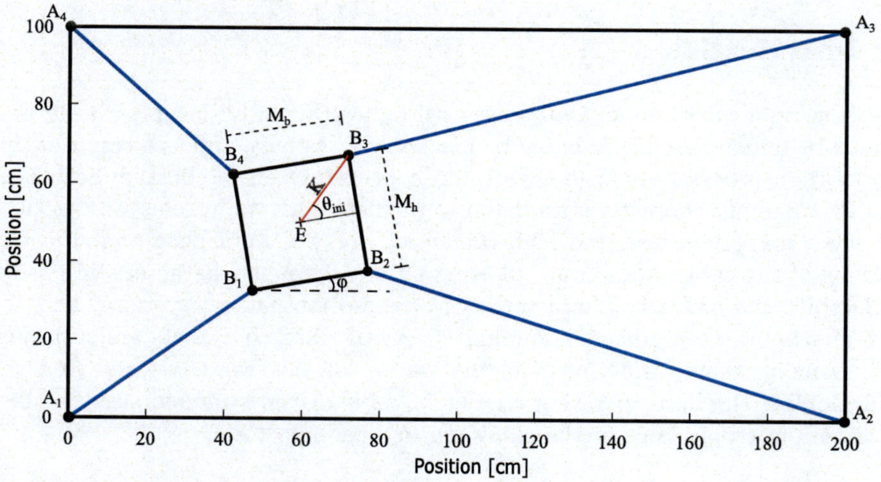

Fig. 1. Topology of the studied cable-driven system

This work applied the Davies' method [14,15] in order to obtain the static model for the studied cable-driven system, despite admittedly the same results can be obtained by applying other methods like the Jacobian based method or the vectorial method, the authors have chosen this method due to the facility to obtain all the inner variables and the potential for an easy expansion in future researches.

The obtained static model for the studied system is shown in Eqs. (1), (2) and (3). These equations represent the forces along the axes x and y and the moment around the axis z. Notice that in those equations appear the term $\theta_n \ \forall \ n = 1...4$ representing the angle's direction of the vector obtained from the joints $B_n \ \forall \ n = 1...4$ to the joints $A_n \ \forall \ n = 1...4$ and the terms $K = (K_{n_x}, K_{n_y}) \ \forall \ n = 1, ..., 4$ representing vectors from E to $B_n \ \forall \ n = 1...4$. Also notice that Eqs. 1, 2 and 3 can be represented in a matrix form as shown in Eq. 4 [11].

$$F_x = F_{1x}+F_{2x}+F_{3x}+F_{4x} = T_1 \cdot cos(\theta_1)+T_2 \cdot cos(\theta_2)+T_3 \cdot cos(\theta_3)+T_4 \cdot cos(\theta_4) \quad (1)$$

$$F_y = F_{1y}+F_{2y}+F_{3y}+F_{4y} = F_Y = T_1 \cdot sin(\theta_1)+T_2 \cdot sin(\theta_2)+T_3 \cdot sin(\theta_3)+T_4 \cdot sin(\theta_4) \quad (2)$$

$$\begin{aligned} M_z &= M_{1z} + M_{2z} + M_{3z} + M_{4z} \\ &= ...F_1 \cdot (K_{1x} \cdot sin(\theta_1) - K_{1y} \cdot cos(\theta_1)) + F_2 \cdot (K_{2x} \cdot sin(\theta_2) - K_{2y} \cdot cos(\theta_2)) \\ &+ ...F_3 \cdot (K_{3x} \cdot sin(\theta_3) - K_{3y} \cdot cos(\theta_3)) + F_4 \cdot (K_{4x} \cdot sin(\theta_4) - K_{4y} \cdot cos(\theta_4)) \end{aligned} \quad (3)$$

$$\begin{bmatrix} F_x \\ F_y \\ M_z \end{bmatrix} = \begin{bmatrix} G_{1,1} & G_{1,2} & G_{1,3} & G_{1,4} \\ G_{2,1} & G_{2,2} & G_{2,3} & G_{2,4} \\ G_{3,1} & G_{3,2} & G_{3,3} & G_{3,4} \end{bmatrix} \cdot \begin{bmatrix} T_1 \\ T_2 \\ T_3 \\ T_4 \end{bmatrix} \quad (4)$$

4 Proposed Method

The proposed method in this paper starts with the determination of four vectorial subspaces, obtained from Eq. (4) and assuming the fact that just three cables can be actuated simultaneously. In this way, each vectorial subspace is composed within a generic matrix A as shown in Eqs. (5), (6), (7) and (8).

$$S_1 = \begin{bmatrix} a_{1,1} & a_{1,2} & a_{1,3} \\ a_{2,1} & a_{2,2} & a_{2,3} \\ a_{3,1} & a_{3,2} & a_{3,3} \end{bmatrix} = \begin{bmatrix} G_{1,1} & G_{1,2} & G_{1,3} \\ G_{2,1} & G_{2,2} & G_{2,3} \\ G_{3,1} & G_{3,2} & G_{3,3} \end{bmatrix} \quad (5)$$

$$S_2 = \begin{bmatrix} a_{1,1} & a_{1,2} & a_{1,3} \\ a_{2,1} & a_{2,2} & a_{2,3} \\ a_{3,1} & a_{3,2} & a_{3,3} \end{bmatrix} = \begin{bmatrix} G_{1,1} & G_{1,2} & G_{1,4} \\ G_{2,1} & G_{2,2} & G_{2,4} \\ G_{3,1} & G_{3,2} & G_{3,4} \end{bmatrix} \quad (6)$$

$$S_3 = \begin{bmatrix} a_{1,1} & a_{1,2} & a_{1,3} \\ a_{2,1} & a_{2,2} & a_{2,3} \\ a_{3,1} & a_{3,2} & a_{3,3} \end{bmatrix} = \begin{bmatrix} G_{1,1} & G_{1,3} & G_{1,4} \\ G_{2,1} & G_{2,3} & G_{2,4} \\ G_{3,1} & G_{3,3} & G_{3,4} \end{bmatrix} \quad (7)$$

$$S_4 = \begin{bmatrix} a_{1,1} & a_{1,2} & a_{1,3} \\ a_{2,1} & a_{2,2} & a_{2,3} \\ a_{3,1} & a_{3,2} & a_{3,3} \end{bmatrix} = \begin{bmatrix} G_{1,2} & G_{1,3} & G_{1,4} \\ G_{2,2} & G_{2,3} & G_{2,4} \\ G_{3,2} & G_{3,3} & G_{3,4} \end{bmatrix} \tag{8}$$

Once determined the four vectorial subspaces S_n $(n = 1...4)$, the novelty of this paper is that it is possible to calculate three critical saturation indices by using Eq. (9) for each vectorial subspace. Notice that this equation is expressed in terms of the imposed moment at the end effector (M_d), the maximum acceptable tension in each cable $(\tau_{i_{max}})$, and, the elements $a_{i,1}$, $a_{i,2}$ and $a_{i,3}$ $(i = 1, 2, 3)$ from each vectorial subspace. The maximum isotropic force of each vectorial subspace is obtained as the minimum value for the three calculated critical saturation indices as shown in Eq. (10) [12,13]. Equations 9 and 10 were proposed by the authors of this paper as a mathematical closed solution for this problem and its obtaining constitutes one the main contribution in this paper.

$$\psi_i = \frac{\tau_{i_{max}} - a_{i,3} \cdot M_d}{\sqrt{(a_{i,1})^2 + (a_{1,2})^2}} \qquad i = 1, 2, 3 \tag{9}$$

$$\Psi_{S_n} = min(\psi_i) \quad n = 1, ..., 4; \quad i = 1, 2, 3 \tag{10}$$

The maximum isotropic force calculated by Eq. 10 can be interpreted as the maximum minimal force that the studied manipulator can apply, a polar representation of this magnitude can be seen as a circumference as shown in red in Fig. 2. In the same figure can be seen in blue the maximum force the manipulator could apply in each single direction. Analysing Fig. 2 can also be seen that the maximum isotropic force is represented as an inner circumscribed circle to the polar representation of the maximum force obtained in each single direction.

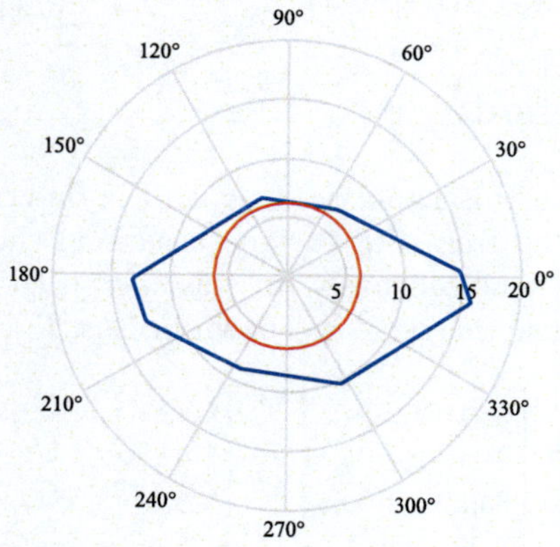

Fig. 2. Polar representation of the local forces (the forces are expressed in [N])

5 Simulation Results

In order to validate the proposed method, four simulations were executed maintaining the original topology for the mobile platform of the studied manipulator, but changing the vertices of the fixed platform. In the firs study case the vertices of the fixed platform were placed at $A_1 = (0, 0)$ [cm], $A_2 = (2000, 0)$ [cm], $A_3 = (2000, 1000)$ [cm] and $A_4 = (0, 1000)$ [cm] representing a rectangular workspace. In the second study case the vertices of the fixed platform were placed at $A_1 = (0, 0)$ [cm], $A_2 = (2000, 0)$ [cm], $A_3 = (1700, 1000)$ [cm] and $A_4 = (0, 1000)$ [cm] representing a rectangular trapezoidal workspace. In the third study case the vertices of the fixed platform were placed at $A_1 = (0, 0)$ [cm], $A_2 = (2000, 0)$ [cm], $A_3 = (1700, 1000)$ [cm] and $A_4 = (300, 1000)$ [cm] representing a trapezoidal workspace, and finally, In the fourth study case the vertices of the fixed platform were placed at $A_1 = (300, 0)$ [cm], $A_2 = (2000, 0)$ [cm], $A_3 = (1700, 1000)$ [cm] and $A_4 = (0, 1000)$ [cm] representing a parallelogramic workspace. For all the studied cases the fixed platform was considered as a square of side 25 [cm] and the maximum length for the cables was not restricted. The maximum admitted tension for each cable is 10 [N]. By applying the proposed method for each discretized point into the workspace we can obtain the complete map for the maximum isotropic force in each studied case as shown in Figs. 3, 4, 5 and 6.

In Figs. 3, 4, 5 and 6, it is possible to realize that there exist central regions in which the maximum isotropic forces are higher than in the external regions. These regions with maximum isotropic force are characterized in yellow and represent the best operation areas for each studied case. In these figures is also possible to notice that depending on the position of the fixed platform's vertices, the shapes for the best operation regions in the studied planar cable driven system, also changes.

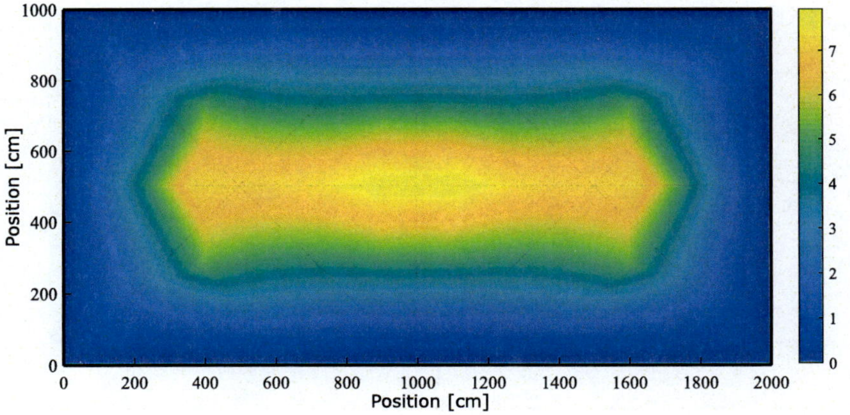

Fig. 3. First study case: maximum isotropic force map in a rectangular workspace

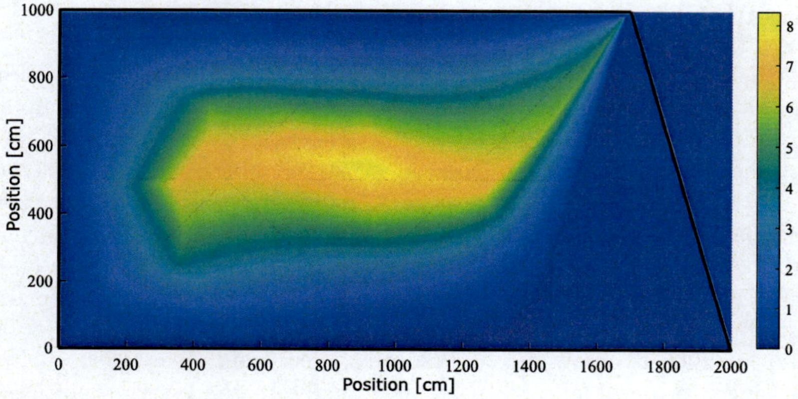

Fig. 4. Second study case: maximum isotropic force map in a rectangular trapezoidal workspace

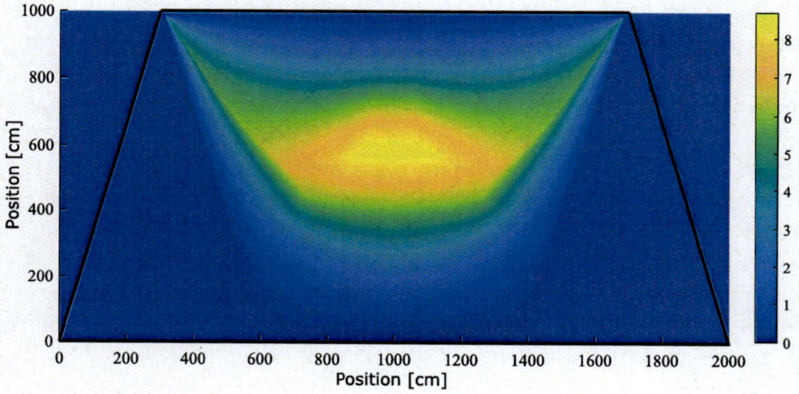

Fig. 5. Third study case: maximum isotropic force map in a trapezoidal workspace

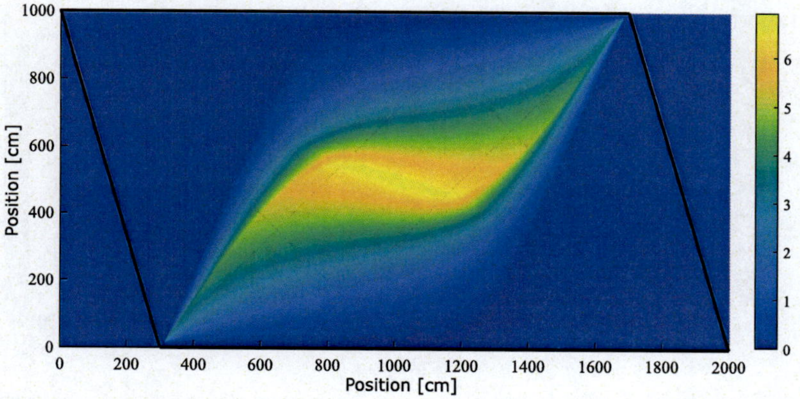

Fig. 6. Fourth study case: maximum isotropic force map in a parallelogramic workspace

6 Conclusions

This paper presented a new method to obtain the best operation regions in a planar cable driven system based on the calculation of the maximum isotropic force at all the discrete points within the manipulator's workspace.

The influence of including variations in the topology of the fixed platform in the manipulator was shown in four study cases, opening the possibility to visualize the best regions where the manipulator can act in order to improve its efficiency.

The proposed method to obtain analytically the maximum isotropic force with a prescribed moment is based on a closed mathematical formulation and uses a general static model and a generic kinematic basis. Authors used the formalism presented by Davies as the primary mathematical tool to analyse the mechanisms statically.

This work may be extended in various ways, for example, studying manipulators with different kinematic chains, DOF's and including their dynamic behaviour. The study of the influence of variations on the imposed moment may be considered in future researches.

Acknowledgements. Authors would like to thank to the Federal University of Santa Catarina which has made the present work possible.

References

1. Pott, A.: Cable-Driven Parallel Robots: Theory and Application. Siciliano, B., Khatib, O. (eds.) Springer, Cham (2018). https://doi.org/10.1007/978-3-319-76138-1
2. Bruckmann, T., Mikelsons, L., Brandt, T., Hiller, M., Schramm, D.: Wire robots part I: kinematics, analysis and design (2008)
3. Viana de Oliveira, M.: Force capability of planar cable-driven robots. UFSC, Florianópolis, SC, Brazil (2018)
4. Muraro, T., Martins, D., Sacht, L.K.: Vector analysis of the cable tension conditions. In: Carvalho, J.C.M., Martins, D., Simoni, R., Simas, H. (eds.) MuSMe 2017. MMS, vol. 54, pp. 39–48. Springer, Cham (2018). https://doi.org/10.1007/978-3-319-67567-1_4
5. Merlet, J.-P.: Wire-driven parallel robot: open issues. In: Romansy 19-Robot, Design, Dynamics and Control, pp. 3–10. Springer, Cham (2013). https://doi.org/10.1007/978-3-7091-1379-0_1
6. Goselin, C.: Cable-driven parallel mechanisms: state of the art and perspectives. Mech. Eng. Rev. **1**(1), DSM0004–DSM0004 (2014). The Japan Society of Mechanical Engineers
7. Verhoeven, R.: Analysis of the workspace of tendon-based Stewart platforms. Thesis (Doctorate) - Universitat Duisburg-Essen, Ingenieurwissenschaften Maschinenbau und Verfahrenstechnik (2004)
8. Weihmann, L.: Force capabilities of kinematically redundant planar parallel manipulators. In: 13th World Congress in Mechanism and Machine Science, vol. 1. sn (2011)

9. Mejia, L., Simas, H., Martins, D.: Force capability in general 3 DoF planar mechanisms. Mech. Mach. Theory **91**, 120–134 (2015)
10. Mejia, L., Simas, H., Martins, D.: Wrench capability in redundant planar parallel manipulators with net degree of constraint equal to four, five or six. Mech. Mach. Theory **105**, 58–79 (2015)
11. Frantz, F.: Capacidade de força em robôs atuados por cabos. UFSC, Blumenau, SC, Brazil (2021)
12. Mejia, L., Ponce, D., Herrera, J.C., Simas, H., Martins, D.: Influence of the working mode on the maximum isotropic force capability maps for a $3RRR$ planar parallel manipulator. In: Carvalho, J.C.M., Martins, D., Simoni, R., Simas, H. (eds.) MuSMe 2017. MMS, vol. 54, pp. 151–159. Springer, Cham (2018). https://doi.org/10.1007/978-3-319-67567-1_14
13. Mejia, L., Simoni, R., Herrera, J.C., Simas, H., Martins, D.: Maximum isotropic force capability maps in redundant planar cooperative systems. Int. J. Mech. Robot. Syst. **4**(4), 323–338 (2018)
14. Davies, T.H.: Mechanical networks I: passivity and redundancy. Mech. Mach. Theory **18**, 95–105 (1983)
15. Davies, T.H.: Mechanical networks III: wrenches on circuit screws. Mech. Mach. Theory **18**, 107–112 (1983)

Stiffness and Transparency of a Collaborative Cable-Driven Parallel Robot

Marceau Métillon[1], Camilo Charron[2], Kévin Subrin[3], and Stéphane Caro[1(✉)]

[1] Centre National de la Recherche Scientifique (CNRS), LS2N, 44300 Nantes, France
{marceau.metillon,stephane.caro}@ls2n.fr

[2] Université de Rennes 2, 35043 Rennes, France
camilo.charron@ls2n.fr

[3] Nantes Université, 44470 Carquefou, France
kevin.subrin@ls2n.fr

Abstract. The subject of this paper is about the relationship between the stiffness and the transparency of Cable-Driven Parallel Robots (CDPRs) used as human-machine interfaces in object comanipulation tasks. An index quantifying the transparency of a CDPR is first introduced. The stiffness of the robot is determined in simulation which parameters have been experimentally identified. Particular attention is paid to the effect of the Moving-Platform pose and cable tension management on CDPR stiffness. Then, the relationship between the stiffness and the transparency is analysed. Finally, the transparency index is traced throughout the constant-orientation static workspace and throughout the cable tension feasibility polygon for a given MP pose.

Keywords: Cable-driven parallel robot · Stiffness · Collaboration · Transparency

1 Introduction

A CDPR is a particular type of parallel robot. A Moving-Platform (MP) is connected to a rigid base frame using cables. The cables are coiled around motorized winches which control the cable lengths therefore translating and orientating the MP. Cable routing is made from the winches to the anchor points on the MP via exit points located on pulleys attached to the base frame. CDPRs are advantageous for industrial applications such as intra-logistics manipulation [4], large scale 3D printing [1], manipulation of large and/or heavy parts [14] and search-and-rescue operations [11].

When an external wrench is applied to the MP, cables exhibit an elastic behaviour and the MP tends to move further from its static equilibrium pose. The stiffness analysis of a CDPR is crucial to predict the MP displacement due to an external wrench resulting from its weight or a collision with its environment. Several papers deal with CDPR stiffness analysis [2,16] for CDPR design [3] or

O. Altuzarra and A. Kecskémethy (Eds.): ARK 2022, SPAR 24, pp. 101–109, 2022.
https://doi.org/10.1007/978-3-031-08140-8_12

control [5,13]. In the field of Human-Robot Interactions, the interaction transparency has different meanings depending on the nature of the interaction. Considering non-physical interactions between a human user agent and a robotic agent, the transparency describes the reciprocal knowledge of the other agent intentions [10]. Considering physical Human-Robot Interactions (pHRI) in the field of bilateral teleoperation, the transparency index is defined as the system capacity of rendering the interacting environment impedance [9]. In the field of comanipulation, where the robotic system is granting a manipulation assistance to the user, the transparency of the interaction denotes the capacity of the robot to moves in the direction desired by the user [8]. This definition is used in this paper to characterize the transparency. The paradigm of pHRI involving a human user and a CDPR is considered as shown in Fig. 1(a).

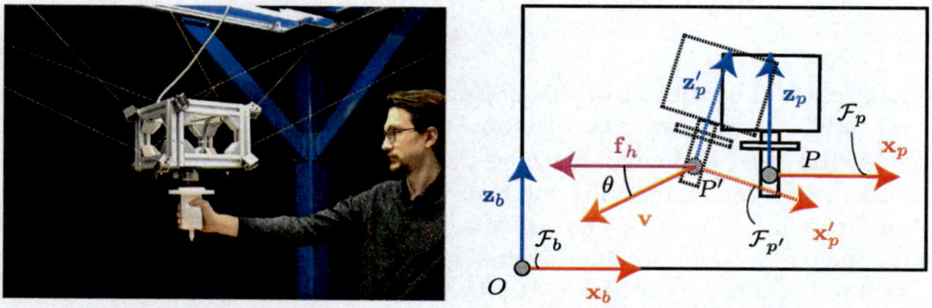

(a) Human user interacting with a CDPR

(b) Transparency

Fig. 1. Physical Human-Robot Interaction with a Cable-Driven Parallel Robot - LS2N, Nantes, France

The MP is equipped with a handle that the user grabs and exert wrench onto to infer motion intention. A force/torque sensor mounted between the MP and the handle measures the wrench exerted by the user. An admittance control strategy converts the measured wrench into a desired acceleration of the MP for the robot to servo [7]. As the user is exerting wrench on the MP the latter is subject to displacements that can affect the wrench measurement therefore impacting the motion direction of the robot thus affecting the transparency. Works have been done on the link between stiffness and transparency in teleoperation paradigms [15] but to the best of the authors knowledge, no study has been conducted on the effect of the stiffness of a CDPR on its transparency when used as a human-robot interface in a comanipulation task. As a consequence, this paper deals with the analysis of the effect of: (*i*) the cable tension distribution and the MP pose on the CDPR stiffness; (*ii*) the CDPR stiffness on its transparency when used as an interface in a human-robot comanipulation task. Section 2 describes the stiffness models of the CDPR under study. Section 3 presents the results on the stiffness analysis. A novel transparency index is introduced and analysed in Sect. 4 as well as the effect of stiffness on transparency. Conclusions and future work are drawn in Sect. 5.

2 CDPR Stiffness Modelling

From [2] and [13], the Cartesian stiffness matrix of a CDPR can be expressed as:

$$\mathbf{K} = \mathbf{K}_a + \mathbf{K}_p \tag{1}$$

with $\mathbf{K}_a$ being the active stiffness matrix and $\mathbf{K}_p$ being the passive stiffness matrix.

The passive stiffness matrix is dependent on the MP pose and the cable elasticity so that:

$$\mathbf{K}_p = \sum_{i=1}^{m} k_i \begin{bmatrix} {}^b\mathbf{u}_i {}^b\mathbf{u}_i^\top & {}^b\mathbf{u}_i {}^b\mathbf{u}_i^\top {}^b\hat{\mathbf{b}}_i^\top \\ {}^b\hat{\mathbf{b}}_i {}^b\mathbf{u}_i {}^b\mathbf{u}_i^\top & {}^b\hat{\mathbf{b}}_i {}^b\mathbf{u}_i {}^b\mathbf{u}_i^\top {}^b\hat{\mathbf{b}}_i^\top \end{bmatrix} \tag{2}$$

where k_i is the i-th cable elasticity, ${}^b\mathbf{u}_i$ is the unit cable vector denoting the i-th cable direction and ${}^b\hat{\mathbf{b}}_i$ is the cross-product matrix of the anchor point vector ${}^b\mathbf{b}_i$ expressing the coordinates of the MP anchor points in the base frame. The cable elasticity is given by $k_i = ES/l_i$ where E is the Young's modulus of the cable material, S is the cable cross-sectional area and l_i is the uncoiled cable length from the winch to the MP anchor points.

The active stiffness matrix is dependent on the MP pose, the cable lengths and the cable tensions so that:

$$\mathbf{K}_a = -\sum_{i=1}^{m} \tau_i \begin{bmatrix} -\frac{1}{l_i}\left(\mathbf{I}_3 - {}^b\mathbf{u}_i {}^b\mathbf{u}_i^\top\right) & \frac{1}{l_i}\left(\mathbf{I}_3 {}^b\mathbf{u}_i {}^b\mathbf{u}_i^\top\right) {}^b\hat{\mathbf{b}}_i^\top \\ -\frac{1}{l_i} {}^b\hat{\mathbf{b}}_i \left(\mathbf{I}_3 - {}^b\mathbf{u}_i {}^b\mathbf{u}_i^\top\right) & \left[{}^b\hat{\mathbf{u}}_i + \frac{{}^b\hat{\mathbf{b}}_i}{l_i}\left(\mathbf{I}_3 - {}^b\mathbf{u}_i {}^b\mathbf{u}_i^\top\right)\right] {}^b\hat{\mathbf{b}}_i^\top \end{bmatrix} \tag{3}$$

where τ_i is the i-th cable tension.

3 Analysis of Factors Influencing Stiffness

In this section, the influence of the MP pose and tension distribution on CDPR stiffness is carried out in simulation in order to further analyse the impact on the transparency. The CDPR considered is the CRAFT prototype, located at LS2N, Nantes, of size 3.75 m × 4.34 m × 2.78 m. The prototype is equipped with eight motorized winches which have lower and upper cable tensions capacities τ_{min} and τ_{max} equal to 0 N and 100 N, respectively. The cables used are synthetic cables VECT070LE composed of 8 braids of Vectran® fibres[1]. The cable diameter is 0.7 mm. The cable elasticity was experimentally determined using a Universal Tensile Machine during a stress-strain analysis experiment. The identified elasticity is found to be $ES = 17318.1$ N.

To assess the evolution of the stiffness for the CRAFT prototype, the displacements of the MP under an external wrench are determined and compared.

[1] Acquired from Corderie Lancelin, https://www.lancelin.com/en/.

The external wrench is the wrench exerted by the user and is expressed as $\mathbf{w}_h = [\mathbf{f}_h \ \mathbf{m}_h]^\top$ with $\mathbf{f}_h = [-15 \ 0 \ 0]^\top$ N and $\mathbf{m}_h = \mathbf{0}_3$. The human operator exerts a pure force $\mathbf{f}_h$ at point P being located on the handle. The small-displacement screw $\delta\mathbf{X} = [\delta\mathbf{p} \ \boldsymbol{\varphi}]^\top$ of the MP due to force exerted by the human operator is the following:

$$\delta\mathbf{X} = \mathbf{K}^{-1}\mathbf{w}_h \tag{4}$$

with $\delta\mathbf{X}$ being the displacement of the MP in term of translation and orientation so that $\delta\mathbf{X} = [\delta\mathbf{p} \ \boldsymbol{\varphi}]^\top$ with $\delta\mathbf{p} = [\delta p_x \ \delta p_y \ \delta p_z]^\top$ and $\boldsymbol{\varphi} = [\varphi_x \ \varphi_y \ \varphi_x]^\top$.

3.1 Influence of the MP Pose on the Manipulator Stiffness

From Eq. (2) and Eq. (3), both the passive and active stiffness matrices are dependent on the unit cable vectors $\mathbf{u}_i$, therefore, the stiffness of the robot varies depending of the pose. Here the influence of the pose on the stiffness is simulated. The constant and null orientation Static Workspace of the robot is determined using the Capacity Margin index [6]. Then, the static workspace is discretized and the robot stiffness is computed at each point. In this scenario, the cable tension distribution considered corresponds to the barycentre of the cable tension feasibility polygon of each pose considering the static equilibrium of the MP [12]. The displacement of the MP under the external wrench exerted by the human user is computed using the matrix $\mathbf{K}$ at each pose.

Figure 2 shows the influence of the MP pose on the robot stiffness. Figure 2(a) shows the translational displacement of the MP δp_x along $\mathbf{x}_b$ axis and Fig. 2(b) shows the MP rotational displacement φ_y around $\mathbf{y}_b$ axis. It can be seen that the robot stiffness decreases when the MP altitude is increasing and when the MP moves away from the vertical centreline of the workspace.

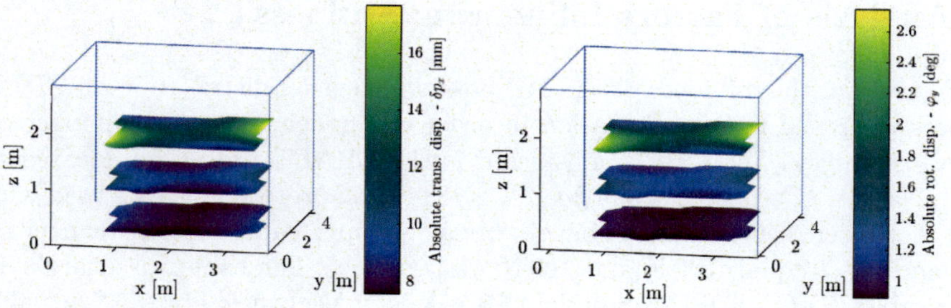

(a) Translational displacement along $\mathbf{x}_b$ (b) Rotational displacement around $\mathbf{y}_b$

Fig. 2. Displacement of the MP subject to an external wrench through the manipulator static workspace

3.2 Influence of the Cable Tension Distribution

The static equilibrium of the MP is written as

$$\mathbf{W}\boldsymbol{\tau} + \mathbf{w}_h + \mathbf{w}_g = \mathbf{0}_8 \tag{5}$$

where $\mathbf{W}$ is the pose-dependent wrench matrix associated to the MP, $\boldsymbol{\tau}$ being the cable tension vector, $\mathbf{w}_h$ being the wrench applied by the human user on the MP and $\mathbf{w}_g$ being the gravity wrench.

The number of cables of the CDPR under study is larger than its number of degrees of freedom. Therefore an infinite number of cable tension sets exists to ensure the static equilibrium of the MP. It can be seen from Eq. (3) that the cable tensions have an influence on the overall robot stiffness. In this section the influence of the cable tension distribution is studied. The methodology to compute the cable tension feasible polygon detailed in [13] is used hereafter. The cable tension feasible polygon Λ_f is the convex polytope of $\boldsymbol{\lambda} = [\lambda_1\ \lambda_2]^\mathsf{T}$ that satisfies the static equilibrium of the MP and the cable tension limits so that:

$$\Lambda_f = \left\{\boldsymbol{\lambda} \in \mathbb{R}^2 \mid \boldsymbol{\tau}_{min} \leq \boldsymbol{\tau}_0 + \mathbf{N}\boldsymbol{\lambda} \leq \boldsymbol{\tau}_{max}\right\} \tag{6}$$

with $\boldsymbol{\tau}_0 = -\mathbf{W}^\dagger\left(\mathbf{w}_g + \mathbf{w}_h\right)$, $\mathbf{N} = \text{null}(\mathbf{W})$ being the null space of the wrench matrix. $\boldsymbol{\tau}_{min}$ and $\boldsymbol{\tau}_{max}$ are the lower and upper bound cable tension vectors, respectively. To study the influence of the cable tension distribution, the space Λ_f was numerically obtained for a given MP pose of the MP, in this case $\mathbf{p} = [1.87\ 2.15\ 0.5]^\mathsf{T}$ m. The feasible polygon is then discretized and the displacement of the MP under the external wrench is computed for each discrete point. Figure 3 shows the influence of cable tension distribution on the robot stiffness. Figure 3(a) shows the translational MP displacement δp_x along $\mathbf{x}_b$ axis and Fig. 3(b) shows the MP rotational displacement φ_y around $\mathbf{y}_b$ axis. The robot stiffness is maximized along $\mathbf{x}_b$ axis and around $\mathbf{y}_b$ axis for $\boldsymbol{\lambda} = [35.54\ \ -41.9]^\mathsf{T}$.

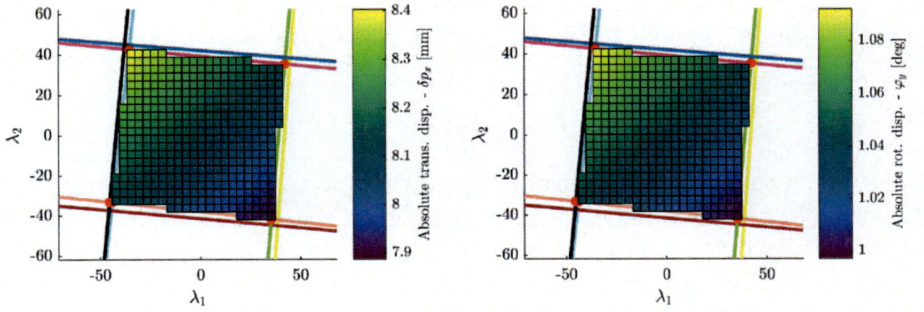

(a) Translational displacement along $\mathbf{x}_b$ (b) Rotational displacement about $\mathbf{y}_b$

Fig. 3. Effect of the cable tension distribution on the MP displacement subject to an external wrench.

4 Transparency

From [8], the transparency index describes the capacity of the robot to follow the human intention in terms of desired displacement. The user desired motion direction relies in the force direction. For the CDPR used as an assistive device, the force exerted on the sensor induces a desired acceleration of the MP based on the force direction. The transparency index μ is expressed as:

$$\mu = \mathbf{v}_n^\top \mathbf{f}_{hn} \tag{7}$$

with $\mathbf{v}_n$ being the unit vector of the MP translational velocity vector and $\mathbf{f}_{hn}$ being the unit vector of the human force $\mathbf{f}_h$. From Eq. (7), μ amounts to the cosine of angle $\theta = \angle(\mathbf{v}_n,\ \mathbf{f}_{hn})$ between vectors $\mathbf{v}_n$ and $\mathbf{f}_{hn}$

It should be noted that the CDPR stiffness affects the transparency index when used in comanipulation, especially when using an admittance control strategy based on wrenches applied on the MP. The parameters used in this section are depicted in Fig. 1(b). Equation (4) denotes the displacement of the MP under the external force due to the pose-dependant robot stiffness $\mathbf{K}$. $\mathbf{X}'$ denotes the new MP pose after small displacement. Due to this displacement, the force measurement in $\mathcal{F}_p$ is distorted due to the small rotational motion of the MP. The expression of the force measured in the sensor frame considering the displacement of the MP is obtained as:

$$\mathbf{f}'_h =^b \mathbf{R}_{p'} \mathbf{f}_h \tag{8}$$

with ${}^b\mathbf{R}_{p'}$ being the rotation matrix from frame $\mathcal{F}_b = (O, \mathbf{x}_b, \mathbf{y}_b, \mathbf{z}_b)$ and $\mathcal{F}_{p'} = (P', \mathbf{x}_{p'}, \mathbf{y}_{p'}, \mathbf{z}_{p'})$ where $\mathbf{x}_{p'} =^b \mathbf{R}_{p'}\mathbf{x}_b$, $\mathbf{y}_{p'} =^b \mathbf{R}_{p'}\mathbf{y}_b$ and $\mathbf{z}_{p'} =^b \mathbf{R}_{p'}\mathbf{z}_b$.

Using the measured force as seen by the sensor, the admittance controller defines a desired Cartesian speed set-point for the robot to reach $\mathbf{t} = [\mathbf{v}, \boldsymbol{\omega}]^\top$, with $\mathbf{v} = [v_x\ v_y\ v_z]^\top$ and $\boldsymbol{\omega} = [\omega_x\ \omega_y\ \omega_z]^\top$ being the MP translational velocity and angular velocity respectively. In this specific case only translational motions are generated using the admittance and angular velocity are kept null so that $\boldsymbol{\omega} = \mathbf{0}_3$. The speed set-point $\mathbf{t}$ is then used in the robot controller to compute the cable speed set-point $\dot{\mathbf{l}}$ using the kinematic forward Jacobian matrix $\mathbf{A}(\mathbf{X})$ such as:

$$\dot{\mathbf{l}} = \mathbf{A}(\mathbf{X})\mathbf{t} \tag{9}$$

In order to compute the transparency as the dot product of the unit vectors of the velocity and the force applied on the MP, it is necessary to determine the effective velocity of the MP by taking into account that the robot controller does not have prior knowledge of the MP displacement. Therefore, the force measured and the cable speed defined using the admittance controller will not lead to the correct robot motion. The MP velocity is obtained using the kinematic forward Jacobian matrix corresponding to the displaced pose $\mathbf{A}(\mathbf{X}')$ and the cable speed set-point issued by the controller $\dot{\mathbf{l}}$ as:

$$\mathbf{t}' = \mathbf{A}^\dagger(\mathbf{X}')\dot{\mathbf{l}} \tag{10}$$

with $\mathbf{A}^{\dagger}(\mathbf{X}')$ being the pseudo-inverse of the matrix $\mathbf{A}(\mathbf{X}')$ and $\mathbf{t}' = [\mathbf{v}' \ \boldsymbol{\omega}']^{\mathrm{T}}$ being the MP twist accounting for the MP displacement under external wrenches. Therefore, the transparency index μ' accounting for the CDPR stiffness can be defined as:

$$\mu' = \mathbf{v}_n'^{\mathrm{T}} \ \mathbf{f}_{hn} \tag{11}$$

where $\mathbf{v}_n'$ is the unit vector of the MP linear velocity vector, $\mathbf{v}'$.

The results on the stiffness analysis carried out in Sect. 3 are used to compute the transparency index throughout the manipulator workspace. It is then possible to apprehend the influence of the MP pose and the cable tension distribution on the transparency index. It should be noted that index μ' is bounded between -1 and 1. In order to have an index bounded between 0 and 1 and have a better assessment of the effect of the orientation error of the MP on the human user's feeling, a novel transparency index, named ν', is expressed as:

$$\nu' = e^{-k \frac{1-\mu'}{2(\mu'+1)}} \tag{12}$$

with k being a scaling factor to be tuned based on the task and the CDPR performance. ν' is bounded between 0 and 1. The higher ν', the better the transparency of the comanipulation task. Figure 4 shows the effect of the MP pose and cable tension distribution on transparency with a scaling factor $k = 500$. Figure 4(a) shows that the cable tension distribution has a negligible effect on the transparency. Due to the cable tension limits, the cable tension distribution offers a small range of controllability of the stiffness and therefore a small range of controllability of the transparency. Figure 4(b) shows that the lower the MP, the better the transparency and that the MP pose has a preponderant effect on the transparency over the cable tension distribution.

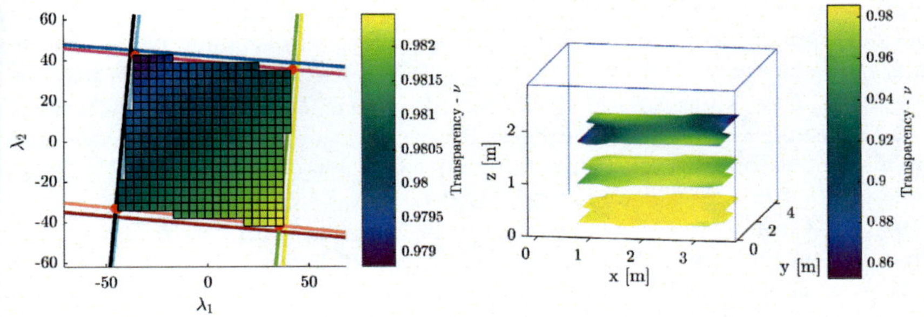

(a) Influence of the cable tension distribution

(b) Influence of the MP pose

Fig. 4. Influence of stiffness on transparency

5 Conclusion and Future Work

In this paper, the stiffness of the CDPR prototype CRAFT was analysed and compared based on MP pose and cable tension management. The displacement of the MP under an external pure force was studied and its impact on the robot transparency was investigated. A novel transparency index was introduced and it turns out that the transparency of the CDPR is clearly a function of its stiffness. The higher the robot stiffness, the better the transparency. Future work will focus on additional elements influencing the robot stiffness such as couplings between the motor gearboxes and the winches. The transparency analysis will be extended to multiple wrenches exerted by a human-operator on the MP and control solutions towards transparency improvement. The effect of the communication time from the force sensor to the robot control bay on the transparency will also be investigated.

Acknowledgements. This work was supported by the ANR CRAFT project, grant ANR-18-CE10-0004. https://anr.fr/Project-ANR-18-CE10-0004.

References

1. Barnett, E., Gosselin, C.: Large-scale 3D printing with a cable-suspended robot. Addit. Manuf. **7**, 27–44 (2015). https://doi.org/10.1016/j.addma.2015.05.001
2. Behzadipour, S., Khajepour, A.: Stiffness of cable-based parallel manipulators with application to stability analysis. J. Mech. Des. **128**(1), 303–310 (2006). https://doi.org/10.1115/1.2114890
3. Bolboli, J., Khosravi, M.A., Abdollahi, F.: Stiffness feasible workspace of cable-driven parallel robots with application to optimal design of a planar cable robot. Robot. Auton. Syst. **114**, 19–28 (2019). https://doi.org/10.1016/j.robot.2019.01.012
4. Bruckmann, T., Lalo, W., Nguyen, K., Salah, B.: Development of a storage retrieval machine for high racks using a wire robot. In: Proceedings of the ASME International Design Engineering Technical Conferences and Computers and Information in Engineering Conference-2012, pp. 771–780. American Society of Mechanical Engineers, New York, NY (2012). https://doi.org/10.1115/DETC2012-70389
5. Cui, Z., Tang, X.: Analysis of stiffness controllability of a redundant cable-driven parallel robot based on its configuration. Mechatronics **75**, 102519 (2021). https://doi.org/10.1016/j.mechatronics.2021.102519
6. Guay, F., Cardou, P., Cruz-Ruiz, A.L., Caro, S.: Measuring how well a structure supports varying external wrenches. In: Petuya, V., Pinto, C., Lovasz, E.C. (eds.) New Advances in Mechanisms, Transmissions and Applications, Mechanisms and Machine Science, vol. 17, pp. 385–392. Springer, New York (2013). https://doi.org/10.1007/978-94-007-7485-8_47
7. Hogan, N.: Impedance control: an approach to manipulation: part I–theory. J. Dyn. Syst. Meas. Contr. **107**(1), 1–7 (1985). https://doi.org/10.1115/1.3140702
8. Jarrasse, N., Paik, J., Pasqui, V., Morel, G.: How can human motion prediction increase transparency? In: 2008 IEEE International Conference on Robotics and Automation, ICRA 2008, pp. 2134–2139. IEEE Xplore, Pitscataway, NJ (2008). https://doi.org/10.1109/ROBOT.2008.4543522

9. Lawrence, D.A.: Stability and transparency in bilateral teleoperation. IEEE Trans. Robot. Autom. **9**(5), 624–637 (1993). https://doi.org/10.1109/70.258054
10. Lyons, J.B.: Being transparent about transparency: a model for human-robot interaction. In: Trust and Autonomous Systems, Technical Report. Association for the Advancement of Artificial Intelligence SS, AAAI Press, Palo Alto, California (2013)
11. Merlet, J.P., Daney, D.: A portable, modular parallel wire crane for rescue operations. In: 2010 IEEE International Conference on Robotics and Automation, pp. 2834–2839. IEEE, Piscataway, NJ (2010). https://doi.org/10.1109/ROBOT.2010.5509299
12. Mikelsons, L., Bruckmann, T., Hiller, M., Schramm, D.: A real-time capable force calculation algorithm for redundant tendon-based parallel manipulators. In: 2008 IEEE International Conference on Robotics and Automation, ICRA 2008, pp. 3869–3874. IEEE Xplore, Pitscataway, NJ (2008). https://doi.org/10.1109/ROBOT.2008.4543805
13. Picard, E., Caro, S., Plestan, F., Claveau, F.: Stiffness oriented tension distribution algorithm for cable-driven parallel robots. In: Lenarčič, J., Siciliano, B. (eds.) ARK 2020. SPAR, vol. 15, pp. 209–217. Springer, Cham (2021). https://doi.org/10.1007/978-3-030-50975-0_26
14. Picard, E., Plestan, F., Tahoumi, E., Claveau, F., Caro, S.: Control strategies for a cable-driven parallel robot with varying payload information. Mechatronics **79**, 102648 (2021). https://doi.org/10.1016/j.mechatronics.2021.102648
15. Roche, L., Saint-Bauzel, L.: High stiffness in teleoperated comanipulation: necessity or luxury? In: Lynch, K. (ed.) 2018 IEEE International Conference on Robotics and Automation (ICRA), pp. 477–483 (2018). https://doi.org/10.1109/ICRA.2018.8461005
16. Yuan, H., Courteille, E., Deblaise, D.: Static and dynamic stiffness analyses of cable-driven parallel robots with non-negligible cable mass and elasticity. Mech. Mach. Theory **85**, 64–81 (2015). https://doi.org/10.1016/j.mechmachtheory.2014.10.010

An Approach for Predicting the Calibration Accuracy in Planar Cable-Driven Parallel Robots

Bozhao Wang[1], Philippe Cardou[2], and Stéphane Caro[1(✉)]

[1] CNRS, Laboratoire des Sciences du Numerique de Nantes, UMR CNRS 6004, 1, rue de la Noë, 44321 Nantes, France
{Bozhao.Wang,stephane.caro}@ls2n.fr

[2] Laboratoire de robotique, Departement de génie mécanique, Université Laval, Québec, QC G1V 0A6, Canada
Philippe.Cardou@gmc.ulaval.ca

Abstract. This paper presents a simulation of the calibration of a 3-DoF, 2-cable, planar cable-driven parallel robot (CDPR). The calibration is realized with the combination of a laser displacement sensor and an inclinometer attached to the moving platform. The actual accuracies of the sensors are tested at first for higher calibration quality. Through simulation, with more measurement poses used, the system variable identification errors are reduced, and have decreasing dispersion, finally form plateaus. The effect of each sensor on the calibration quality is studied. Based on the sensors considered in this work, the system variable errors are all within ±9 mm, and most are within ±5 mm for 5.209 m-span CDPR.

1 Introduction

Cable-driven parallel robots (CDPRs) are a group of parallel robots that are actuated through flexible cables instead of rigid links. Compared with serial robots, this particular type of robot benefits from its high flexibility, reconfigurability, and potentially large translational workspace. In most CDPRs, the moving platform (MP) is usually connected by several cables, then through pulleys, winches and then linked to the base frame. There are two configurations according to the number and spatial position of cables used: suspended and fully-constrained CDPRs [2].

Parallel mechanisms, because of their large numbers of links and passive joints [6], may not necessarily have a high accuracy. Therefore, kinematic calibration is important for such robot architectures. Daney et al. [8] used a method based on interval arithmetics for kinematic parallel robot calibration. In [11], a high precision and robustness iterative calibration method is proposed to significantly improve the end-effector position errors. For CDPRs, most of previous works relied on non-linear least square (NLLS) methods for parameter identification, as it applies conveniently to the minimization of the cable length residuals. In [3], the NLLS method is used on a 6-cable, 3-degrees of freedom (DoF)

O. Altuzarra and A. Kecskémthy (Eds.): ARK 2022, SPAR 24, pp. 110–121, 2022.
https://doi.org/10.1007/978-3-031-08140-8_13

CDPR, with a parallelogram, which is able to reach a larger workspace. The application is proven effective by simulations and experiments. The authors also proposed an algorithm to select optimal measurement poses. The authors of [1] used NLLS method for a CDPR called TBot and also considered pulley kinematics. They proposed a measurement pose optimization method, which consists in minimizing the identification matrix condition number. In [4], several identification methods derived from NLLS are proposed and tested. The other similar approach is orthogonal distance regression (ODR) [14].

Sensors used for robot calibration can be divided into two categories: exteroceptive and proprioceptive, which can be combined or not. Proprioceptive motor position sensors are used in [12] for auto-calibration. Zhang et al. [11] used only external measurement device to perform the iterative calibration method. Martin et al. [13] used laser-based cable length measurement sensor to improve calibration quality. Little research has been done on the combination of different types of sensors in CDPRs. Calibration with exteroceptive sensors has drawbacks such as low ratios between measurement accuracy and measurement volume, and the difficulty to access the end-effector in certain cases [5]. The use of proprioceptive sensors on the MP can mitigate those disadvantages. Renaud et al. [5] proposed to perform the kinematic calibration of a parallel mechanism by observing its legs with a camera, which can make the legs be observed more easily. The method combines the advantages of both exteroceptive and proprioceptive sensors. Andreff et al. [7] proposed kinematic calibration of a H4 parallel robot with computer vision. In Ref. [8], Daney et al. used internal sensors to provide the leg length differences of a parallel mechanism.

This paper focuses on the simulation of the kinematic calibration for a 3-DoF, 2-cable, planar CDPR, using the combination of a laser displacement sensor and an inclinometer. Pulley kinematics is considered in the modelling. The intrinsic measurement errors of both sensors are examined beforehand to improve the calibration quality. Then the system variable identification errors are simulated based on the sensor measurement error after curve fitting. Besides, the effect of each sensor on the calibration quality is studied. The NLLS method is applied to minimize the cable length residual in order to identify the system parameters.

Section 2 focuses on the CDPR kinematic modelling and on accuracy. Section 3 describes the identification methodology of the current study. The obtained simulation results are discussed in Sect. 4, and then the conclusions are given.

2 CDPR Modelling

Figure 1(a) shows the CDPR base frame $\mathcal{F}_1$ and the real 3-DoF moving platform. Two pulleys in the diagonal direction are used, forming the CDPR workspace plane. The workspace frame is defined as $\mathcal{F}_2$, with a 5.2 m by 2.8 m size. $\mathcal{F}_2$ is built based on exit points in $\mathcal{F}_1$ that are measured accurately with a laser tracker. Cables are supposed to be inelastic, straight and massless. The robot geometry inside $\mathcal{F}_2$ can be seen in Fig. 1(b). The MP is a 0.5 m-long straight bar. It is suspended by the two cables, which go through the two pulleys located near the cable exit points A_1 and A_2. The cables are then led to the motor winches fixed on the corresponding bottom corners in the base frame. The horizontal movement of the winch exit point because of cable coiling is neglected.

The MP will be equipped with an inclinometer and a laser displacement sensor connected to the MP bar through a revolute joint. Thanks to this revolute joint, the laser sensor always points vertically to the ground, to directly measure the MP height. The total MP mass is 2.2 kg. The ground is assumed to be flat, and is able to reflect enough laser light to perform proper measurements. Motors control the cable lengths, and are equipped with encoders that measure the motor angular position, and thus the cable-length variations.

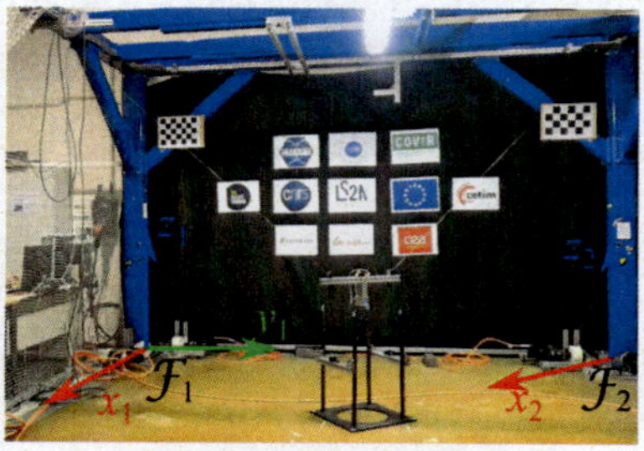

(a) MP in $\mathcal{F}_1$ and $\mathcal{F}_2$

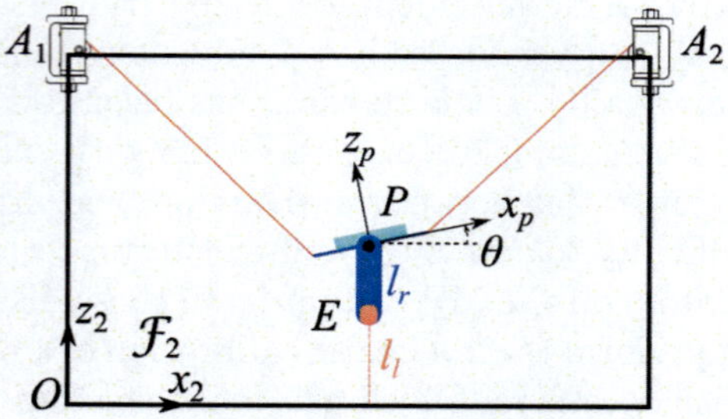

(b) CDPR geometry inside $\mathcal{F}_2$

Fig. 1. CDPR under study equipped with a laser displacement sensor and an inclinometer

The MP has three DoF: one rotational and two translational DoF. However it is held by only two cables, which makes the robot under-constrained. With a certain set of cable lengths, the MP will always have one degree of freedom to move, but will eventually stay at the unique position where its gravitational potential energy is the smallest. The 50 poses used in the simulation and generated randomly are illustrated in Fig. 2.

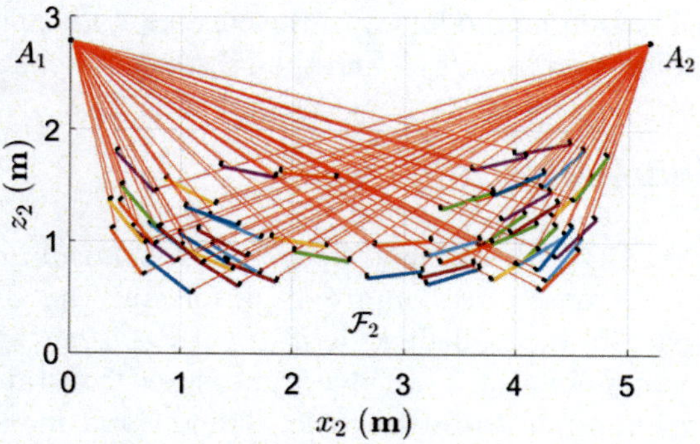

Fig. 2. Measurement poses randomly generated

2.1 Kinematic Modelling

Figure 4 shows the ith loop of the current CDPR. The vectors pointing from the workspace origin O to the ith pulley exit point and from O to the MP center are $\mathbf{a}_i$ and $\mathbf{p}_j$, respectively; with $i = 1, \ldots, m$ and $j = 1, \ldots, n$ are the numbers of each cable and measurement poses, respectively. $m = 2$ and n is the total number of measurement poses. Detailed pulley kinematic modelling is described in [2,9] and is illustrated in Fig. 3. The vector from the pulley exit point A'_i to the cable anchor point B_i is $\mathbf{l}_{ci}$. The jth MP rotational angle compared with the horizontal plane is θ_j. As a result, the cable length from the cable exit point A_i to cable anchor point B_i is:

$$l_{ti} = l_{pi} + l_{ci} \tag{1}$$

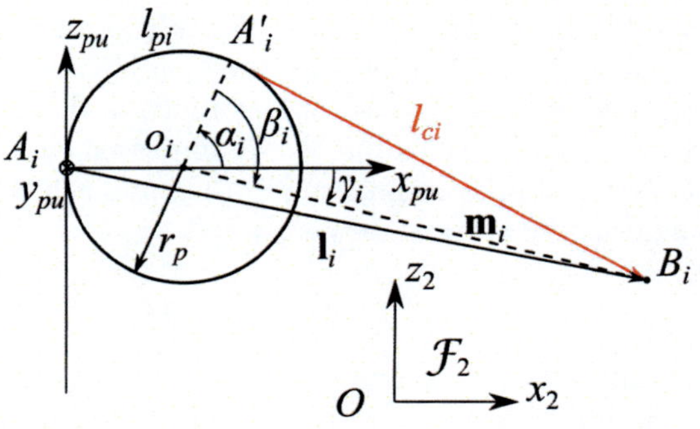

Fig. 3. Parameterization of the ith pulley

Following the ith loop of the CDPR geometry, l_{ti} can be expressed as:

$$l_{ti} = r_p \Big[\pi - \beta_i - \gamma_i\Big] + \sqrt{\mathbf{m}_i \mathbf{m}_i^T - r_p^2} \tag{2}$$

where r_p is the pulley radius, $\tan(\beta_i) = \frac{\sqrt{\mathbf{m}_i \mathbf{m}_i^T - r_p^2}}{r_p}$, $\sin(\gamma_i) = \frac{a_{iz} - b_{iz}}{||\mathbf{m}_i||_2}$, $\mathbf{m}_i = \mathbf{p} + \mathbf{b}_i - \mathbf{a}_i + r_p \, {}^b\mathbf{R}_i \mathbf{x}_{pu}$ is the vector pointing from the pulley center to the anchor point, $\mathbf{x}_{pu}$ is the unit vector along the x-axis of the pulley frame, and ${}^b\mathbf{R}_i$ is the rotation matrix from base frame to the pulley frame.

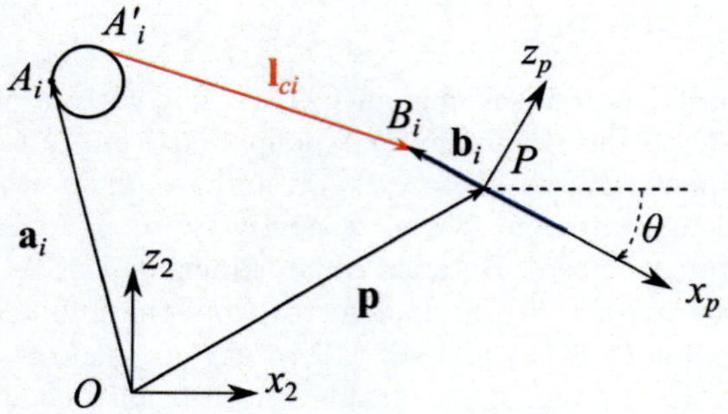

Fig. 4. The ith loop of the CDPR under study

2.2 Sensors Accuracy Test

Two sensors are considered in the simulation: the WitMotion BWT61CL inclinometer and the SICK DT50-2 laser displacement sensor. The calibration method in this work relies on the sensor measurements. Better knowledge on the measurements will certainly contribute to an improved calibration accuracy. Sensor measurements from both sensors are recorded and compared to the standard predefined angles and distances, respectively:

$$e_\theta = \theta_s - \theta_r, \qquad e_d = d_s - d_r \tag{3}$$

where e_θ and e_d are the actual measurement errors of the inclinometer and the laser displacement sensor, θ_s and d_s are the sensor output measurements from both sensors, θ_r and d_r are the standard predefined measurement values for both sensors.

Measurement curve fitting for both sensors are conducted. The $\sec(\theta_j)$ with a constant ratio between lateral and gravitational acceleration $\frac{da}{g} = -0.24$ fits e_θ the best. As e_d is roughly uniformly distributed, its mean value $e_{d,fit}$ is used over its full range. Regarding the realistic poses, the ranges lower than 60° for the inclinometer and less than 3000 mm mm for the displacement sensor are considered. From the test, the inclinometer accuracy is determined as ±0.1° with 0.05° repeatability, and the laser displacement sensor accuracy is ±6 mm with a 1.5 mm repeatability (Fig. 5).

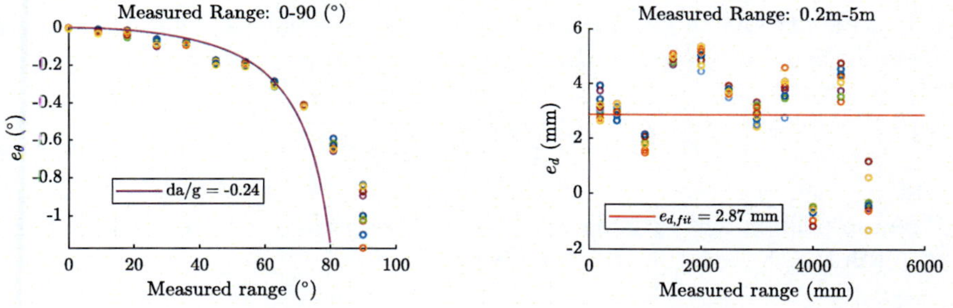

(a) Inclinometer accuracy (b) Laser displacement sensor accuracy

Fig. 5. WitMotion BWT61CL inclinometer and SICK DT50-2 laser displacement sensor accuracy measurements

3 Identification Methodology

The robot calibration is achieved with the information from all the sensors throughout all the MP measurement poses. At the initial position, the cable lengths are denoted as l_{i0}. When the MP is moved to all the n different poses within the workspace, the cable length variations Δl_{ij} are recorded by the encoders. Thus, the actual cable lengths l_{ij} are calculated by:

$$l_{ij} = \Delta l_{ij} + l_{i0} \tag{4}$$

The CDPR system variables are the outputs of the problem, including the Cartesian coordinates of cable exit points a_{ix}, a_{iz}, the initial cable lengths l_{i0}, the Cartesian coordinates of the MP center p_{jx}, p_{jz} and the MP rotation angle θ_j. Among these variables, the laser displacement sensor measures p_{jz}, and the inclinometer measures θ_j. All the sensor measurements are the inputs of the identification problem. The rest of the output variables are considered to be the problem unknowns, included in a $3m + n$ dimensional vector $\mathbf{x}$:

$$\mathbf{x} = [a_{1x}\, a_{1z}\, a_{2x}\, a_{2z}\, l_{10}\, l_{20}\, p_{1x} \dots p_{nx}]^T \tag{5}$$

Between the real cable length l_{ij} from Eq. (4) and the estimated cable lengths l_{ti} from Eq. (2), a system of mn equations can be obtained [2,10]:

$$f_{ij}(\mathbf{x}) = (l_{pi} + l_{ci})^2 - (\Delta l_{ij} + l_{i0})^2, \quad i = 1, 2 \quad, j = 1, \dots, n \tag{6}$$

To solve the nonlinear system of equations defined by Eq. (6), the number of inputs must be larger than or equal to the number of unknowns:

$$mn + 2n \geq 2m + m + 3n \tag{7}$$

As the CDPR has $m = 2$ cables, the least number of measurement poses is $n = 6$. A similar simulation process is used in [9,10]. Arbitrary errors are added on $\mathbf{x}$ to simulate the approximately known system variables. The identification problem is then formulated as the non-linear least square problem in Eq. (8):

$$\min_{\mathbf{x}} \left(\sum_{i=1}^{m} \sum_{j=1}^{n} f_{ij}^2 \right) \tag{8}$$

The pre-defined real variable values $\mathbf{x}_r$ are compared with the identified ones $\mathbf{x}^*$, to evaluate the identification quality:

$$\delta\mathbf{x}_k = \mathbf{x}_k^* - \mathbf{x}_{r,k}, \qquad k = 1, 2, ..., 3m + n \tag{9}$$

where $\delta\mathbf{x}_k$ is the difference between the real and identified variable values.

4 Simulation of 3-DoF, 2-Cable Planar CDPR

4.1 Results of the Simulated Identification Problem

Simulations with 6 to 50 measurement poses are performed. For each pose, the simulation is repeated 500 times with different sensor measurement errors. Based on the accuracy test results given in Sect. 2, the applied accuracy and repeatability are generated as normally distributed values, assuming the sensor accuracy and repeatability ranges as three times the standard deviation, and zero distribution mean.

The dispersion or standard deviations of the obtained results $\delta\mathbf{x}_k$ is defined as:

$$\sigma = std(\delta\mathbf{x}_k), \qquad k = 1, 2, ..., 3m + n \tag{10}$$

Figure 6 shows σ of the x-coordinate of the 2nd cable exit point, the initial cable length of the 1st cable and the x-coordinate of the 4th MP pose. It is apparent that the higher the number of poses, the lower the σ, the better the geometric calibration quality. It should be noted that σ does not decrease after 50 measurement poses.

Figures 7, 8 and 9 are the summary of three examples of the identification errors of the system variables with different number of measurement poses used, and the σ of the identified values are shown. The results are plotted as probability density function (PDF). It can be seen that the higher the number of poses (from 6 to 50), the lower the system variable identification errors. When only six poses are used, the probability density function plots are quite flat, the errors are bounded between −50 and 50 mm. Afterwards, the PDF plots tend towards normal distributions, the main part of the identification errors is distributed in the lower value area. Besides, the σ values keep decreasing as the number of measurement poses increases, which means that the identification error dispersion becomes lower and the identification quality is higher.

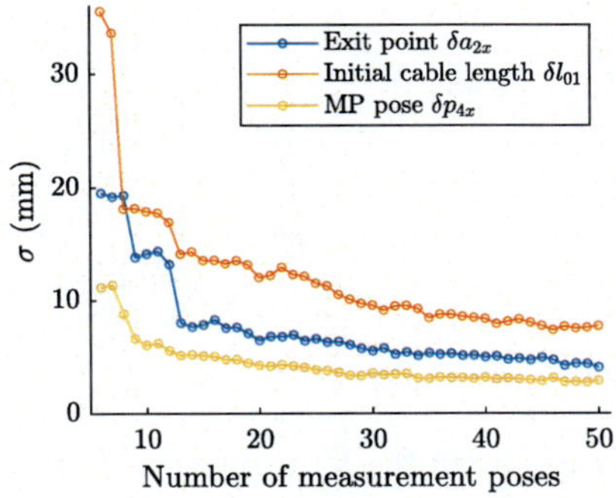

Fig. 6. The σ results of 3 examples of the system variables

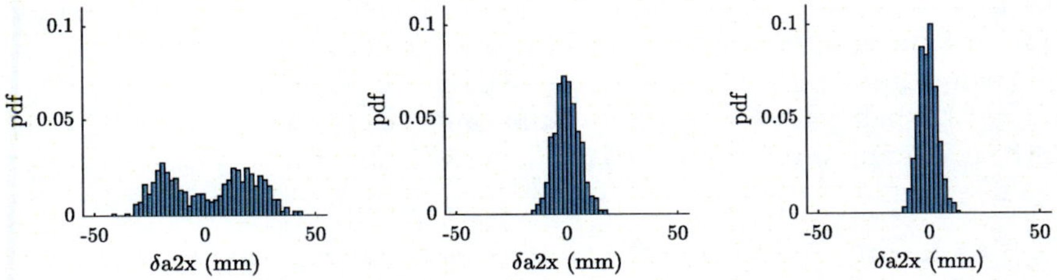

(a) 6 poses, σ=19.5003 mm (b) 30 poses, σ=5.5474 mm (c) 50 poses, σ=4.0962 mm

Fig. 7. Identification error results in the x-coordinates of the 2nd cable exit point, with different number of measurement poses and σ values

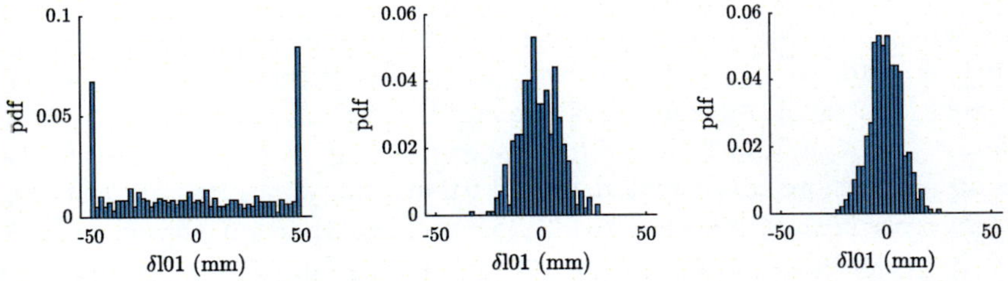

(a) 6 poses, σ=35.4669 mm (b) 30 poses, σ=9.4841 mm (c) 50 poses, σ=7.6688 mm

Fig. 8. Identification error results of the initial length in the 1st cable, with different number of measurement poses and σ values

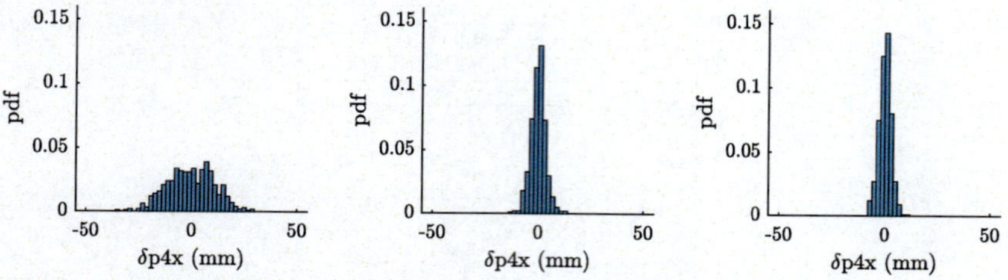

(a) 6 poses, σ=11.0545 mm (b) 30 poses, σ=3.4903 mm (c) 50 poses, σ=2.8290 mm

Fig. 9. Identification error results of the x-coordinates of the 4th MP position vector, with different number of measurement poses and σ values

In general, from the identification error results, the variables of MP coordinates have the lowest errors, with nearly all the values below 10 mm and the minimum σ being 2.8290 mm. The errors in the coordinates of exit point come next, the maximum identification errors slightly exceeding 15 mm, and the minimum σ value being 4.0962 mm. The initial cable lengths have the largest errors, with the maximum values around 25 mm and minimum σ value being 7.6688 mm.

4.2 Simulation of Sensor Effects on Identification Errors

The two sensors provide different types of quantities (lengths and angles), which makes the effect of each sensor on the overall calibration quality difficult to discern. Therefore, the simulations when eliminating each of the sensor measurement errors one at a time are carried out. If one sensor is eliminated, its measurement error will be set to zero, and the simulation will be processed by considering the errors in the other sensor only.

An example of this set of results for the 2nd exit point x-coordinate a_{2x} is shown in Fig. 10. As a result, when the SICK sensor measurement errors are eliminated, the identification errors are much reduced, the dispersion of the 500 results are significantly reduced, as shown in Fig. 10(c). The identification errors are much more concentrated around 0 mm, and within ±6 mm. And the σ values decreased from 7.4969 mm to 0.2145 mm. In comparison, the errors for the x-coordinates of initial cable lengths are even more affected, with identification errors inside ±1 mm range. It should be noted that the errors in the x-coordinates of the MP positions are less affected, reduced from ±10 mm to ±5 mm.

On the other hand, when the WitMotion inclinometer is not considered, the identification errors do not really change as shown in Fig. 10(b). The elimination of inclinometer measurement errors does not improve much the identification quality. It means that the identification quality of the geometric parameters of the planar CDPR, the initial cable lengths and the moving-platform poses are not very sensitive to measurement errors in WitMotion inclinometer.

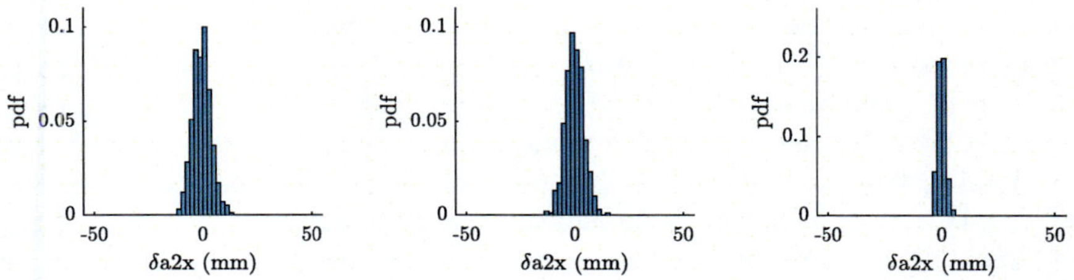

(a) Both sensors, σ=4.0962mm (b) SICK sensor, σ=4.1247mm (c) Inclinometer, σ=1.6193mm

Fig. 10. Identification error results in the x-coordinates of the 2nd exit point when 50 poses are used, comparison of sensor effects

4.3 Overall Simulation

The previous sections explained in detail different factors that affect overall identification quality. In this section, the overall simulation regarding the realistic experiment process is discussed. Similar to previous simulations, the same sensor measurement errors are applied. One single calibration using all the 50 measurement poses is performed. Identification errors $\delta\mathbf{x}$, i.e. the differences between the identified and real system variables are calculated. MP movement is also simulated, the control scheme code that will be able to move the real robot is used in the simulation. The overall simulation result is shown in Fig. 11. All identification errors fall within the range of ± 9 mm, which is consistent with the previous detailed simulation. Besides, among the identification errors, 89.8% are less than ± 5 mm.

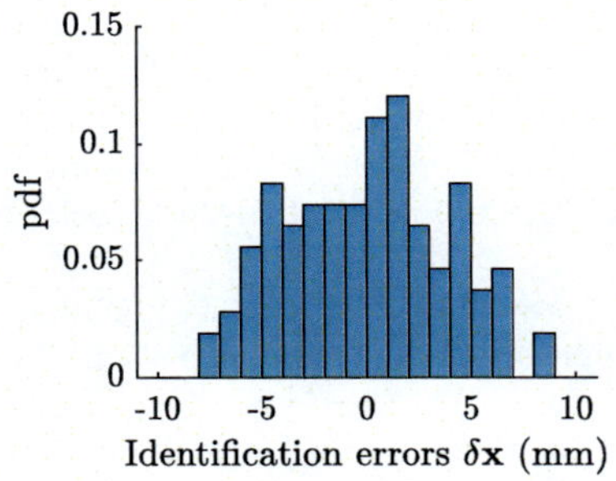

Fig. 11. Identification errors in all system variables

5 Conclusions and Future Work

This paper presented the simulation of the calibration of a 3-DoF, 2-cable, planar CDPR. In addition to motor encoders, the calibration relies on the combination of a laser displacement sensor and an inclinometer embedded on the moving platform. Detailed pulley kinematics are considered. The actual accuracies of the sensors are examined beforehand. The simulations showed that with more measurement poses used, the identification errors of the exit point, pose Cartesian coordinates and the initial cable lengths were reduced, and were less dispersed. It turns out that the laser displacement sensor has a much larger influence than the inclinometer on the identification errors. Furthermore, amongst all system variables, the errors in initial cable lengths are the most sensitive to the laser displacement sensor measurement errors whereas the errors in the MP position coordinates are the least sensitive ones. Based on the sensor considered in this work, the system variable errors are all below ± 9 mm, and most are below ± 5 mm.

Later on, experimental validations will be carried out. In addition to pulley kinematics, cable elasticity and mass will be considered to improve calibration quality. Future work will also deal with the determination of optimal measurement poses for CDPR calibration.

Acknowledgements. This work was supported by the ANR CRAFT project, grant ANR-18-CE10-0004, https://anr.fr/Project-ANR-18-CE10-0004. Bozhao Wang is grateful for the support of China Scholarship Council (CSC Grant No. 202008070051).

References

1. Zhang, Z., Xie, G., Shao, Z., Gosselin, C.: Kinematic calibration of cable-driven parallel robots considering the pulley kinematics. Mech. Mach. Theory **169**, 104648 (2022)
2. Picard, E., Caro, S., Claveau, F., Plestan, F.: Pulleys and force sensors influence on payload estimation of cable-driven parallel robots. In: 2018 IEEE/RSJ International Conference on Intelligent Robots and Systems (IROS 2018) (2018). https://hal.archives-ouvertes.fr/hal-01862015
3. Qian, S., Bao, K., Zi, B., Wang, N.: Kinematic calibration of a cable-driven parallel robot for 3D printing. Sensors (Switzerland) **18**, 2898 (2018)
4. Sandretto, J., Daney, D., Gouttefarde, M.: Calibration of a fully-constrained parallel cable-driven robot. CISM Int. Centre Mech. Sci. Courses Lect. **544**, 77–84 (2013)
5. Renaud, P., Andreff, N., Martinet, P., Gogu, G.: Kinematic calibration of parallel mechanisms: a novel approach using legs observation. IEEE Trans. Rob. **21**, 529–538 (2005)
6. Wang, J., Masory, O.: On the accuracy of a Stewart platform. I. The effect of manufacturing tolerances. In: [1993] Proceedings IEEE International Conference on Robotics and Automation, vol. 1, pp. 114–120 (1993)
7. Andreff, N., Renaud, P., Martinet, P., Pierrot, F.: Vision-based kinematic calibration of an H4 parallel mechanism: practical accuracies. Ind. Robot. **31**, 273–283 (2004)
8. Daney, D., Papegay, Y., Neumaier, A.: Interval methods for certification of the kinematic calibration of parallel robots. In: Proceedings - IEEE International Conference on Robotics and Automation, vol. 2004, pp. 1913–1918 (2004)
9. Wang, B., Caro, S.: Exit point, initial length and pose self-calibration method for cable-driven parallel robots. Mech. Mach. Sci. **103**, 90–101 (2021)

10. Fortin-Côté, A., Cardou, P., Gosselin, C.: An admittance control scheme for haptic interfaces based on cable-driven parallel mechanisms. In: Proceedings - IEEE International Conference on Robotics and Automation, pp. 819–825 (2014)
11. Zhang, F., Shang, W., Li, G., Cong, S.: Calibration of geometric parameters and error compensation of non-geometric parameters for cable-driven parallel robots. Mechatronics **77**, 102595 (2021). https://doi.org/10.1016/j.mechatronics.2021.102595
12. Miermeister, P., Pott, A., Verl, A.: Auto-calibration method for overconstrained cable-driven parallel robots. In: 7th German Conference on Robotics, ROBOTIK 2012, pp. 301–306 (2012)
13. Martin, C., Fabritius, M., Stoll, J., Pott, A.: A laser-based direct cable length measurement sensor for CDPRS. Robotics **10**, 1–11 (2021)
14. Boggs, P., Byrd, R., Schnabel, R.: A stable and efficient algorithm for nonlinear orthogonal distance regression. SIAM J. Sci. Stat. Comput. **8**, 1052–1078 (1987). https://doi.org/10.1137/0908085

A Panorama of Methods for Dealing with Sagging Cables in Cable-Driven Parallel Robots

J.-P. Merlet[1(✉)] and R. Tissot[2]

[1] INRIA, 2004 Route des Lucioles, 06902 Sophia-Antipolis, France
Jean-Pierre.Merlet@inria.fr

[2] INRIA/3IA Côte d'Azur, 2004 Route des Lucioles, 06902 Sophia-Antipolis, France
Romain.Tissot@inria.fr

Abstract. We are considering cable-driven parallel robot (CDPR), where the legs of the robot are constituted of cables that can be independently coiled/uncoiled. We show that whatever the size of the CDPR is we may have slack cables so that using a sagging cable model that takes into account both the mass and elasticity of the cables will improve the positioning accuracy. Being able to solve the inverse and direct kinematics (IK/DK) with sagging cables is crucial for kinematic analysis while being quite complex as both IK/DK may have multiple solutions. We present a panorama of solving methods for the IK/DK with their advantages and drawbacks.

Keywords: Cable-driven parallel robot · Cable sagging · Kinematics

1 Cable-Driven Parallel Robots and Sagging Cables

Cable-driven parallel robots (CDPR) are now quite well known as a variant of classical parallel robots that use cables instead of rigid legs with the advantage of a large workspace. We will assume that the winch output point of cable j is A_j while the other extremity of this cable is connected to the platform at point B_j. Cables have the drawback of being only able to have a pulling action on the platform. Therefore managing cable tensions is of importance and two classes have to be considered:

- suspended CDPR where B_j is lower than A_j for all cables
- fully constrained CDPR for which some cable(s), denoted the *pulling cables*, have a A_j that is lower than the B_j, the other cables being called *lifting cables*

In the first case only the action of gravity, that cannot be controlled, may lead to have cables under tension while in the second case both gravity and the downward action of the pulling cables influence the tension in the lifting cables, possibly allowing for optimizing the cable tension distribution. Fully constrained

O. Altuzarra and A. Kecskémthy (Eds.): ARK 2022, SPAR 24, pp. 122–130, 2022.
https://doi.org/10.1007/978-3-031-08140-8_14

CDPRs have the drawback that the pulling cables are obstacles below the platform (which may be troublesome for some tasks) and do not contribute to supporting the platform weight. Furthermore it may be conjectured that for a given geometry the workspace of a suspended CDPR will be larger than the one of a fully constrained CDPR with the same number of cables.

Another important issue for the kinematic analysis of CDPR is the cable model i.e. the relationship between cable shape, tension and length. Many works on CDPR use the *ideal cable* model assuming no elasticity, no cable mass and a straight line shape of the cable as soon it is under tension. Some works use different variants of cable model with the purpose of taking elasticity and/or sagging into account. Sagging is important for suspended CDPR even for ideal cables as it has been shown that in general the number of taught cables cannot be larger than the number of d.o.f. of the platform as all cables may be under tension only temporarily.In that case an efficient control requires to identify which cable(s) are slack and to evaluate their slackness using a sagging model is more efficient than trying to measure reliably and accurately the cable tensions (and only them). Furthermore at the actuation level slack cable may leads to winding problems, possibly even leading to reverting the normal coiling process [20].

Simple sagging model may not be sufficient: for example approximating the shape of the slack cable by a parabola is valid only if the cable mass is negligible compared to the tension in the cable which is, by essence, not true for slack cable. Hence a more accurate model must be used and the Irvine textbook model [11], see Eq. (1,2), is appropriate as it has been experimentally proven to be valid for usual CDPR [23] (more complex model are available [9,25]).

For improving the accuracy of CDPR taking into account the sagging in the inverse kinematics (IK) and direct kinematics (DK) problems is of high importance. As soon as the CDPR velocities are small and the initial state of the CDPR has been determined real-time kinematics based on a certified Newton scheme to determine the current state of the CDPR is efficient and very fast [18,21]. In our latest prototype, that has be run continuously for several months, we use redundant measurements [20] for determining the pose of the platform:

- a real-time DK solver that uses a full cable sagging model and a pulley model
- two horizontal and one vertical lidars that measure distance to the surrounding wall and to the ground for calculating very accurately the platform pose at a sampling frequency of 1/2 Hz (horizontal) 1 Hz (vertical)
- an accelerometer on each cable located close to the platform that measure the direction of the cable tangent. Combined with a sagging model these measurements allows one to estimate the slackness of each cable [17] with the safety advantage of allowing to detect cable failures
- an adaptive coiling model that uses the other measurements to estimate the current drum radius of the multi-layer winch

If managing sagging cables is not a problem with these measurements a difficulty will appear in the design phase where a kinematic analysis has to be performed for evaluating the performances of a given CDPR over its workspace.

2 Sagging Cables Model

The non-algebraic Irvine model is a planar one with a planar frame in which A, B have respectively (0,0), (x_b, z_b) as coordinates. The model relates x_b, z_b to the cable length at rest L_0 and to the horizontal and vertical forces $F_x > 0, F_z$ applied on the cable at B. They are different formulations to describe this model and one of them is:

$$x_b = F_x\left(\frac{L_0}{EA_0} + \frac{sinh^{-1}(\frac{F_z}{F_x}) - sinh^{-1}(\frac{F_z - \mu g L_0}{F_x})}{\mu g}\right) \tag{1}$$

$$z_b = \frac{F_z L_0}{EA_0} - \mu g L_0^2/2 + \frac{\sqrt{F_x^2 + F_z^2} - \sqrt{F_x^2 + (F_z - \mu g L_0)^2}}{\mu g} \tag{2}$$

where E is the Young modulus of the cable material, μ its linear density and A_0 the area of a cross-section of the cable. Let us consider the IK/DK for a spatial CDPR with n cables based on the Irvine model. Regarding equations we have $2n$ equations coming from the cable model and 6 equations that characterize the mechanical equilibrium of the platform so that we have a total of $2n + 6$ equations. For the IK we have $3n$ unknowns, namely the L_0, F_x, F_z for each cable, leading to a square system only if $n = 6$. For the DK the unknowns are the $2n$ F_x, F_z for each cable and the 6 pose parameters, namely 3 coordinates in the reference frame of a specific point on the platform and 3 rotation angles that will characterize its orientation (the coordinates x_b, z_b of cable model can be derived from the pose parameters). Hence we have $2n+6$ unknowns and the DK system is always square. The important point is that with this sagging model both the IK and DK may have multiple solutions and we will address in Sect. 4 the possible methods that may be used for finding these solutions. But we will first emphasize the importance of being able to solve the IK/DK for kinematics analysis in the design phase.

3 Sagging and Kinematics Analysis

Kinematic analysis is used in the design phase to check the performances of a given CDPR design. For example it is of interest to

1. being able to calculate cross-sections of the workspace assuming for example an upper limit L_0^M for the L_0
2. calculate the maximal cable tensions over a specified workspace for a given load

In many works the workspace problem is approximately solved by using a discretisation of the workspace: a regular grid is used and at each node of the grid the IK is solved for determining if the node is in or out of the workspace. A faster and more accurate method consists in determining only the workspace border. Being given the non-algebraic nature of the Irvine equations it seems to be difficult to find an analytic description of the border curve so that we have to rely on a numerical method. For that purpose an efficient approach [13] is to start from a pose $\mathbf{X_a}$, that may be arbitrarily chosen as soon as it lies in the workspace, and from one of its IK solutions S_a^{ik}. We then moves incrementally the platform along an arbitrary unit vector $\mathbf{N}$ by setting the pose $\mathbf{X}$ as $\mathbf{X}] = \mathbf{X_a} + \lambda\mathbf{N}$ and solving its IK for a given λ by using the previous IK solution as guess for the Newton method, until we obtain a pose $\mathbf{X_s}$ such that its current IK solution S_s^{ik} has a L_0^j very close to L_0^M so that $\mathbf{X_s}$ is close to the border. We then find a pose $\mathbf{X_c}(x_c, y_c)$ that is arbitrarily close to the border by solving with Newton a new square IK system $\mathcal{S}$ obtained by setting $L_0^j = L_0^M$ and λ as new unknown. Note that by using classical mathematical tools we are able to calculate an upper bound on the smallest distance between $\mathbf{X_b}$ and the border and adjust the accuracy of the Newton scheme so that this bound is lower than a fixed threshold. Then we follow the border by setting the pose parameters to $x_c + \epsilon, y_c$ where ϵ is a small increment and solving with Newton the IK system with $L_0^j = L_0^M$ and ϵ as new unknown. By repeating this process successive poses on the border are obtained, leading to an approximation of the border by a polygonal line. Note that this border is specific to one of the IK solutions at $\mathbf{X_a}$ so that we obtain a border curve for each IK solutions at $\mathbf{X_a}$. Finding the pose $\mathbf{X_s}$ may fail if a singularity is encountered when moving from $\mathbf{X_a}$ so that several initial poses $\mathbf{X_a}$ and/or $\mathbf{N}$ may have to be used for obtaining a closed region and to find all closed components of the workspace.

The workspace example illustrates the necessity of being able to find all the IK solutions for a finite set of poses. Now if we are willing to determine the maximum of the cable tensions over the CDPR articular workspace we have to consider all solutions of the DK as we cannot predict the history of the robot motion.

4 Sagging and Solving the IK/DK

As finding all solutions of the IK/DK is crucial for the kinematic analysis we will now present possible solving methods that may provide all solutions for the IK/DK. Numerous works have addressed the DK problem with ideal cables [1,4, 6,10] but much less works have considered Irvine-based sagging cables [7,22,24]. To the best of the author knowledge they are only 3 available methods for dealing with solving the IK/DK having sagging cables and aiming at finding **all** solutions. We present them for the Irvine model but they may be used with any cable model as soon as it is analytic:

Interval analysis (IA): this method looks for solutions within a search space defined by intervals for the unknowns that have to be carefully selected. For the IK with 6 sagging cables the unknowns are the $L_0 > 0, F_x > 0, F_z$. As IK solutions with very large L_0, F_x can be found we have to choose a large upper bound $\overline{L_0}, \overline{F_x}$ for the L_0, F_x intervals. If M is the platform mass, then a lower bound for the F_z interval may be $-Mg - (n-1)\mu g\overline{L_0}$ while its upper bound may be set to $\mu g\overline{L_0}$. For the DK the unknowns are the pose parameters and the $F_x > 0, F_z$. The translational part of the pose parameters may easily be bounded being given the geometry of the CDPR while the rotation angles intervals may be set to $[0, 2\pi]$. Intervals for F_z may be chosen in the same way than for the IK. As for the upper bound of the F_x intervals we choose a very large value as a DK solution may have cables close to the horizontal. As we have unknowns with very large intervals the IA branch-and-bound process may lead to a large computation time (hours for the IK and days for the DK for a 6-cables CDPR [16]). For both the IK and DK we may miss solution(s) because of the unknown upper bound limit on the F_x.

Continuation [2]: if $E \rightarrow \infty$ and $\mu \rightarrow 0$, then Irvine equations reflect the behavior of an ideal cable. For the DK we first assume ideal cables and calculate all DK solutions $\{S_k^i\}$ for all cable configurations having from 6 to 1 cables under tension, the other cables being slack and therefore disregarded. Then we set E to a large value E_l and μ to a small value μ_l and use the Newton scheme with each S_k^i as guess for deriving the DK solutions C_k of the CDPR with sagging cables. If E_0, μ_0 are the real E, μ values we define a single parameter *continuation path* as

$$E = E_l + \lambda(E_0 - E_l) \qquad \mu = \mu_l + \lambda(\mu_0 - \mu_l) \tag{3}$$

The continuation parameter λ is such that C_k is a DK solution for $\lambda = 0$ while the sought DK solutions will be obtained for $\lambda = 1$. Starting from $\lambda = 0$ we increase λ by an increment ϵ that is automatically calculated at each step and solve the DK problem for the current λ using the previous solution as guess for Newton. We then repeat the process until $\lambda = 1$. Continuation is in general much faster than IA but the continuation path (3) may end-up in a singularity while another path (e.g. adjusting first E then μ) may avoid any singularity and leads to a DK solution. Therefore we may also miss DK solution(s) with the continuation approach [14]. For the IK the process is similar: we consider all configurations of IK with $m = 6$ to $m = 1$ ideal cables under tension (therefore having m fixed pose parameters among the desired pose). Then we use first a continuation on E, μ and if $m < 6$ a second continuation on the $6 - m$ free pose parameters so that they reach their assigned values [19]. Here again the continuation may end up in a singularity aand therefore miss an IK solution.

Neural networks: a *multi-layer perceptron (MLP)* [8] is an universal estimator that is theoretically able to provide an estimation $\mathbf{V}$ of any set of functions $\mathbf{F}(\mathbf{U})$ with input $\mathbf{U}$, even if $\mathbf{F}$ is not known. For constructing a MLP we need a large training set of pairs $(\mathbf{U_i}, \mathbf{F}(\mathbf{U_i}))$. In our case $\mathbf{V}$ will be the solutions of the IK/DK problems while $\mathbf{U}$ will be the pose parameters for the IK and the L_0s for the DK.

A major problems with a MLP is that it produces a single estimation $\mathbf{V}$ of the IK/DK while we may have several solutions whose number cannot be predicted. Furthermore the estimation error should be acceptable for being used in the kinematic analysis. A MLP has been used for the IK of a redundant planar CDPR with 4 cables by fixing the minimal cable tension and therefore having a single IK solution [5]. The prediction error on the L_0 was acceptable (less than 0.1) for the kinematic analysis but not for control while the errors on the F_x, F_z where not acceptable (more than 10 N for a load of 1 kg).

We have investigated the use of MLP for CDPR with 2 and 3 cables for both the IK and DK that have a single solution in this case [3]. The training set is obtained by solving with IA or continuation the IK/DK for a limited number of inputs P_j and using continuation to calculate the IK/DK solutions for $P_j + \lambda \mathbf{N_{jk}}$ where $\mathbf{N_{jk}}$ are randomly chosen unit vectors (the verification set is obtained in the same manner but with different $\mathbf{N_{jk}}$) Beside the training set a MLP requires to fix several parameters: number of layers, number of neurons per layer, type of activation functions and the cost function. As the learning time is small (mean value: 7 min) we have used a systematic brute force approach for creating MLPs with between 1 and 6 layers, 10 to 200 neurons per layer and various combinations of activation functions. We have then systematically calculated statistics (mean, max, variance, ...) of the estimation errors on a large verification set. It appears that none of the MLPs were producing estimation whose accuracy was sufficient for kinematic analysis. However in some cases we have got an exact solution by using the MLP estimation as guess for the Newton scheme. Furthermore the statistical analysis have shown that some MLPs were able to provide a reasonable prediction for part of the unknowns and poor one for the others. Hence we have devised a solving strategy that is based on various MLPs whose estimation is used as guess for Newton and if necessary we combine the good MLPs estimations to create new estimations that are fed to Newton. We have then tested this strategy on another large verification set with 100% of success. The interest of this approach is that the computation time is very low (less than 5 ms while the IA/continuation require about one minute)). Still we cannot guarantee that we will always find the solution for the IK/DK.

It remains to determine if MLPs can manage the case of multiple IK/DK solutions. We are currently investigating an approach for that purpose. A first step is to establish **all** IK/DK solutions for m inputs. A training set will be established for each solution by using a continuation process. This training set will be used to build 12 different MLPs with different parameters. If we get n_j solutions for input j and define $k = \sum_{j=1}^{j=m} n_j$, then $12k$ MLPs will be built. The number m of inputs must be chosen large enough so that k is large as we cannot get more than k IK/DK solutions. Being given the MLPs we use the same solving strategy than in the previous example.

Building the MLPs is computer intensive but has to be done only once for possibly getting a very fast IK/DK solver so that the pre-processing time will be small compared to the time necessary for performing the kinematic analysis. Still to reduce the pre-processing time we are currently investigating the use of specific processors devoted to neural networks learning with a large number of GPUs. In parallel we plan to use also these processors for speeding up IA solving as this approach is intrinsically appropriate for a distributed implementation.

In **summary** IA is guaranteed to find all solutions within a search domain in a large computation time but will miss non realistic solutions (e.g. for the IK the one having one or several very large cable length) while continuation, that is usually faster than IA, may miss even realistic solution(s) because of the singularity problem. Regarding MLP it is currently unclear if all solutions may be determined but possibly may determine all realistic solutions in a very short computation time.

Another possible solving approach is to use a *cable lumped model* that is a much simpler alternate to Irvine cable model. It has been tested for a 3-cables CDPR with satisfactory experimental results [12] but this CDPR is relatively simple. A large number of issues has still to be solved for this approach to determine if this model will allow for faster IA/continuation solvers such as how to choose the number of elements, their elasticity and mass in order to get the best compromise between computation time of the solver and errors on the kinematic parameters (considering that the model estimation may be used as Newton guess for getting the exact result).

5 Conclusion

In this paper we have addressed different issues for taking sagging into account for CDPR in the design phase and during control. We have shown that the ideal cable model limits the possibility of accurate control and is not sufficient for kinematic analysis. We then have emphasized that using a sagging model is not an issue for real-time kinematics and that sagging may also be measured with additional sensors for improving control and safety. For kinematic analysis it is crucial to have efficient solving methods for both the IK and DK (that usually have multiple solutions) and we have presented a panorama of existing methods with their advantages and drawback and the alternate lumped mass model that seems to be worth investigating.

A kinematic issue is not addressed in this paper: singularity. Indeed CDPR with sagging cables may exhibit singularities beside classical parallel robots one [15] but this issue will be addressed in a devoted paper.

Acknowledgements. Through the 3IA Côte d'Azur with the reference number ANR-19-P3IA-0002 and ANR CRAFT, grant ANR-18-CE10-0004. The authors want to thank the reviewer that has provided the interesting Ref. [12].

References

1. Abbasnejad, G., Carricato, M.: Direct geometrico-static problem of underconstrained cable-driven parallel robots with n cables. IEEE Trans. on Rob. **31**(2), 468–478 (2015)
2. Allgower, E.: Numerical continuation methods. Springer-Verlag (1990)
3. Ben Yahia, I., Merlet, J.P., Papegay, Y.: Mixing neural networks, newton method and inverval analysis for the kinematics of cable-driven parallel robots with sagging cables. In: ICAR, Ljulbjana (2021)
4. Berti, A., Merlet, J.P., Carricato, M.: Solving the direct geometrico-static problem of underconstrained cable-driven parallel robots by interval analysis. Int. J. of Rob. Res. **35**(6), 723–739 (2016)

5. Chawla, I., et al.: Neural network-based inverse kineto-static analysis of cable-driven parallel robot considering cable mass and elasticity. In: 5th International Conference on Cable-Driven Parallel Robots, CableCon, virtual, 7–9 July 2021
6. Filipovic, M., Djuric, A., Kevac, L.: Contribution to the modeling of cable-suspended parallel robot hanged on the four points. In: IEEE International Conference on Intelligent Robots and Systems, IROS, Vilamoura, 7–12 October 2012
7. Gouttefarde, M., Nguyen, D., Baradat, C.: Kinetostatics analysis of cable-driven parallel robots with consideration of sagging and pulleys. In: ARK, Ljulbjana, pp. 213–221, 29 June-3 July 2014
8. Haykin, S.: Neural networks: a comprehensive foundation. Prentice Hall PTR (1994)
9. Hussein, H., Gouttefarde, M., Pierrot, F.: Static modeling of sagging cables with flexural rigidity and shear forces. In: ARK, Bologna, 1–5 July 2018
10. Husty, M., Schadlbauer, J., Zsombor-Murray, P.: A new approach to the direct geometrico-static problem of cable suspended robots using kinematic mapping. In: 3rd International Conference on Cable-Driven Parallel Robots, CableCon, Québec (2017)
11. Irvine, H.M.: Cable Structures. MIT Press (1981)
12. Lambert, C., Nahon, M.: Study of a multitethered aerostat system: experimental observations and model validation. J. Aircr. **43**(4), 1182–1189 (2006)
13. Merlet, J.P.: Computing cross-sections of the workspace of cable-driven parallel robots with 6 sagging cables. In: Computational Kinematics, Poitiers (2017)
14. Merlet, J.P.: Preliminaries of a new approach for the direct kinematics of suspended cable-driven parallel robot with deformable cables. In: Eucomes, Nantes, 20–23 September 2016
15. Merlet, J.P.: Singularity of cable-driven parallel robot with sagging cables: preliminary investigation. In: IEEE International Conference on Robotics and Automation, Montréal, 20–24 May 2019
16. Merlet, J.P.: The forward kinematics of cable-driven parallel robots with sagging cables. In: 2nd International Conference on Cable-Driven Parallel Robots, CableCon, Duisburg, pp. 3–16, 24–27 August 2014
17. Merlet, J.P.: Direct kinematics of CDPR with extra cable orientation sensors: the 2 and 3 cables case with perfect measurement and sagging cables. In: IEEE International Conference on Intelligent Robots and Systems, IROS, Vancouver, 24–28 September 2017
18. Merlet, J.P.: On the real-time calculation of the forward kinematics of suspended cable-driven parallel robots. In: 14th IFToMM World Congress on the Theory of Machines and Mechanisms, Taipei, 27–30 October 2015
19. Merlet, J.P.: A new generic approach for the inverse kinematics of cable-driven parallel robot with 6 deformable cables. In: ARK, Grasse, 27–30 June 2016
20. Merlet, J.P., Papegay, Y., Gasc, A.V.: The Prince's tears, a large cable-driven parallel robot for an artistic exhibition. In: IEEE International Conference on Robotics and Automation, Paris, 31 May-31 August 2020
21. Pott, A.: An algorithm for real-time forward kinematics of cable-driven parallel robots. In: ARK, Piran, pp. 529–538, 28 June-1 July 2010
22. Pott, A., Tempel, P.: A unified approach to forward kinematics for cable-driven parallel robots based on energy. In: ARK, Bologna, 1–5 July 2018
23. Riehl, N., et al.: Effects of non-negligible cable mass on the static behavior of large workspace cable-driven parallel mechanisms. In: IEEE International Conference on Robotics and Automation, Kobe, pp. 2193–2198, 14–16 May 2009

24. Sridhar, D., Williams, R., II.: Kinematics and statics including cable sag for large cable suspended robots. Global J. Res. Eng. H Rob. Nano-Tec **17**(1), 11 pages (2017)
25. Talke, K., De Oiveira, M., Bewley, T.: Catenary tether shape analysis for a UAV - USV team. In: IEEE International Conference on Intelligent Robots and Systems, IROS, Madrid, 1–5 October 2018

Sensitivity of the Direct Kinematics of Underactuated Cable-Driven Parallel Robots to Redundant Sensor-Measurement Errors

Sara Gabaldo, Edoardo Ida(✉), and Marco Carricato

Department of Industrial Engineering, University of Bologna, Viale del Risorgimento, 2, 40136 Bologna, Italy
sara.gabaldo2@studio.unibo.it, {edoardo.ida2,marco.carricato}@unibo.it

Abstract. Underactuated cable-driven parallel robots (*UACDPRs*) displace a 6-degree-of-freedom end-effector (*EE*) with less than 6 cables. As a consequence, the *EE* is underconstrained and, for assigned cable lengths, the *EE* pose cannot be obtained from geometrical constraint equations only. In this paper, we study the performance of a direct-kinematics algorithm using redundant sensor measurements for a 4-cable *UACDPR*. The proposed method measures two orientation parameters of the *EE* besides cable lengths, in order to determine the other four pose variables, namely 3 position coordinates and one additional orientation parameter. A methodology for the computation of the sensitivity of the direct-kinematics solution to measurement errors is finally proposed.

1 Introduction

A cable-driven parallel robot (*CDPR*) is underactuated if the number n of actuated cables is smaller than the degrees of freedom (*DoFs*) of the end-effector (*EE*). Underactuated *CDPRs* (*UACDPRs*) are always underconstrained since the *EE* preserves some freedoms when actuators are locked [1,2]. Such robots are usually designed in a suspended configuration, with cables being routed towards the *EE* from an elevated position, so that gravity can keep them taut. Because of underactuation, the n geometric constraint equations of the system are insufficient to determine the 6 pose parameters of the *EE* when cable lengths are assigned. If static equilibrium is sought, the problem can be tackled by also considering static-equilibrium equations in the solution process, thus establishing a so-called direct geometrico-static problem [3–5]. Alternatively, one can consider additional sensors directly measuring some pose parameters or other model variables [6–8]. Swivel-pulley angles are good extra-measurement candidates, due to their simplicity of implementation, and general good accuracy. On the other hand, when the cable robot is large, or motion is highly dynamic [2],

O. Altuzarra and A. Kecskémthy (Eds.): ARK 2022, SPAR 24, pp. 131–138, 2022.
https://doi.org/10.1007/978-3-031-08140-8_15

these measures may become unreliable. Merlet [7] experimentally showed that measuring some orientation parameters of the *EE* with static inclinometers (i.e. accelerometers) provides acceptable results for static postures and quasi-static motions.

Few studies are available on the *EE* pose sensitivity to control or measurement errors. In [9] the maximum rotation and maximum point-displacement sensitivity were defined for robotic manipulators. A sensitivity analysis was carried out in [10] for planar overconstrained *CDPRs* and in [11] for spatial suspended *UACDPRs*. Finally, Ref. [12] evaluated how the pose of a translational *CDPR* changes as some errors of known magnitude are introduced in cable lengths.

In this paper, we study the theoretical sensitivity of the direct-kinematics solution to measurement errors in cable lengths and pitch-and-roll orientation parameters, on a 4 cable *UACDPR*. This set of measurements allows the *EE* position and yaw angle to be determined according to direct kinematics only. Technologically speaking pitch and roll measurements needs to be obtained by accurate *vertical reference units* [13]. Following the approach of [9], sensitivities are obtained through robot geometrical constraints differential analysis. Additionally, maximum point-displacement and yaw errors are computed throughout the workspace for given upper bounds on cable-length, pitch, and roll errors.

The rest of the paper is structured as follows. Section 2 recalls the kinematic model of a *UACDPR*. Section 3 defines the point-displacement and yaw angle sensitivities to cable length, roll and pitch measurement errors. The computation of the latter through the workspace of a 4–cable *UACDPR* is presented in Sect. 4. Finally, conclusions and future works are discussed in Sect. 5.

2 Kinematic Model

A 6-*DoF* *UACDPR* consists of a mobile platform coupled to the base by n cables, with $n < 6$. For the sake of simplicity, cables are modeled as massless and inextensible straight line segments.

$Oxyz$ is an inertial frame, whereas $Px'y'z'$ is a mobile frame attached to the platform in point P. The pose $\boldsymbol{\zeta} = \left[\mathbf{p}^T\, \boldsymbol{\epsilon}^T\right]^T$ of the *EE* is described by the position vector $\mathbf{p}$ of P and by roll-pitch-yaw (RPY) angles $\boldsymbol{\epsilon} = [\phi, \theta, \chi]^T$. The latters allow one to compute the *EE* rotation matrix $\mathbf{R}$, such that $\mathbf{R} = \mathbf{R}_z(\chi)\mathbf{R}_y(\theta)\mathbf{R}_x(\phi)$. Each cable enters the workspace through an eyelet at point B_i, whose position is denoted by vector $\mathbf{b}_i$, and is then attached to the platform at point A_i, whose position in the mobile frame is denoted by the vector ${}^P\mathbf{a}'_i$ (Fig. 1a). The coordinates of A_i in the inertial frame can be computed as:

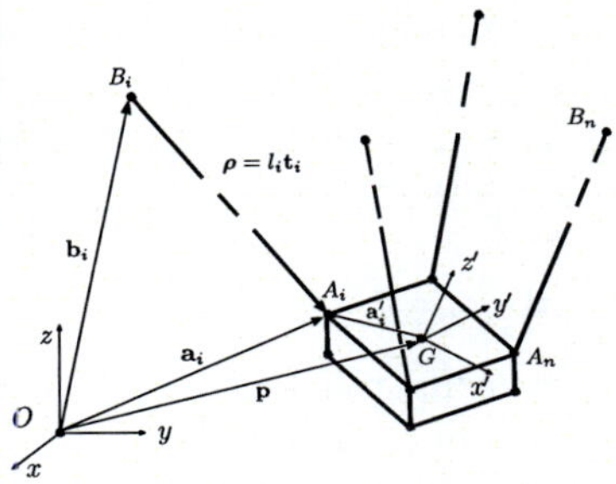

(a) Geometrico-static model

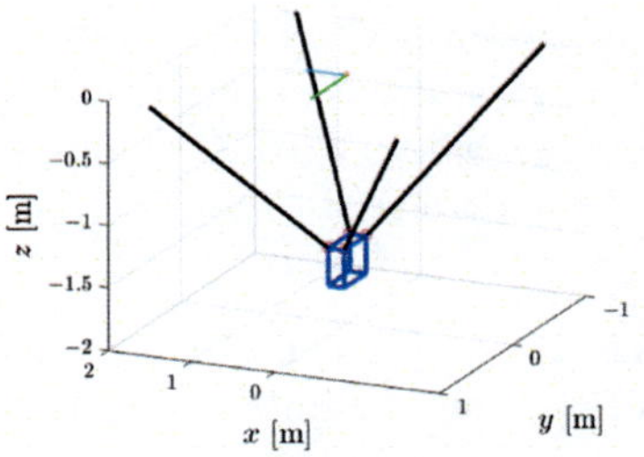

(b) *UACDPR* schematics for simulations

Fig. 1. *UACDPR* model.

$$\mathbf{a}_i = \mathbf{p} + \mathbf{a}'_i = \mathbf{p} + \mathbf{R}\,{}^P\mathbf{a}'_i, \qquad \mathbf{a}'_i = \mathbf{R}\,{}^P\mathbf{a}'_i \tag{1}$$

If l_i is the cable length and $\boldsymbol{\rho} = \mathbf{b}_i - \mathbf{a}_i$ is the cable vector, we have:

$$\boldsymbol{\rho}_i = \|\boldsymbol{\rho}_i\|\mathbf{t}_i = l_i\mathbf{t}_i, \qquad l_i = \|\boldsymbol{\rho}_i\|, \qquad \mathbf{t}_i = \boldsymbol{\rho}/l_i \tag{2}$$

Thus, the geometrical constraint imposed by the cables on the *EE* pose is:

$$\boldsymbol{\rho}_i \cdot \boldsymbol{\rho}_i - l_i^2 = 0 \tag{3}$$

Differentiating Eq. (3), for $i = 1, \ldots n$, yields [11]:

$$\mathbf{J}\mathrm{d}\boldsymbol{\zeta} = \mathrm{d}\mathbf{l}, \qquad \mathbf{J} = \boldsymbol{\Xi}\mathbf{D}, \tag{4}$$

$$\boldsymbol{\Xi} = \left[\boldsymbol{\xi}_1 \cdots \boldsymbol{\xi}_n\right], \qquad \boldsymbol{\xi}_i = \begin{bmatrix} \mathbf{t}_i \\ \mathbf{a}'_i \times \mathbf{t}_i \end{bmatrix}, \qquad \mathbf{D} \in \mathbb{R}^{6\times 6}$$

where $\boldsymbol{\omega}$ is the *EE* angular velocity, $\mathbf{v} = [\dot{\mathbf{p}}^T \boldsymbol{\omega}^T]^T$ is the *EE* twist and matrix $\mathbf{D}$ depends on the particular choice of orientation representation $\boldsymbol{\epsilon}$ so that:

$$\mathbf{v} = \mathbf{D}(\boldsymbol{\epsilon})\dot{\boldsymbol{\zeta}} = \mathbf{D}\dot{\boldsymbol{\zeta}} \tag{5}$$

3 Sensitivity of Direct Kinematics with Redundant Measurements

In this Section, the theoretical point-displacement and yaw sensitivities to cable length, roll and pitch errors are introduced. A 6-*DoF* *EE* suspended by 4 cables is considered, so that the direct kinematics with extra roll and pitch measurements is well-posed, namely it comprises a set of 4 nonlinear equations, the geometrical constraints in Eq. (3), in 4 unknowns, the *EE* position and yaw.

In the following, the *EE* pose is divided in *known-parameters* $\boldsymbol{\zeta}_k$, the pitch and the roll, and the *unknown-parameters* $\boldsymbol{\zeta}_u$, the *EE* position $\mathbf{p}$ and the yaw angle ψ, to be computed by means of direct kinematics. In order to evaluate the sensitivity of the solution $\boldsymbol{\zeta}_u$ to errors on measurements (cable lengths $\mathbf{dl}$ and $\mathrm{d}\boldsymbol{\zeta}_k$), differential kinematics must be considered. If we re-order pose parameters $\boldsymbol{\zeta} = [\boldsymbol{\zeta}_u^T, \boldsymbol{\zeta}_k^T]^T$ and, accordingly, we re-order the columns of the Jacobian matrix $\mathbf{J} = [\mathbf{J}_u \quad \mathbf{J}_k]$, where $\mathbf{J}_u \in \mathbb{R}^{4\times4}$ and $\mathbf{J}_k \in \mathbb{R}^{4\times2}$, the differential kinematics of Eq. (4) can be written as:

$$\mathbf{J}_u \mathrm{d}\boldsymbol{\zeta}_\mathbf{u} + \mathbf{J}_k \mathrm{d}\boldsymbol{\zeta}_\mathbf{k} = \mathrm{d}\mathbf{l} \tag{6}$$

By assuming to have no errors on sensors measuring roll and pitch, that is, $\mathrm{d}\boldsymbol{\zeta}_\mathbf{k} = \mathbf{0}$, we may compute $\boldsymbol{\zeta}_u$ sensitivities to errors on cable lengths:

$$\mathbf{J}_u \mathrm{d}\boldsymbol{\zeta}_\mathbf{u} = \mathrm{d}\mathbf{l} \tag{7}$$

Out of singularities, matrix $\mathbf{J}_u$ is full rank, and thus invertible. If $\mathbf{K} = \mathbf{J}_u^{-1}$, we have:

$$\mathrm{d}\boldsymbol{\zeta}_\mathbf{u} = \mathbf{K}\mathrm{d}\mathbf{l} \tag{8}$$

To compute dimensionally consistent sensitivity indices, it is necessary to decouple unknown pose-parameters: point-position vector $\mathbf{p}$ and yaw angle ψ. Matrix $\mathbf{K}$ can be decoupled in two dimensionally-consistent parts so that Eq. (8) becomes:

$$\begin{bmatrix} \mathrm{d}\mathbf{p} \\ \mathrm{d}\psi \end{bmatrix} = \begin{bmatrix} \mathbf{K}_p \\ \mathbf{K}_\psi \end{bmatrix} \mathrm{d}\mathbf{l} \tag{9}$$

where $\mathbf{K}_p \in \mathbb{R}^{3\times4}$ and $\mathbf{K}_\psi \in \mathbb{R}^{1\times4}$, such that $\mathbf{K} = [\mathbf{K}_p^T \quad \mathbf{K}_\psi^T]^T$.

Finally, following the approach described in [12], we can define the *point-displacement kinematic sensitivity to errors in cable-length measurements* as:

$$\sigma_{\infty,2}^{p,l} = \max_{\|\mathrm{d}\mathbf{l}\|_\infty=1} \|\mathrm{d}\mathbf{p}\|_2 \tag{10}$$

where $\|\cdot\|_\infty$ and $\|\cdot\|_2$ are the ∞-norm and the 2-norm of $(\cdot)$. In this paper, according to observations presented in [12], we chose the matrix norms induced by $\|\cdot\|_2$ and $\|\cdot\|_\infty$, since they attain the clearest physical meaning. In fact, the infinity norm states that the absolute value of the measurements errors are independently bound by 1, while the 2-norm indicates the maximum distance error (the 2- and ∞-norm of the scalar ψ are equivalent). Substituting the first part of Eqs. (9) in (10), we get:

$$\sigma_{\infty,2}^{p,l} = \max_{\|\mathrm{d}\mathbf{l}\|_\infty=1} \|\mathbf{K}_p \mathrm{d}\mathbf{l}\|_2 \tag{11}$$

that is, by definition [12], the matrix $\mathbf{K}_p$ norm induced by $\|\cdot\|_2$ and $\|\cdot\|_\infty$:

$$\sigma_{\infty,2}^{p,l} = \|\mathbf{K}_p\|_{\infty,2} \tag{12}$$

Similarly, *yaw sensitivity to errors in cable-length measurements* can be defined as:

$$\sigma_{\infty,2}^{\psi,l} = \max_{\|\mathrm{d}\mathbf{l}\|_\infty=1} \|\mathrm{d}\psi\|_2 \tag{13}$$

and, substituting the second part of Eq. (9) therein, we show that yaw angle sensitivity is equal to the matrix $\mathbf{K}_\psi$ norm induced by $\|\cdot\|_2$ and $\|\cdot\|_\infty$, namely:

$$\sigma_{\infty,2}^{\psi,l} = \|\mathbf{K}_\psi\|_{\infty,2} \tag{14}$$

In order to compute $\boldsymbol{\zeta}_u$ sensitivity to errors on roll and pitch measurements we assume $\mathrm{d}\mathbf{l} = \mathbf{0}$; Eq. (6) thus becomes:

$$\mathbf{J}_u \mathrm{d}\boldsymbol{\zeta}_u = -\mathbf{J}_k \mathrm{d}\boldsymbol{\zeta}_\mathbf{k} \tag{15}$$

Out of singularities, matrix $\mathbf{J}_u$ is full rank, and thus invertible. Thus, we have:

$$\mathrm{d}\boldsymbol{\zeta}_u = -\mathbf{F}\mathrm{d}\boldsymbol{\zeta}_\mathbf{k} \tag{16}$$

where $\mathbf{F} = \mathbf{J}_u^{-1}\mathbf{J}_k \in \mathbb{R}^{4\times 2}$. Matrix $\mathbf{F}$ can be decoupled in two dimensionally-consistent parts so that Eq. (16) becomes:

$$\begin{bmatrix} \mathrm{d}\mathbf{p} \\ \mathrm{d}\psi \end{bmatrix} = \begin{bmatrix} \mathbf{F}_p \\ \mathbf{F}_\psi \end{bmatrix} \mathrm{d}\boldsymbol{\zeta}_\mathbf{k} \tag{17}$$

where $\mathbf{F}_p \in \mathbb{R}^{3\times 2}$ and $\mathbf{F}_\psi \in \mathbb{R}^{1\times 2}$ such that matrix $\mathbf{F} = -[\mathbf{F}_p^T \quad \mathbf{F}_\psi^T]^T$.

Again, it is possible to define the *point-displacement sensitivity to errors in roll-and-pitch measurements* as:

$$\sigma_{\infty,2}^{p,\zeta_k} = \max_{\|\mathrm{d}\boldsymbol{\zeta}_\mathbf{k}\|_\infty = 1} \|\mathrm{d}\mathbf{p}\|_2 \tag{18}$$

then, substituting the first part of Eq. (17) therein, we obtain the matrix $\mathbf{F}_p$ norm induced by $\|\cdot\|_2$ and $\|\cdot\|_\infty$:

$$\sigma_{\infty,2}^{p,\zeta_k} = \|\mathbf{F}_p\|_{\infty,2} \tag{19}$$

Similarly, we define the *yaw sensitivity to errors in roll-and-pitch measurements* as:

$$\sigma_{\infty,2}^{\psi,\zeta_k} = \max_{\|\mathrm{d}\boldsymbol{\zeta}_\mathbf{k}\|_\infty = 1} \|\mathrm{d}\psi\|_2 \tag{20}$$

and substituting the second part of Eq. (17) therein, we obtain the matrix $\mathbf{F}_\psi$ norm induced by $\|\cdot\|_2$ and $\|\cdot\|_\infty$:

$$\sigma_{\infty,2}^{\psi,\zeta_k} = \|\mathbf{F}_\psi\|_{\infty,2} \tag{21}$$

4 Application on a 4-*UACDPR*

In this section, we present the application of the proposed sensitivity indices throughout the static workspace [14] of a 4-cable *UACDPR* (Fig. 1b) with parameters:

$$^{P}\mathbf{a}_1' = \begin{bmatrix} 0.1 \\ 0.15 \\ 0.15 \end{bmatrix} m, \quad ^{P}\mathbf{a}_2' = \begin{bmatrix} 0.1 \\ -0.15 \\ 0.15 \end{bmatrix} m, \quad ^{P}\mathbf{a}_3' = \begin{bmatrix} -0.1 \\ -0.15 \\ 0.15 \end{bmatrix} m, \quad ^{P}\mathbf{a}_4' = \begin{bmatrix} -0.1 \\ 0.15 \\ 0.15 \end{bmatrix} m,$$

$$\mathbf{b}_1 = \begin{bmatrix} 1.5 \\ 1 \\ 0 \end{bmatrix} m, \quad \mathbf{b}_2 = \begin{bmatrix} 1.5 \\ -1 \\ 0 \end{bmatrix} m, \quad \mathbf{b}_3 = \begin{bmatrix} -1.5 \\ -1 \\ 0 \end{bmatrix} m, \quad \mathbf{b}_4 = \begin{bmatrix} -1.5 \\ 1 \\ 0 \end{bmatrix} m, \quad m = 1Kg$$

Point-displacement and yaw sensitivities to errors in cable-length measurements are computed according to Eqs. (12), (14); the matrix norms induced by $\|\cdot\|_2$ and $\|\cdot\|_\infty$ are calculated by way of the algorithm proposed in [15]. The results range from 1.78 to 6.19 mm/mm for $\sigma^{p,l}_{\infty,2}$, and from 0.77 to 1.51 °/mm for $\sigma^{\psi,l}_{\infty,2}$. **p** is sensitive to errors in cable lengths in the upper central part of the workspace, whereas ψ is sensitive to errors in cable lengths in the lower side parts (Fig. 2).

Point-displacement and yaw sensitivities to errors in roll-and-pitch measurements are computed according to Eqs. (19), (21); in the given workspace, the results range from 0.64 to 3.89 mm/° for $\sigma^{p,\zeta_k}_{\infty,2}$, and from 0.01 to 1.81 °/° for $\sigma^{\psi,\zeta_k}_{\infty,2}$. **p** is more sensitive to errors in pitch and roll in the lower and upper parts of the workspace, whereas ψ is more sensitive to errors in pitch and roll in the lateral parts (Fig. 3).

In addition, if maximum errors in cable lengths $\mathrm{d}l_M$ and in roll-and-pitch angles $\mathrm{d}\zeta_{k,M}$ are known, the aforementioned sensitivity indices can be employed to evaluate the maximum combined point-displacement and yaw errors:

$$\|\mathrm{d}\mathbf{p}_{max}\|_2 = \sigma^{p,l}_{\infty,2}\mathrm{d}l_M + \sigma^{\psi,l}_{\infty,2}\mathrm{d}\zeta_{k,M} \tag{22}$$

$$\|\mathrm{d}\psi_{max}\|_2 = \sigma^{p,\zeta_k}_{\infty,2}\mathrm{d}l_M + \sigma^{\psi,\zeta_k}_{\infty,2}\mathrm{d}\zeta_{k,M} \tag{23}$$

For the exemplary 4-cable *UACDPR*, assuming $\mathrm{d}l_M = 1$ mm, which may be due to cable elasticity, and $\mathrm{d}\zeta_{k,M} = 0.4°$, which is compatible with accurate vertical reference units, the resulting accuracy on the determination of *EE* position and yaw orientation angle, computed throughout the workspace, range from 2.13 mm to 7.46 mm for $\mathrm{d}\mathbf{p}$ and from 0.77° to 2.12° for $\mathrm{d}\psi$ (Fig. 4). Higher accuracies are obtained, for both **p** and ψ, in the central part of the workspace, while the major inaccuracies are obtained in the upper part for **p** and in the side parts for ψ.

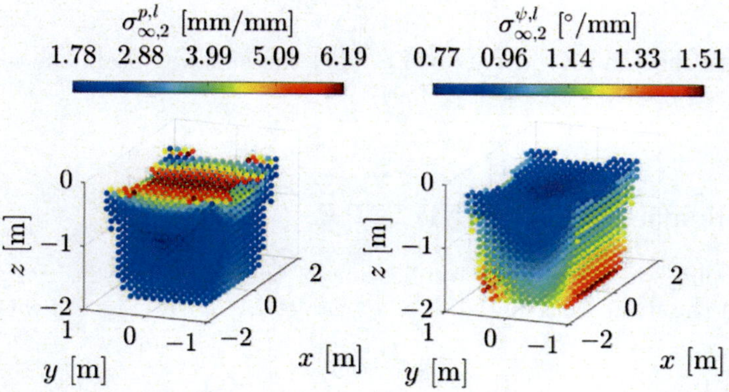

Fig. 2. Point-displacement and yaw sensitivities to errors in cable-length measurements

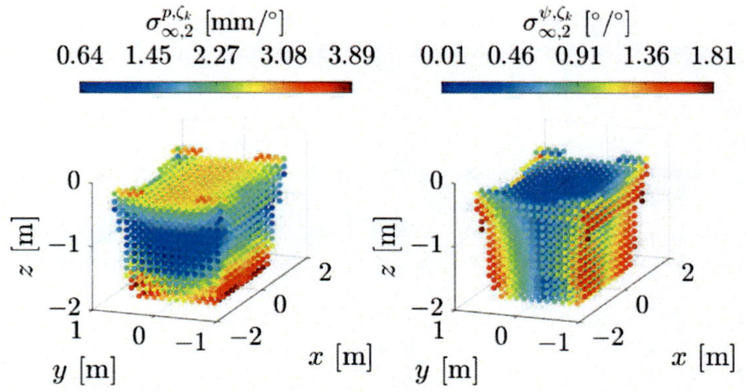

Fig. 3. Point-displacement and yaw sensitivities to errors in roll-and-pitch measurements

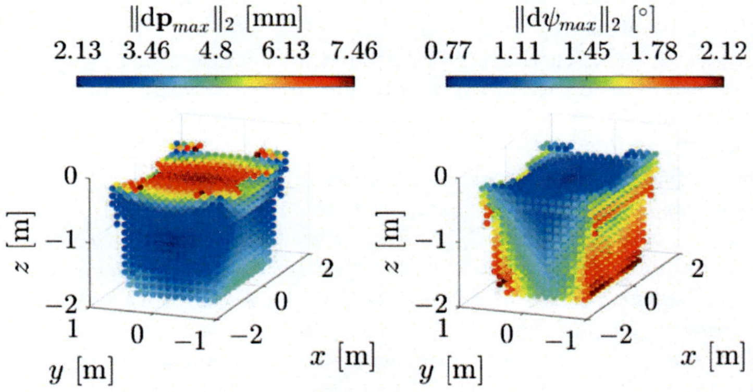

Fig. 4. Point-displacement and yaw accuracies for $\mathrm{d}l_M = 1$ mm and $\mathrm{d}\zeta_{k,M} = 0.4°$

5 Conclusions

In this paper, the point-displacement and yaw sensitivities to cable-length and roll-and-pitch errors were investigated. The application of sensitivity indices to an exemplary 4-cable robot showed that the yaw angle is limitedly influenced by errors, whereas the end-effector position is more affected. In the future, sensitivity indices of pose variables to extra measurements provided by different types of sensors, such as encoders on the swivel axes of pulleys, will be investigated.

References

1. Idà, E., Briot, S., Carricato, M.: Natural oscillations of underactuated cable-driven parallel robots. IEEE Access **9**, 71660–71672 (2021)
2. Idà, E., Bruckmann, T., Carricato, M.: Rest-to-rest trajectory planning for underactuated cable-driven parallel robots. IEEE Trans. Rob. **35**(6), 1338–1351 (2019)

3. Carricato, M.: Direct geometrico-static problem of underconstrained cable-driven parallel robots with three cables. ASME. J. Mechanisms and Robotics **5**(3), 031008 (2013)
4. Carricato, M., Abbasnejad, G.: Direct geometrico-static analysis of under-constrained cable-driven parallel robots with 4 cables. In: Bruckmann, T., Pott, A. (eds.) Cable-Driven Parallel Robots, pp. 269–285, Springer (2013), https://doi.org/10.1007/978-3-642-31988-4_17
5. Abbasnejad, G., Carricato, M.: Direct geometrico-static problem of undercon-strained cable-driven parallel robots with *n* cables. IEEE Trans. Rob. **31**(2), 468–478 (2015)
6. Merlet, J.-P.: Direct kinematics of CDPR with extra cable orientation sensors: the 2 and 3 cables case with perfect measurement and ideal or elastic cables. In: Gosselin, C., Cardou, P., Bruckmann, T., Pott, A. (eds.) Cable-Driven Parallel Robots. MMS, vol. 53, pp. 180–191. Springer, Cham (2018). https://doi.org/10.1007/978-3-319-61431-1_16
7. Merlet, J.P.: An experimental investigation of extra measurements for solving the direct kinematics of cable-driven parallel robots. In: 2018 IEEE Int. Conf. on Robotics and Automation, pp. 6947–6952 (2018)
8. Idá, E., Merlet, J.-P., Carricato, M.: Automatic self-calibration of suspended under-actuated cable-driven parallel robot using incremental measurements. In: CableCon 2019. MMS, vol. 74, pp. 333–344. Springer, Cham (2019). https://doi.org/10.1007/978-3-030-20751-9_28
9. Cardou, P., Bouchard, S., Gosselin, C.: Kinematic-sensitivity indices for dimension-ally nonhomogeneous jacobian matrices. IEEE Trans. Rob. **26**(1), 166–173 (2010)
10. Mattioni, V., Idà, E., Carricato, M.: Force-distribution sensitivity to cable-tension errors: a preliminary investigation. In: Gouttefarde, M., Bruckmann, T., Pott, A. (eds.) CableCon 2021. MMS, vol. 104, pp. 129–141. Springer, Cham (2021). https://doi.org/10.1007/978-3-030-75789-2_11
11. Idà, E., Carricato, M.: A new performance index for underactuated cable-driven parallel robots. In: Gouttefarde, M., Bruckmann, T., Pott, A. (eds.) CableCon 2021. MMS, vol. 104, pp. 24–36. Springer, Cham (2021). https://doi.org/10.1007/978-3-030-75789-2_3
12. Mottola, G., Gosselin, C., Carricato, M.: Effect of actuation errors on a purely-translational spatial cable-driven parallel robot. In: 9th Int. Conf. on CYBER Technology in Automation, Control, and Intelligent Systems, pp. 701–707 (2019)
13. Mecheri, H., Robert-Lachaine, X., Larue, C., Plamondon, A.: Evaluation of eight methods for aligning orientation of two coordinate systems. ASME. J. Biomech. Eng. **138**(8), 084501 (2016)
14. Kumar, A.A., Antoine, J.-F., Zattarin, P., Abba, G.: Workspace analysis of a 4 cable-driven spatial parallel robot. In: Arakelian, V., Wenger, P. (eds.) ROMANSY 22 – Robot Design, Dynamics and Control. CICMS, vol. 584, pp. 204–212. Springer, Cham (2019). https://doi.org/10.1007/978-3-319-78963-7_27
15. Johnston, N., QUETLAB: A MATLAB toolbox for quantum entanglement, version 0.9 (2016). http://qetlab.com

Stacked Tensegrity Mechanism for Medical Application

Dhruva Khanzode[1,2], Ranjan Jha[1,2], Emilie Duchalais[4], and Damien Chablat[3](✉)

[1] BioMedical Application Division, CSIR-Central Scientific Instruments Organisation, Chandigarh, India
{khanzode.dhruva,ranjan.jha}@csio.res.in
[2] Academy of Scientific and Innovative Research (AcSIR), Ghaziabad, Uttar Pradesh 201002, India
[3] Laboratoire des Sciences du Numérique de Nantes, UMR CNRS 6004, Nantes, France
damien.chablat@cnrs.fr
[4] CHU Nantes-Centre Hospitalier Universitaire de Nantes, Nantes, France
emilie.dassonneville@chu-nantes.fr

Abstract. In this article a multi-segmented planar tensegrity mechanism was presented. This mechanism has a three-segment structure with each segment residing on top of another. The size of the segments may decrease proportionally from base to top, resulting in a tapered shape from base to tip like an elephant trunk. The system was mechanically formulated as having linear springs and cables functioning as actuators. The singularities, as well as the stability of the parallel mechanism, were analyzed by using the principle of minimum energy. Optimization was also done to obtain the greatest angular deflection for a segment according to a ratio between the size of the base and the moving platform of the robotic system. The result of this work is a family of mechanisms that can generate the same workspace for different stability properties.

1 Introduction

Robotics is one of the most rapidly emerging branches in the world of technology. From heavy-duty applications to the most sensitive microscopic application, robotic systems are present everywhere. Therefore, with the rise of applications of robotic systems, grew the demand for facilitating critical tasks and requirements of high dexterity applications. To fulfill these demands, a separate branch of robotics emerged coined as flexible robotic. A field in which manipulators and actuators of hyper redundant mechanisms were developed. Hyper redundant systems are robotic systems having a very high number of degree of freedom [1]. These flexible robotic systems are very versatile in their applications. These flexible robotic systems are generally employed at a place with convoluted

O. Altuzarra and A. Kecskémethy (Eds.): ARK 2022, SPAR 24, pp. 139–148, 2022.
https://doi.org/10.1007/978-3-031-08140-8_16

workspace and restricted maneuverability [2]. Such kind of flexible robotic systems can be abundantly seen as probing and inspection drones, surgical robots, military UGVs, etc. [3,4]. The flexible robotic systems also called snake/worm-like robotic systems can be divided into two groups, based on their mechanical structure, discrete systems, and continuous systems. Discrete systems have finite discrete rigid elements joint together [5] while the continuous systems have soft or compliant materials used in the robotic system [6]. But in recent times, the line between these two categories has been blurred as newer designs being developed have both of the features. Another classification of hyper redundant robotic systems can be made based on how the robot is used. It can be classified either as a manipulator or a locomotor. If a robotic system is attached to a fixed base, then it can be called a manipulator [7], but if the system can move freely (such as a work or a snake), then it is to be called a locomotor [8].

The flexible robotic systems are generally tendon actuated, pneumatically, or hydraulically actuated or are based on the shape memory feature of the material. Tendon-based mechanisms are one of the most used mechanisms for flexible manipulators. The manipulators are actuated by pushing or pulling a tendon to facilitate bending or curvature [9]. These tendons-driven mechanisms are sometimes also referred to as tensegrity mechanisms because of the design principle wherein compression and expansion of lengths of opposite tendons takes place simultaneously [10,11]. A study was conducted to analyze and compare the kinematic properties of discrete tendon-driven flexible robots, continuous tendon-driven flexible robots, and concentric tube manipulators [12]. The results of this study indicated that in the case of dexterity, the discrete system was the finest followed by continuous and concentric tube type manipulators. From a scaling point of view, concentric tube manipulators are better for miniaturization whereas discrete tendon-driven systems were more suitable for scaling up [13].

This flexible robotic system has found some very interesting applications in the field of medical sciences [14]. Even before the conceptualization of robotic surgeries, endoscopes were used to investigate the insides of the human gastrointestinal tract, and it was done manually [15,16]. With the medical science turning into minimally invasive surgeries (MIS), several surgical tools were required to be redesigned for MIS such as clippers, scalpels, cauterizers, catheters, etc. [17,18]. One such essential surgical tool is the surgical stapler [19], which is generally used to avoid any spillage of intestine content onto the wounds of the patient and prevent any infection during the surgery. Commonly five kinds of staplers are used, namely TA, Thoracic-Abdominal; GIA, Gastro-Intestinal Anastomosis; Endo GIA, Endoscopic Gastro-Intestinal Anastomosis; EEA, End-End Anastomosis and Skin Stapler. The staplers are generally equipped with a separating knife to cut the tissues after the stapling is done. Endo GIA staplers are most commonly used for MIS. The Endo GIA staplers are available in three forms: passive articulated wrist type (PAW) [20], active articulated wrist type (AAW) [21], and radial reload type staplers (RR). The RR has a fixed "U" shaped jaw which requires a huge incision into the abdomen for insertion. Therefore there is a dire need for a surgical stapler that can enter through the laparoscopic openings but can work as an RR-type stapler inside the body. Our study aims to design a stapler for laparoscopic rectal cancer surgery where conventional tools cannot be easily accessed. Two mechanisms will have to be positioned on each side of the colon to insert two lines of staples and then carry out its cutting.

The article is organized as follows. In Sect. 2, the kinematics of the studied mechanism is presented. Section 3 reports the different computational results to study the singularities associated with the mechanism, then in Sect. 4, the study of stability based on the principle of minimum energy, and finally, in Sect. 5, an optimization of the dimensions of the mechanism is presented.

2 Mechanism Design

In this article, the robotic system designed is a multi-segment planar robotic system. As depicted in Fig. 1, the robotic system comprises of 3 trapezoidal segments A, B, and C stacked one above the other. Each segment may have a larger base plate and a smaller moving plate. Both the plates are connected by a central serial spine linkage, comprising of 3 links. The system is only composed of 3 sections to facilitate the position control of the end effector. Since the system is being controlled by only 2 tendons, hence it is under-actuated. Therefore, not more than 3 segments can be controlled effectively.

For example, in section A, the central spinal linkages are A_0B_0, B_0C_0 and C_0D_0. The link A_0B_0 and C_0D_0 are rigidly fixed and are perpendicular to their respective plate A_1A_2 and D_1D_2. These central spinal links are connected by two revolute joints with identical rotation angles. To achieve this coupling, the joints can have a sliding surface, similar to a knee joint in humans, or can also have an X-shape tensegrity module [10,11], or a gear train to couple the movement of the two revolute joints.

The base plate and the moving plates are also attached with the help of two cables ρ_1 and ρ_2, present on either side of the central spine and two springs of stiffness k_1 and k_2 between (A_1 B_1) and (A_2 B_2), respectively.

The contraction of the cables will stimulate angular displacement in the revolute joints of the central spine and replication of this phenomenon in each of the three segments will eventually facilitate the bending of the robotic structure. The length of the cables is measured as ρ_i and the angular deviation of the revolute joints is measured as α_i where

$$\alpha_1 = \alpha_2, \quad \alpha_3 = \alpha_4, \quad \alpha_5 = \alpha_6, \tag{1}$$

because of the coupling inside each segment.

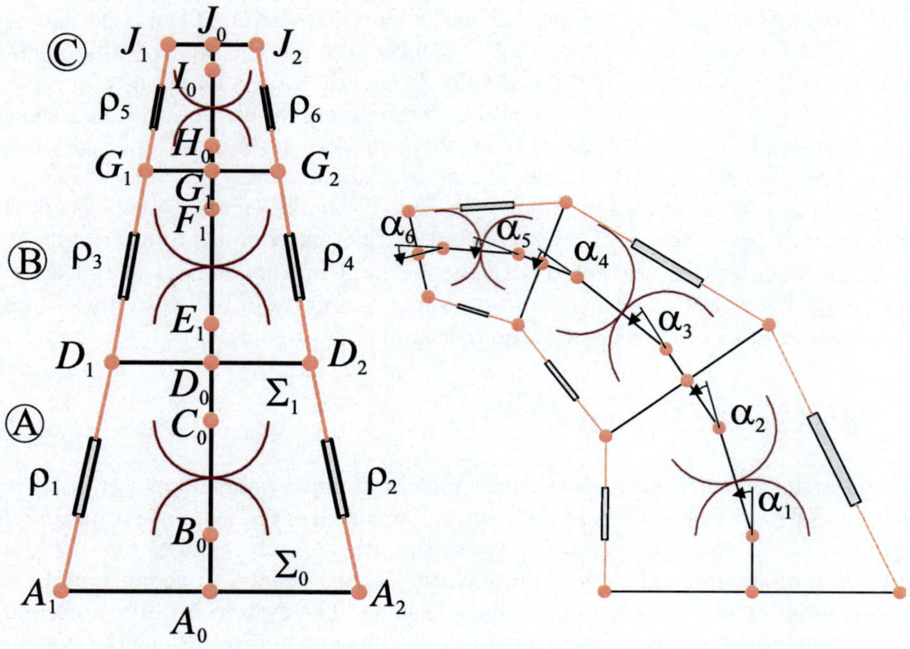

Fig. 1. The tensegrity mechanism understudy with three segments stacked, named A, B, C for $\alpha_i = 0$ (home pose) in the left and $\alpha_i = 0.3$ in the right

The fixed coordinate frame of the base is represented by Σ_0 with the origin at A_0. The distances between the coordinates of each points are $||\mathbf{a_1} - \mathbf{a_0}|| = \mathbf{l_1}$, $||\mathbf{a_2} - \mathbf{a_0}|| = \mathbf{l_1}$, $||\mathbf{b_1} - \mathbf{a_0}|| = \mathbf{h_1}$, $||\mathbf{c_0} - \mathbf{b_0}|| = \mathbf{h_2}$, $||\mathbf{d_0} - \mathbf{c_0}|| = \mathbf{h_3}$, $||\mathbf{d_1} - \mathbf{d_0}|| = \mathbf{l_2}$ and $||\mathbf{d_2} - \mathbf{d_0}|| = \mathbf{l_2}$. The vector co-ordinates for the base mounting points are given by

$$\mathbf{a}_1 = [-l_1, 0] \quad \mathbf{a}_2 = [l_1, 0] \tag{2}$$

The kinematic chain between the base and the platform of the first segment is defined by the following points

$$\mathbf{b}_0 = \begin{bmatrix} 0 \\ h_1 \end{bmatrix}, \mathbf{c}_0 = \begin{bmatrix} -h_2 \sin(\alpha_1) \\ h_1 + h_2 \cos(\alpha_1) \end{bmatrix}, \mathbf{d}_0 = \begin{bmatrix} -h_2 \sin(\alpha_1) - h_3 \sin(2\alpha_1) \\ h_1 + h_2 \cos(\alpha_1) + h_3 \cos(2\alpha_1) \end{bmatrix} \tag{3}$$

The moving co-ordinate frame of the first segment is represented by Σ_1 with its origin at D_0. The spring mounting points are represented by D_1 and D_2.

$$\mathbf{d}_1 = \begin{bmatrix} (-2h_3 \cos(\alpha_1) - h_2) \sin(\alpha_1) - 2l_2 \left(\cos^2(\alpha_1)\right) + l_2 \\ 2h_3 \left(\cos^2(\alpha_1)\right) + (-2l_2 \sin(\alpha_1) + h_2) \cos(\alpha_1) + h_1 - h_3 \end{bmatrix} \tag{4}$$

$$\mathbf{d}_2 = \begin{bmatrix} (-2h_3 \cos(\alpha_1) - h_2) \sin(\alpha_1) + 2l_2 \left(\cos^2(\alpha_1)\right) - l_2 \\ 2h_3 \left(\cos^2(\alpha_1)\right) + (2l_2 \sin(\alpha_1) + h_2) \cos(\alpha_1) + h_1 - h_3 \end{bmatrix} \tag{5}$$

The inverse kinematic model for a segment is used to determine the length of the springs and the cables between the base and the moving platform and the moving platform. The equations are given by

$$||\mathbf{a}_1 - \mathbf{d}_1|| = \rho_1, \quad ||\mathbf{a}_2 - \mathbf{d}_2|| = \rho_2 \tag{6}$$

These two equations can also be written as,

$$\begin{aligned}&\left((-2h_3\cos(\alpha_1) - h_2)\sin(\alpha_1) - 2l_2\left(\cos^2(\alpha_1)\right) + l_2 + l_1\right)^2\\&+\left(2h_3\left(\cos^2(\alpha_1)\right) + (-2l_2\sin(\alpha_1) + h_2)\cos(\alpha_1) + h_1 - h_3\right)^2 = \rho_1^2\end{aligned} \tag{7}$$

$$\begin{aligned}&\left((-2h_3\cos(\alpha_1) - h_2)\sin(\alpha_1) + 2l_2\left(\cos^2(\alpha_1)\right) - l_2 - l_1\right)^2\\&+\left(2h_3\left(\cos^2(\alpha_1)\right) + (2l_2\sin(\alpha_1) + h_2)\cos(\alpha_1) + h_1 - h_3\right)^2 = \rho_2^2\end{aligned} \tag{8}$$

3 Singularity Analysis

For determining the singularities of the multi-segment planar mechanism, only section A was taken into consideration. As all other sections are similarly constructed, analysis of singularities in one section will lead to similar results in all other sections, and sections B and C will also have similar singularities.

Two closed-loop mechanisms can be described by (A_1, A_0, B_0, C_0, D_0, D_1) and (A_2, A_0, B_0, C_0, D_0, D_2). For the first closed-loop mechanism, the singular configurations with α_1 as Cartesian values and ρ_1 as the input values are computed by differentiating with respect to time the Eq. 7, as follow,

$$\begin{aligned}&-8h_3^2\cos^3(\alpha_1)\sin(\alpha_1) + 8h_3^2\cos(\alpha_1)\sin(\alpha_1) - 8l_2^2\cos^3(\alpha_1)\sin(\alpha_1)\\&+8l_2^2\cos(\alpha_1)\sin(\alpha_1) - 4h_3\sin(\alpha_1)\cos^2(\alpha_1)\,h_2 - 4h_3\sin^3(\alpha_1)\,h_2\\&-4h_3\cos^2(\alpha_1)\,l_1 + 4h_3\sin^2(\alpha_1)\,l_1 - 4l_2\cos^2(\alpha_1)\,h_1 + 4l_2\sin^2(\alpha_1)\,h_1\\&-8h_3^2\sin^3(\alpha_1)\cos(\alpha_1) - 2h_2\cos(\alpha_1)\,l_2 - 2h_2\cos(\alpha_1)\,l_1 + 8l_2\cos(\alpha_1)\,l_1\sin(\alpha_1)\\&-8l_2^2\sin^3(\alpha_1)\cos(\alpha_1) - 8h_3\cos(\alpha_1)\,h_1\sin(\alpha_1) - 2h_2\sin(\alpha_1)\,h_1\\&+2h_2\sin(\alpha_1)\,h_3 = 0\end{aligned} \tag{9}$$

The singularities are the roots of a 4th-degree equation. When $l_1 \neq l_2$, only a numerical method allows us to calculate them. In our case, we use the "RootFinding:-Isolate" function of Maple which computes all the roots after a substitution by the half-angle of α_1 to obtain an algebraic equation [22].

For the second closed-loop, the singularity locus is for opposite values of α_1. For each closed loop, there can be up to four singular positions. The absolute value of the smallest angle α_i is called α_{sing} and it represents the largest travel that the mechanism can achieve. This value must be maximized.

As an example, for $h_1 = 1$, $h_2 = 1$, $h_3 = 1$, $l_1 = 1$, $l_2 = 1$, the singularity locus are

$$\alpha_1 = -\frac{\pi}{4}, \alpha_1 = \frac{3\pi}{4}, \alpha_1 = \arctan\left(\frac{\frac{1}{4} + \frac{\sqrt{7}}{4}}{-\frac{1}{4} + \frac{\sqrt{7}}{4}}\right), \alpha_1 = \arctan\left(\frac{\frac{1}{4} - \frac{\sqrt{7}}{4}}{-\frac{1}{4} - \frac{\sqrt{7}}{4}}\right) - \pi \tag{10}$$

Only two singular configurations are close to the home pose as shown in Fig. 2. The smallest absolute value of α_1 defines the range of motion of the segment in both directions.

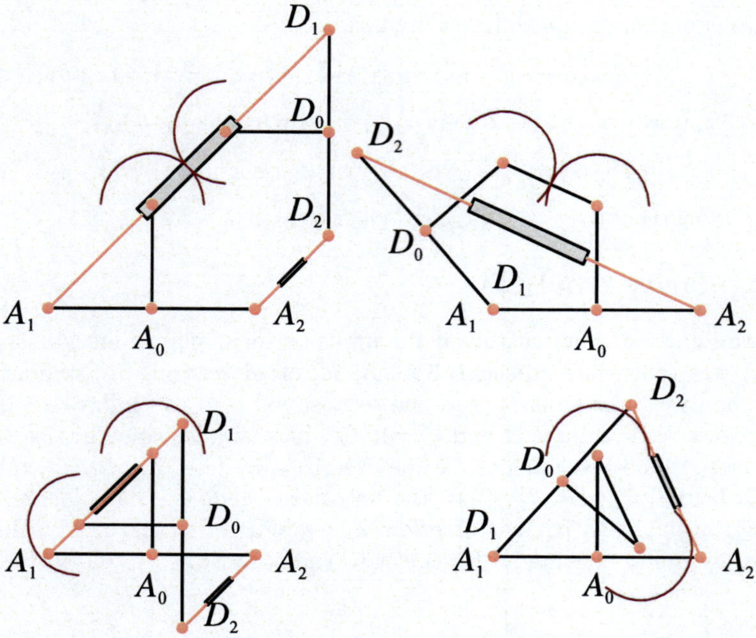

Fig. 2. Four singular configurations for $h_1 = 1, h_2 = 1, h_3 = 1, l_1 = 1$ and $l_2 = 1$

4 Stability Analysis

The stability of the robotic system is one of the major components in its performance evaluation [23]. Generally, when a system is in equilibrium and is displaced by an external force if the system comes back to its equilibrium state, it is called a stable equilibrium. If the system does not return to equilibrium, it is considered an unstable system and if the system does not affect itself, then the system is known as a neutral system [24]. A robotic system is always desired to be a stable system in its operational capacity. If a robotic system becomes unstable at any point in its trajectory, then it may cause substantial errors during its control [25]. For this multi-segment planar robotic system, the stability is evaluated by using the principle of minimum internal energy.

For section A, two springs are connected between A_1D_1 and A_2D_2 with a no-load length equal to l_0. This length is chosen as 40% of the length in the home pose. This value will be proportional to the size of the second and third segments if we want to characterize the stability of the complete mechanism.

For segment A, the energy E for a given α inside the mechanism is

$$E = \frac{1}{2}\left(k_1(\rho_1 - l_0)^2 + k_2(\rho_2 - l_0)^2\right) \tag{11}$$

To compare two mechanisms, we introduce the total energy as $E_t = \int_{-\alpha_{sing}}^{\alpha_{sing}} E$.

Depending on the sizes of the robot, several stability schemes can appear either stable around $\alpha = 0$ up to the singularities (Fig. 3(left)) or unstable in the position for $\alpha = 0$ (Fig. 3(right)).

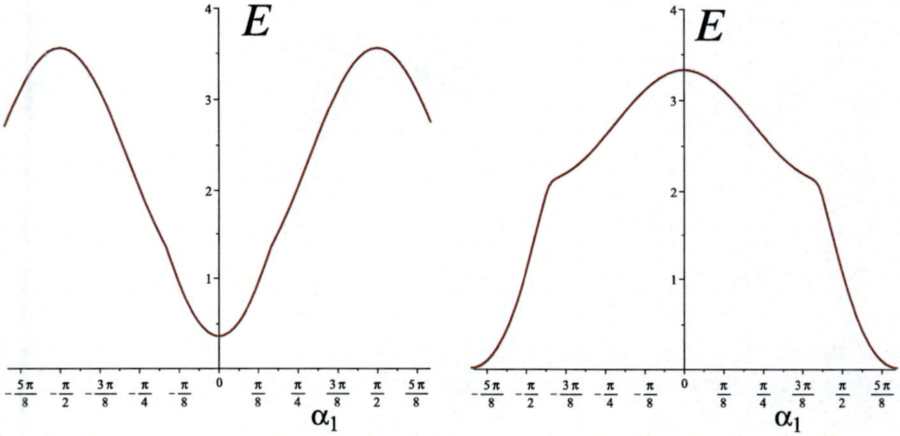

Fig. 3. Variation of the energy as a function of α_1 for a stable home pose for $h_1 = 0$, $h_2 = 1$, $h_3 = 0$, $l_1 = 1$, $l_2 = 1$, $k = 1$ (left) and unstable home pose for $h_1 = 1$, $h_2 = 1$, $h_3 = 1$, $l_1 = 1$, $l_2 = \frac{1}{2}$, $k = 1$

5 Optimization of One Segment

The objective of our optimization is to find the design parameters that allow the greatest angular deflection and then to study the stability of the solutions found. A ratio λ between the base and the platform is also studied to see if elephant trunk structures, with sections of decreasing sizes, are relevant [26].

So, the objective function is defined as

$$f(\mathbf{x}) = \alpha_{sing} \rightarrow \max \tag{12}$$

where $\mathbf{x}$ is the set of the design parameters $[h_1, h_2, h_3, l_1, \lambda]$. To simplify the optimization, we set $h_3 = h_1$, $l_2 = \lambda l_1$ and we defined the following constraints:

$$0 < l_1 < 4.5 \quad 0 \leq h_1 \leq 1 \quad 0 \leq h_2 \leq 2 \quad 1/20 \leq \lambda \leq 1 \tag{13}$$

The solution to this optimization problem can be achieved by several methods. As the dimension of the problem is small, a discretization of the parameter space has been performed to find the optimal solutions.

The results of the optimization show that the maximum value of α is $\pi/2$ and this for any value of λ and for $l_2 = 2$, i.e. the maximum bound, and for $h_1 = 0$ for the minimum bound. Figure 4 shows the evolution of l_1 and l_2 as a function of λ.

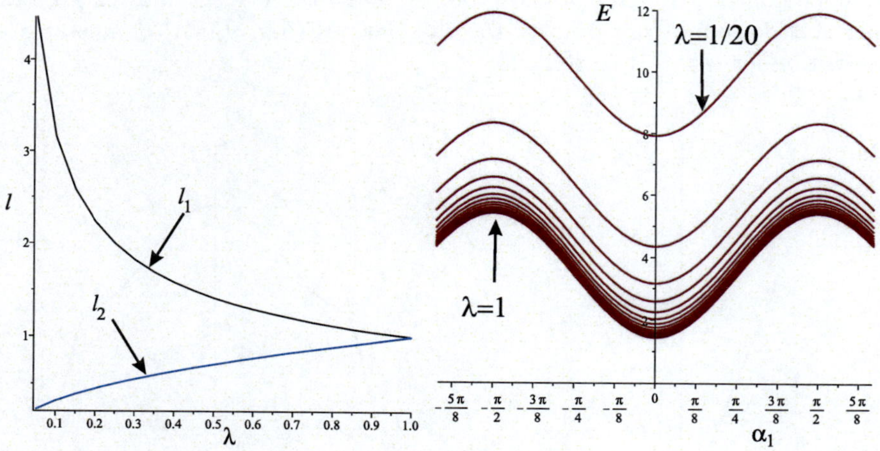

Fig. 4. Variation of l_1 in black and l_2 in blue as a function of λ (left) and the stability analysis of the optimal solutions as a function of λ

For all solutions, the energy has its maximum value for α_{sing} and has a local minimum for $\alpha = 0$. If we observe the value of $E(\alpha = 0)$ and $E(\alpha = \pi/2)$, the difference is always the same. The total energy decreases when λ tends to 1. This means that the actuation forces will be smaller when $\lambda = 1$. Conversely, the smaller λ is, the greater the external forces must be to move the mechanism from its home position. Moreover, to have a regular stack of modules, module C is smaller than module B which is smaller than module A if λ is different from 1.

6 Conclusions and Future Work

To summarise, in this article a multi-segmented planar tensegrity mechanism was designed consisting of a stack of three tensegrity mechanisms. The mechanism was mechanically formulated as having linear springs and cables for actuation. The singularities of the parallel mechanism were analyzed as well as the stability by using the principle of minimum energy. Optimization was done to obtain the greatest angular deflection according to a ratio between the size of the base and the platform.

We observe that the angle α can vary between $-\pi/2$ and $\pi/2$, that all optimal solutions are stable up to the singularity, and that when $\lambda = 1$, the total energy in a module is minimal.

Further research is in progress to control the three modules with only two cables going from the base to the third segment and to know the stability of the mechanism against external forces. We also need to estimate the radii of curvature required to encompass the patients' colon from CT scan studies and encapsulate the mechanism with an elastomer with the clamps and knife to section the colon.

References

1. Yeshmukhametov, A., Koganezawa, K., Yamamoto, Y.: A novel discrete wire-driven continuum robot arm with passive sliding disc: design, kinematics and passive tension control. Robotics **8**(3), 1–18 (2019). https://doi.org/10.3390/ROBOTICS8030051
2. Degani, A., Choset, H., Zubiate, B., Ota, T., Zenati, M.: Highly articulated robotic probe for minimally invasive surgery. In: Proceeding of 30th Annual International Conference of the IEEE Engineering in Medicine and Biology Society EMBS 2008 - Personalized Healthcare through Technology, pp. 3273–3276, May (2008). https://doi.org/10.1109/iembs.2008.4649903
3. Wang, H.C., Cui, S.H., Wang, Y., Song, C.L.: A hybrid electromagnetic and tendon-driven actuator for minimally invasive surgery. Actuators **9**(3), 92 (2020). https://doi.org/10.3390/act9030092
4. Hwang, M., Kwon, D.S.: K-FLEX: a flexible robotic platform for scar-free endoscopic surgery. Int. J. Med. Robot. Comput. Assist. Surg. **16**(2), e2078 (2020). https://doi.org/10.1002/rcs.2078
5. Yeshmukhametov, A., Koganezawa, K., Yamamoto, Y.: Design and kinematics of cable-driven continuum robot Arm with universal joint backbone. In: 2018 IEEE International Conference on Robotics and Biomimetics, ROBIO 2018, pp. 2444–2449 (2018). https://doi.org/10.1109/ROBIO.2018.8665186
6. Suzumori, K., Iikura, S., Tanaka, H.: Development of flexible microactuator and its applications to robotic mechanisms. In: Proceedings of International Conference on Robotics and Automation, vol. 2, pp. 1622–1627, April 1991. https://doi.org/10.1109/robot.1991.131850
7. Xu, K., Zhao, J., Member, S., Fu, M.: Development of the SJTU Unfoldable Robotic System (SURS) for single port laparoscopy. IEEE/ASME Trans. Mechatronics **20**(5), 2133–2145 (2014)
8. Whitman, J., Zevallos, N., Travers, M., Choset, H.: Snake robot urban search after the 2017 Mexico City earthquake. In: 2018 International Symposium on Safety, Security and Rescue Robotics, SSRR 2018, pp. 7–12 (2018). https://doi.org/10.1109/SSRR.2018.8468633
9. Li, Z., Feiling, J., Ren, H., Yu, H.: A novel tele-operated flexible robot targeted for minimally invasive robotic surgery. Engineering **1**(1), 073–078 (2015). https://doi.org/10.15302/J-ENG-2015011
10. Furet, M., Lettl, M., Wenger, P.: Kinematic analysis of planar tensegrity 2-X manipulators. In: Lenarcic, J., Parenti-Castelli, V. (eds.) ARK 2018. SPAR, vol. 8, pp. 153–160. Springer, Cham (2019). https://doi.org/10.1007/978-3-319-93188-3_18
11. Wenger, P., Chablat, D.: Kinetostatic analysis and solution classification of a class of planar tensegrity mechanisms. Robotica **37**(7), 1214–1224 (2019). https://doi.org/10.1017/S026357471800070X

12. Li, Z., Wu, L., Ren, H., Yu, H.: Kinematic comparison of surgical tendon-driven manipulators and concentric tube manipulators. In: Mechanism and Machine Theory, vol. 107, pp. 148–165, June 2016 (2017). https://doi.org/10.1016/j.mechmachtheory.2016.09.018
13. Le, H.M., Do, T.N., Cao, L., Phee, S.J.: Towards active variable stiffness manipulators for surgical robots. In: Proceeding of International Conference on Robotics and Automation, pp. 1766–1771 (2017). https://doi.org/10.1109/ICRA.2017.7989209
14. Shang, J., et al.: A single-port robotic system for Transanal microsurgery-design and validation. IEEE Robot. Autom. Lett. **2**(3), 1510–1517 (2017). https://doi.org/10.1109/LRA.2017.2668461
15. Haber, G.P., et al.: SPIDER surgical system for urologic procedures with laparoendoscopic single-site surgery: from initial laboratory experience to first clinical application. Eur. Urol. **61**(2), 415–422 (2012). https://doi.org/10.1016/j.eururo.2010.12.033
16. Abbott, D.J., Becke, C., Rothstein, R.I., Peine, W.J.: Design of an endoluminal NOTES robotic system. In: EEE International Workshop on Intelligent Robots and Systems, pp. 410–416 (2007). https://doi.org/10.1109/IROS.2007.4399536
17. Roh, S.G. et al.: Development of the SAIT single-port surgical access robot slave arm based on RCM mechanism. In: Proceeding of Annual International Conference of the IEEE Engineering in Medicine and Biology Society, EMBS, vol. 2015, pp. 5285–5290, November 2015. https://doi.org/10.1109/EMBC.2015.7319584
18. Hu, X., Chen, A., Luo, Y., Zhang, C., Zhang, E.: Steerable catheters for minimally invasive surgery: a review and future directions. Comput. Assist. Surg. **23**(1), 21–41 (2018). https://doi.org/10.1080/24699322.2018.1526972
19. Akopov, A., Artioukh, D.Y., Molnar, T.F.: Surgical staplers: the history of conception and adoption. Ann. Thorac. Surg. **112**, 1716–1721 (2021)
20. Bolanos, H., Norwalk, E., Sherts, C.R., Southport, A.T. Pel- letier, Wallingford.: ENDOSCOPICSTAPLER, US Patent 690269 (1997)
21. Milliman, L., et al.: Surgical Stapling Apparatus, US Patent 586531 (1999)
22. Rouillier, F.: Solving zero-dimensional systems through the rational univariate representation. Appl. Algebra Eng. Commun. Comput. **9**(5), 433–461 (1999)
23. Behzadipour, S., Khajepour, A.: Stiffness of cable-based parallel manipulators with application to stability analysis. J. Mech. Des. Trans. ASME **128**(1), 303–310 (2006). https://doi.org/10.1115/1.2114890
24. Chadefaux, T.: The Triggers of war: disentangling the spark from the powder keg. SSRN Electron. J. 49 pages April 2014. https://doi.org/10.2139/ssrn.2409005
25. Zhao, W., Pashkevich, A., Klimchik, A., Chablat, D.: Elastostatic modeling of multi-link flexible manipulator based on two-dimensional dual-triangle tensegrity mechanism. J. Mech. Robot. **14**(2), 1–31 (2022). https://doi.org/10.1115/1.4051789
26. Yang, J., Pitarch, E.P., Potratz, J., Beck, S., Abdel-Malek, K.: Synthesis and analysis of a flexible elephant trunk robot. Adv. Robot. **20**(6), 631–659 (2006). https://doi.org/10.1163/156855306777361631

Kinematic and Static Analysis of a Cable-Driven 2-X Tensegrity Manipulator for Two Actuation Strategies

Vimalesh Muralidharan(✉), Philippe Wenger, and Christine Chevallereau

Laboratoire des Sciences du Numérique de Nantes (LS2N), CNRS, 44321 Nantes, France
{Vimalesh.Muralidharan,Philippe.Wenger,Christine.Chevallereau}@ls2n.fr

Abstract. This paper analyzes the instantaneous kinematic and static performance of a two degree-of-freedom serial manipulator composed of anti-parallelogram (or X-) tensegrity joints and remotely driven with cables. Two actuation schemes with 4 cables and 3 cables, respectively, are considered for this manipulator. The physical limitations on the velocities and forces of the actuating cables are mapped onto the task space of the manipulator, to quantify the corresponding velocity and force-application capabilities of the end-effector. A comparative study is carried out between the 4-cable and 3-cable actuation schemes in terms of their velocity and force performance.

Keywords: Tensegrity manipulator · Anti-parallelogram joint · Velocity kinematics · Static analysis

1 Introduction

Cable-driven parallel robots have gained immense popularity over the past two decades, due to their large workspaces, high payload to weight ratio, and fast dynamics. A vast fraction of literature on these robots has been directed towards the study of their wrench-feasibility [1,2]. The twist feasibility analysis of such robots has been conducted in [3]. Both wrench and twist capabilities of these robots are usually represented by convex polytopes [2,3]. Two performance measures, namely, manipulability, and condition index, have been extended from ellipsoids to polytopes in [4]. These measures will be used in this paper.

Unlike cable-driven parallel robots, the literature dedicated to cable-driven serial robots [5] is relatively sparse. A difficult challenge that exists in these robots is to find a suitable cable routing scheme [6]. The force-closure analysis of cable-driven open chains has been performed using reciprocal screw theory in [7]. Most of these papers consider robots composed of simple revolute joints only. As an alternative, it is possible to use an anti-parallelogram joint (referred as X-joint) [8]. The X-joint has been used for modeling bio-mechanical systems such as the knee [9], and inter-vertebrae joints of the bird neck [10]. The kinematic

O. Altuzarra and A. Kecskémthy (Eds.): ARK 2022, SPAR 24, pp. 149–159, 2022.
https://doi.org/10.1007/978-3-031-08140-8_17

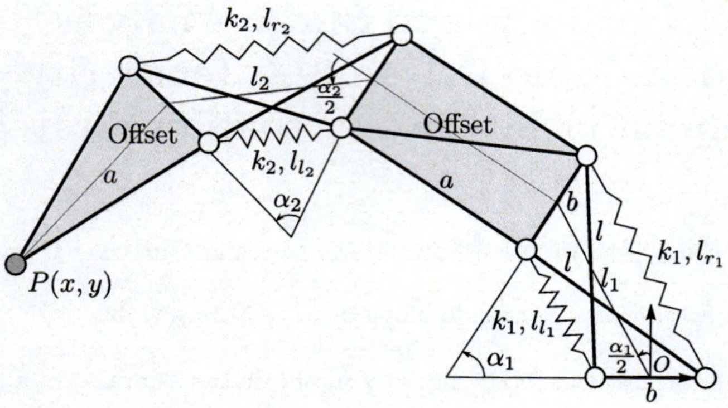

Fig. 1. Schematic of the 2-X tensegrity manipulator with link offsets, $\alpha_i \in]-\pi, \pi[$.

analysis of a 2-X manipulator with link offsets has been performed in [11], where it was shown that this manipulator has a larger workspace than a conventional 2-revolute manipulator with the same maximal reach and joint limits.

This paper studies a two degree-of-freedom (DoF) manipulator composed of two X-joints arranged in series with rigid offsets between them (see Fig. 1). Two different actuation strategies with 4 cables and 3 cables, respectively, are proposed for this manipulator. The 4-cable scheme is a direct extension of the antagonistic actuation of a single X-joint [12] to two of them, while the 3-cable scheme uses the minimum number of cables required to fully actuate this 2-DoF system [7]. One of the objectives of this study is to see the impact of reducing the number of actuators and therefore the cost of the associated robot. It might seem obvious that reducing the number of actuators will reduce the performance of the robot, but we will show that it is not so direct.

In the following, a complete description of the 2-X manipulator, 4-cable and 3-cable actuation schemes are presented. The performance of these two schemes is studied in terms of instantaneous velocity and force transmission. Finally, this performance is compared inside the respective workspaces of the two schemes.

2 Manipulator Description and Kinematic Relations

A schematic of the 2-X manipulator with two X-joints and link offsets is shown in Fig. 1. Each X-joint is composed of a base and top bar of length b and two crossed bars of length l, such that $(l > b)$ for its assembly. Two rigid offsets of length a are introduced between the two X-joints and between the second joint and the end-effector point $P(x, y)$, as shown in the schematic (shaded parts).

For the ith X-joint, the orientation of the top bar relative to its base bar, is given by α_i. This joint is equipped with identical springs of stiffness k_i (zero free-length) on the two sides, ensuring that it remains in equilibrium at $\alpha_i = 0$, in the absence of forces. Since the flat-singularities at $\alpha_i = \pm\pi$ limit the movement of this joint, the study of this manipulator is carried out only for $\alpha_i \in$

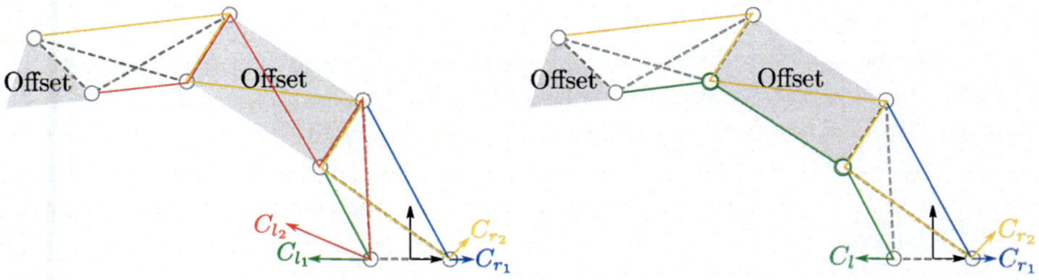

Fig. 2. Actuation scheme with 4 cables (left) and 3 cables (right).

$]-\pi, \pi[$, $i = 1, 2$. Within this range, it is possible to unambiguously write the direct kinematics of this manipulator as follows (see [11] for details):

$$\begin{cases} l_i(\alpha_i) = \sqrt{l^2 - b^2\cos^2(\alpha_i/2)}, \quad i = 1,2 \\ x = -l_1(\alpha_1)\sin(\alpha_1/2) - a(\sin\alpha_1 + \sin(\alpha_1+\alpha_2)) - l_2(\alpha_2)\sin(\alpha_1+\alpha_2/2) \\ y = l_1(\alpha_1)\cos(\alpha_1/2) + a(\cos\alpha_1 + \cos(\alpha_1+\alpha_2)) + l_2(\alpha_2)\cos(\alpha_1+\alpha_2/2) \end{cases} \tag{1}$$

Differentiating the above equations w.r.t. time yields :

$$\begin{bmatrix}\dot{x}\\ \dot{y}\end{bmatrix} = \boldsymbol{J}\begin{bmatrix}\dot{\alpha}_1\\ \dot{\alpha}_2\end{bmatrix}, \boldsymbol{J} = \begin{bmatrix}\frac{\partial x}{\partial \alpha_1} & \frac{\partial x}{\partial \alpha_2}\\ \frac{\partial y}{\partial \alpha_1} & \frac{\partial y}{\partial \alpha_2}\end{bmatrix} = \begin{bmatrix}J_{11} & J_{12}\\ J_{21} & J_{22}\end{bmatrix}, \text{is a Jacobian matrix.} \tag{2}$$

The side lengths l_{l_i}, l_{r_i} with $i = 1, 2$ (see Fig. 1) can be expressed as follows [10]:

$$\begin{cases} l_{l_i} = -b\sin(\alpha_i/2) + \sqrt{l^2 - b^2\cos^2(\alpha_i/2)} \implies \dot{l}_{l_i} = -Z_{l_i}\dot{\alpha}_i \\ l_{r_i} = b\sin(\alpha_i/2) + \sqrt{l^2 - b^2\cos^2(\alpha_i/2)} \implies \dot{l}_{r_i} = -Z_{r_i}\dot{\alpha}_i, \text{where,} \\ Z_{l_i} = -\frac{\mathrm{d}l_{l_i}}{\mathrm{d}\alpha_i} = b\cos(\alpha_i/2)\left(1 - \frac{b\sin(\alpha_i/2)}{\sqrt{l^2-b^2\cos^2(\alpha_i/2)}}\right) \\ Z_{r_i} = -\frac{\mathrm{d}l_{r_i}}{\mathrm{d}\alpha_i} = -b\cos(\alpha_i/2)\left(1 + \frac{b\sin(\alpha_i/2)}{\sqrt{l^2-b^2\cos^2(\alpha_i/2)}}\right) \end{cases} \tag{3}$$

Note that $Z_{l_i} > 0$ and $Z_{r_i} < 0$ when $l > b$ and $\alpha_i \in]-\pi, \pi[$. The numerical data that will be used in this paper for the purpose of illustration are: $b = 0.05$ m, $l = 0.1$ m, $a = 0.2$ m, $k_1 = 600$ N/m, $k_2 = 300$ N/m. All the bars are assumed to be made of Aluminum material with a circular cross-section of radius 0.005 m.

3 Cable Actuation Schemes

The 2-X manipulator will be remotely actuated by motors fixed in the base, using cables as transmission elements. Two such actuation schemes are considered in this study. The first one, referred as the 4-cable scheme, involves four actuators/cables (see Fig. 2, left), namely, $C_{l_1}, C_{r_1}, C_{l_2}, C_{r_2}$. The cables C_{l_1} and C_{r_1} actuate the first joint *antagonistically* without any effect on the second one.

The other two cables C_{l_2} and C_{r_2} are routed through the rigid links with pulleys, following the *strut-routed* scheme [8], to actuate the second joint independent of the first one. The advantage of this scheme is the *independence* in actuation of the two joints, but it uses one cable more than the minimum required [7].

The second actuation scheme, referred as the 3-cable scheme, uses only three actuators/cables (see Fig. 2, right), the minimum number required for this system [7]. Cables C_{r_1} and C_{r_2} have the same connections as in the 4-cable scheme. In contrast, the cable C_l is routed along the left side of both the joints in a *side-routed* scheme [10]. It is wound completely around two of the pulleys (at the top left of the first joint and bottom left of the second joint) before being connected to the top left vertex of the second X-joint. This way, the cable C_l can control the movement of both the joints, albeit in a dependent manner.

In this paper, the cables are assumed to be inextensible and the pulleys are approximated as massless points, i.e., with zero radius. The velocity and static force analysis for the two actuation schemes are presented in the following. The velocity analysis is useful for designing robots for manipulation tasks like palletizing, while the force analysis is relevant for machining tasks like drilling.

4 Instantaneous Velocity Analysis

In a cable-driven system, the peak speed of the cables and the maximum force transmitted by them, are limited by the actuators driving them. The aim of this section and the next one is to map these limitations onto the task space and evaluate the capabilities of the manipulator.

Let us assume that all the cables are actuated by identical motors, the one used in [10]. Using the data sheet of the specified motor and the drum characteristics, a value for maximum cable speed can be found as $\dot{l}_{\max} = 0.423$ m/s. From Fig. 2, it is apparent that the velocities of the cables C_{l_i}, C_{r_i}, when they are wound, are equal to the negative time-derivatives of l_{l_i}, l_{r_i}, respectively. Using Eq. (3), the associated constraints on the joint velocities can be obtained as:

$$\dot{C}_{l_i} \in \left[-\dot{l}_{\max}, \dot{l}_{\max}\right] \implies -\dot{l}_{l_i} \in \left[-\dot{l}_{\max}, \dot{l}_{\max}\right] \implies -\frac{\dot{l}_{\max}}{Z_{l_i}} \leq \dot{\alpha}_i \leq \frac{\dot{l}_{\max}}{Z_{l_i}} \tag{4}$$

$$\dot{C}_{r_i} \in \left[-\dot{l}_{\max}, \dot{l}_{\max}\right] \implies -\dot{l}_{r_i} \in \left[-\dot{l}_{\max}, \dot{l}_{\max}\right] \implies -\frac{\dot{l}_{\max}}{-Z_{r_i}} \leq \dot{\alpha}_i \leq \frac{\dot{l}_{\max}}{-Z_{r_i}} \tag{5}$$

for $i = 1, 2$. The velocity of cable C_l (see Fig. 2, right) is $-(\dot{l}_{l_1} + \dot{l}_{l_2})$, leading to:

$$\dot{C}_l \in \left[-\dot{l}_{\max}, \dot{l}_{\max}\right] \implies -\dot{l}_{\max} \leq (Z_{l_1}\dot{\alpha}_1 + Z_{l_2}\dot{\alpha}_2) \leq \dot{l}_{\max} \tag{6}$$

For the 4-cable actuation scheme, the constraints in Eqs. (4) and (5) apply simultaneously, resulting in a total of eight constraints. For the 3-cable scheme, there are only six constraints defined by Eqs. (5) and (6). As an illustration, these constraints are shown in Fig. 3a (resp. Fig. 3b) for the 4-cable (resp. 3-cable) scheme, when the manipulator is at $(\alpha_1, \alpha_2) = (-1.65, 1.18)$ radians. The

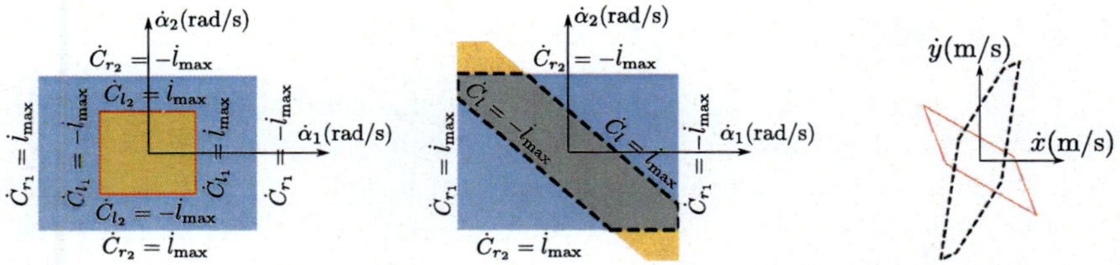

(a) Joint velocity(4-cable) (b) Joint velocity(3-cable) (c) Task velocity polytopes

Fig. 3. Joint velocity polytope and task velocity polytope for the 4-cable (red continuous) and 3-cable (black dashed) schemes at $(\alpha_1, \alpha_2) = (-1.65, -1.18)$ rad.

region respecting the constraints formed by the left cables (Eqs. (4), (6)) is shown in orange shade and by the right cables (Eq. (5)) in blue shade. The polytope that lies at the intersection of all the constraints represents the set of feasible joint velocities and is referred to as the joint velocity polytope. This polytope is highlighted by red continuous line for the 4-cable scheme and by black dashed line for the 3-cable scheme. This style is followed throughout the paper.

The corresponding polytope in the task velocity space $(\dot{x}, \dot{y})$ can be obtained by mapping the vertices of joint velocity polytope using the Jacobian matrix in Eq. (2). The task velocity polytopes for the two actuation schemes are superimposed in Fig. 3c. Interestingly, the polytope of the 3-cable scheme is larger, and its direction of maximum velocity is nearly perpendicular to its counterpart.

5 Static Force Analysis

The equation of static equilibrium for the 2-X manipulator can be written as:

$$\boldsymbol{J}^\top \boldsymbol{f} + \boldsymbol{G} = \boldsymbol{\Gamma} \tag{7}$$

where $\boldsymbol{J}$ is the Jacobian matrix defined in Eq. (2), $\boldsymbol{f} = [f_x, f_y]^\top$ represents the forces applied by the robot end-effector on the environment, $\boldsymbol{G} = [G_1, G_2]^\top$ denotes the wrench due to the springs and gravity effects, and $\boldsymbol{\Gamma} = [\Gamma_1, \Gamma_2]^\top$ is the wrench applied by the cables, which depends on the actuation scheme.

For the 4-cable scheme, the cables C_{l_i}, C_{r_i} apply forces F_{l_i}, F_{r_i} on the ith joint, respectively. Thus, the actuation wrench components are given by $\Gamma_i = Z_{l_i} F_{l_i} + Z_{r_i} F_{r_i}, i = 1, 2$ (see [8]), which can be written component-wise as:

$$\begin{bmatrix} J_{11} & J_{21} \\ J_{12} & J_{22} \end{bmatrix} \begin{bmatrix} f_x \\ f_y \end{bmatrix} + \begin{bmatrix} G_1 \\ G_2 \end{bmatrix} = \begin{bmatrix} Z_{l_1} F_{l_1} + Z_{r_1} F_{r_1} \\ Z_{l_2} F_{l_2} + Z_{r_2} F_{r_2} \end{bmatrix} \tag{8}$$

The limits on actuation forces shall be imposed as: $F_{l_i}, F_{r_i} \in [F_{\text{min}}, F_{\text{max}}]$ with numerical values $F_{\text{min}} = 5$ N, $F_{\text{max}} = 155$ N. Recalling that $Z_{l_i} > 0$ and $Z_{r_i} < 0$, the lower (resp. upper) bound of the actuation wrench Γ_i occurs when $F_{l_i} = F_{\text{min}}, F_{r_i} = F_{\text{max}}$ (resp. $F_{l_i} = F_{\text{max}}, F_{r_i} = F_{\text{min}}$). Thus, from Eq. (8), the following inequalities must hold:

$$Z_{l_i}F_{\min} + Z_{r_i}F_{\max} - G_i \leq (J_{1i}f_x + J_{2i}f_y) \leq Z_{l_i}F_{\max} + Z_{r_i}F_{\min} - G_i \quad (9)$$

for $i = 1, 2$. The region in the task force space (f_x, f_y) that satisfies these four conditions represents the force polytope for the 4-cable scheme. This is shown by the red continuous boundary for an example in Fig. 4.

In the 3-cable scheme, the side-routed cable C_l applies the same force F_l on the left side of both the joints (see Fig. 2, right), while the cable C_{r_i} applies a force F_{r_i} only on the ith joint. Thus, the actuation wrench components are given by $\Gamma_i = Z_{l_i}F_l + Z_{r_i}F_{r_i}$. The associated equilibrium equation resembles the one in Eq. (8), with F_{l_1}, F_{l_2} replaced by F_l. Thus, proceeding in a similar manner, one obtains the same conditions in Eq. (9) for the 3-cable scheme also. Additionally, the constraints due to coupling of the side-routed cable C_l must also be imposed. This is obtained by firstly eliminating F_l from the two equilibrium equations:

$$J'_{11}f_x + J'_{21}f_y + G' = Z'_{l_1}F_{r_2} + Z_{r_1}F_{r_1} \quad (10)$$

where $J'_{i1} = \left(J_{i1} - \frac{Z_{l_1}}{Z_{l_2}}J_{i2}\right), i = 1, 2,\ G' = \left(G_1 - \frac{Z_{l_1}}{Z_{l_2}}G_2\right),\ Z'_{l_1} = -\left(\frac{Z_{l_1}}{Z_{l_2}}Z_{r_2}\right)$. It can be shown that $Z'_{l_1} > 0$ in the above equation. Thus, the wrench bounds can be obtained as above, and the final conditions on (f_x, f_y) are:

$$Z'_{l_1}F_{\min} + Z_{r_1}F_{\max} - G' \leq J'_{11}f_x + J'_{21}f_y \leq Z'_{l_1}F_{\max} + Z_{r_1}F_{\min} - G' \quad (11)$$

The force polytope for the 3-cable scheme lies at the intersection of all the six conditions in Eqs. (9) and (11), as illustrated by the dashed boundary in Fig. 4.

Evidently, the force polytope of 3-cable scheme lies inside that of the 4-cable scheme, as it satisfies two additional conditions in Eq. (11). Thus, it can be concluded that the 4-cable actuation scheme permits the end-effector to apply more forces than the 3-cable scheme. From Fig. 4, it is observed that the force polytopes are off-centered w.r.t. the origin. This is an effect of gravity and springs.

6 Global Performance Inside the Task Space

The objective of this section is to use the velocity and force polytopes defined above, to assess the performance of the manipulator in the task space. In order to do this, it is firstly necessary to qualify the points (x, y) in the task space, to define a *workspace* for this manipulator. Several such definitions exist in the literature of cable-driven robots, e.g., wrench-closure workspace [6], wrench-feasible workspace [8]. In this study, a stable wrench-feasible workspace will be used. It is defined as the set of points (x, y) that can be reached by the end-effector, while satisfying the conditions of static equilibrium (only under self-weight), stability, and the bounds on actuation forces of the cables. It can be computed by a scanning method analogous to the one used in [8]. The details are not discussed here due to lack of space. A plot of this workspace is shown in Figs. 5 and 6 for the 4-cable (left) and 3-cable (right) schemes. As expected, the workspace obtained for the 4-cable scheme is larger than that of the 3-cable scheme.

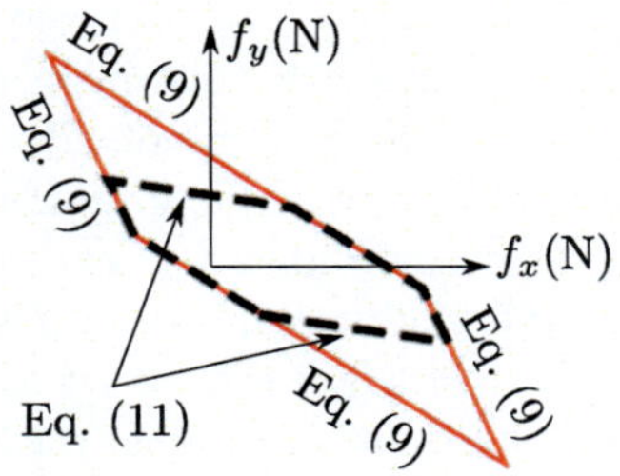

Fig. 4. Force polytopes and associated constraints of the 4-cable (red continuous) and 3-cable (black dashed) schemes at $(\alpha_1, \alpha_2) = (-1.65, -1.18)$ rad.

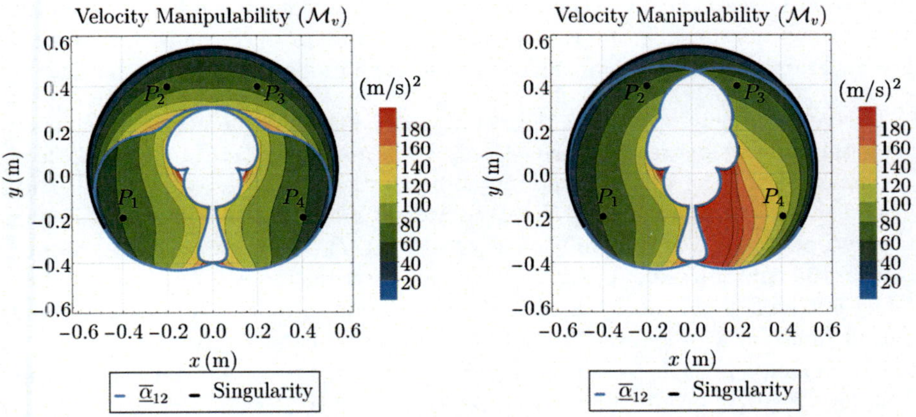

Fig. 5. Contour plot of velocity manipulability index inside the stable wrench-feasible workspace for the 4-cable (left) and 3-cable (right) actuation schemes.

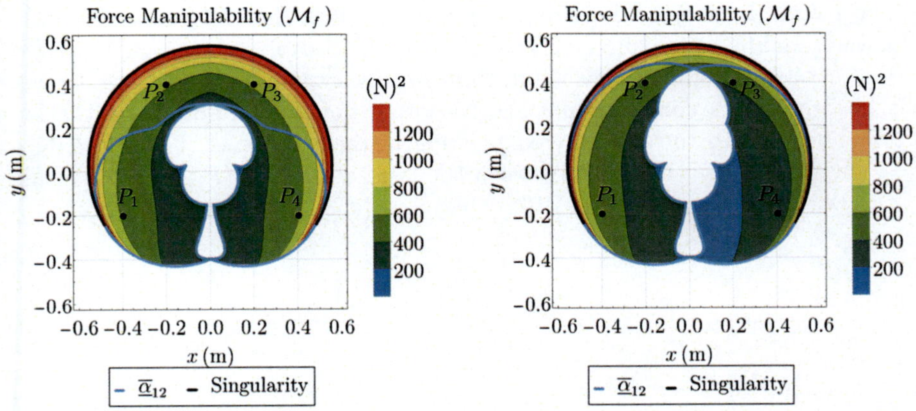

Fig. 6. Contour plot of force manipulability index inside the stable wrench-feasible workspace for the 4-cable (left) and 3-cable (right) actuation schemes.

Four points: P_1, P_2, P_3, P_4 (see Fig. 5) that are common to the workspaces of 4-cable and 3-cable actuation schemes have been chosen, to compare their velocity and force performance locally. The velocity and force polytopes are visualized at these points in Table 1. In order to quantify the size and shape of these polytopes, two performance measures, namely, manipulability index ($\mathcal{M}$) and condition index, have been introduced in [4]. In this paper, the inverse of the condition index (κ) will be used instead of the condition index, as it is a bounded measure. Geometrically, the manipulability index measures the surface area of the polytope and the inverse condition index is the ratio of maximum performance (velocity/force) that can be achieved in *all* directions to the maximum performance inside the polytope. The distribution of velocity and force manipulability indices have been plotted for the two schemes inside their workspaces in Figs. 5 and 6, respectively. When there is more than one feasible configuration for the manipulator, the one that results in maximum performance measure is chosen. From these plots and Table 1, the following observations can be made:

- The velocity polytopes are identical at point P_1, since the active constraints on the joint velocities are formed only by the cables C_{r_1}, C_{r_2} in both cases. Consequently, the velocity manipulability indices ($\mathcal{M}_v$) are also equal. But, for the 3-cable scheme, it is slightly lower at P_2, equal at P_3 (with different polytope shape), and significantly greater at P_4. This can also be verified from the contour plots in Fig. 5, where the distribution is nearly the same on the left half of the workspaces, but the value is significantly greater (about 1.53 times on average) on the right half for the 3-cable scheme.
- The velocity inverse condition index (κ_v) is greater for the 3-cable scheme at (P_2, P_3, P_4). Inside the workspace, $(\kappa_v)_4 \in [0, 0.5]$ (4-cable) and $(\kappa_v)_3 \in [0, 0.7]$ (3-cable). Its distribution is not shown here due to lack of space.
- The force manipulability ($\mathcal{M}_f$) must be necessarily greater for the 4-cable scheme as the force polytope of 3-cable scheme must be contained inside that of the 4-cable one (see Sect. 5). The difference $((\mathcal{M}_f)_4 - (\mathcal{M}_f)_3)$ increases as one moves from P_1 to P_4. From Fig. 6, it is found that $\mathcal{M}_f$ is significantly lower (about 1/1.56 times on average) for the 3-cable scheme on the right half of the workspace, converse to that of velocity manipulability.
- The force inverse condition index (κ_f) is low, in general, for both the actuation schemes, due to the off-centered position of the polytopes. There is only a small difference in its value between the two schemes in $(P_1, \ldots, P_4)$. Inside the workspace $(\kappa_v)_4 \in [0, 0.14]$ (4-cable) and $(\kappa_v)_3 \in [0, 0.18]$ (3-cable).

Table 1. Comparison of velocity and force polytopes for 4-cable and 3-cable actuation schemes at discrete points in their workspaces. The joint velocity constraints imposed by the left cables (C_{l_1}, C_{l_2}, C_l) are shown in orange shade, and those imposed by the right cables (C_{r_1}, C_{r_2}) in blue shade, for both the schemes. The resulting polytopes obtained for the 4-cable (resp. 3-cable) scheme are shown in red continuous (resp. black dashed) boundary. Symbols $(\mathcal{M}_v, \kappa_v)$ and $(\mathcal{M}_f, \kappa_f)$ denote (manipulability, inverse condition index) for velocity and force polytopes, respectively. A measure $(\cdot)$ associated with 4-cable scheme is given by $(\cdot)_4$ and with 3-cable scheme by $(\cdot)_3$. All the numerical values have been rounded off to 2 decimal places.

Position [m]/ Configuration [rad]	Joint velocity polytopes $(\dot{\alpha}_1, \dot{\alpha}_2)$ (4-cable) (3-cable)	Task velocity polytopes $(\dot{x}, \dot{y})$	Task force polytopes (f_x, f_y)	$\mathcal{M}_v$ [(m/s)^2], $\mathcal{M}_f$ [N^2] κ_v, κ_f [no units]
$P_1 =$ $(-0.4, -0.2)$; $(\alpha_1, \alpha_2) =$ $(1.65, 1.18)$	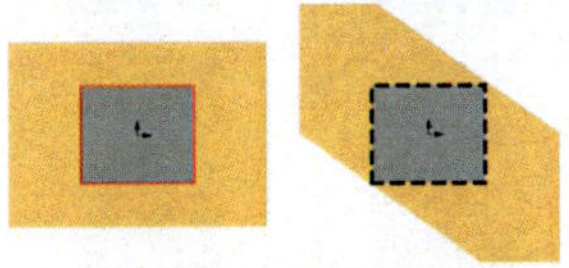		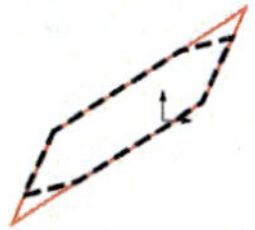	$(\mathcal{M}_v)_4 = 61.18$, $(\mathcal{M}_v)_3 = 61.18$ $(\kappa_v)_4 = 0.18$, $(\kappa_v)_3 = 0.18$ $(\mathcal{M}_f)_4 = 579.68$, $(\mathcal{M}_f)_3 = 518.45$ $(\kappa_f)_4 = 0.04$, $(\kappa_f)_3 = 0.04$
$P_2 =$ $(-0.2, 0.4)$; $(\alpha_1, \alpha_2) =$ $(-0.2, 1.57)$	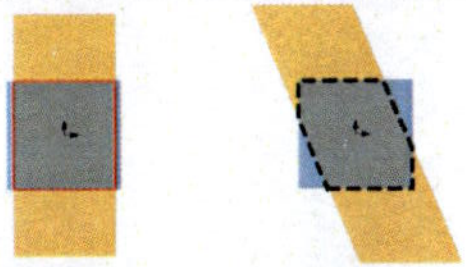	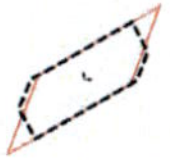	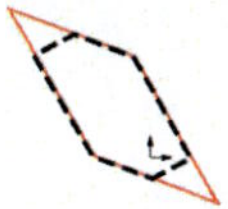	$(\mathcal{M}_v)_4 = 67.35$, $(\mathcal{M}_v)_3 = 63.26$ $(\kappa_v)_4 = 0.25$, $(\kappa_v)_3 = 0.35$ $(\mathcal{M}_f)_4 = 657.14$, $(\mathcal{M}_f)_3 = 549.08$ $(\kappa_f)_4 = 0.09$, $(\kappa_f)_3 = 0.12$
$P_3 =$ $(0.2, 0.4)$; $(\alpha_1, \alpha_2) =$ $(0.2, -1.57)$	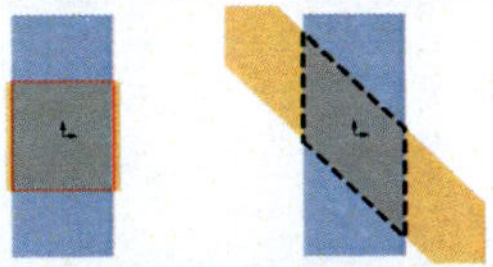	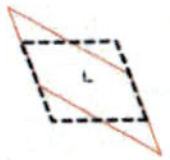	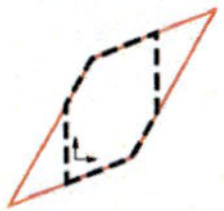	$(\mathcal{M}_v)_4 = 67.35$, $(\mathcal{M}_v)_3 = 67.35$ $(\kappa_v)_4 = 0.25$, $(\kappa_v)_3 = 0.54$ $(\mathcal{M}_f)_4 = 657.14$, $(\mathcal{M}_f)_3 = 443.68$ $(\kappa_f)_4 = 0.09$, $(\kappa_f)_3 = 0.06$
$P_4 =$ $(0.4, -0.2)$; $(\alpha_1, \alpha_2) =$ $(-1.65, -1.18)$	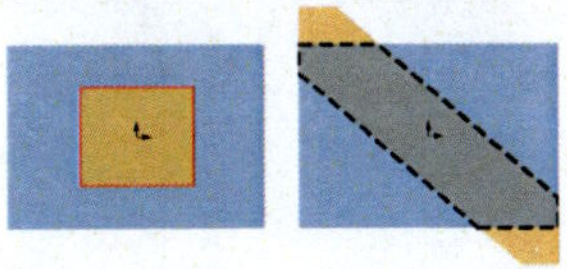	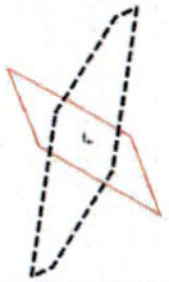	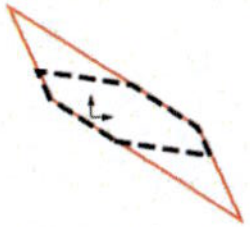	$(\mathcal{M}_v)_4 = 61.18$, $(\mathcal{M}_v)_3 = 109.71$ $(\kappa_v)_4 = 0.18$, $(\kappa_v)_3 = 0.22$ $(\mathcal{M}_f)_4 = 579.68$, $(\mathcal{M}_f)_3 = 316.38$ $(\kappa_f)_4 = 0.04$, $(\kappa_f)_3 = 0.05$

7 Conclusion

The instantaneous velocity and static force analysis have been performed for the 2-X tensegrity manipulator for two different cable actuation schemes. The first scheme involves four cables, while the second one involves only three. Practical limits have been considered on the cable velocities and the actuation forces. The effect of these limits on the end-effector velocity and force-application capabilities have been studied for the two actuation schemes, in terms of polytopes.

A comparative study has been carried out between the velocity and force polytopes of the 4-cable and 3-cable schemes, in terms of their manipulability and inverse condition indices. It is observed that the velocity manipulability of the 3-cable scheme is significantly larger than its counterpart on the right half of the workspace. However, its force manipulability is less than that of the 4-cable scheme throughout its workspace, and significantly lower on the right half. The velocity inverse condition index is greater for the 3-cable scheme, but there is no significant difference in the force inverse condition index between the two schemes. In summary, the 3-cable scheme has a smaller stable wrench-feasible workspace, lesser force capabilities, but larger velocity capabilities in some parts of the workspace. In the future, these velocity and force indices will be used to design tensegrity manipulators for manipulation and machining applications.

References

1. Gouttefarde, M., Daney, D., Merlet, J.: Interval-analysis-based determination of the wrench-feasible workspace of parallel cable-driven robots. IEEE Trans. Rob. **27**(1), 1–13 (2011)
2. Bouchard, S., Gosselin, C., Moore, B.: On the ability of a cable-driven robot to generate a prescribed set of wrenches. ASME J. Mech. Rob **2**(1), 011010 (2010)
3. Lessanibahri, S., Gouttefarde, M., Caro, S., Cardou, P.: Twist feasibility analysis of cable-driven parallel robots. In: Gosselin, C., Cardou, P., Bruckmann, T., Pott, A. (eds.) Cable-Driven Parallel Robots. MMS, vol. 53, pp. 128–139. Springer, Cham (2018). https://doi.org/10.1007/978-3-319-61431-1_12
4. Lee, J.: A study on the manipulability measures for robot manipulators. In: Proceedings of the 1997 IEEE/RSJ International Conference on Intelligent Robot and Systems, IROS 1997, vol. 3, pp. 1458–1465 (1997)
5. Rezazadeh, S., Behzadipour, S.: Workspace analysis of multibody cable-driven mechanisms. ASME. J. Mech. Rob. **3**(2), 021005 (2011)
6. Ramadoss, V., Lau, D., Zlatanov, D., Zoppi, M.: Analysis of planar multilink cable driven robots using internal routing scheme. In: Proceedings of the ASME 2020 International Design Engineering Technical Conferences and Computers and Information in Engineering Conference (2020)
7. Mustafa, S.K., Agrawal, S.K.: On the force-closure analysis of n-DOF cable-driven open chains based on reciprocal screw theory. IEEE Trans. Rob. **28**(1), 22–31 (2012)
8. Furet, M., Wenger, P.: Kinetostatic analysis and actuation strategy of a planar tensegrity 2-X manipulator. ASME J. Mech. Rob. **11**(6), 060904 (2019)

9. Hamon, A., Aoustin, Y., Caro, S.: Two walking gaits for a planar bipedal robot equipped with a four-bar mechanism for the knee joint. Multibody Sys. Dyn. **31**(3), 283–307 (2013). https://doi.org/10.1007/s11044-013-9382-7
10. Fasquelle, B., et al.: Identification and control of a 3-X cable-driven manipulator inspired from the Bird's neck. ASME J. Mech. Rob. **14**(1), 011005 (2022)
11. Wenger, P., Furet, M.: Kinematic analysis of a planar manipulator with anti-parallelogram joints and offsets. In: Lenarčič, J., Siciliano, B. (eds.) ARK 2020. SPAR, vol. 15, pp. 319–326. Springer, Cham (2021). https://doi.org/10.1007/978-3-030-50975-0_39
12. Furet, M., van Riesen, A., Chevallereau, C., Wenger, P.: Optimal design of tensegrity mechanisms used in a Bird neck model. In: Corves, B., Wenger, P., Hüsing, M. (eds.) EuCoMeS 2018. MMS, vol. 59, pp. 365–375. Springer, Cham (2019). https://doi.org/10.1007/978-3-319-98020-1_43

Computational Kinematics

Complete Inverse Geometric Model Computation of the Quattro Parallel Kinematic Robot

Hélène Chanal[1](✉), Rani Rizk[2], Nicolas Bouton[1], and Benoît Blaysat[1]

[1] Clermont Auvergne INP, CNRS, Institut Pascal, Clermont Auvergne Université, 63000 Clermont-Ferrand, France
helene.chanal@sigma-clermont.fr

[2] School of Engineering, Lebanese University, CRSI, Roumieh, Lebanon

Abstract. In this paper, a formalism based on a vector modeling of joint invariants is applied to the Quattro robot with 3° of freedom. The aim is to compute analytically the complete inverse kinematic model, i.e. both the value of active and passive joints regarding the mobile platform position. This formalism has been already applied to a parallel kinematic machine with revolute joints; in this work, we discuss its extension to parallel kinematic robots with spherical joints.

1 Introduction

Different kinds of industrial machine or robot architectures exist [1]. Considering the influence of robot architecture choice on moving mass for a performance level of a given task has been studied in the literature [2]. However, a ground-breaking in the robotic architecture design can bring a high benefit in the mass/stiffness ratio. For example, parallel kinematic robots feature high stiffness compare to serial manipulators with the same moving part weight [3].

The use of over-constrained mechanism is another option to increase stiffness or decrease moving part weight [3]. An over-constrained parallel kinematic manipulator (OPKM) is defined as parallel manipulators with common or redundant constraints that can be removed without changing the kinematics of the mechanism [4]. A stiffness comparison between Tricept robot and Exechon robot (an over-constrained robot equivalent to Tricept in term of architecture and motion) shows a decrease of 0.177 mm on average for the linear displacement induced by a 450 N load which can be important in the context of machining [5].

OPKM have some other advantages such as avoiding kinematic singularities, improving kinematic characteristics and eliminating backlash [6]. However, an OPKM shall satisfy strict geometric conditions during assembly and use [7]. Thus, there is a challenge to determine the geometric constraints induced by the over-constrained systems. The first step is based on the establishment of an inverse nominal kinematic model to determine active and passive joint values [8]. The chosen formalism will allow to introduce, in a next time, the second order defect exhaustively. We previously proposed such formalism on a SCARA robot, which relies on the vector modelling of kinematic joint invariant

O. Altuzarra and A. Kecskémthy (Eds.): ARK 2022, SPAR 24, pp. 163–170, 2022.
https://doi.org/10.1007/978-3-031-08140-8_18

[9]. It improves the final accuracy after its identification compared to a model based on the usual or modified Denavit-Hartenberg (DH) formalism. Our formalism requires the same number of parameters than DH but it avoids two DH drawbacks: (i) the numerical instabilities in the case of two adjacent joints with collinear or intersecting axes due to the sensibility of geometric parameter values to small variations in axis alignment [10], and (ii) the joint axis orientation constraints imposed by the previous frame.

The proposed formalism has been previously applied to an OPKM with over-constrained leg assembly and with only revolute passive joints [11]. However, two types of OPKM are introduced in the literature: OPKM with redundant legs and OPKM with over-constrained leg assembly. Thus, in this paper, we propose to extend this formalism for the modelling of the Complete Inverse Kinematic Model (CIKM) of an OPKM with redundant legs with spherical joints, a Quattro robot. Note that the aim of this work is not to establish an Inverse Kinematic Model (IKM) for controlling robot motion like in [1], but is to define a CIKM to identify the assembly constraints of OPKM in a next stage.

First, the architecture of the Quattro robot is presented. A formalism which describes the invariant of spherical joint is then proposed. This formalism is finally applied to determine the CIKM of a Quattro robot.

2 Presentation of the Quattro Robot

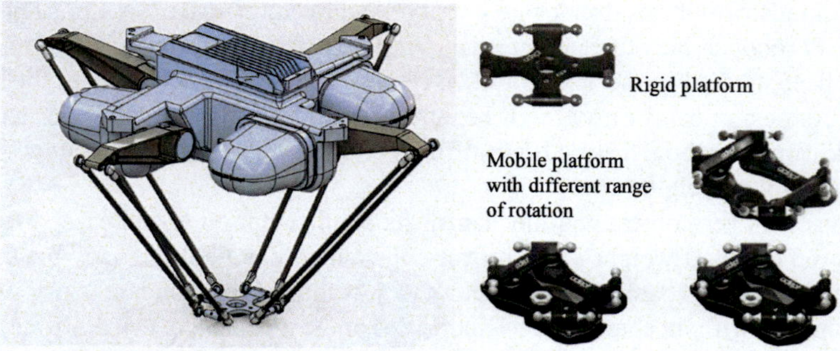

Fig. 1. Quattro robot of Omron and the four types of possible platform.

The Quattro robot is a parallel robot based on Par4 [12] (Fig. 1). It consists of 4° of freedom (dof), 3 translations and 1 rotation [13]. It is used for pick-and-place. Each leg is composed of an active revolute joint and four passive spherical joints. The rotation is realized according to revolute joints on the mobile platform.

To increase the maximum payload, a mobile platform without rotation can be chosen (Fig. 1). With this platform, the system is a 3-dof robot with over-constrained and a redundancy actuation which justifies a study to identify the induced geometric constraints. In this article, the focus is made on the first step of this study, the elaboration of the CIKM of the 3-dof Quattro robot (i.e. the three translations).

3 Formalism Based on Joint Invariant

To establish the CIKM, the first step requires the definition of each frame linked to each joint. The formalism relies on a vector description of the joint invariant. The modelling of joint kinematic should be consistent with the real OPKM to have the ability to add representative joint defects.

The invariant of the rotational joint is a straight- line (rotational joint axis) [14]. Therefore, 4 parameters are introduced for describing this feature, i.e. 2 parameters to define the straight-line orientation and 2 parameters to define a point of this straight-line. The invariant of a spherical joint is a point (the center of the spherical joint). 3 parameters are introduced for defining the point position.

For a link $i-1$ and a link i connected with a revolute joint J_i, the frame definition is realized as in [11]. The rotational joint axis is along the vector $\mathbf{z}_i$ and the axis $\mathbf{x}_i$ is normal to $\mathbf{z}_i$ and $\mathbf{z}_{i-1}$ (Fig. 2). Then, the mapping between frame $\mathcal{R}_{i-1}$ to frame $\mathcal{R}_i$ is realized according to two transformation matrices. The first one, $\mathbf{D}_{J_i}^{R_{i-1}}$, defines a rigid transformation fixed in frame $\mathcal{R}_{i-1}$. The second one, $\mathbf{R}_{R_i}^{J_i}$, defines the movement of the revolute joint. These matrices are given in Eq. (1).

$$\mathbf{X}_{R_{i-1}} = \mathbf{D}_{J_i}^{R_{i-1}} \mathbf{R}_{R_i}^{J_i} \mathbf{X}_{R_i} \tag{1}$$

With $\mathbf{R}_{R_i}^{J_i} = \begin{bmatrix} c\theta_i & -s\theta_i & 0 & 0 \\ s\theta_i & \theta_i & 0 & 0 \\ 0 & 0 & 1 & 0 \\ 0 & 0 & 0 & 1 \end{bmatrix}$ with $s. = sin(.)$ and $c. = cos(.)$

and $\mathbf{D}_{J_i}^{R_{i-1}} = \begin{bmatrix} -\frac{J_i}{\sqrt{I_i^2+J_i^2}} & -\frac{I_iK_i}{\sqrt{I_i^2+J_i^2}} & I_i & a_i \\ \frac{I_i}{\sqrt{I_i^2+J_i^2}} & -\frac{J_iK_i}{\sqrt{I_i^2+J_i^2}} & J_i & b_i \\ 0 & -\frac{I_i^2+J_i^2}{\sqrt{I_i^2+J_i^2}} & K_i & c_i \\ 0 & 0 & 0 & 1 \end{bmatrix}$ if $\mathbf{z}_i$ and $\mathbf{z}_{i-1}$ are not colinear else $\mathbf{x}_i$ is chosen along the straight line which connecting the center of joint J_{i-1} and joint J_i.

For a spherical joint, since the invariant is a point, there is no constraint on the associated basis. The frame is chosen according to the type of previous joint J_{i-1} to simplify the expression of the matrix $\mathbf{D}_{J_i}^{R_{i-1}}$ (Fig. 2). In the case of two successive spherical joints, the axis $\mathbf{z}_i$ is along the straight line which crossed the two joint center points. In this case, an undetermined rotation appears about the axis $\mathbf{z}_i$. The matrix $\mathbf{S}_{R_i}^{J_i}$, which defines the movement of a spherical joint, is expressed with three angles according to the Euler angles $(\psi_i, \theta_i, \varphi_i)$ (Eq. (2)).

$$\mathbf{S}_{R_i}^{J_i} = \begin{bmatrix} c\psi_i c\varphi_i - s\psi_i c\theta_i s\varphi_i & -c\psi_i s\varphi_i - s\psi_i c\theta_i c\varphi_i & s\psi_i s\theta_i & 0 \\ s\psi_i c\varphi_i + c\psi_i c\theta_i s\varphi_i & -s\psi_i s\varphi_i + c\psi_i c\theta_i c\varphi_i & -c\psi_i s\theta_i & 0 \\ s\theta_i s\varphi_i & s\theta_i c\varphi_i & c\theta_i & 0 \\ 0 & 0 & 0 & 1 \end{bmatrix} \tag{2}$$

Figure 2 illustrates the application of this formalism to the Quattro robot with 3 dofs. $\mathcal{R}_f$ is the fixed platform frame and $\mathcal{R}_m$ is the mobile platform frame. Each leg is composed of two arms (A_iB_i) and $(A_i'B_i')$. The notations of frames and angles attached to arm $(A_i'B_i')$ are the same of arm (A_iB_i) with "$'$".

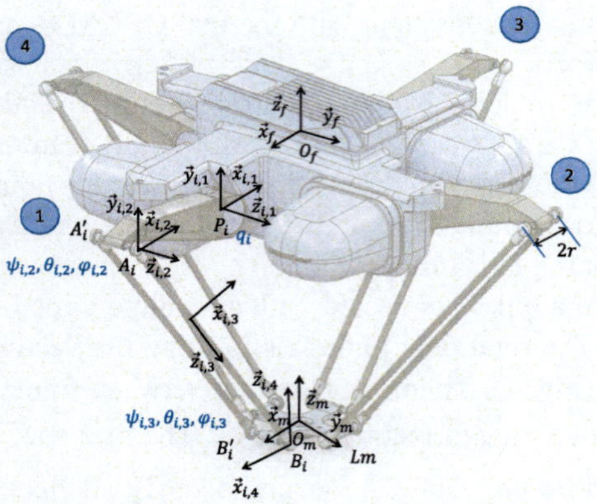

Fig. 2. Frame definitions of the Quattro robot

Note that for each leg i, there is one active joint parameter q_i associated to a revolute joint and 12 passive joint parameters $(\psi_{i,2}, \theta_{i,2}, \varphi_{i,2}, \psi_{i,3}, \theta_{i,3}, \varphi_{i,3}, \psi_{i,2}', \theta_{i,2}', \varphi_{i,2}', \psi_{i,3}', \theta_{i,3}', \varphi_{i,3}')$ associated to the spherical joints.

4 CIKM of the Quattro Robot

The determination of the geometric constraint due to over-constraints imposed to establish the CIKM. In this section, we introduce the well-known Inverse Kinematic Model (IKM) of a 3-dof Quattro robot. Second, according to the formalism introduced in the previous section, its complete inverse kinematic model is determined. This allows the computation of the value of the passive joints regarding the active joints position and the position of point O_m in frame $\mathcal{R}_f$.

4.1 Inverse Kinematic Model

In this section, the well-known IKM of the 3-dof Quattro robot is detailed [15]. Figure 3 illustrates the geometric parameters. Points P_i are on the active revolute joint axis. Points A_i, A_i', B_i, B_i' are the center of the spherical joints (Fig. 2). α_i corresponds to the orientation of the first leg revolute joint axis around $\mathbf{z}_f$. l is the length of arms ($\overrightarrow{A_iB_i} = l\mathbf{z}_3$). The coordinates of point O_m in frame $\mathcal{R}_f$ are noted (X, Y, Z).

As the distance from A_i to B_i is constant $\overrightarrow{A_iB_i} = \overrightarrow{O_fOm} + \overrightarrow{OmB_i} - \overrightarrow{O_fP_i} - \overrightarrow{P_iA_i}$. This equation writes:

$$C_{i,1}\sin(q_i) + C_{i,2}\cos(q_i) + C_{i,3} = 0 \tag{3}$$

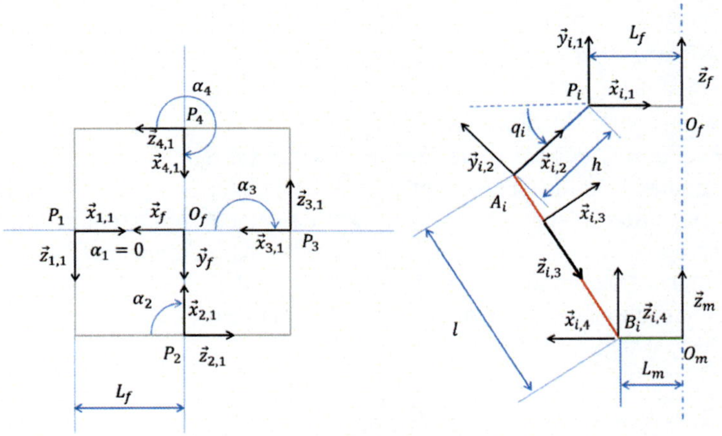

Fig. 3. Quattro geometrical parameters

where $C_{i,1} = 2Zh$, $C_{i,2} = 2h(L_f - L_m) - 2hX\cos(\alpha_i) - 2hY\sin(\alpha_i)$ and $C_{i,3} = (L_f - L_m)^2 - 2\cos(\alpha_i)X(L_f - L_m) - 2\sin(\alpha_i)Y(L_f - L_m) + X^2 + Y^2 + Z^2 + h^2 - l^2$. One obtains:

$$\tan\left(\frac{q_i}{2}\right) = \frac{-C_{i,1} - \sqrt{C_{i,1}^2 - C_{i,3}^2 + C_{i,2}^2}}{C_{i,3} - C_{i,2}} \tag{4}$$

This IKM ensures the computation of the active joint values q_i regarding the mobile platform position. The determination of the passive joint values should be realized according to a CIKM.

4.2 Complete Inverse Kinematic Model

The CIKM is obtained from Eq. (5) by looping each leg with a virtual leg.

$$\mathbf{T}_i = \mathbf{D}_{J_{i,1}}^{R_f} \mathbf{R}_{R_{i,1}}^{J_{i,1}} \mathbf{D}_{J_{i,2}}^{R_{i,1}} \mathbf{S}_{R_{i,2}}^{J_{i,2}} \mathbf{D}_{J_{i,3}}^{R_{i,2}} \mathbf{S}_{R_{i,3}}^{J_{i,3}} \mathbf{D}_{R_m}^{R_{i,3}} = \begin{bmatrix} 1 & 0 & 0 & X \\ 0 & 1 & 0 & Y \\ 0 & 0 & 1 & Z \\ 0 & 0 & 0 & 1 \end{bmatrix} \tag{5}$$

where $\mathbf{D}_{J_{i,1}}^{R_f} = \mathbf{D}_{J_i}^{R_{i-1}}$ with $I_i = -\sin(\alpha_i)$, $J_i = \cos(\alpha_i)$, $K_i = 0$, $a_i = L_f\cos(\alpha_i)$, $b_i = L_f\sin(\alpha_i)$, $c_i = 0$, $\mathbf{D}_{J_{i,2}}^{R_{i,1}} = \begin{bmatrix} 1 & 0 & 0 & -h \\ 0 & 1 & 0 & 0 \\ 0 & 0 & 1 & c_r \\ 0 & 0 & 0 & 1 \end{bmatrix}$, $\mathbf{D}_{J_{i,3}}^{R_{i,2}} = \begin{bmatrix} 1 & 0 & 0 & 0 \\ 0 & 1 & 0 & 0 \\ 0 & 0 & 1 & l \\ 0 & 0 & 0 & 1 \end{bmatrix}$, $\mathbf{D}_{R_m}^{R_{i,3}} =$

$$\begin{bmatrix} \cos(\alpha_i) & -\sin(\alpha_i) & 0 & -L_m \\ \sin(\alpha_i) & \cos(\alpha_i) & 0 & b_r \\ 0 & 0 & 1 & 0 \\ 0 & 0 & 0 & 1 \end{bmatrix}$$, with $c_r = r$ and $b_r = -r$ for arm (A_iB_i), and $c_r = -r$ and $b_r = r$ for arm $(A_i^{'}B_i^{'})$ ($2r$ is the distance $A_iA_i^{'}$, and $B_iB_i^{'}$).

The computation of the CIKM is illustrated for leg 1 for arm (A_iB_i), i.e. with $\alpha_i = 0$. In this case, 6 passive joint parameters should be computed from 20 equations. Note that $\mathbf{T}_i(m, n)$ represents the value of the mth line and n^{th} column of the matrix $\mathbf{T}_i$.

With a first analysis of matrix $\mathbf{T}_1$, we can conclude that $\varphi_{1,2}$ and $\psi_{1,3}$ are around the same axis and cannot be separated. Thus, we choose to have $\varphi_{1,2} = 0$.

The principal steps of resolution are:

a) $\mathbf{T}_1(2,3) = 0 \Rightarrow \cos(\psi_{1,3}) = \frac{\cos(\theta_{1,2})\cos(\theta_{1,3})}{\sin(\theta_{1,2})\sin(\theta_{1,3})}$.

b) $\mathbf{T}_1(2,1) = 0 \Rightarrow \sin(\psi_{1,3}) = -\frac{\cos(\theta_{1,2})\sin(\varphi_{1,3})}{\sin(\theta_{1,3})\cos(\varphi_{1,3})\sin(\theta_{1,2})}$

c) $\mathbf{T}_1(1,3) = 0 \Rightarrow \cos(\theta_{1,2}) = \frac{\cos(\theta_{1,3})\sin(\psi_{1,2}+q_1)\cos(\varphi_{1,3})}{\cos(\psi_{1,2}+q_1)\sin(\varphi_{1,3})}$

d) $\mathbf{T}_1(3,3) = 1 \Rightarrow \sin(\theta_{1,2}) = -\frac{\cos(\theta_{1,3})}{\cos(\psi_{1,2}+q_1)}$

e) $\mathbf{T}_1(2,2) = 1 \Rightarrow \sin(\varphi_{1,3}) = \frac{\cos(\theta_{1,3})\sin(\psi_{1,2}+q_1)}{\cos(\psi_{1,2}+q_1)\sin(\theta_{1,3})}$

f) $\mathbf{T}_1(3,1) = 0 \Rightarrow \cos^2(\varphi_{1,3}) = \frac{\cos^2(\theta_{1,3})-\cos^2(\psi_{1,2}+q_1)}{\cos^2(\psi_{1,2}+q_1)\cos^2(\theta_{1,3})-\cos^2(\psi_{1,2}+q_1)}$.

g) If $\varphi_{1,3} \in \left[-\frac{\pi}{2}; \frac{\pi}{2}\right]$ then $\cos(\varphi_{1,3}) = \sqrt{\frac{\cos^2(\theta_{1,3})-\cos^2(\psi_{1,2}+q_1)}{\cos^2(\psi_{1,2}+q_1)\cos^2(\theta_{1,3})-\cos^2(\psi_{1,2}+q_1)}}$

h) $\mathbf{T}_1(2,4)^2 = Y^2 \Rightarrow \cos^2(\theta_{1,3}) = \cos^2(\psi_{1,2}+q_1)\left(1-\frac{Y^2}{l^2}\right)$.

i) If $\theta_{1,3} \in \left[-\frac{\pi}{2}; \frac{\pi}{2}\right]$ then $\cos(\theta_{1,3}) = \cos(\psi_{1,2}+q_1)\sqrt{1-\frac{Y^2}{l^2}}$

j) $\mathbf{T}_1(2,4) = Y \Rightarrow sin(\theta_{1,3}) = \frac{\sqrt{(Y^2-l^2)\cos^2(\psi_{1,2}+q_1)+l^2}}{l}$

k) $\mathbf{T}_1(3,4) = Z \Rightarrow \cos^2(\psi_{1,2}+q_1) = -\frac{(Z+h\sin(q_1))^2}{(Y^2-l^2)}$.

l) If $(\psi_{1,2}+q_1) \in \left[-\frac{\pi}{2}; \frac{\pi}{2}\right]$ then $\cos(\psi_{1,2}+q_1) = \frac{Z+h\sin(q_1)}{\sqrt{(l^2-Y^2)}}$

m) $\mathbf{T}_1(1,4) = X \Rightarrow \sin(\psi_{1,2}+q_1) = \frac{X-(L_f-L_m+h\cos(q_1))}{\sqrt{(l^2-Y^2)}}$

This methodology can be applied to leg 2, 3 and 4. Similar expressions are thus obtained for angles $\varphi_{i,2}$, $\psi_{i,3}$, $\theta_{i,2}$, $\varphi_{i,3}$, $\varphi_{i,2}^{'}$, $\psi_{i,3}^{'}$, $\theta_{i,2}^{'}$ and $\varphi_{i,3}^{'}$. The expression of $\theta_{i,3}$ and $\theta_{i,3}^{'}$ depend on the considered leg.

Finally, an analytical complete inverse geometric model of a Quattro robot is written by adding Eq. (4) which is the Quattro robot IKM.

5 Conclusion

In this article, the application of a formalism based on a vector modelling of joint invariants is proposed. This methodology is applied to a 3-dof Quattro robot for determining the CIKM. This model allows determining analytically the passive joint, the active joint, and the mobile platform positions.

The next step consists of the introduction of second order defect to identify the geometric constraint due to over-constrained.

References

1. Khalil, W., Dombre, E.: Modeling Identification and Control of Robots. Butterworth-Heinemann (2004). https://doi.org/10.1016/B978-1-903996-66-9.X5000-3
2. Lajili, M., Chanal, H., Bouzgarrou, B.C.: Machine tool architecture selection at the preliminary design stage: application to hard material machining. J. Comput. Des. Eng. **7**, 657–667 (2020). https://doi.org/10.1093/jcde/qwaa049
3. Huang, T., Zhao, X., Whitehouse, D.J.: Stiffness estimation of a tripod-based parallel kinematic machine. IEEE Trans. Robot. Autom. **18**, 50–58 (2002). https://doi.org/10.1109/70.988974
4. Liu, W.-L., Xu, Y.-D., Yao, J.-T., Zhao, Y.-S.: Methods for force analysis of overconstrained parallel mechanisms: a review. Chinese J. Mech. Eng. **30**(6), 1460–1472 (2017). https://doi.org/10.1007/s10033-017-0199-9
5. Bonnemains, T., Chanal, H., Bouzgarrou, C., Ray, P.: Definition of a new static model of parallel kinematic machines: highlighting of overconstraint influence. In: IEEE/RSJ International Conference on Intelligent Robots Systems, IROS 2008, pp. 2416–2421 (2008). https://doi.org/10.1109/IROS.2008.4650957
6. Liu, X., Xu, Y., Yao, J., Xu, J., Wen, S., Zhao, Y.: Control-faced dynamics with deformation compatibility for a 5-DOF active over-constrained spatial parallel manipulator 6PUS-UPU. Mechatronics **30**, 107–115 (2015). https://doi.org/10.1016/j.mechatronics.2015.06.014
7. Mabire, A., Serré, P., Moinet, M., Rameau, J.F., Clément, A.: Computing clearances and deviations in over-constrained mechanisms. Procedia CIRP. **75**, 238–243 (2018). https://doi.org/10.1016/j.procir.2018.04.070
8. Guyon, J.-B., Boudon, B., Chanal, H., Blaysat, B.: Toward explicit determination of kinematic model of Exechon like PKM. In: ECCOMAS Thematic Conference on Multibody Dynamic, Duisburg, Germany (2019)
9. Chanal, H., et al.: Geometrical defect identification of a SCARA robot from a vector modeling of kinematic joints invariants. Mech. Mach. Theory. **162**, 104339 (2021). https://doi.org/10.1016/j.mechmachtheory.2021.104339
10. Everett, L.J., Driels, M., Mooring, B.W.: Kinematic modelling for robot calibration. In: IEEE International Conference on Robotics Automation, pp. 183–189 (1987)
11. Altuzarra, O., Campa, F.J.: On singularity and instability in a planar parallel continuum mechanism. In: Lenarčič, J., Siciliano, B. (eds.) Advances in Robot Kinematics 2020. Springer Proceedings in Advanced Robotics, vol. 15, pp. 327–334. Springer, Cham (2021). https://doi.org/10.1007/978-3-030-50975-0_40
12. Pierrot, F., Nabat, V., Company, O., Krut, S., Poignet, P.: Optimal design of a 4-DOF parallel manipulator: from academia to industry. IEEE Trans. Robot. **25**, 213–224 (2009). https://doi.org/10.1109/TRO.2008.2011412
13. Özgür, E., Bouton, N., Andreff, N., Martinet, P.: Dynamic control of the quattro robot by the leg edges. In: Proceeding of IEEE International Conference on Robotics and Automation, pp. 2731–2736 (2011). https://doi.org/10.1109/ICRA.2011.5979733

14. Clément, A., Valade, C., Rivière, A.: The TTRS: 13 oriented constraints for dimensioning, tolerancing and inspection. Adv. Math. Tools Metrol. II **I**, 24–42 (1997)
15. Corbel, D., Nabat, V., Maurine, P.: Geometrical calibration of the high speed hobot Par4 using a laser tracker. In: 12th IEEE International Conference on Methods Models in Automation and Robotics, pp. 687–692 (2006)

A New Approach to Forward Kinematics for a SILS Robotic Orientation Platform Based on Perturbation Theory

Iosif Birlescu, Calin Vaida(✉), Alexandru Pusca, Gabriela Rus, and Doina Pisla

CESTER, Technical University of Cluj-Napoca, Cluj-Napoca, Romania
{iosif.birlescu,calin.vaida,alexandru.pusca,gabriela.rus, doina.pisla}@mep.utcluj.ro

Abstract. The paper presents a mathematical model for the forward kinematics of an orientation platform integrated within a robotic system for Single Incision Laparoscopic Surgery (SILS). The authors propose the use of perturbation theory as a method for solving the mechanism. The approximation accuracy is assessed for several active joints value ranges, being compared with Newton-Raphson (NR) method and Maple solver validating this new approach.

Keywords: Forward kinematics · Orientation platform · Parallel robot · Perturbation theory · SILS

1 Introduction

Concerning parallel robots, a common statement refers to the fact that the inverse kinematic problem provides a simple analytic solution (most of the time unique) while the forward kinematics has multiple solutions, and for some structures the equations require numerical methods to generate an approximate answer. While in many applications, inverse kinematics alone is sufficient to control the pose of a parallel mechanism (isolating a single working mode), in medical robotics, the increased safety conditions require a fast and accurate solution for the forward kinematics to verify/validate the pose of the robot.

The aim of this paper is to explore a possible alternative to numerical methods, namely perturbation theory, which represents a set of methods to find approximate solutions to a problem, normally unsolvable analytically. Refined by Carl Bender, this technique [1] is used normally in quantum mechanics, but it has applications in robotics, such as control [2, 3], flexible robot modelling [4] and dynamic gait locomotion of humanoid robots [5]. The authors aim to show, that perturbation theory can be a powerful, fast, reliable, and accurate alternative for solving the forward kinematics of parallel mechanisms. A case study is presented, solving the forward kinematics of the orientation mechanism positioned on the mobile platform of a surgical robot for Single Incision Laparoscopic Surgery (SILS), which is a type of minimally invasive surgery (MIS) that uses a single trocar to insert the medical instruments within the operating field [6, 7]. To enable

O. Altuzarra and A. Kecskémthy (Eds.): ARK 2022, SPAR 24, pp. 171–178, 2022.
https://doi.org/10.1007/978-3-031-08140-8_19

the SILS advantages (e.g., better cosmetics, faster recovery) over classical MIS, the procedure challenges [6] can be overcome by using robotic systems. One such robotic system (Fig. 1) consisting of: i) a 6-degrees of freedom (DOF) parallel robot guiding a mobile platform that contains ii) two 3-DOF orientation platforms on which the SILS active instruments are mounted on [8].

The paper presents a new model for solving the forward kinematics of a parallel orientation platform based on perturbation theory. Section 2 describes the 3-DOF platform for SILS and its input-output equations. Section 3 presents a short introduction to the perturbation theory methodology that will be used to solve the equations from Sect. 2. Section 4 presents the perturbation theory algorithm used to solve the forward kinematics of the independent orientation platforms for the active instruments, and some numerical results, which based on the data accuracy validate the new mathematical model proposed by the authors.

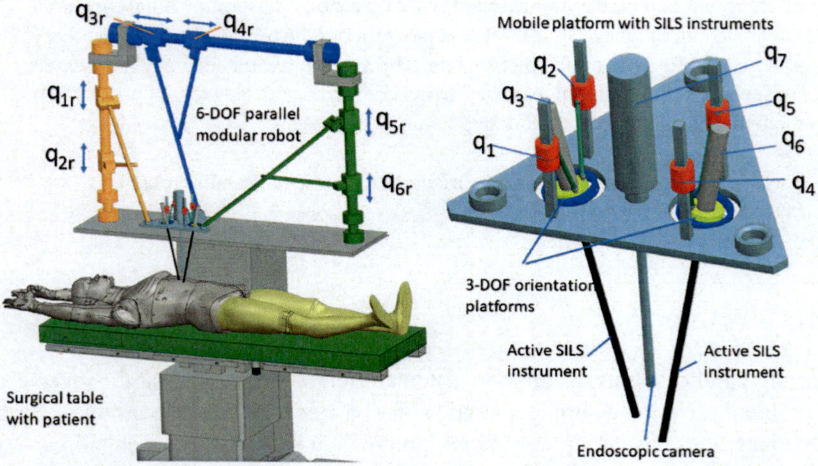

Fig. 1. Parallel robotic system for SILS: (left) the robotic system positioned on the side of the surgical table; (right) detailed view of the mobile platform.

2 The 3-DOF Parallel Orientation Platform for SILS

The kinematic scheme of the orientation platform is presented in Fig. 2, along with the envisioned workspace of the three SILS instruments. The workspace was generated for active instruments with bends [9] with the orientations θ, $\psi \in [-35\ 35]$ [°] (defined with the help of medical experts based on a medical protocol), to avoid instruments crossover at the insertion point and a rigid endoscopic camera.

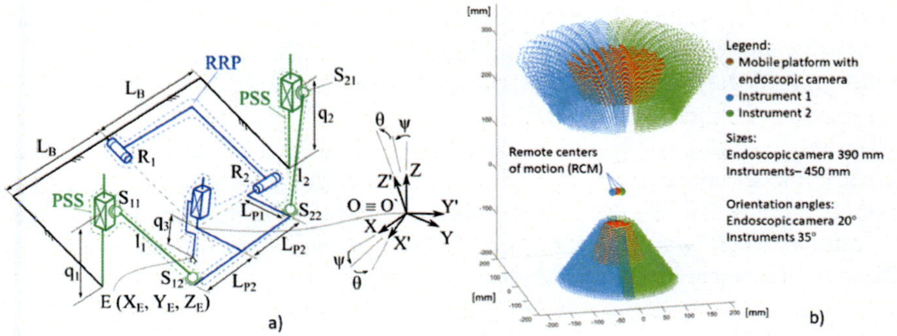

Fig. 2. Mobile orientation platform for SILS: a) kinematic scheme; b) envisioned workspace for the SILS instruments.

The 3-DOF orientation platform for one active instrument consists of:

- Two PSS chains actuated by the prismatic joints q_1 respectively and q_2 (positioned at a distance L_B and $-L_B$ from the origin of the fixed frame *OXYZ*), which move on trajectories parallel with the *OZ*. These kinematic chains integrate two pairs of passive spherical joints S_{11}, S_{12} and S_{21}, S_{22} connected by the links l_1 and l_2;
- One RRP chain with two passive rotational joints R_1 and R_2 with orthogonal axes and an active translational joint q_3 (parallel with the *O'Z'* of the mobile coordinate frame *O'X'Y'Z'*) along the longitudinal axis of the instrument for its insertion/retraction. With respect to the origin O', the passive spherical joints S_{12} and S_{22} are linked at a distance L_{P1} on *O'X'* direction and at L_{P2} (and $-L_{P2}$) on *O'Y'* direction, respectively.

In [10], the authors derived the input-output equations for a parallel robot that uses a similar topology with the parallel mechanism shown in Fig. 2a. The input-output equations were obtained for given numerical values for the geometric parameters $\{l_1 = 5, L_B = 1, L_{P1} = 4, L_{P2} = 1\}$ (dimensions are considered 20 × mm). The input-output equations for the orientation platform are [10]:

$$\begin{cases} \left(\alpha^2+1\right)\left(\beta^2+1\right)q_1^2 - \left(4\alpha^2\beta + 16\alpha\beta^2 + 16\alpha - 4\beta\right)q_1 - 5\left(\alpha^2+1\right)\left(\beta^2 + \frac{9}{5}\right) = 0 \\ \left(\alpha^2+1\right)\left(\beta^2+1\right)q_2^2 + \left(4\alpha^2\beta - 16\alpha\beta^2 - 16\alpha - 4\beta\right)q_2 - 5\left(\alpha^2+1\right)\left(\beta^2 + \frac{9}{5}\right) = 0 \end{cases} \tag{1}$$

where α and β are the tangents of the half-angle of θ, and ψ, respectively. In addition, the equations that describe the relative position between the fixed coordinate system and point E (X_E, Y_E, Z_E), characterizing the tip of the instrument are:

$$\begin{cases} X_E\left(\alpha^2+1\right)\beta^2 - 2q_3\left(\alpha^2+1\right)\beta + X_E\left(\alpha^2+1\right) = 0 \\ Y_E\alpha^2 - 2q_3\alpha + Y_E = 0 \\ Z_E\left(\beta^2+1\right)\alpha^2 + q_3\left(\beta^2-1\right)\left(\alpha^2-1\right) + Z_E\left(\beta^2+1\right) = 0 \end{cases} \tag{2}$$

Equation (2) has closed form solutions for both the forward (i.e., X_E, Y_E, Z_E) and the inverse (i.e., α, β, q_3) kinematic model, respectively. However, Eq. (1) does not have closed form solutions for the forward geometric model (i.e., for α, β); a general solution for α, β, computed with solve command in Maple (which is not presented due to its length) contains the roots of polynomials of degree 8, 7, 6, and 5 (which have no closed form solutions for their roots).

The motivation of finding an explicit (yet approximate) solution for the forward geometric model comes from its requirement of the future development of safe and robust robot control, where the actuators positions are used to determine the instruments tips positions (via forward kinematics), and a classical numerical method would be very difficult to implement in a real-time control system.

3 Perturbation Theory Background

Perturbation theory can offer approximate solutions for otherwise difficult to solve (exactly) mathematical problems. However, it is important to note that perturbation theory is not a numerical method. The approach is to split an "infinitely hard problem" (which cannot be solved exactly) into infinitely many solvable problems. The algorithm in its general form is the following [1, 11]:

Step (1) Define the unsolvable problem P and insert a small parameter ε such that $P(\varepsilon)$ has an exact solution for $\varepsilon = 0$;

Step (2) Assume that the solution for P (denoted SOL) has the form of a Taylor like series:

$$SOL(\varepsilon) = \sum_{n=0}^{\infty} a_n \varepsilon^n \tag{3}$$

Step (3) Substitute $SOL(\varepsilon)$ into $P(\varepsilon)$, compute the coefficients and add up the series for $\varepsilon = 1$. Each computed term in the series better approximates the solution, meaning that perturbation theory provides an arbitrary accurate solution but not an exact solution.

4 Solving the Forward Kinematics

Equation (1) shows that the input-output equations are functions of (α, β, q_1), and (α, β, q_2) respectively, which after applying perturbation theory will lead to a series in the form:

$$\begin{cases} \alpha = a_0(q_1, \beta) + a_1(q_1, \beta) + \ldots + a_n(q_1, \beta) \\ \beta = b_0(q_2, \alpha) + b_1(q_2, \alpha) + \ldots + b_n(q_2, \alpha) \end{cases} \tag{4}$$

but for the forward kinematic model, the goal is to obtain a series in the form:

$$\begin{cases} \alpha = a_0(q_1, q_2) + a_1(q_1, q_2) + \ldots + a_n(q_1, q_2) \\ \beta = b_0(q_1, q_2) + b_1(q_1, q_2) + \ldots + b_n(q_1, q_2) \end{cases} \tag{5}$$

where a_i and b_i $(i = 1 \ldots n)$ are the coefficients of the perturbation series. The input-output equations found in Eq. (1) were rewritten such that the input-output equations are functions of (α, q_1, q_2) and (β, q_1, q_2). To achieve this, two Groebner bases were computed using Maple (from Eq. (1)) to obtain univariate polynomials (in α and β) using pure lexicographic orderings, $\alpha < \beta < q_1 < q_2$, and $\beta < \alpha < q_1 < q_2$, respectively. The first equations of the two Groebner bases (both being univariate and degree 8) have the general form:

$$\begin{cases} P_1 : D_1\alpha^8 + D_2\alpha^7 + D_3\alpha^6 + D_4\alpha^5 + D_5\alpha^4 + D_4\alpha^3 + D_3\alpha^2 + D_2\alpha + D_1 \\ P_2 : T_1\beta^8 + T_2\beta^6 + T_3\beta^4 + T_4\beta^2 + T_5 \end{cases} \tag{6}$$

with:

$$\begin{aligned}
&T_1 = 4(q_1q_2 + 5)^2(q_1 - q_2)^2,\ T_2 = 17q_1^4q_2^2 - \left(30q_2^3 - 182q_2\right)q_1^3 \\
&\quad + \left(17q_2^4 - 660q_2^2 + 585\right)q_1^2 + \left(182q_2^3 - 1070q_2\right)q_1 + 585q_2^2, \\
&T_3 = 26q_1^4q_2^2 - \left(44q_2^3 - 308q_2\right)q_1^3 + \left(26q_2^4 - 1240q_2^2 + 1234\right)q_1^2 \\
&\quad + \left(308q_2^3 - 2108q_2\right)q_1 + 1234q_2^2,\ T_4 = 17q_1^4q_2^2 - \left(30q_2^3 - 238q_2\right)q_1^3 \\
&\quad + \left(17q_2^4 - 804q_2^2 + 1089\right)q_1^2 + \left(238q_2^3 - 1854q_2\right)q_1 + 1089q_2^2, \\
&T_5 = 4(q_1q_2 + 9)^2(q_1 - q_2)^2,\ D_1 = (q_1 + q_2)^2\left(\left(q_2^2 + 1\right)q_1^2 - 16q_1q_2 + q_2^2 + 45\right), \\
&D_2 = -32(q_1 + q_2)\left(2q_1^2q_2^2 + q_1^2 - 16q_1q_2 + q_2^2\right),\ D_3 = 4q_1^4 + \left(1016q_2^2 + 256\right)q_1^2 \\
&\quad - 512q_1q_2 + 4q_2^2 + 256q_2^2,\ D_4 = 32(q_1 + q_2)\left(2q_1^2q_2^2 - 3q_1^2 - 8q_1q_2 - 3q_1^2\right),
\end{aligned} \tag{7}$$

$$\begin{aligned}
D_5 = &\left(6 - 2q_2^2\right)q_1^4 - \left(4q_2^3 - 28q_2\right)q_1^3 - (2q_2^4 + 2004q_2^2 + 422)q_1^2 \\
&+ \left(28q_2^3 - 1204q_2\right)q_1 + 6q_2^4 + 422q_2^2
\end{aligned}$$

A numerical check for various values of q_1 and q_2 reveals that the set of solutions of Eq. (1) is included in the set of solutions of Eq. (6). Even though P_2 has exact solutions (being a quadric in β^2), the solutions are unpractical for the robot control due to: i) computation speed (the achieved perturbation solution evaluates approx. 10 times faster while remaining accurate); ii) numerical evaluation of the explicit solution from P_2 evaluates to NaN for $q_1 = q_2 = 3$ which should be $\beta = 0$ (the perturbation series evaluates to 0).

Step 1 of the perturbation theory algorithm is to insert a small parameter ε into the problem such that when $\varepsilon = 0$, the problem has an exact solution. The following model is proposed, to reduce P_1 to a quadratic equation and P_2 to a quadric one:

$$\begin{aligned}
&P_1(\varepsilon) : \varepsilon\left(D_1\alpha^8 + D_2\alpha^7 + D_3\alpha^6 + D_4\alpha^5 + D_5\alpha^4 + D_4\alpha^3\right) + D_3\alpha^2 + D_2\alpha + D_1 = 0 \\
&P_2(\varepsilon) : \varepsilon\left(T_1\beta^8 + T_2\beta^6\right) + T_3\beta^4 + T_4\beta^2 + T_5 = 0
\end{aligned} \tag{8}$$

For $\varepsilon = 0$, the simplified input-output equations have exact solutions $\{\alpha_{P1}, \beta_{P2}\}$ (one quadratic and one quadric), which represent the coefficients a_0 and b_0 in the perturbation series. The exact solutions that will converge to the forward kinematic solution that describes the intended working mode are (two solutions for b_0 are needed to describe positive and negative values for β, whereas only one solution for b_0 is needed for both positive and negative values of α):

$$a_0 = \frac{1}{2}\frac{D_2 - \sqrt{D_2^2 - 4D_1D_3}}{D_3}, \quad b_0 = \frac{\sqrt{\pm 2T_3\left(T_4 + \sqrt{T_4^2 - 4T_3T_5}\right)}}{2T_3} \tag{9}$$

Two series are defined (Eq. 10), but only six terms are used for the computation (up to ε^5), which will provide good approximations for α and β:

$$\begin{aligned} S_1 &= a_0 + a_1\varepsilon + a_2\varepsilon^2 + a_3\varepsilon^3 + a_4\varepsilon^4 + a_5\varepsilon^5 + \ldots + a_n\varepsilon^n \sim \alpha, \ (n \to \infty) \\ S_2 &= b_0 + b_1\varepsilon + b_2\varepsilon^2 + b_3\varepsilon^3 + b_4\varepsilon^4 + b_5\varepsilon^5 + \ldots + b_n\varepsilon^n \sim \beta, \ (n \to \infty) \end{aligned} \tag{10}$$

Substituting $\alpha = S_1$ and $\beta = S_2$ into Eq. (8) (Step 3) yields a series in powers of ε. Step 2 in Sect. 2 assumed that the solutions are Taylor-like series. Consequently, the equations' lefthand and righthand sides must be equal term by term. Therefore, each coefficient of ε^i (i = 1…n) must vanish, and the terms a_i, b_i are computed by solving (linearly) the coefficients of ε^i. The terms a_1, b_1, b_2 are shown (the other coefficients are lengthy and not presented):

$$\begin{aligned} a_1 &= -\frac{a_0^3\left(a_0^5D_1 + a_0^4D_2 + a_0^3D_3 + a_0^2D_4 + a_0D_5 + D_4\right)}{2\alpha_0D_3 + D_2}, \quad b_1 = -\frac{b_0^5\left(b_0^2T_1 + T_2\right)}{2\left(2b_0^2T_3 + T_4\right)}, \\ b_2 &= \frac{b_0^9\left(b_0^2T_1 + T_2\right)\left(26b_0^4T_1T_3 + 15b_0^2T_1T_4 + 18b_0^2T_2T_3 + 11T_2T_4\right)}{8\left(2b_0^2T_3 + T_4\right)^3} \end{aligned} \tag{11}$$

The final step is to substitute a_i, b_i into Eq. (10) and sum the series for $\varepsilon = 1$.

To assess the accuracy and computation speed of the proposed model several trajectories (based on α and β parameters) were studied (Fig. 3). Based on defined numerical intervals of q_1, q_2 the values of α and β were computed with the perturbation series, the Newton-Raphson (NR) algorithm (on Eq. 1) and using the solve function in Maple (on Eq. 1). The largest errors (compared to the Maple solver, as NR had negligible errors with respect to Maple solver $\approx 10^{-11}$) for α were obtained at the boundary of the operational workspace ($\alpha \approx \pm 0.35$ – corresponding to an angle $\theta \approx 38.5°$) with a maximum error of $4 \cdot 10^{-3}$ (angular error $\approx 0.48°$). For β, the maximum error was $1.5 \cdot 10^{-6}$. The computation time of the perturbation series was on average about 60 times faster than Newton-Raphson with 5 iterations. Considering an insertion depth of 200 [mm] for the tip of the instrument (using Eq. (2)) and the maximum error for α, β the maximum positioning error computed by the root of the mean squared (RMS) for point $E\,(X_E, Y_E, Z_E)$ was 0.5 [mm].

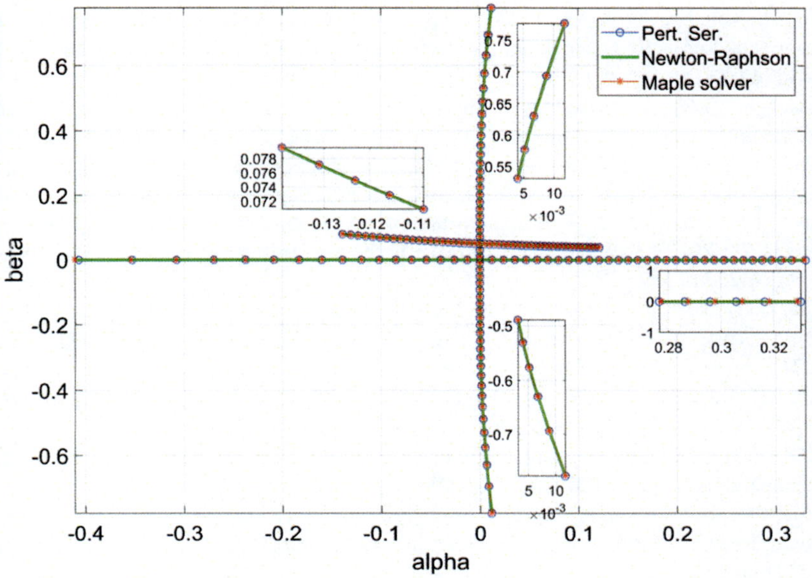

Fig. 3. Accuracy assessment of the proposed model for various trajectories.

5 Conclusions

To avoid numerical methods, the authors used perturbation theory to develop a model that solves (approximatively) the forward kinematics of the SILS platform. The initial numerical evaluation shows good accuracy in the operational workspace and even though NR algorithm is more accurate (maximum error $\approx 10^{-11}$) than the perturbation series approach (maximum error $\approx 10^{-3}$), the proposed method was about 60 times faster, and does not require an initial value for the solution. For the SILS robot control, the better accuracy of NR may be insignificant for a real time control whereas the computation speed is.

Future work aims to analyze how the proposed model behaves in special configurations (e.g., near singularities) and implement it into the control of the parallel orientation platform.

Acknowledgments. This work was supported by a grant of the Ministry of Research, Innovation and Digitization, CNCS/CCCDI – UEFISCDI, project number PCE171/2021 - Challenge within PNCDI III, and project POCU/380/6/13/123927–ANTREDOC, "Entrepreneurial competencies and excellence research in doctoral and postdoctoral studies programs", a project co-funded by the European Social Fund through the Human Capital Operational.

References

1. Bender, C.M., Orszag, S.A.: Advanced Mathematical Methods for Scientists and Engineers I: Asymptotic Methods and Perturbation Theory. Springer, New York (1999). https://doi.org/10.1007/978-1-4757-3069-2

2. Kim, D., Koh, K., Cho, G.-R., Zhang, L.-Q.: A robust impedance controller design for series elastic actuators using the singular perturbation theory. IEEE/ASME Trans. Mechatron. **25**(1), 164–174 (2020)
3. Vossoughi, G.R., Karimzadeh, A.: Impedance control of a two degree-of-freedom planar flexible link manipulator using singular perturbation theory. Robotica **24**(2), 221–228 (2006)
4. Garg, M., Swarup, A., Kanth, A.R.: Application of singular perturbation theory in modeling and control of flexible robot arm. Int. J. Adv. Technol. Eng. Exp. **3**(24), 176–181 (2016)
5. Sentis, L., Fernandez, B.: Perturbation theory to plan dynamic locomotion in very rough terrains. In: 2011 IEEE/RSJ International Conference on Intelligent Robots and Systems, pp. 2267–2273 (2011)
6. Pisla, D., et al.: New approach to hybrid robotic system application in single incision laparoscopic surgery. Acta Technica Napocensis Ser. Appl. Math. Mech. Eng. **24**(3), 369–378 (2021)
7. Rao, P.P., Rao, P.P., Bhagwat, S.: Single-incision laparoscopic surgery - current status and controversies. J. Minim. Access Surg. **7**(1), 6–17 (2011)
8. Pisla, D., et al.: Family of parallel modular robots with active translational joints for SILS, patent pending, OSIM A00733/03.12.2021
9. Vaida, C., et al.: Preliminary control design of a single-incision laparoscopic surgery robotic system. In: 2021 25th International Conference on System Theory, Control and Computing, pp. 384–389 (2021)
10. Birlescu, I., et al.: Motion parameterization of parallel robots used in lower limb rehabilitation. In: Lenarčič, J., Siciliano, B. (eds.) ARK 2020. SPAR, vol. 15, pp. 57–64. Springer, Cham (2021). https://doi.org/10.1007/978-3-030-50975-0_8
11. Pakdemirli, M., Yurtsever, H.: Estimating roots of polynomials using perturbation theory. Appl. Math. Comput. **188**(2), 2025–2028 (2007)

Forward Kinematics of a Novel 6-DoF Spatial Hybrid Manipulator

Shashank Ramesh[1], Pranathi Golla[2], Prem Kumar Prasad[2], and Sandipan Bandyopadhyay[2](✉)

[1] Department of Mechanical Engineering, Indian Institute of Technology Madras, Chennai, India

[2] Department of Engineering Design, Indian Institute of Technology Madras, Chennai 600 036, India

sandipan@iitm.ac.in

Abstract. This paper presents a novel six-degrees-of-freedom spatial hybrid manipulator and the analysis of its forward kinematics. The proposed manipulator has three limbs of identical architecture connecting a hexagonal fixed platform to a moving platform in the form of an equilateral triangle. The limbs are spaced symmetrically w.r.t. the fixed platform. Each limb consists of a planar five-bar mechanism at its base, the tip of which connects to the moving platform via a passive RS link. The forward kinematic problem of the manipulator is reduced to that of a circle intersecting an oct-circular curve of degree 16 in its own plane. As expected, this leads to a maximum of 16 isolated (complex) solutions, the reals among which lead to poses with a certain mirror-symmetry. Consequently, a polynomial of only degree 8 has to be solved eventually. Coefficients of this polynomial have been found as closed-form symbolic expressions in the architecture parameters and input variables. The theoretical results are demonstrated and validated numerically using an example.

Keywords: Hybrid manipulator · Forward kinematic problem · Symbolic computation · Closed-form expressions · Oct-circular curve

1 Introduction

Hybrid manipulators are becoming increasingly popular in research since they afford reasonable compromises between the parallel characteristics, e.g., rigidity and better payload-to-self-weight ratio, and the serial ones, e.g., larger workspaces. A prominent class among these are platform-type robots having three limbs and all the six spatial *degrees-of-freedom* (DoF). Monsarrat et al. introduced a six-DoF hybrid manipulator in [1], whose moving platform (MP) connects to a fixed platform (FP) via three identical limbs consisting of individual five-bar mechanisms. The closed-form forward kinematics of another three-limbed manipulator with six-DoF was presented by Tahmasebi and Tsai in [2].

O. Altuzarra and A. Kecskémthy (Eds.): ARK 2022, SPAR 24, pp. 179–187, 2022.
https://doi.org/10.1007/978-3-031-08140-8_20

The limbs of this "minimanipulator" are driven by five-bar manipulators placed in a plane parallel to the FP. The manipulator presented in [3] also has three limbs actuated by four-bar mechanisms lying in the base plane. Long and Collins, in [4], used five-bar driver mechanisms placed in different moving planes for force-reflected teleoperation.

This paper proposes a new architecture in this class of robots, wherein each limb, placed in a symmetric manner on the fixed platform has two-DoF and an identical architecture. Each limb forms a serial combination of an actuated planar five-bar sub-chain at its base and a passive RS sub-chain linking it to the moving platform. Unlike in the minimanipulator described in [2], here the five-bar mechanisms are contained in planes forming an angle (denoted by α in Fig. 1b) to the base plane, and also, a separation is introduced (marked as $\boldsymbol{b}_{l_1}\boldsymbol{b}_{r_1}$ in the same figure) between the actuators situated at the base of each five-bar mechanism. These features enhance the application potential of the manipulator by introducing new parameters in its design, without complicating the kinematic analysis (as can be seen in Sect. 3). For instance, these architecture parameters can be tuned to attain desired combinations of workspace dimensions and static/dynamic performance suited to any particular application. Moreover, a judicious choice of α allows the designer to critically apportion the payload-induced forces (along the *RS* links) into components to be supported *actively*, i.e., via letting them reflect on the actuators, and to be resisted *passively* in the *structural mode*, i.e., letting the links support the loads via elastic deformation, without transmitting these to the actuators.

The forward kinematic problem (FKP) is studied comprehensively in the closed-form using symbolic computations, via a *computer algebra system* (CAS), namely, `Mathematica` [5]. The kinematics of the planar five-bar mechanisms are both decoupled and trivial, as opposed to the rest of the manipulator. However, performing an inversion in which the MP is considered fixed, the problem is rendered equivalent to that of the FKP of the 3-R$\underline{\text{P}}$S manipulator (more precisely, a 3-RS chain), which is solved following the geometric approach described in [6]. The problem is finally reduced to the solution of a univariate polynomial equation of degree 8, whose coefficients have been obtained as closed-form symbolic expressions in the actuator inputs and the architecture parameters of the manipulator. However, since each of the five-bar mechanisms can have two configurations for a given input to them, and that the poses of the upper portion of the manipulator maintain a pair-wise mirror symmetry w.r.t. the MP, the total number of possible isolated solutions in the complex plane is $2 \times 2 \times 2 \times 8 \times 2 = 128$.

The rest of the paper is organised as follows: the geometry of the manipulator is described in Sect. 2. Section 3 presents the formulation and solution procedure of the loop-closure equations. The results are numerically validated in Sect. 4. The conclusions and the scope for potential extensions are presented in Sect. 5.

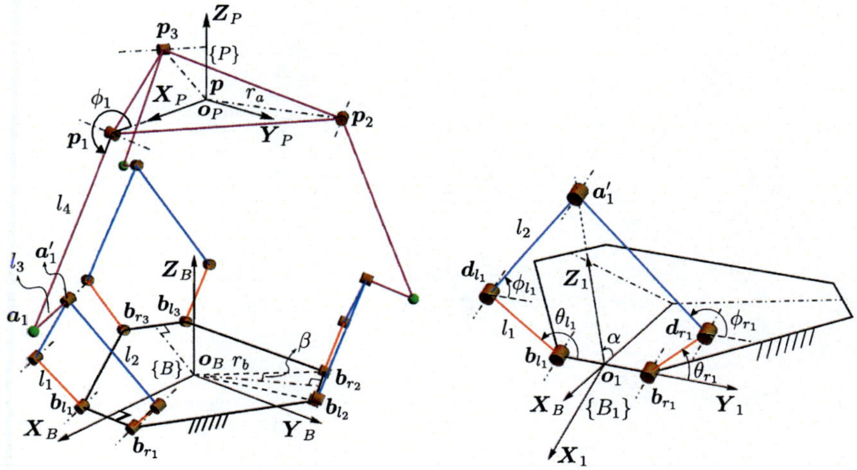

(a) Schematic of the hybrid manipulator

(b) Five-bar sub-chain of the first limb

Fig. 1. Kinematic description of the proposed hybrid manipulator

2 Geometry of the Manipulator

The kinematic sketch of the proposed hybrid manipulator is depicted in Fig. 1a. It is composed of three limbs connecting the FP to the MP. The vertices[1] of the MP, namely, $\boldsymbol{p}_1, \boldsymbol{p}_2$ and $\boldsymbol{p}_3$, form an equilateral triangle whose circumradius is r_a. The vertices of the hexagonal FP, namely, $\boldsymbol{b}_{l_1}, \boldsymbol{b}_{r_1}, \boldsymbol{b}_{l_2}, \boldsymbol{b}_{r_2}, \boldsymbol{b}_{l_3}$ and $\boldsymbol{b}_{r_3}$, lie on a circle of radius r_b, such that alternate sides of the hexagon have equal length. Consequently, the angular spacing between $\boldsymbol{b}_{l_i}$ and $\boldsymbol{b}_{r_i}$ is 2β, $i = 1, 2, 3$. The MP is driven by actuating the three five-bar sub-chains attached to its base. The five-bar sub-chains lie in their respective planes all of which are inclined at an angle α with the FP. The axes of all of its rotary joints make an angle of $(\alpha - \pi/2)$ with the FP.

A global reference frame $\boldsymbol{o}_B$-$\boldsymbol{X}_B\boldsymbol{Y}_B\boldsymbol{Z}_B$, denoted by $\{B\}$, is fixed to the centroid of the FP such that $\boldsymbol{X}_B$ bisects the line segment $\boldsymbol{b}_{l_1}\boldsymbol{b}_{r_1}$. Similarly, the reference frame $\{P\}$, constituted of $\boldsymbol{o}_P$-$\boldsymbol{X}_P\boldsymbol{Y}_P\boldsymbol{Z}_P$, is attached to the centroid of the MP, $\boldsymbol{p}$, with the axis $\boldsymbol{X}_P$ passing through $\boldsymbol{p}_1$. The axes $\boldsymbol{Z}_P$ and $\boldsymbol{Z}_B$ define the normals to the MP and FP, through their respective origins. A reference frame $\{B_1\}$, specified by $\boldsymbol{o}_1$-$\boldsymbol{X}_1\boldsymbol{Y}_1\boldsymbol{Z}_1$, is defined with its origin at the midpoint of $\boldsymbol{b}_{l_1}\boldsymbol{b}_{r_1}$ and $\boldsymbol{Y}_1$-axis along $\boldsymbol{b}_{l_1}\boldsymbol{b}_{r_1}$. The axis $\boldsymbol{Z}_1$ is inclined at an angle α with the FP as seen in Fig. 1b. In a similar fashion, frames $\{B_2\}$ and $\{B_3\}$ are attached to the edges $\boldsymbol{b}_{l_2}\boldsymbol{b}_{r_2}$ and $\boldsymbol{b}_{l_3}\boldsymbol{b}_{r_3}$, respectively (not shown in the figure to avoid clutter). The five-bar sub-chain of the ith limb lies in the plane $\boldsymbol{Z}_i\boldsymbol{Y}_i$, actuated by the

[1] In this paper, ${}^F\boldsymbol{p}$ would indicate the point $\boldsymbol{p}$ being expressed in the frame $\{F\}$, and so on. An absence of any leading superscript would imply the point being in frame $\{B\}$.

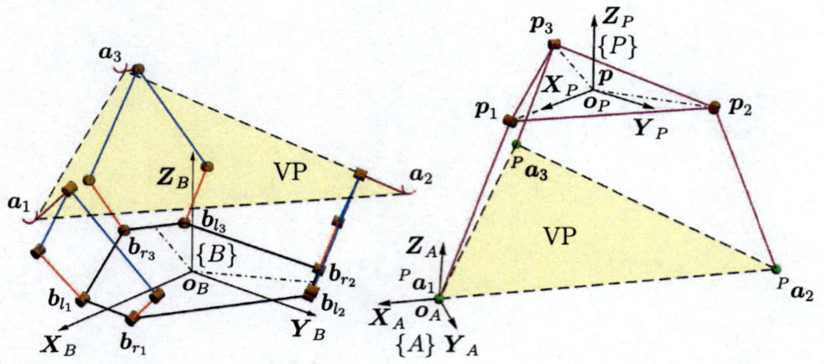

Fig. 2. The manipulator bisected hypothetically by the *virtual platform* (VP)

motors located at $\boldsymbol{b}_{l_i}$ and $\boldsymbol{b}_{r_i}$. The links attached to the actuators, namely, $\boldsymbol{b}_{l_i}\boldsymbol{d}_{l_i}$ and $\boldsymbol{b}_{r_i}\boldsymbol{d}_{r_i}$, are each of length l_1 and inclined to $\boldsymbol{Y}_i$-axis at angles θ_{l_i} and θ_{r_i}, respectively, which define the active/actuated variables. The passive variables ϕ_{l_i} and ϕ_{r_i} define the orientation of the passive links of the five-bar sub-chain, i.e., $\boldsymbol{d}_{l_i}\boldsymbol{a}'_i$ and $\boldsymbol{d}_{r_i}\boldsymbol{a}'_i$, respectively, w.r.t. the axis $\boldsymbol{Y}_i$. Both the passive links are of length l_2. A rigid link of length l_3 directed along $\boldsymbol{X}_i$ connects the ith five-bar sub-chain to a link of length l_4, i.e., $\boldsymbol{p}_i\boldsymbol{a}_i$, via a spherical joint at $\boldsymbol{a}_i$. The other end of $\boldsymbol{p}_i\boldsymbol{a}_i$ is connected to the MP by a revolute joint at $\boldsymbol{p}_i$, whose axis lies in the plane of the MP and is perpendicular to the line joining $\boldsymbol{p}$ and $\boldsymbol{p}_i$. The angle between $\boldsymbol{p}_i\boldsymbol{a}_i$ and $\boldsymbol{p}_i\boldsymbol{p}$ is denoted by ϕ_i. Hence, the six-DoF motion of the moving platform is realised by actuating all the three five-bar sub-chains.

3 The Forward Kinematic Problem

The objective of the FKP is to find the pose of the MP in the frame of the FP defined by $\boldsymbol{p} = [x_p, y_p, z_p]^\top$ and the orientation defined by the unit quaternion $\boldsymbol{q} = [q_0, q_1, q_2, q_3]^\top$, respectively, given the actuator inputs, denoted by $\boldsymbol{\theta} = [\theta_{l_1}, \theta_{r_1}, \theta_{l_2}, \theta_{r_2}, \theta_{l_3}, \theta_{r_3}]^\top$, and the architecture parameters $\boldsymbol{\mu} = [l_1, l_2, l_3, l_4, r_a, r_b, \alpha, \beta]^\top$. It is solved by locating the *virtual platform* (denoted by VP in Fig. 2) in the FP and MP through the following steps.

1. **Decomposition into two kinematic sub-chains**
   As shown in Fig. 2, the manipulator can be decomposed into two parts by hypothetically removing the spherical joints at $\boldsymbol{a}_i$: the lower part, starting from the FP and leading up to the VP, and the upper part, starting from the MP and coming down to the same VP. This decomposition allows for modular computation of the forward kinematics of both the parts.
2. **Determination of the pose of the VP w.r.t. the FP (frame $\{B\}$)**
   The FKP of the lower part is rather trivial, as it involves a decoupled analysis of three five-bar sub-chains, namely, $\boldsymbol{b}_{l_i}\boldsymbol{d}_{l_i}\boldsymbol{a}'_i\boldsymbol{d}_{r_i}\boldsymbol{b}_{r_i}$, $i = 1, 2, 3$. The external revolute inputs, $\boldsymbol{\theta}$, are converted to the locations of the tips of

the five-bar sub-chains, $\boldsymbol{a}'_i$, through an intermediary step of computing the passive angles $\boldsymbol{\phi} = [\phi_{l_1}, \phi_{r_1}, \phi_{l_2}, \phi_{r_2}, \phi_{l_3}, \phi_{r_3}]^\top$. Consequently, the vertices of the VP, $\boldsymbol{a}_i$, are defined in the frame $\{B\}$. A new frame of reference, given by $\boldsymbol{o}_A$-$\boldsymbol{X}_A\boldsymbol{Y}_A\boldsymbol{Z}_A$ and denoted by $\{A\}$, is attached to $\boldsymbol{a}_1$. The axes are chosen/computed as:

$$\boldsymbol{X}_A := \frac{\boldsymbol{a}_1 - \boldsymbol{a}_2}{\|\boldsymbol{a}_1 - \boldsymbol{a}_2\|}, \quad \boldsymbol{Z}_A := \frac{\boldsymbol{X}_A \times (\boldsymbol{a}_1 - \boldsymbol{a}_3)}{\|\boldsymbol{X}_A \times (\boldsymbol{a}_1 - \boldsymbol{a}_3)\|}, \quad \boldsymbol{Y}_A := \boldsymbol{Z}_A \times \boldsymbol{X}_A. \quad (1)$$

Together, these axes and $\boldsymbol{a}_1$ define the homogeneous transformation matrix ${}^B_A\boldsymbol{T} \in \mathbb{SE}(3)$, that describes the pose of the VP w.r.t. the fixed frame $\{B\}$. Since each five-bar sub-chain can have two configurations for a given set of inputs, the total number of solutions for this part is $2 \times 2 \times 2 = 8$.

3. **Determination of the pose of the VP w.r.t. the MP (frame $\{B\}$)**
The upper part of the manipulator forms a 3-SR passive chain connecting the VP to the MP, which is harder to analyse. Therefore, without any loss of generality, a kinematic inversion is performed, in which the MP is considered to be fixed, and the VP moves w.r.t. the same, forming an equivalent 3-RS chain with identical leg lengths, l_4. The FKP of this "manipulator" is solved following the geometric approach described in [6] and [7], in the contexts of the 3-R$\underline{\text{P}}$S and the 3-R$\underline{\text{P}}$RS manipulators, respectively. The process is described briefly for completeness.
The equivalent 3-RS chain is again decomposed hypothetically into two kinematic sub-chains by removing the spherical joint at ${}^P\boldsymbol{a}_1$, splitting the point ${}^P\boldsymbol{a}_1$ into two physically coincident but notionally distinct points: ${}^P\boldsymbol{a}_{s_1}$, the *coupler point* of the *RSSR* chain ${}^P\boldsymbol{p}_2{}^P\boldsymbol{a}_2{}^P\boldsymbol{a}_3{}^P\boldsymbol{p}_3$; and ${}^P\boldsymbol{a}_{c_1}$, the tip of the revolute link ${}^P\boldsymbol{p}_1{}^P\boldsymbol{a}_1$. The locus of ${}^P\boldsymbol{a}_{s_1}$ has been described as the *spin surface* in [8]. The locus of ${}^P\boldsymbol{a}_{c_1}$ is a circle in the plane $y = 0$, due to the presence of the revolute joint at ${}^P\boldsymbol{p}_1$:

$$C(x, z) := (x - r_a)^2 + z^2 - l_4^2 = 0. \quad (2)$$

In order to find the intersections of this circle with the spin surface, the latter is first sliced by the plane containing the circle, i.e., $y = 0$. This leads to the curve $S(x, z) = 0$, which decomposes into two components:

$$S(x, z) := \xi_1(x, z)\tau(x, z) = 0, \quad \text{where} \quad \xi_1(x, z) := \left((x + 2r_a)^2 + 4z^2\right)^2. \quad (3)$$

Further details on the derivation of Eq. (3) may be found in [6], wherein the special case related to the vanishing of $\xi_1(x, z)$ is also explained. Ignoring that case, the FKP of the 3-RS sub-chain now reduces to the computation of the points of intersection between the curve $\tau = 0$ and the circle $C = 0$. The curve $\tau = 0$ is of total degree $m_S = 16$ in x, z and has a circularity of $c_S = 8$ as mentioned in [7]. For $C = 0$, the degree $m_C = 2$ and circularity $c_C = 1$. Thus, the maximum number of isolated points of intersection in the complex plane, given by n, is calculated as the modified Bézout's number (see, e.g., [8]):

$$n := m_S m_C - 2c_S c_C = 16 \times 2 - 2 \times 8 \times 1 = 16. \quad (4)$$

Furthermore, it is apparent from the geometry of the manipulator (and seen in Fig. 3) that the real poses of the VP appear pairwise as mirror images of each other about the MP, i.e., $z = 0$ in $\{P\}$. This is observed algebraically as well, i.e., when the symbolic expression of $\tau(x, z)$ is derived in the closed-form in terms of the architecture parameters $\boldsymbol{\mu}$ and the inputs $\boldsymbol{\theta}$, the coefficients of the odd powers of z in it vanish identically upon simplification. Therefore, the variable z is eliminated from Eqs. (2, 3), by a direct substitution of z^2 from the former into the latter. This results in a polynomial equation of degree 16 in x, whose leading coefficients vanish upon simplification (as might be expected), resulting in the *forward kinematic univariate* (FKU) equation of degree 8 in x:

$$f(x) := a_0x^8 + a_1x^7 + a_2x^6 + a_3x^5 + a_4x^4 + a_5x^3 + a_6x^2 + a_7x + a_8 = 0. \quad (5)$$

The coefficients a_i, $i = 0, \ldots, 8$ of the FKU are also obtained as exact functions of $\boldsymbol{\mu}$ and $\boldsymbol{\theta}$ in their symbolic forms. Hence, the solution procedure and the number of solutions are valid for the most general architecture and all possible actuator inputs of the manipulator. The total *size*[2] of the FKU is 4.222 MB.
For each value of x satisfying Eq. (eq:FKU), two solutions for z are obtained from $C = 0$, thus, leading to a total number of 16 solutions[3], matching up to the predicted modified Bézout number. Using the real and accurate (see Sect. 4) ones among these, the vertices of the VP, ${}^P\boldsymbol{a}_i$, $i = 1, 2, 3$ are computed. Finally, in a manner analogous to that used in Step 2, the pose of the VP is described in the frame $\{P\}$ via the transformation matrix, ${}^P_A\boldsymbol{T}$. The details are skipped to avoid repetition.

4. **Closing the kinematic loop by joining the two parts separated at the VP**
This step reverses Step 1 to reconstitute the manipulator, enforcing the loop-closure constraints in the process. This is done by equating the two descriptions of the pose of the VP: one obtained directly in the frame $\{B\}$, and the other originally obtained in the frame $\{P\}$, but transformed back into $\{B\}$, i.e., ${}^B_A\boldsymbol{T} = {}^B_P\boldsymbol{T}{}^P_A\boldsymbol{T}$. From this, the desired final result, i.e., ${}^B_P\boldsymbol{T}$, may be obtained trivially. Thus, the FKP is seen to have a total of $8 \times 16 = 128$ solutions.

[2] In this paper, the "size" of an expression refers to the amount of memory used internally by the CAS `Mathematica`, as seen through the built-in command `ByteCount`.
[3] These solutions, obtained in the VP frame, divide equally among two *operation modes* (see, e.g., [9]).

Table 1. Poses of the MP corresponding to the real and accurate solutions of Eq. (6). Each pose is described via the coordinates of its centroid, $\boldsymbol{p} = [x_p, y_p, z_p]^\top$ expressed in frame $\{B\}$ and the orientation of the associated frame $\{P\}$ w.r.t. frame $\{B\}$, expressed in terms of the unit quaternion $\boldsymbol{q} = [q_0, q_1, q_2, q_3]^\top$.

Solution no.	x_p	y_p	z_p	q_0	q_1	q_2	q_3	$e \times 10^{-11}$
1	0.2511	−1.5998	5.0430	0.0934	0.2976	−0.9435	0.1115	1.1182
2	−0.5033	0.2051	4.4713	0.1606	−0.1562	0.0741	0.9717	1.1396
3	−0.3851	−0.0346	5.9941	0.9844	−0.0203	−0.0717	−0.1590	0.0022
4	−1.6622	1.1489	4.6518	0.4097	0.9108	0.0484	−0.0027	0.0013
5	0.7973	−1.6682	−0.4506	0.1826	−0.3083	0.9309	0.0699	1.1184
6	−0.0505	0.1485	−0.0833	0.1544	0.0573	−0.0776	0.9832	1.1646
7	0.3644	−0.1284	−1.5466	0.9869	0.0237	−0.0279	−0.1570	0.0021
8	−1.1489	1.0846	−0.5115	0.3916	−0.9111	−0.0887	−0.0922	0.0014

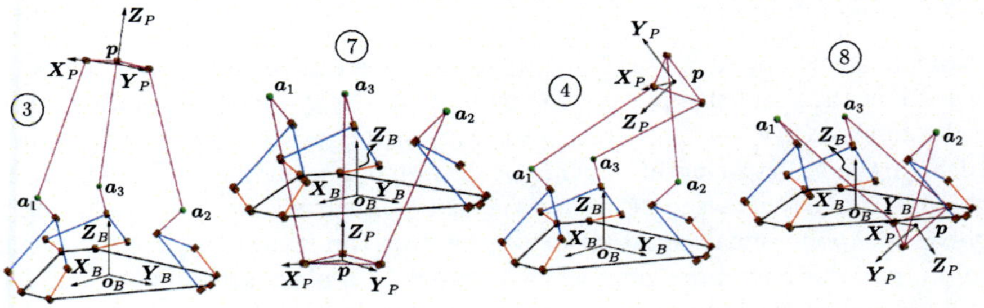

Fig. 3. Configurations of the manipulator depicting the mirror symmetry of the solution pairs 3, 7 and 4, 8 (in Table 1) about the MP. It may also be seen that the MP and the RS links attached to it are reflected together about the VP in each pair.

4 Numerical Example

This section illustrates the theoretical developments presented in the previous section using a numerical example. Choosing the architecture parameters as $\boldsymbol{\mu} = [1, 2, 1, 4, 1, 3, \frac{\pi}{4}, \frac{\pi}{18}]^\top$ and the actuator inputs[4] as $\boldsymbol{\theta} = \frac{\pi}{36}[30, 18, 34, 14, 33, 10]^\top$, the monic form of Eq. (5) becomes:

$$x^8 - 6.3797x^7 - 33.9720x^6 + 186.9577x^5 + 533.0284x^4 \\ - 1669.4662x^3 - 3943.7132x^2 + 3943.9716x + 8774.8695 = 0. \tag{6}$$

[4] All angles in this paper are measured in radians. The length parameters, i.e., $l_1, l_2, l_3, l_4, r_a, r_b$, are rendered unitless by normalising them w.r.t. the MP circumradius, i.e., r_a.

The function `NSolve` of `Mathematica` is used to compute the roots of this polynomial using the default numerical precision. The accuracy of the final solutions are ascertained by enforcing the condition $e := \|[e_1, e_2, e_3]^\top\| \leq \epsilon$, where:

$$e_i = (\boldsymbol{p}_i - \boldsymbol{p}_j) \cdot (\boldsymbol{p}_i - \boldsymbol{p}_j) - 3r_a^2, \quad i = 1, 2, 3, \ j = (i+1) \mod 3, \text{and } \epsilon = 10^{-10}.$$

The passive angles for the *elbow-up* configurations of all the five-bar sub-chains are given by $\boldsymbol{\phi} = [1.3114, 1.8379, 1.2118, 1.7500, 1.0193, 1.5673]^\top$. These configurations yield 8 real solutions for the pose of the MP w.r.t. the frame $\{B\}$, as listed in Table 1. The configurations of the 3-RS sub-chain corresponding to the FKP solutions 5–8 are the mirror reflections of those associated with solutions 1–4, respectively, about the plane containing the MP. This can be seen in Fig. 3 for the mirror-symmetric FKP solution pairs 3, 7 and 4, 8.

5 Conclusions

A novel six-DoF spatial hybrid manipulator is presented and analysed for its forward kinematics in this paper. The manipulator may be conceptually decomposed into three five-bar mechanisms situated at its base, the tips of which meet a 3-RS sub-chain, starting backwards from its moving platform. The kinematic problem is solved via a geometric formulation, which reduces to the solution of a univariate polynomial equation of degree 8 at the last stage. For the whole manipulator the maximum number of isolated (complex) solutions is found to be 128. However, these may be computed accurately since the coefficients of the said 8-degree polynomial have been obtained *exactly* in their symbolic forms, for the general architecture of the manipulator and for generic actuator inputs. This enables parametric studies of the manipulator with regard to its kinematic and static/dynamic performance without incurring huge computational costs. In the present paper, the solution procedure is demonstrated through a numerical example and the results are validated.

References

1. Monsarrat, B., Gosselin, C.M.: Workspace analysis and optimal design of a 3-Leg 6-DOF parallel platform mechanism. IEEE Trans. Robot. Autom. **19**(6), 954–966 (2003)
2. Tahmasebi, F., Tsai, L.-W.: Closed-form direct kinematics solution of a new parallel minimanipulator. J. Mech. Des. **116**(4), 1141–1147 (1994)
3. Han, C., Hudgens, J.C., Tesar, D., Traver, A.E.: Modeling, synthesis, analysis, and design of high resolution micromanipulator to enhance robot accuracy. In: Proceedings IROS: IEEE/RSJ International Workshop on Intelligent Robots and Systems, vol. 2, pp. 1157–1162 (1991)
4. Collins, C., Long, G.: The singularity analysis of an in-parallel hand controller for force-reflected teleoperation. IEEE Trans. Robot. Autom. **11**(5), 661–669 (1995)
5. Wolfram Research, Inc.: Mathematica, Version 12.0, Champaign, IL (2019)

6. Mamidi, T.K., Baskar, A., Bandyopadhyay, S.: A novel geometric analysis of the kinematics of the 3-RPS manipulator. In: Zeghloul, S., Romdhane, L., Laribi, M. (eds.) Computational Kinematics. Mechanisms and Machine Science, vol. 50, pp. 483–490. Springer, Cham (2018). https://doi.org/10.1007/978-3-319-60867-9_55
7. Santosh Kumar, S.P., Bandyopadhyay, S.: Forward kinematics of the 3-RPRS parallel manipulator using a geometric approach. In: Badodkar, D., Dwarakanath, T. (eds.) Machines, Mechanism and Robotics. LNME. Springer, Singapore (2019). https://doi.org/10.1007/978-981-10-8597-0_14
8. Hunt, K., Primrose, E.: Assembly configurations of some in-parallel-actuated manipulators. Mech. Mach. Theory **28**(1), 31–42 (1993)
9. Schadlbauer, J., Walter, D., Husty, M.: The 3-RPS parallel manipulator from an algebraic viewpoint. Mech. Mach. Theory **75**, 161–176 (2014)

Structural and Dimensional Synthesis of Overconstraint Symmetric 3T2R Parallel Robots Using Tait-Bryan-Angle Kinematic Constraints

Moritz Schappler(✉)

Leibniz University Hannover, Institute of Mechatronic Systems, Hanover, Germany
moritz.schappler@imes.uni-hannover.de

Abstract. The basic parallel robotics principle of defining kinematic constraints as vector loops is transferred from the general 3T3R case to the 3T2R case by applying a nonlinear Tait-Bryan-angle rotation constraint using the intrinsic Z-Y'-X'' convention. This presents an alternative way to the deduction of differential inverse kinematics by the theory of linear transformations. The resulting formulation is used in a permutational combined structural and dimensional synthesis. The modular approach allows to combine databases of serial and parallel robots without manual intervention. The validation shows the reproducibility of existing kinematic structures using the new kinematic formulation. The optimization scheme allows to obtain suitably dimensioned symmetric 3T2R parallel robots for a given task.

Keywords: Parallel robot · Parallel manipulator · Overconstraint · 3T2R · Kinematic constraints · Tait-Bryan angles · Euler angles · Dimensional synthesis

1 Introduction and State of the Art

A special case of parallel robots or manipulators (PMs) are those with five structural degrees of freedom (DoF): three translational and two rotational (3T2R). These PMs provide an interesting kinematic structure e.g. for machining tasks. Symmetric 3T2R PMs with identical limbs were shown to exist rather late [5,9]. Since then, works on analysis, modeling and synthesis of these types of PMs have increased [7,16].

The *analysis* of symmetric 3T2R PMs can be performed by several methods. Most prominent is screw theory [5,7,10,13], which can be extended by other algebraic and geometric concepts like Grassmann-Cayley algebra and Grassmann geometry [1] or algebraic geometry (Study parameters and Gröbner bases) [13]. For the related 2T3R PMs Lie groups of displacements were used in [11]. The concept of linear transformations presents an approach of less mathematical

O. Altuzarra and A. Kecskémthy (Eds.): ARK 2022, SPAR 24, pp. 188–197, 2022.
https://doi.org/10.1007/978-3-031-08140-8_21

complexity and has also been used in the context of general 3T2R PMs [8]; in [16] together with a geometrical analysis. The existing works focus on different aspects like forward kinematics [13], workspace [14], singularity analysis [1,12] or compliance modeling [3].

The *synthesis* of this type of PM has been performed by screw theory in combination with a constraint method [9] or a virtual chain approach [10]. In [4] screw theory is used for synthesis together with a systematic deduction and naming scheme to build up a database of symmetric and asymmetric PMs. In [8] the synthesis is performed by evolutionary morphology and the theory of linear transformations, which was taken up in similar form in [16]. Works on robot synthesis give high focus on the *structural* synthesis of the leg chains and mainly provide rules for the synthesis of complete PMs together with some selected examples. However, the performance of parallel robots is strongly depending on the *dimensioning* of the parameters [15].

To obtain PMs for a given purpose—even for an academic example—the *dimensional* synthesis has to be performed—either manually or automatically. Therefore, the structural synthesis should be *combined* with a dimensional synthesis [6]. The kinematics parameter optimization of a five-DoF PM with complex kinematic structure but few parameters is performed in [18]. Other works on dimensional synthesis like [6] focus on different architectures, but are also transferable to the 3T2R case.

To further the combined synthesis of 3T2R PMs, the paper's contributions are

- an *alternative approach* to the *inverse kinematics* model of symmetric 3T2R PMs with a deduction similar to the theory of linear transformations from [8],
- an optimization scheme suitable for combined structural and dimensional synthesis of 3T2R PMs with *less mathematical complexity* than established methods,
- the reproduction of symmetric 3T2R PMs from literature with the new method,
- an open-source Matlab toolbox for the kinematics model, the structural and dimensional synthesis toolchain and a serial chain and parallel robot database.

The remainder of the paper is structured as follows. Section 2 introduces the kinematic model for symmetric 3T2R parallel robots. The synthesis of these robots is discussed in Sect. 3, followed by results in Sect. 4 and a conclusion in Sect. 5.

2 Inverse Kinematic Model for 3T2R Parallel Robots

The constraints equations for parallel robots are usually defined on or equivalent to the velocity level when using screw theory [10] or the theory of linear transformations [8]. The following section presents an inverse kinematics model based on full kinematic constraints equations, which can be seen as an alternative deduction to the latter method. Approaching the kinematics problem from the nonlinear position level can avoid difficulties regarding the reference frame of angular velocities. A method from [17] for 3T3R parallel robots with functional redundancy for 3T2R tasks is transferred to the case of 3T2R robots without redundancy. It should be kept in mind that the method is dedicated to a numeric evaluation and only parts of the expressions are derived symbolically. An elimination of passive joint coordinates or the use for the forward kinematics problem is not feasible with the proposed approach. This does not present a disadvantage for the combined synthesis or in simulation.

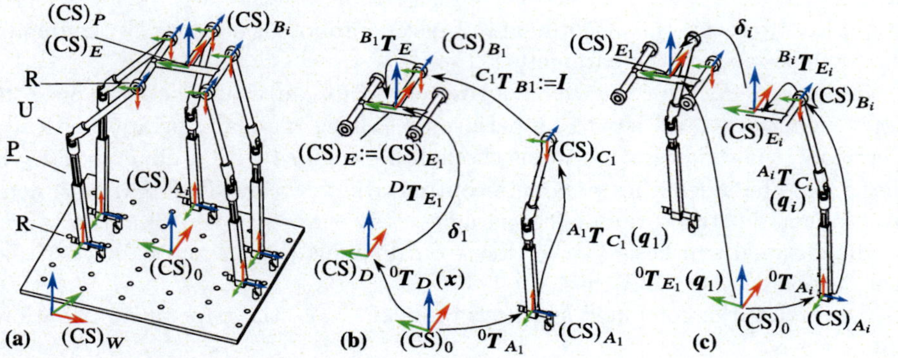

Fig. 1. Sketch of the kinematics model at the example of the modified 5-RPUR image of [12]. (**a**) Full model with all coordinate frames, (**b**) constraints for first leg chain and (**c**) other leg chains $i \neq 1$

A set of coordinate systems (CS), shown in Fig. 1a, is used to model the parallel robot with $m = 5$ leg chains $n = 5$ platform DoF, extending the model of [2]. The robot base frame $(\mathrm{CS})_0$ is fixed regarding the world frame $(\mathrm{CS})_W$. Each leg chain has a virtual base frame $(\mathrm{CS})_{A_i}$ and a virtual end frame $(\mathrm{CS})_{C_i}$. This allows a modular use of models for serial kinematic leg chains. The cut joint frames at the platform are $(\mathrm{CS})_{B_i}$, corresponding to the leg's $(\mathrm{CS})_{C_i}$. The desired end effector frame $(\mathrm{CS})_D$ is used in the derivation of the model and $(\mathrm{CS})_E$ corresponds to the actual frame of the end effector. Geometric parameters are given by the SE(3) matrices ${}^0\boldsymbol{T}_{A_i}$ for the base coupling joint frames, ${}^P\boldsymbol{T}_{B_i}$ for the platform coupling joint frames and ${}^P\boldsymbol{T}_E$ for the (for now neglected) end effector (e.g. tool) frame on the platform.

The translational component of the operational space coordinates $\boldsymbol{x}_\mathrm{t} = {}_{(0)}\boldsymbol{r}_D$ is the vector to the desired end effector frame relative to the base frame. The rotational component marks the pointing direction of the end effector frame with the two angles $\boldsymbol{x}_\mathrm{r}^\mathrm{T} = [\varphi_x, \varphi_y]$, as established in literature [1, 7, 14, 16, 18] with $\varphi_z = 0$. To allow general (tool) frame definitions, using X-Y'-Z'' Tait-Bryan angles for the full orientation of the end effector frame gives the SO(3) rotation matrix

${}^0\boldsymbol{R}_D = \boldsymbol{R}_x(\varphi_x)\boldsymbol{R}_y(\varphi_y)\boldsymbol{R}_z(\varphi_z)$. The third angle $\varphi_z = \text{const}$ (depending on frame definitions) corresponds to a rotation around the Z-axis of $(\text{CS})_E$ and—as a dependent variable for 3T2R PMs—is not included in the five-DoF minimal coordinate $\boldsymbol{x}^\mathsf{T} = [\boldsymbol{x}_\mathrm{t}^\mathsf{T}, \boldsymbol{x}_\mathrm{r}^\mathsf{T}]$.

The forward kinematics for a leg chain i with (active and passive) joint coordinates $\boldsymbol{q}_i$ are defined as

$$ {}^0\boldsymbol{T}_{E_i}(\boldsymbol{q}_i) = {}^0\boldsymbol{T}_{A_i}\,{}^{A_i}\boldsymbol{T}_{C_i}(\boldsymbol{q}_i)\,{}^{C_i}\boldsymbol{T}_{B_i}\,{}^{B_i}\boldsymbol{T}_{E}. \tag{1} $$

The kinematic constraints for the first leg chain $i = 1$ are expressed as residual regarding $(\text{CS})_E$ from chain 1 and the desired platform frame $(\text{CS})_D$, giving

$$ \delta_{\mathrm{t},i}(\boldsymbol{q}_i, \boldsymbol{x}) = {}_{(0)}\boldsymbol{r}_{D,E}(\boldsymbol{q}_i, \boldsymbol{x}) = -\boldsymbol{x}_\mathrm{t} + {}_{(0)}\boldsymbol{r}_{E_i}(\boldsymbol{q}_i) \in \mathbb{R}^3 \tag{2} $$

for translation, as depicted in Fig. 1b. The rotational part is

$$ \delta_{\mathrm{r},i}(\boldsymbol{q}_i, \boldsymbol{x}) = \begin{bmatrix}\alpha_y & \alpha_x\end{bmatrix}^\mathsf{T} = \boldsymbol{\alpha}_\mathrm{red}\left({}^D\boldsymbol{R}_{E_i}(\boldsymbol{x}, \boldsymbol{q}_i)\right) = \boldsymbol{\alpha}_\mathrm{red}\left({}^0\boldsymbol{R}_D^\mathsf{T}(\boldsymbol{x}_\mathrm{r})\,{}^0\boldsymbol{R}_{E_i}(\boldsymbol{q}_i)\right) \in \mathbb{R}^2. \tag{3} $$

The function $\boldsymbol{\alpha}(\boldsymbol{R})$ computes the intrinsic Z-Y'-X'' Tait-Bryan angles $[\alpha_z, \alpha_y, \alpha_x]$ from the rotation matrix $\boldsymbol{R}$ and $\boldsymbol{\alpha}_\mathrm{red}(\boldsymbol{R}) = [\alpha_y, \alpha_x]^\mathsf{T}$ does the same without the Z angle. Using proper Euler angles like Z-X'-Z'' instead would be impractical due to the singularity for $\boldsymbol{\alpha} = 0$. Therefore, the general term "Euler angles" is omitted. A convention with first rotation around Z should be used for independence of φ_z, [17].

The nonlinear function (3) allows a minimal coordinate representation of the inverse kinematics problem of 3T2R robots (or tasks) and is not dependent on the uncontrollable (or redundant) variable φ_z, as elaborated in more detail in [17].

The rotational constraints of further leg chains are modeled following the first leg chain. See e.g. [2,8] for similar approaches on velocity level. As sketched in Fig. 1c, this is expressed relative to the first leg's rotation matrix ${}^0\boldsymbol{R}_{E_1}(\boldsymbol{q}_1)$ from (1) as

$$ \delta_{\mathrm{r},i}(\boldsymbol{q}_i, \boldsymbol{q}_1) = \boldsymbol{\alpha}\left({}^0\boldsymbol{R}_{E_1}^\mathsf{T}(\boldsymbol{q}_1)\,{}^0\boldsymbol{R}_{E_i}(\boldsymbol{q}_i)\right) \in \mathbb{R}^3 \quad \text{for} \quad i = 2, \ldots, m. \tag{4} $$

The translational component for further leg chains is linear and therefore used identically as in (2). The constraints δ_i for each leg chain and δ for the full robot are

$$ \delta_i^\mathsf{T} = \begin{bmatrix}\delta_{\mathrm{t},i}^\mathsf{T} & \delta_{\mathrm{r},i}^\mathsf{T}\end{bmatrix} \quad \text{and} \quad \delta^\mathsf{T} = \begin{bmatrix}\delta_1^\mathsf{T} & \delta_2^\mathsf{T} & \cdots & \delta_m^\mathsf{T}\end{bmatrix}. \tag{5} $$

A symmetric 3T2R PM has $n_{\boldsymbol{q}} = 25$ active and passive joint DoFs and the constraints are $\delta(\boldsymbol{q}, \boldsymbol{x}) = \boldsymbol{0} \in \mathbb{R}^{29}$, which descriptively shows the overconstraint of degree four.

The first-order inverse kinematics can be obtained by partial derivatives as

$$ \frac{\mathrm{d}}{\mathrm{d}t}\delta(\boldsymbol{q}, \boldsymbol{x}) = \delta_{\partial\boldsymbol{q}}\dot{\boldsymbol{q}} + \delta_{\partial\boldsymbol{x}}\dot{\boldsymbol{x}} = \boldsymbol{0} \quad \text{with} \quad \delta_{\partial\boldsymbol{q}} := \frac{\partial}{\partial\boldsymbol{q}}\delta \quad \text{and} \quad \delta_{\partial\boldsymbol{x}} := \frac{\partial}{\partial\boldsymbol{x}}\delta. \tag{6} $$

The gradients in (6) can be obtained in a closed form only depending on known expressions such as the geometric Jacobian of the leg chains and Euler angle transformation matrices [17]. The "direct kinematic matrix" [8] $\delta_{\partial q}$ is rectangular and the linear relation $\dot{q} = \tilde{J}^{-1}\dot{x}$ can be obtained numerically from (6) e.g. using a QR solver. The tilde sign is used to demarcate the Jacobian corresponding to all joint coordinates (including passive and coupling joints). For analysis of the robot the (inverse) Jacobian matrix J^{-1} corresponding to the active joints q_{a} is needed. The appropriate rows are selected with the matrix P_{a} giving $q_{\mathrm{a}} = P_{\mathrm{a}}q$ and $J^{-1} = P_{\mathrm{a}}\tilde{J}^{-1}$.

This approach has several properties which makes it favorable to use for a combined structural and dimensional synthesis of symmetric 3T2R PMs:

- The inverse kinematics problem can be solved using the Newton-Raphson algorithm with $\delta(q^{k+1}, x) = \delta(q^k, x) + \delta_{\partial q}(q, x)|_{q^k}(q^{k+1} - q^k) = 0$ for a step k.
- Only mathematical concepts in the scope of textbooks like [15] are necessary, avoiding the explicit use of screw theory and Grassmann or Lie algebra.
- A completely numeric and modular implementation allows setting up the kinematics model automatically for creation and use of a database of robot structures.

The model (5) has the property of overconstraint, which shows in a rectangular direct kinematic matrix $\delta_{\partial q}$. This can be avoided by using the reduced orientation residual (3) on all leg chains. This geometric elimination of the overconstraint is only permitted if the kinematic constraints can be met, i.e. $\delta = 0$. This reduced model is written with letter ψ instead of δ to easier distinguish the two. The translational part stays unchanged with $\psi_{\mathrm{t},i} = \delta_{\mathrm{t},i}$ and the rotational part is

$$\psi_{\mathrm{r},i}(q_i, x) = \left[\alpha_y \ \alpha_x\right]^{\mathrm{T}} = \alpha_{\mathrm{red}}\left({}^{D}R_{E_i}(x_{\mathrm{r}}, q_i)\right) \in \mathbb{R}^2 \quad \text{for} \quad i = 1, \ldots, m. \tag{7}$$

The reduced dimension leads to the non-overconstraint $\psi \in \mathbb{R}^{25}$ and $\psi_{\partial q} \in \mathbb{R}^{25\times 25}$.

The Jacobian can be obtained numerically by formulating (6) with ψ instead of δ using standard solvers for square matrices like the LU solver. For evaluating the mobility of a parallel robot first the full constraints δ have to be regarded. Otherwise a false positive $\psi = 0$ can result while the full constraints are not met, giving an infeasible $\delta \neq 0$. The leg chain's coupling joints then perform an inconsistent rotation around the end effector's Z-axis with feasible ${}_{(0)}r_{E_i}$ and last column in ${}^{0}R_{E_i}$.

3 Combined Structural and Dimensional Synthesis of 3T2R PMs

The kinematic model from the previous section is used in a combined structural and dimensional synthesis of symmetric parallel mechanisms (PMs) with

3T2R DoF. In the *combined synthesis* all possible structures are evaluated by an algorithm which first creates a PM database in a structural synthesis (using the dimensional synthesis implementation) and then performs the dimensional synthesis on the database.

The *dimensional synthesis* is performed for an exemplary task by the kinematics simulation of a representative end effector trajectory. A more general, but computationally more expensive solution would include a subsequent workspace analysis. The kinematic parameters are subject to a particle swarm optimization (PSO), as outlined in Fig. 2. The fitness function $\boldsymbol{f}$ for evaluating a set of dimensional parameters $\boldsymbol{p}$ begins with the solution of the first-order inverse kinematics (IK) on position level for reference points in the workspace. If successful, a second-order inverse kinematics is solved for the trajectory using the Jacobian relation. The process is repeated for all the PM's IK configurations, found numerically. Within the fitness function evaluation several *constraints* are checked in a hierarchical manner with decreasing priority, meaning that violation of a constraint leads to the abortion of the evaluation and a penalty corresponding to constraint priority. Some of the constraints (checked in structural "S" or dimensional "D" synthesis) are (in this order)

- (S/D) geometric plausibility (e.g. leg chain lengths vs base/platform dimensions),
- (S/D) success of the inverse kinematics (first and second order, based on (7), (6)),
- (S) validity of full kinematic constraints $\boldsymbol{\delta} = \boldsymbol{0}$ from (5),
- (D) self-collisions (with elementary geometry such as spheres and capsules),
- (D) installation space (e.g. robot has to be in a cylinder with reasonable radius),
- (D) joint angle ranges and velocities (to be in a technically feasible range of values),
- (D) singularities of type I (of $\boldsymbol{\delta}_{\partial \boldsymbol{q}}$) and II (of $\boldsymbol{J}$) (by a condition-number threshold).

If all constraints are met, the *optimization objective* is evaluated, as discussed next.

In the *structural synthesis* mode of the toolchain the *objective* is to obtain the mobility based on the Jacobian $\boldsymbol{J}$. By this framework there is no need of a complex geometric analysis as done by other authors. The underlying assumption is that a numeric evaluation of the PM mobility is possible if (and only if) the kinematic constraints can be met. While many implementation details are different, the general leg chain synthesis approach is similar to [8] and the database approach is similar to [4]. Serial kinematic leg chains are obtained by permutation of Denavit-Hartenberg parameters and successive elimination of isomorphisms which leaves 213 possible serial kinematic leg chains with five joints consisting of revolute (R) or prismatic (P) joints. Substituting some RR-subchains with universal (U) joints further gives 180 variants. The geometric characteristics of possible symmetric parallel robots are obtained from literature. This leads to several possible joint alignments relative to the fixed base (transformation ${}^{0}\boldsymbol{T}_{A_i}$) or the moving platform (transformation ${}^{P}\boldsymbol{T}_{B_i}$). The synthesis is performed by permutation of all implemented possibilities for leg chain, base coupling, platform coupling and actuation with additional filtering for feasibility (e.g. no passive prismatic joints, proximal actuation). Eleven PMs with four joints (classes 5-PRUR, 5-RPUR, 5-RRUR, 5-RUPR, 5-RURR) and 16 PMs with five joints (classes 5-PRRRR, 5-RPRRR, 5-RRPRR, 5-RRRRR) are generated.

4 Exemplary Results of the Combined Synthesis for 3T2R PMs

These structures are validated for an exemplary task with middle position of $[r_{\mathrm{Tx}}, r_{\mathrm{Ty}}, r_{\mathrm{Tz}}] = \boldsymbol{x}_{\mathrm{T,t}}^{\mathrm{T}} = [0, 0, 1500\,\mathrm{mm}]$ in the world frame. The tilting angles $[\varphi_{\mathrm{Tx}}, \varphi_{\mathrm{Ty}}] = \boldsymbol{x}_{\mathrm{T,r}}^{\mathrm{T}} = [20°, 20°]$ of the tool axis have to be non-zero to avoid singularities. The end effector pose is then changed consecutively by ±300 mm for each position component of $\boldsymbol{x}_{\mathrm{t}}$ and ±10° for the orientation $\boldsymbol{x}_{\mathrm{r}}$. A further required singularity [12] and workspace [14] analysis beyond the specific trajectory is out of this paper's scope.

The PM is floor-mounted and the base position $\boldsymbol{r}_0^{\mathrm{T}} = [r_{0x}, r_{0y}, r_{0z}]$ is *subject to optimization* with $-600\,\mathrm{mm} < r_{0x}, r_{0y} < 600\,\mathrm{mm}$ and $0\,\mathrm{mm} < r_{0z} < 800\,\mathrm{mm}$. The base and platform diameter are *optimization variables* as well with $1400\,\mathrm{mm} < d_{\mathrm{B}} < 3000\,\mathrm{mm}$ and $200\,\mathrm{mm} < d_{\mathrm{P}} < 800\,\mathrm{mm}$. Other variables are one scaling parameter, one platform coupling joint alignment angle and one to seven Denavit-Hartenberg parameters [2], representing the link's lengths and alignment angle. The latter *parameter's bounds* are computed from the task dimensions and are not vital for the results.

This leads to nine to 15 *optimization variables* in total, depending on the kinematic structure of the robot. The high number of dimensional parameters facilitates finding a non-singular setting of the robot. *Minimization objectives* in the *multi-objective* PSO are the condition number and the summed lengths of the leg chains as (debatable) indicators for general feasibility. A further elaboration on the optimization details is omitted for the sake of brevity but can be reconstructed from the published source code. The optimization with up to 200 generations and 100 individuals took 7 h to 10 h per robot (with time limit of 10

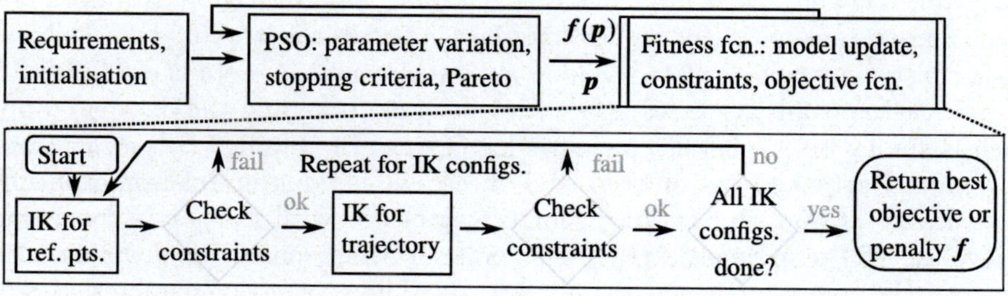

Fig. 2. Overall procedure for the dimensional synthesis of a robot with hierarchical constraints

h) on a state-of-the-art Intel Xeon computing cluster system, running in parallel using a MATLAB implementation and MEX-compiled functions.

The results are summarized in a Pareto diagram shown in Fig. 3a. The robots in Fig. 3b are named using the established PM notation [15] together with the kinematic chain notation from [10], where joints with the same accent on R̀ or Ŕ are parallel to each other and underlined R̲ or P̲ are actuated. In addition, P̲̀R̀ and Ṕ̲Ŕ subchains are used to distinguish the structures as this provides more information than C̲ in [14].

Known structures like the 5-RP̲UR (no. 6) [1,7,13] and 5-P̲RUR (no. 4) [14] can be reproduced. The 5-R̀P̲R̀ŔŔ (no. 5) emerges when separating the DoF of the universal joint of no. 6. The same goes for no. 3 emerging from no. 4. Structures like this may not be popular in literature due to the higher number of kinematic parameters, difficulty of analysis and the necessity of the dimensional synthesis beyond manual parameter tuning. Some structures are unconventional

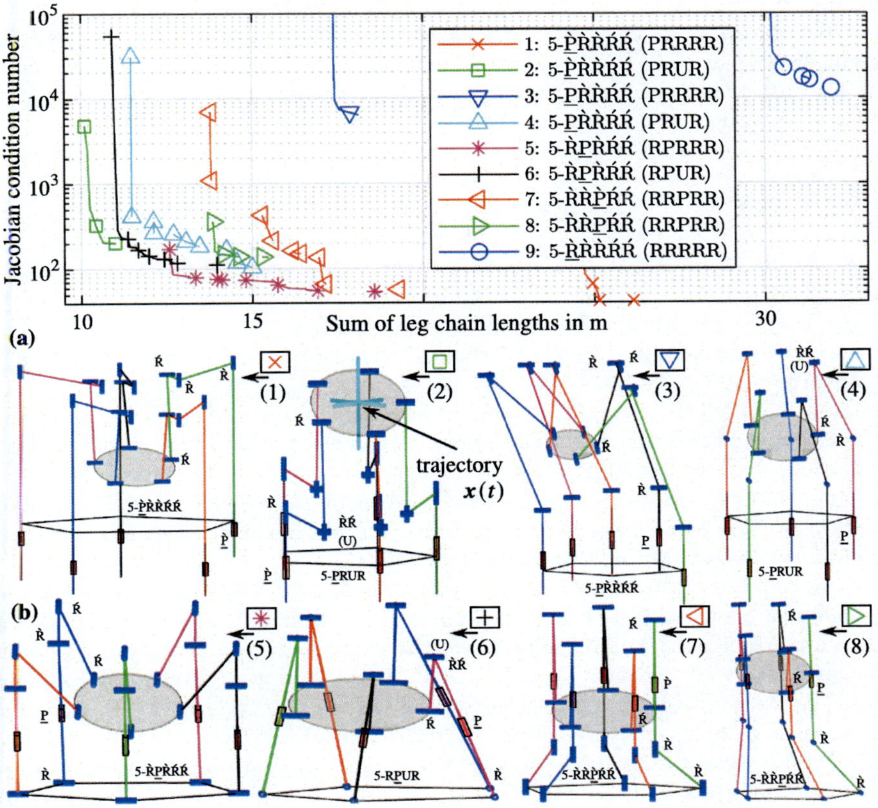

Fig. 3. Results of the synthesis: **(a)** Pareto front, **(b)** robot examples with active prismatic base coupling joint (numbers 1–4), and with revolute base joint and active prismatic joint (no. 5–8)

like 5-$\underline{\grave{\text{P}}}\grave{\text{R}}\grave{\text{R}}\acute{\text{R}}\acute{\text{R}}$ (no. 1 and 2) similar to the $\underline{\text{C}}$UR chains of the Pentapteron in [14]. Structure no. 7 is disadvantageous for technical realization due to the distal position of the actuation in the chain, despite its acceptable performance. Robots with only revolute joints like the 5-RRUR reported in [3] were not successful due to self collisions with the long limbs required for the large-scale platform motion as can be seen by high values for no. 9 in the Pareto diagram.

5 Conclusion

The presented inverse kinematics model for 3T2R PMs can be regarded as a variation of the theory of linear transformations. Using the model in a combined structural and dimensional synthesis allows to reproduce relevant symmetric 3T2R parallel robots. The practical application of symmetric 3T2R robots will be promoted by the open-source MATLAB tool which generates feasibly-dimensioned robot structures.

Acknowledgements. The author acknowledges the support by the Deutsche Forschungsgemeinschaft (DFG) under grant number 341489206. MATLAB code to reproduce the results is available at GitHub under free license at github.com/SchapplM/robotics-paper_ark2022_3T2R.

References

1. Amine, S., Masouleh, M.T., Caro, S., Wenger, P., Gosselin, C.: Singularity analysis of 3T2R parallel mechanisms using Grassmann-Cayley algebra and Grassmann geometry. Mech. Mach. Theory **52**, 326–340 (2012)
2. Staicu, S.: Dynamics of Parallel Robots. PRTA, Springer, Cham (2019). https://doi.org/10.1007/978-3-319-99522-9
3. Cao, W., Ding, H., Yang, D.: A method for compliance modeling of five degree-of-freedom overconstrained parallel robotic mechanisms with 3T2R output motion. J. Mech. Rob. **9**, 011011 (2017)
4. Ding, H., Cao, W., Cai, C., Kecskeméthy, A.: Computer-aided structural synthesis of 5-DOF parallel mechanisms and the establishment of kinematic structure databases. Mech. Mach. Theory **83**, 14–30 (2015)
5. Fang, Y., Tsai, L.W.: Structure synthesis of a class of 4-DoF and 5-DoF parallel manipulators with identical limb structures. Int. J. Rob. Res. **21**(9), 799–810 (2002)
6. Frindt, M., Krefft, M., Hesselbach, J.: Structure and type synthesis of parallel manipulators. In: Schütz, D., Wahl, F.M. (eds.) Robotic Systems for Handling and Assembly. Springer Tracts in Advanced Robotics, vol. 67, pp. 17–37. Springer, Heidelberg (2010). https://doi.org/10.1007/978-3-642-16785-0_2
7. Gallardo-Alvarado, J., Abedinnasab, M.H., Islam, M.N.: A simple method to solve the instantaneous kinematics of the 5-RPUR parallel manipulator. Robotica **37**(6), 1143–1157 (2019)
8. Gogu, G.: Structural Synthesis of Parallel Robots, Part 1: Methodology. Solid Mechanics and Its Applications, vol. 866. Springer, Cham (2008). https://doi.org/10.1007/978-1-4020-5710-6

9. Huang, Z., Li, Q.: Type synthesis of symmetrical lower-mobility parallel mechanisms using the constraint-synthesis method. Int. J. Rob. Res. **22**(1), 59–79 (2003)
10. Kong, X., Gosselin, C.M.: Type Synthesis of Parallel Mechanisms. Springer, Heidelberg (2007). https://doi.org/10.1007/978-3-540-71990-8
11. Li, Q., Huang, Z., Hervé, J.M.: Type synthesis of 3R2T 5-DOF parallel mechanisms using the Lie group of displacements. IEEE Trans. Rob. Autom. **20**, 173–180 (2004)
12. Masouleh, M.T., Gosselin, C.: Singularity analysis of 5-RPUR parallel mechanisms (3T2R). Int. J. Adv. Manuf. Technol. **57**(9–12), 1107–1121 (2011)
13. Masouleh, M.T., Gosselin, C., Husty, M., Walter, D.R.: Forward kinematic problem of 5-RPUR parallel mechanisms (3T2R) with identical limb structures. Mech. Mach. Theory **46**(7), 945–959 (2011)
14. Masouleh, M.T., Saadatzi, M.H., Gosselin, C., Taghirad, H.D.: A geometric constructive approach for the workspace analysis of symmetrical 5-PRUR parallel mechanisms (3T2R). In: ASME 2010 International Design Engineering Technical Conferences, pp. 1335–1344 (2010)
15. Merlet, J.P.: Parallel Robots, Solid Mechanics and Its Applications, vol. 128, 2nd edn. Springer, Cham (2006). https://doi.org/10.1007/1-4020-4133-0
16. Motevalli, B., Zohoor, H., Sohrabpour, S.: Structural synthesis of 5 DoFs 3T2R parallel manipulators with prismatic actuators on the base. Rob. Auton. Syst. **58**(3), 307–321 (2010)
17. Schappler, M., Tappe, S., Ortmaier, T.: Modeling parallel robot kinematics for 3T2R and 3T3R tasks using reciprocal sets of Euler angles. MDPI Robot. **8**(3), 68 (2019)
18. Song, Y., Lian, B., Sun, T., Dong, G., Qi, Y., Gao, H.: A novel five-degree-of-freedom parallel manipulator and its kinematic optimization. ASME J. Mech. Rob. **6**(4), 041008 (2014)

Inverse Kinematics for Functional Redundancy of Symmetric 3T1R Parallel Manipulators Using Tait-Bryan-Angle Kinematic Constraints

Moritz Schappler(✉)

Institute of Mechatronic Systems, Leibniz University Hannover, Hannover, Germany
moritz.schappler@imes.uni-hannover.de

Abstract. Functional redundancy for parallel manipulators (PM) with 3T1R degrees of freedom (DoF) presents an untreated niche regarding a general and systematic kinematic description. For an efficient formulation of the inverse kinematics problem (IKP) an existing approach using intrinsic Z-Y'-X'' Tait-Bryan angles for the rotational kinematic constraints is transferred from 3T3R PMs to 3T1R PMs. The adaption of the kinematics model for the five-DoF leg chains of symmetric 3T1R PMs is elaborated in detail. The presented application in a Newton-Raphson IK scheme with nullspace projection is validated within a dimensional synthesis of such PMs, already exploiting the redundancy. The framework is able to reproduce existing PMs from literature. Results show the dimensioning of PMs for an exemplary task.

Keywords: Parallel robot · Parallel manipulator · 3T1R · Functional redundancy · Kinematic constraints · Tait-Bryan angles · Euler angles · Dimensional synthesis

1 Introduction and State of the Art

Many parallel robot structures for different degrees of freedom (DoF) of the moving platform already exist [6,9,10]. While some few parallel manipulators (PM) are commercialized for motion with three translational DoF and fixed rotation (3T0R), such as the *Delta robot*, structures for the Schoenflies motion with one rotational DoF (3T1R) are less common. Industrially used parallel robots mainly have symmetric leg chains with motors fixed to the robot base, as this reduces costs and complexity [13]. One seeming example of a commercial 3T1R PM is the *Adept Quattro robot*, which has four kinematic leg chains. However, kinematically the robot is a 3T0R PM. The rotational DoF is transmitted via an articulated traveling plate over the fourth leg chain [13]. Therefore it can be classified as a parallel-hybrid structure.

The *structural synthesis* of PMs with symmetric leg chains was introduced in [5] by using screw theory for 4-DoF and 5-DoF PMs, which results in several

Code: github.com/SchapplM/robotics-paper_ark2022_3T1R.

O. Altuzarra and A. Kecskéméthy (Eds.): ARK 2022, SPAR 24, pp. 198–207, 2022.
https://doi.org/10.1007/978-3-031-08140-8_22

general rules regarding the properties of leg chains. In [8] screw theory is used with a constraint synthesis method for 3T1R PMs for fully-symmetrical PMs. The *Grübler-Kutzbach criterion* can be used to perform an assessment of the mobility of the synthesized lower-mobility PMs [6,8], but can not provide a synthesis itself.

Extensive frameworks on the structural synthesis have been e.g. proposed by [6] based on *evolutionary morphology* and the *theory of linear transformations* and by [9] based on *screw theory* and the *virtual-chain* approach. The results from [6] also incorporate many asymmetric PMs, e.g. of the Isoglide family, from which some have been built as prototypes. In [9] also many symmetric PMs are proposed. Both works give an extensive overview of possible leg chains and coupling joint alignments for parallel robots with reduced mobility, among others for 3T1R DoF.

The abundance of possible kinematic structures makes selecting the best—or at least a suitable—parallel robot for a 3T1R task a tedious endeavor. Not only the kinematic structure, but also the dimensioning of kinematic parameters strongly influences performance criteria of parallel robots [10].

A major performance criterion for parallel robots is the absence of singularities in the workspace, expressed by the condition number or other characteristics derived from the manipulator Jacobian. A good conditioning can be accomplished by a well-chosen structural design, such as for isotropic PMs [3,6], where each actuator is dedicated to one platform DoF. Another mean to avoid singularities is the use of redundant DoF. The *intrinsic redundancy* reviewed in [7] requires additional joints in at least one leg chain (*kinematic redundancy*) or additional leg chains or actuators (*actuation redundancy*). Exploiting the *functional redundancy* allows to use a DoF of the PM's operational space – unused in the task space – for optimization [4].

Optimization schemes for functional redundancy have been investigated, such as

- interval analysis in [11] for a Gough platform (6-U$\underline{\text{P}}$S Hexapod robot[1]),
- linear and quadratic programming in [12], also for a Hexapod robot,
- sequential quadratic programming in [4] for a combination of translating (3T0R) and spherical (0T3R) parallel machine,
- the gradient-projection method on position level for general 3T3R PMs in [17] and on acceleration level in [1] for a 3-$\underline{\text{R}}$RR and in [16] for a 6-U$\underline{\text{P}}$S robot.

Several approaches exist for serial-link robots to formulate the inverse kinematics problem (IKP) properly for functional redundancy [14]. The problem can be addressed straightforward for first- or second-order differential kinematics (on velocity or acceleration level) by using the linear Jacobian relation. This can be transferred to parallel robots as shown for kinematic redundancy in [7] or explicitly for functional redundancy by [1]. The position-level IKP has to be

[1] The parallel robot notation from [10] is used with prepended number of legs, chains with universal (U), revolute (R), prismatic (P) and spherical (S) joint and underlining for actuation.

handled differently due to the nonlinearity of rotation [17]. For the *explicit* IKP the substitution of the redundant coordinate is necessary for global optimization schemes, such as [11].

If only an *implicit* solution of the position-level IKP is available, a numeric method such as the *Newton-Raphson algorithm* can be used to obtain joint coordinates to solve the kinematic constraints equations. This can be combined with the *nullspace-projection method* and allows a local optimization of the redundant DoF while finding the IK solution. It can also be used for kinematic redundancy, as shown by [15] for a 3-PRRR planar robot in combination with differential dynamic programming. The kinematics formulation has to be handled differently than non-redundant approaches [10] to be able to exclude the redundant coordinate from the equations [17].

The presented references on parallel robot redundancy mainly focus on general spatial manipulators, i.e. 3T3R PMs for 3T2R tasks [4,11,12,17] or planar manipulators [1], i.e. 2T1R PMs for 2T0R tasks. Functional redundancy for 3T1R PMs has not been investigated yet, to the best knowledge of the author. A practical use for functionally redundant 3T1R PMs is not obvious, but can be motivated from axis-symmetric pick-and-place tasks, e.g. when picking parts with arbitrary rotation or with an interval of tolerance for orientation. A technical realization of the robot end effector is e.g. a vacuum gripper with one suction cup. Another possible application are drilling tasks in three-axis-machining or in five-axis machining using a two-DoF table orientation mechanism, similar to [4]. Despite the lack of obtruding specific use cases, the sketched problem may be interesting from an academic point of view.

Using functionally redundant 3T1R PMs with four leg chains instead of 3T0R PMs with three leg chains has the disadvantage of in principle smaller workspace due to additional constraints and more possible self-collisions. Using the task redundancy for singularity and collision avoidance may on the contrary abolish these disadvantages. This may also be beneficial for isotropic PMs of [3,6] when nullspace optimization for collision avoidance enlarges the workspace. In the following, an axis-symmetric 3T0R task is denoted by 3T0*R to distinguish it from fixed-orientation tasks termed with 3T0R. In the latter case e.g. an additional revolute joint at the platform is necessary when the 3T1R PM is operated with redundant orientation.

The performance of the discussed alternatives can be estimated with a *combined structural and dimensional synthesis*, where all possible structures are simulated for the tasks and their kinematic parameters are each optimized. The functional redundancy has to be exploited already in the dimensional synthesis to take it's potential into account. To be able to compare numerous different solutions, a *general inverse kinematics model* is necessary.

These aspects are addressed for 3T1R PMs in this paper by adapting a kinematics model presented previously by the author based on full kinematic constraints and Tait-Bryan angles for rotation [17]. The paper's contributions are

- a novel geometric model for the IKP of task-redundant 3T1R PMs,
- application of the model in a gradient-projection scheme for task redundancy,

- validation of the model in a dimensional synthesis framework.

The remainder of the paper is structured as follows. Section 2 presents the kinematics model which is applied in Sect. 3 for the IK scheme. The model is used in the synthesis framework, with results presented in Sect. 4. Section 5 concludes the paper.

2 Inverse Kinematic Model for 3T1R Parallel Robots

A fully-parallel robot with 3T1R platform mobility is considered. The robot has $m = 4$ leg chains and $n = 4$ platform DoF. It's operational-space coordinates $\boldsymbol{x}^\mathsf{T} = [r_x, r_y, r_z, \varphi_z]$ contain the position $\boldsymbol{r}$ and planar orientation φ_z of the platform-mounted end effector. Leg chains are assumed to have $n_i = 5$ DoF each since six-DoF leg chains provide no constraint and four-DoF chains do not allow a symmetric PM. The kinematic model is assisted by several frames, based on the model in [2].

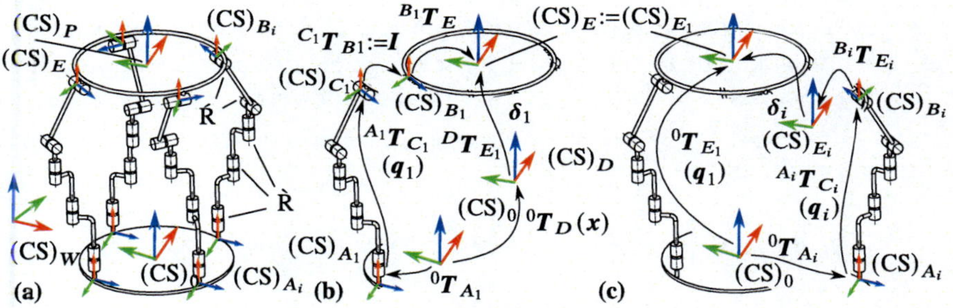

Fig. 1. Sketch of the kinematics model at the example of the modified 4-$\underline{\grave{\mathrm{R}}}\grave{\mathrm{R}}\grave{\mathrm{R}}\acute{\mathrm{R}}\acute{\mathrm{R}}$ image of [9] (p. 155). (**a**) All coordinate frames, (**b**) constraints for first leg chain and (**c**) other leg chains $i \neq 1$. Joints with the same accent on $\grave{\mathrm{R}}$ or $\acute{\mathrm{R}}$ are parallel to each other [9], extending the notation of [10]

Leg chains are modeled with the modified Denavit-Hartenberg parameters [2] and the legs' kinematics can be used modularly to obtain the parallel robot kinematics.

The coordinate systems (CS) are sketched in Fig. 1a. The world frame $(\mathrm{CS})_W$ serves as a reference for the relative position of task and robot, which has a base frame $(\mathrm{CS})_0$. The legs are connected to the fixed base with a base joint coupling frame $(\mathrm{CS})_{A_i}$ and to the moving platform with a coupling joint frame $(\mathrm{CS})_{B_i}$. A corresponding cut joint frame $(\mathrm{CS})_{C_i}$ is attached to the leg. The platform has a body frame $(\mathrm{CS})_P$ and an additional end effector frame $(\mathrm{CS})_E$, which is e.g. helpful for modeling ceiling-mounted robots or additional tool transformations. For the sake of setting up the kinematics model, the desired end effector frame $(\mathrm{CS})_D$ is the reference for operational space coordinate $\boldsymbol{x}$ and each leg chain has a frame $(\mathrm{CS})_{E_i}$ which corresponds to the end effector frame from the perspective of the chain i. If the kinematic constraints $\boldsymbol{\delta}$ are met, i.e. $\boldsymbol{\delta} = \mathbf{0}$, then all $(\mathrm{CS})_{E_i}$ and $(\mathrm{CS})_D$ align.

The forward kinematics of each leg chain, using the leg chain's joint coordinate vector $\boldsymbol{q}_i$, is set up with SE(3) transformation matrices to

$$ {}^0\boldsymbol{T}_{E_i}(\boldsymbol{q}_i) = {}^0\boldsymbol{T}_{A_i}\,{}^{A_i}\boldsymbol{T}_{C_i}(\boldsymbol{q}_i)\,{}^{C_i}\boldsymbol{T}_{B_i}\,{}^{B_i}\boldsymbol{T}_{E}. \tag{1} $$

The full kinematic constraints $\boldsymbol{\delta}$ for all leg chains have the translational part

$$ \boldsymbol{\delta}_{\mathrm{t},i}(\boldsymbol{q}_i, \boldsymbol{x}) = {}_{(0)}\boldsymbol{r}_{D,E_i}(\boldsymbol{q}_i, \boldsymbol{x}) = -\boldsymbol{x}_{\mathrm{t}} + {}_{(0)}\boldsymbol{r}_{E_i}(\boldsymbol{q}_i) \stackrel{!}{=} \mathbf{0} \in \mathbb{R}^3. \tag{2} $$

The rotational part $\boldsymbol{\delta}_{\mathrm{r},1} = \left[\alpha_{z,1}\ \alpha_{y,1}\ \alpha_{x,1}\right]^{\mathsf{T}}$ for the first leg chain $i = 1$ is obtained from the rotation matrix ${}^D\boldsymbol{R}_{E_1}$ corresponding to (2) as

$$ \boldsymbol{\delta}_{\mathrm{r},1}(\boldsymbol{q}_1, \boldsymbol{x}) = \boldsymbol{\alpha}\left({}^D\boldsymbol{R}_{E_1}(\boldsymbol{x}, \boldsymbol{q}_1)\right) = \boldsymbol{\alpha}\left({}^0\boldsymbol{R}_D^{\mathsf{T}}(\boldsymbol{x}_{\mathrm{r}})\,{}^0\boldsymbol{R}_{E_1}(\boldsymbol{q}_1)\right) \stackrel{!}{=} \mathbf{0} \in \mathbb{R}^3 \tag{3} $$

with the function $\boldsymbol{\alpha}(\boldsymbol{R})$, which computes the intrinsic Z-Y'-X'' Tait-Bryan angles $[\alpha_z, \alpha_y, \alpha_x]$ from a given rotation matrix. Proper Euler angles like Z-Y'-Z'' are infeasible due to the singularity for an angle of zero. The approach for the first leg chain is depicted in Fig. 1b. For the following leg chains, as shown in Fig. 1c, the rotational constraints $\boldsymbol{\delta}_{\mathrm{r},i} = \left[\alpha_{z,i}\ \alpha_{y,i}\ \alpha_{x,i}\right]^{\mathsf{T}}$ are expressed w.r.t. the first leg chain as

$$ \boldsymbol{\delta}_{\mathrm{r},i}(\boldsymbol{q}_i, \boldsymbol{q}_1) = \boldsymbol{\alpha}\left({}^0\boldsymbol{R}_{E_1}^{\mathsf{T}}(\boldsymbol{q}_1)\,{}^0\boldsymbol{R}_{E_i}(\boldsymbol{q}_i)\right) \stackrel{!}{=} \mathbf{0} \in \mathbb{R}^3 \quad \text{for} \quad i = 2, ..., m. \tag{4} $$

Assembling the kinematic constraints for each leg chain leads to

$$ \boldsymbol{\delta}_i^{\mathsf{T}} = \left[\boldsymbol{\delta}_{\mathrm{t},i}^{\mathsf{T}}\ \boldsymbol{\delta}_{\mathrm{r},i}^{\mathsf{T}}\right] \in \mathbb{R}^6 \quad \text{and} \quad \boldsymbol{\delta}^{\mathsf{T}} = \left[\boldsymbol{\delta}_1^{\mathsf{T}}\ \boldsymbol{\delta}_2^{\mathsf{T}} \cdots \boldsymbol{\delta}_m^{\mathsf{T}}\right] \in \mathbb{R}^{6m}. \tag{5} $$

This formulation is designed for 3T3R PMs and is overconstraint for 3T1R PMs since six constraint equations are defined for five leg joints in the considered case. However, it is feasible for providing a solution to the IKP. When considering the functionally redundant case, the reason for the selection of the rotational constraints (3), (4) from [17] becomes apparent. The task space coordinates $\boldsymbol{y} = [r_x, r_y, r_z]^{\mathsf{T}}$ can be obtained by removing the platform rotation φ_z. Due to the order of rotations, the components $\alpha_{x,1}$ and $\alpha_{y,1}$ are independent of φ_z [17]. The residual $\alpha_{z,1}$ corresponds to the removed coordinate and can be disregarded. Removing this term from the rotational constraint for the first leg chain of (3) is termed as $\boldsymbol{\delta}_{\mathrm{r},1,\mathrm{red}}$ and accordingly

$$ \boldsymbol{\delta}_{1,\mathrm{red}}^{\mathsf{T}} = [\boldsymbol{\delta}_{\mathrm{t},1}^{\mathsf{T}}, \boldsymbol{\delta}_{\mathrm{r},1,\mathrm{red}}^{\mathsf{T}}] \in \mathbb{R}^5 \quad \text{and} \quad \boldsymbol{\delta}_{\mathrm{red}}^{\mathsf{T}} = \left[\boldsymbol{\delta}_{1,\mathrm{red}}^{\mathsf{T}}\ \boldsymbol{\delta}_2^{\mathsf{T}} \cdots \boldsymbol{\delta}_m^{\mathsf{T}}\right] \in \mathbb{R}^{6m-1}. \tag{6} $$

Since the leg chain has $n_1 = 5$ DoF and the platform has 3T1R mobility, only one DoF of the leg chain remains for the pointing direction of the z axis of $(\mathrm{CS})_{E_i}$. Therefore, for the *functionally redundant case*, the *translational* constraint for *all* legs remains unchanged as $\boldsymbol{\psi}_{\mathrm{t},i} := \boldsymbol{\delta}_{\mathrm{t},i}$. For the *rotational* constraint $\psi_{\mathrm{r},1}$ of the *first* leg chain (unaffected by platform rotation) a function candidate has to be found with

$$\psi_{\mathrm{r},1}(\boldsymbol{q}_1, \boldsymbol{y}) = \begin{cases} 0, & \text{iff } \boldsymbol{\delta}_{\mathrm{r},1,\mathrm{red}}(\boldsymbol{q}_1, \boldsymbol{y}) = \mathbf{0} \quad \text{(uniqueness condition)} \\ \neq 0, & \text{otherwise.} \end{cases} \tag{7}$$

The 2-norm of $\boldsymbol{\delta}_{\mathrm{r},1,\mathrm{red}} = [\alpha_{y,1}, \alpha_{x,1}]^{\mathsf{T}}$ disqualifies since the gradient w.r.t. $\boldsymbol{q}_1$ becomes 0 when approaching $\boldsymbol{\delta}_{\mathrm{r},1,\mathrm{red}} = 0$. Instead, the constraint is selected as the 1-norm

$$\psi_{\mathrm{r},1}(\boldsymbol{q}_1, \boldsymbol{y}) = |\alpha_{x,1}| + |\alpha_{y,1}| \overset{!}{=} 0 \in \mathbb{R}. \tag{8}$$

The rotational constraint $\boldsymbol{\psi}_{\mathrm{r},i}$ for the *following* leg chains also contains the z component $\alpha_{z,i}$, since the leading leg 1 sets the platform orientation. Otherwise, it is constructed with the same considerations like (8) as

$$\boldsymbol{\psi}_{\mathrm{r},i}(\boldsymbol{q}_i, \boldsymbol{x}) = \begin{bmatrix} \alpha_{z,i} \\ |\alpha_{x,i}| + |\alpha_{y,i}| \end{bmatrix} \overset{!}{=} \mathbf{0} \in \mathbb{R}^2 \quad \text{for} \quad i = 2, ..., m. \tag{9}$$

The reduced constraints for the full robot are termed similar to (5) as

$$\boldsymbol{\psi}_i^{\mathsf{T}} = [\boldsymbol{\psi}_{\mathrm{t},i}^{\mathsf{T}}\ \boldsymbol{\psi}_{\mathrm{r},i}^{\mathsf{T}}] \in \mathbb{R}^5 \quad \text{for} \quad i > 1 \quad \text{and} \quad \boldsymbol{\psi}^{\mathsf{T}} = [\boldsymbol{\psi}_1^{\mathsf{T}}\ \boldsymbol{\psi}_2^{\mathsf{T}} \cdots \boldsymbol{\psi}_m^{\mathsf{T}}] \in \mathbb{R}^{5m-1}. \tag{10}$$

The efficient computation of the terms $\boldsymbol{\psi}_{\partial \boldsymbol{q}} := \partial\boldsymbol{\psi}/\partial\boldsymbol{q}$ and $\boldsymbol{\psi}_{\partial \boldsymbol{y}} := \partial\boldsymbol{\psi}/\partial\boldsymbol{y}$, necessary for solving the IKP, is shown in [17]. The differentiation of the absolute value of these terms can be obtained by

$$\frac{\partial|\alpha|}{\partial\boldsymbol{q}} = \mathrm{sgn}(\alpha)\frac{\partial\alpha}{\partial\boldsymbol{q}} \quad \text{with} \quad \mathrm{sgn}(0) \overset{!}{=} 1 \tag{11}$$

to avoid a loss of differentiability. Since the implementation is numeric, the inexactness of the signum function in (11) does not degrade the results.

3 Functional Redundancy and Inverse Kinematics

The advantage of the proposed model of the previous section becomes apparent, when looking at the dimension in (10) which leads for a functionally redundant symmetric 3T1R PM to a fat "direct kinematic matrix" [6] $\boldsymbol{\psi}_{\partial \boldsymbol{q}}$ of dimension 19×20. Using the velocity-level formulation of this matrix from the theory of linear transformations in [6] would in principle also reveal this kind of redundancy. Approaching the formulation on the position level as shown above however is beneficial regarding the generality of the reference frame of the components corresponding to the nullspace. The one-DoF nullspace of the matrix $\boldsymbol{\psi}_{\partial \boldsymbol{q}}$, corresponding to the coordinate φ_z, can be exploited while solving the IKP. Using the task redundancy in the overconstraint equations (6) instead produces a tall 23×20 matrix not feasible for a numeric solution with the pseudo inverse (†). The inverse manipulator Jacobian $\boldsymbol{J}^{-1}$ can be obtained numerically by $\tilde{\boldsymbol{J}}^{-1} = -\boldsymbol{\delta}_{\partial \boldsymbol{q}}^{\dagger} \boldsymbol{\delta}_{\partial \boldsymbol{x}}$ and selecting rows corresponding to active joints.

The first-order IK algorithm on position level is derived with a Taylor series expansion of (10), leading to a step k in the Newton-Raphson algorithm, by

$$\boldsymbol{\psi}(\boldsymbol{q}^{k+1}, \boldsymbol{y}) = \boldsymbol{\psi}(\boldsymbol{q}^{k}, \boldsymbol{y}) + \boldsymbol{\psi}_{\partial \boldsymbol{q}}(\boldsymbol{q}^{k}, \boldsymbol{y})(\boldsymbol{q}^{k+1} - \boldsymbol{q}^{k}) \stackrel{!}{=} \boldsymbol{0}. \tag{12}$$

The task-related increment is obtained using the pseudo inverse as

$$\Delta \boldsymbol{q}_{\mathrm{T}}^{k} = (\boldsymbol{q}^{k+1} - \boldsymbol{q}^{k}) = -\left(\boldsymbol{\psi}_{\partial \boldsymbol{q}}(\boldsymbol{q}, \boldsymbol{y})\right)^{\dagger} \boldsymbol{\psi}(\boldsymbol{q}^{k}, \boldsymbol{y}). \tag{13}$$

The nullspace increment (homogeneous solution) is added, leading to

$$\Delta \boldsymbol{q} = \Delta \boldsymbol{q}_{\mathrm{T}} + \Delta \boldsymbol{q}_{\mathrm{N}} = -\boldsymbol{\psi}_{\partial \boldsymbol{q}}^{\dagger} \boldsymbol{\psi} + \boldsymbol{N} \boldsymbol{v} \quad \text{with} \quad \boldsymbol{N} = \boldsymbol{I} - \boldsymbol{\psi}_{\partial \boldsymbol{q}}^{\dagger} \boldsymbol{\psi}_{\partial \boldsymbol{q}}, \tag{14}$$

where $\boldsymbol{v} = h_{\partial \boldsymbol{q}}$ is the gradient of secondary tasks defined as potential h. The formulation can also be set up on velocity level (first-order differential kinematics), based on the time differentiation of the constraints (10), see e.g. [10] and similar to [1,6,15] or on acceleration level (second order) following [14,16].

4 Dimensional Synthesis of Functionally Redundant 3T1R PMs

The generality of the proposed approach for the IK model of 3T1R PMs is validated with a combined structural and dimensional synthesis. Serial-kinematic leg chains with five joints are generated via permutation of the Denavit-Hartenberg parameters, similar to [6]. A permutation of all possible symmetric alignments of these kinematic chains is performed regarding base and platform coupling joints [2,6,9,10], presenting the *structural synthesis*. Each combination is tested for a reference trajectory to evaluate the PM's mobility using the full model from (5) for 3T1R task DoF. Kinematic parameters $\boldsymbol{p}$ of the leg chains and the parallel alignment are optimized in a dimensional synthesis to obtain a feasible solution for each robot (*combined synthesis*), required for the numeric evaluation in contrast to symbolic schemes.

The reduced model (10) is then used for a *dimensional synthesis* with 3T0*R task DoF, exploiting the functional redundancy of the 3T1R PMs, as sketched in Fig. 2. The proposed kinematic model is mainly used for finding an optimal initial configuration in the position IK. Thereby the following second-order trajectory IK starts in an equilibrium and then optimizes the same criterion, as shown in [16] in more detail. In this example, the kinematic objectives pursued in a Pareto optimization are Jacobian condition number, length of leg chains and installation space. Constraints are e.g. self collisions and joint limit violations.

Examples for the resulting optimized 3T1R PMs and the trajectory are shown in Fig. 3a–c, proving the generality of the model by the variety of alignments. The notation for dashes on P and R joints to mark parallelism is adapted from [9]. The exploitation of the functional redundancy is validated with a heatmap

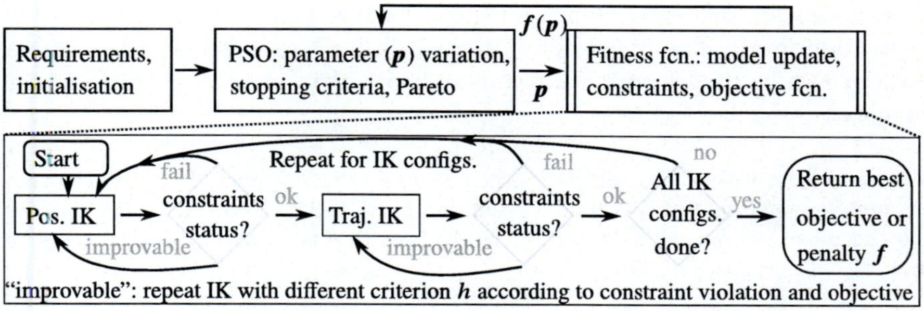

Fig. 2. Optimization scheme with task redundancy within the dimensional synthesis

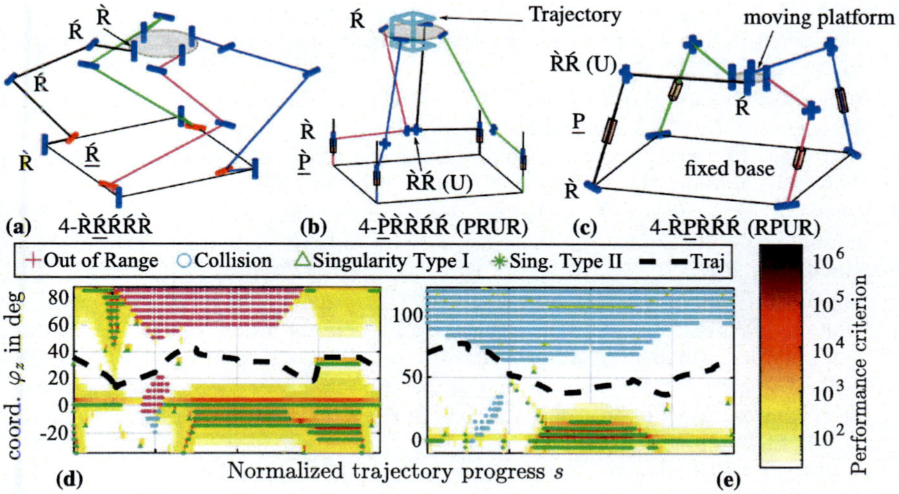

Fig. 3. (**a–c**) Robots optimized for the 3T0*R task and (**d–e**) two exemplary performance maps

of the performance criterion $h = \text{cond}(\boldsymbol{J})$ over the redundant coordinate φ_z and the trajectory progress, shown in Fig. 3d–e for two different robots. The local optimization is sufficient to let the trajectory (dashed line) avoid regions of high condition numbers (dark colors) and therefore of singularities of $\boldsymbol{\psi}_{\partial q}$ (Type I) and of $\boldsymbol{J}$ (Type II) or of collisions and range violations (with markers according to the legend).

5 Conclusion

The exemplary results for the proposed IK model show the necessity of exploiting the task redundancy if a dimensional synthesis for axis-symmetric 3T0*R tasks is performed. Optimal solutions often max out the constraints like self-collisions. This and the problem of singularities can be diminished efficiently by nullspace optimization for reference points (on position level) and for a trajectory (on acceleration level). The open-source framework for structural and dimensional synthesis can be used for practical application and to create variations in academic examples.

Acknowledgements. The author acknowledges the support by the Deutsche Forschungsgemeinschaft (DFG) under grant number 341489206. MATLAB code to reproduce the results is available at GitHub under free license at github.com/SchapplM/robotics-paper_ark2022_3T1R.

References

1. Agarwal, A., Nasa, C., Bandyopadhyay, S.: Dynamic singularity avoidance for parallel manipulators using a task-priority based control scheme. Mech. Mach. Theory **96**, 107–126 (2016)
2. Staicu, S.: Dynamics of Parallel Robots. PRTA, Springer, Cham (2019). https://doi.org/10.1007/978-3-319-99522-9
3. Carricato, M.: Fully isotropic four-degrees-of-freedom parallel mechanisms for Schoenflies motion. Int. J. Rob. Res. **24**(5), 397–414 (2005)
4. Corinaldi, D., Angeles, J., Callegari, M.: Posture optimization of a functionally redundant parallel robot. In: Lenarčič, J., Merlet, J.-P. (eds.) Advances in Robot Kinematics 2016. SPAR, vol. 4, pp. 101–108. Springer, Cham (2018). https://doi.org/10.1007/978-3-319-56802-7_11
5. Fang, Y., Tsai, L.W.: Structure synthesis of a class of 4-DoF and 5-DoF parallel manipulators with identical limb structures. Int. J. Rob. Res. **21**(9), 799–810 (2002)
6. Gogu, G.: Structural Synthesis of Parallel Robots, Part 1: Methodology. Solid Mechanics and Its Applications, vol. 866. Springer, Cham (2008). https://doi.org/10.1007/978-1-4020-5710-6
7. Gosselin, C., Schreiber, L.T.: Redundancy in parallel mechanisms: a review. Appl. Mech. Rev. **70**(1), 010802 (2018)
8. Huang, Z., Li, Q.: Type synthesis of symmetrical lower-mobility parallel mechanisms using the constraint-synthesis method. Int. J. Rob. Res. **22**(1), 59–79 (2003)
9. Kong, X., Gosselin, C.M.: Type Synthesis of Parallel Mechanisms. Springer, Cham (2007). https://doi.org/10.1007/978-3-540-71990-8
10. Merlet, J.P.: Parallel Robots, Solid Mechanics and Its Applications, vol. 128, 2nd edn. Springer, Cham (2006). https://doi.org/10.1007/1-4020-4133-0
11. Merlet, J.P., Perng, M.W., Daney, D.: Optimal trajectory planning of a 5-axis machine-tool based on a 6-axis parallel manipulator. In: Lenarčič, J., Stanišić, M.M. (eds.) Advances in Robot Kinematics, pp. 315–322. Springer, Cham (2000). https://doi.org/10.1007/978-94-011-4120-8_33
12. Oen, K.T., Wang, L.C.T.: Optimal dynamic trajectory planning for linearly actuated platform type parallel manipulators having task space redundant degree of freedom. Mech. Mach. Theory **42**(6), 727–750 (2007)
13. Pierrot, F., Nabat, V., Company, O., Krut, S., Poignet, P.: Optimal design of a 4-DOF parallel manipulator: from academia to industry. IEEE Trans. Rob. **25**(2), 213–224 (2009)
14. Reiter, A., Müller, A., Gattringer, H.: On higher order inverse kinematics methods in time-optimal trajectory planning for kinematically redundant manipulators. IEEE Trans. Ind. Inf. **14**(4), 1681–1690 (2018)

15. Santos, J.C., da Silva, M.M.: Redundancy resolution of kinematically redundant parallel manipulators via differential dynamic programing. ASME J. Mech. Rob. **9**(4), 041016 (2017)
16. Schappler, M., Ortmaier, T.: Singularity avoidance of task-redundant robots in pointing tasks. In: International Conference on Informatics in Control, Automation and Robotics (ICINCO) (2021)
17. Schappler, M., Tappe, S., Ortmaier, T.: Modeling parallel robot kinematics for 3T2R and 3T3R tasks using reciprocal sets of Euler angles. MDPI Robot. **8**(3), 68, 1–24 (2019)

Continuum Mechanisms

Analysis of a Compliant Parallel Manipulator for Torso Balance Rehabilitation

F. J. Campa(✉), D. Diaz-Caneja, M. Diez, and O. Altuzarra

University of the Basque Country UPV/EHU, Escuela de Ingeniería de Bilbao, Plaza Ingeniero Torres Quevedo 1, 48013 Bilbao, Spain
fran.campa@ehu.eus

Abstract. One of the first rehabilitation exercises in a hemiplegic patient is the recovery of the trunk balance, as a previous stage for the gait training. In this paper, a solution to assist this task with a robotic device is proposed and discussed. It is based on a compliant parallel kinematics manipulator of 4 degrees of freedom, 3 rotations and a translation in Z direction, attached to a patient on a standing frame. Thanks to its compliance, the manipulator can move together with the patient in the other two translational degrees of freedom in X and Y directions, although an elastic force that must be absorbed by the patient's body appears. Here, the analysis of those forces in several positions of the manipulator is shown, concluding that it is possible to design the mechanism taking into account those forces to avoid potential damage to the patient, keeping them under a tolerable threshold.

1 Introduction

Population aging has lead in the last decades to an increase of age-related disorders, being stroke one of the most common. One of the consequences of stroke is hemiplegia. In the early stages of rehabilitation, exercises focus on the motor recovery of the torso by means of passive mobilisations, re-education of balance, active assisted movements and muscular strengthening [1]. Here, the use of assistive devices as standing frames is common for safe positioning of the patient. It is of the utmost importance to perform as much repetitions as possible, so the use of robotics has become of interest, given also their capabilities to increase the patients' adherence to the rehabilitation program or to monitor its evolution using sensors [2, 3].

Regarding the robotics for rehabilitation, the use of compliant mechanisms can be of interest due to their flexibility, as it feels less intimidating than a mechanism with rigid elements, and they act as a natural filter that applies forces smoothly and can deflect to reach positions out of their workspace in isolation [4, 5]. After measuring the kinematics requirements for the motion of the trunk, see Fig. 1a, the authors of the present work developed a compliant parallel prototype for the rehabilitation of trunk balance [6]. It has four flexible bars actuated by rotary motors and joined through spherical joints to the backpack that the patient wears to be able to contribute to the patients' trunk rotation around the sagittal, longitudinal and lateral axis. Each bar is formed by: a coupling to the motor, four bars of Nylon 6.6 with a diameter of 6 mm, a coupling to a uniaxial force sensor, and a final coupling to the spherical joints, see Fig. 1b. Regarding the bars, the

O. Altuzarra and A. Kecskémthy (Eds.): ARK 2022, SPAR 24, pp. 211–219, 2022.
https://doi.org/10.1007/978-3-031-08140-8_23

central one is fixed to both couplings and the outer three bars are fixed in the coupling to the sensor but can slide through the motor coupling. They are located at 120° in a circumference of 25 mm radius around the central bar.

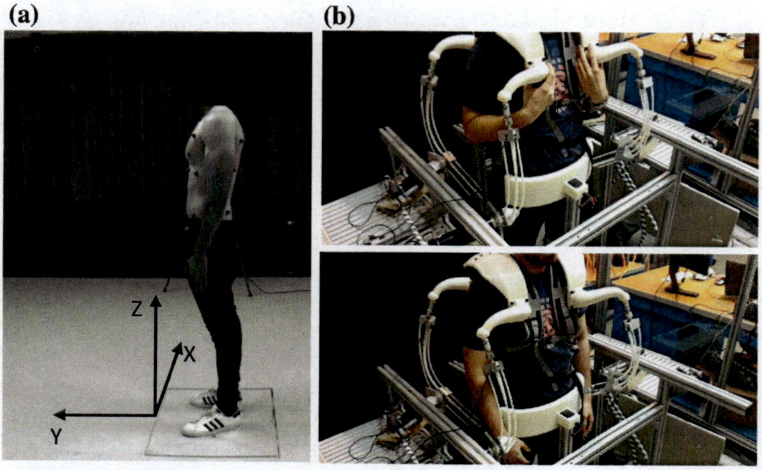

Fig. 1. a) Test layout. b) Images of the prototype.

The main degrees of freedom of the manipulator are the three rotations and a translation along the vertical Z axis. However, the rotation of the trunk is coupled to translations in X and Y axis, which are motions that the manipulator can only perform due to its compliance. This means that an elastic force will appear acting on the patients' trunk that must be calculated to avoid potential damage. In the following sections, we show three basic rehabilitation exercises that have been simulated, the basics of the modelling of the manipulator and the discussion of the results.

2 Kinematic and Dynamic Requirements

A series of exercises typical of physical rehabilitation of patients after stroke were measured on a healthy person (26 years old, 72 kg weight and 1.79 m high) using a vision system with 10 Vicon Systems cameras. Three perimetric rings were arranged at different heights along the torso, each of them with eight reflective markers. The aim was to record the motion of the three main regions of the spinal column: cervical, lumbar and dorsal.

The tests performed by the volunteer included torso rotations in X, Y and Z axis. The data obtained was exported to Matlab to calculate the rotation angles and displacement of the torso. According to the reference system in Fig. 1, the range of rotation specified was from −20° to 5° around X (flexion-extension), from −20° to 20° around Y (lateral bending) and from −10° to 10° around Z (axial rotation).

On the other hand, to calculate the maximum torque that the manipulator must apply to the patient to compensate his balance deficits, a simple static model of the human body on a standing frame was developed [7]. It considers the human parts as rigid bodies linked by spherical joints, actuated to simulate the action of the muscles. The joint reactions would be the forces that the skeleton suffers, due to the weight of the body members. Both the legs and the hip are fixed, simulating the action of the standing frame. The trunk is modelled as two sections and the head, arms, forearms and hands were considered. The weight of each part and its mass center were calculated using formulae in the bibliography. In Fig. 2, the body poses in the extreme position for the flexion-extension, lateral bending and axial rotation are shown.

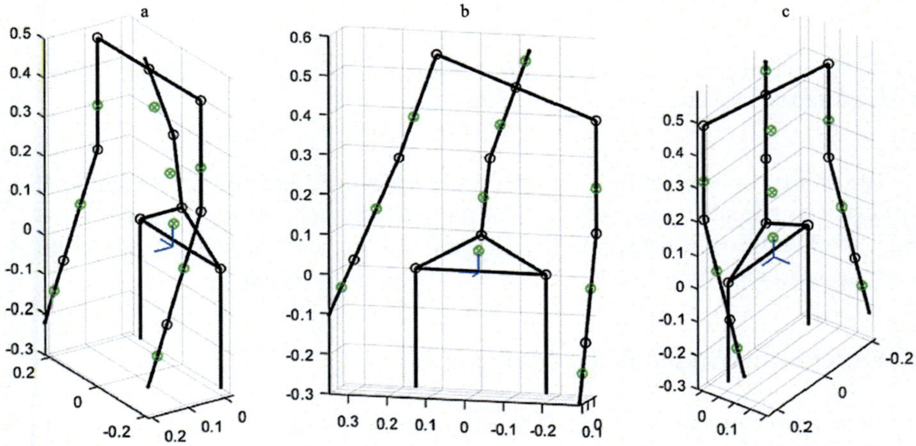

Fig. 2. Simulated body poses: a) Flexion of 20°. b) Lateral bending of 20°. c) Axial rotation of −10°.

The model allowed to calculate the moment that the muscular system must apply on the upper trunk, where the manipulator is attached to the body, to balance the pose. The muscular moment is calculated with respect to the joint that links the two parts of the trunk in the model, see Fig. 2. In Fig. 3, the calculated moment projections in X, Y and Z axis are shown for three exercises in quasi-static conditions. In Fig. 3a, exercise A, a flexion motion from 0° to 20°, where a growing moment in X direction is necessary to balance the body. Note that in 0° there is a moment to balance the body weight, as the mass center of the body has an offset in Y direction with respect to the joint position. In Fig. 3b, exercise B, the moment to perform the lateral bending up to 20° of Fig. 2 is shown. Finally, in Fig. 3c, exercise C, it can be seen the moment to perform an axial rotation up to −10°. There, as the body weight has no effect on the moment around Z, the figure displays the moment due to the passive stiffness in an axial rotation [8].

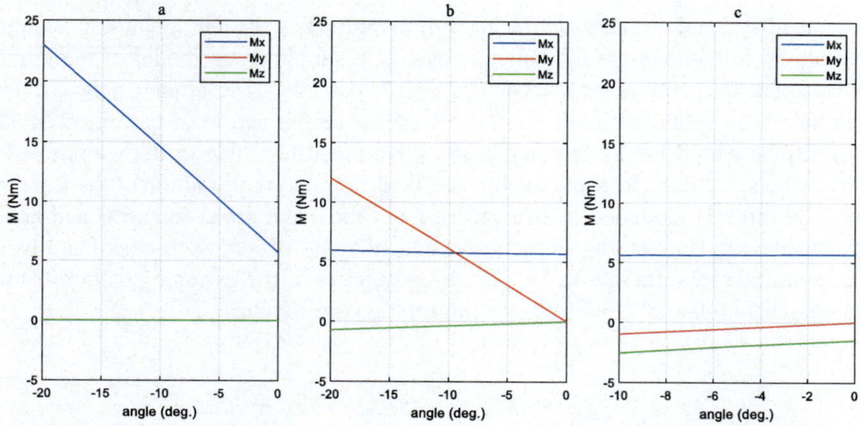

Fig. 3. Muscular moment evolution on upper trunk: a) Flexion up to 20°. b) Lateral bending up to −20°. c) Axial rotation up to −10°.

The moment values in Fig. 3 are the reference in each exercise for the torque that the device must apply on the patient. It must be said, though, that they are indicative as no patient has the same degree of disability after stroke, and the added stiffness due to muscular spasticity, for example, has not been taken into account.

3 Kinematic Model of the Manipulator

The manipulator proposed has the following elements. First, the mobile platform that the patient wears as a backpack. Second, the fixed platform, which is attached to the standing frame. There, four rotary motors axis are disposed in points A_i, see Fig. 4a, in an elliptical configuration with the motor axis tangential to the ellipse. Finally, four flexible bars, fixed to the motors in A_i and joined to the jacket by means of spherical joints in B_i that are in an ellipse parallel to the fixed one. The main degrees of freedom, considering that it is a compliant mechanism, are the 3 rotations around X, Y and Z, and a translation along the Z axis. The reference system has the origin in the centre of the lower ellipse.

To solve the inverse kinematic problem, it is assumed that the mobile platform moves together with the upper torso of the modeled human body, so its position is known, and therefore, the spherical joint positions $\mathbf{OB_i}$. To model the deflection of the bars, they are discretized in N rigid elements joined by spherical joints where the flexural stiffness is modeled as torsion springs between elements [6]. Their mass has been neglected. The position of each element j with respect to the previous one is defined by an azimuthal angle φ_{ij} and a polar angle θ_{ij}, as shown in Fig. 4 right. Being R the radius of the coupling to the motor axis and dL the length of each element, the position of each spherical joint B_i with respect to a local reference system $X_iY_iZ_i$ located in A_i is shown in Eq. 2, where

θ_{i1} is also the position of the i motor.

$$\mathbf{A_iB_i} = \begin{bmatrix} 0 & -1 & 0 \\ c\theta_{i1} & 0 & s\theta_{i1} \\ -s\theta_{i1} & 0 & c\theta_{i1} \end{bmatrix} \begin{Bmatrix} 0 \\ 0 \\ R + dL \end{Bmatrix} + \sum_{j=2}^{N} \begin{bmatrix} c\theta_{ij}c\varphi_{ij} & -s\varphi_{ij} & s\theta_{ij}c\varphi_{ij} \\ c\theta_{ij}s\varphi_{ij} & c\varphi_{ij} & s\theta_{ij}s\varphi_{ij} \\ -s\theta_{ij} & 0 & c\theta_{ij} \end{bmatrix} \begin{Bmatrix} 0 \\ 0 \\ dL \end{Bmatrix} \quad (1)$$

With the corresponding equations, the kinematic condition $\mathbf{OB_i}$ is obtained.

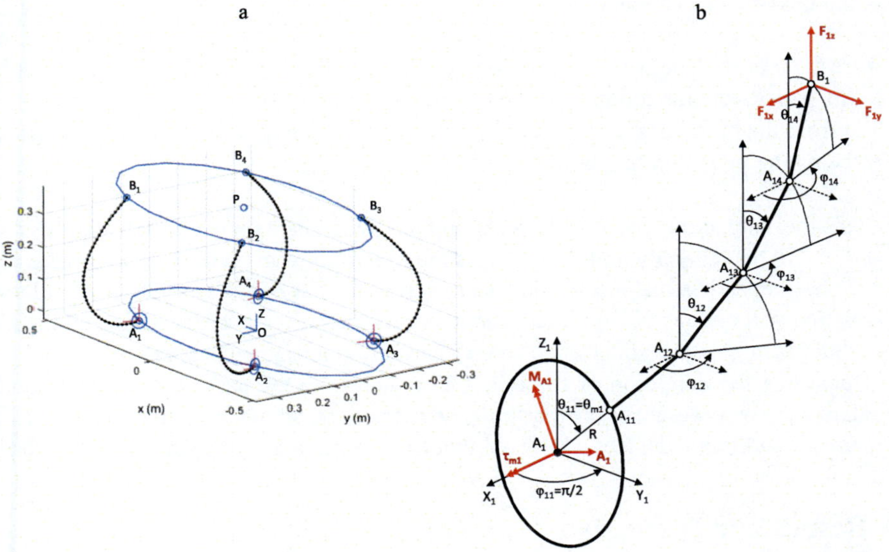

Fig. 4. (a) Schematic of the kinematics of the proposed parallel compliant manipulator. (b) Model of the i = 1 discretized flexible bar for $N = 4$.

In addition, the manipulator must balance the patient applying a moment on the upper trunk. A factor α has been defined to control the moment that the platform must exert on the upper trunk as $\boldsymbol{\tau} = (1-\alpha)\boldsymbol{\tau}_{trunk}$. The factor α ranges from 0 to 1, where 0 represents full patient disability and 1 no disability, that is, when the manipulator must apply all the moment that the patient cannot do, and when it just follows the patient. Hence, initially the four bars must meet a set of 15 conditions, the three cartesian coordinates of each spherical joint B_i and the three components of the moment $\boldsymbol{\tau}$. To solve the kinematics problem, coupled with the statics, there are $5N + 5$ unknowns per bar:

- $2N - 1$ angles θ_{ij} and φ_{ij}, as φ_{i1} is 90°
- 3N joint reactions
- 3 components of the moment in the motor including the motor torque
- 3 components of the force in B_i

Regarding the equations, as there is no torsion due to the spherical joint, the balance of moments is done in each element in the two transversal directions only, except for $j =$

1, which is considered together with the motor and provides 3 equations. Hence, there is a total of $5N + 4$ equations per bar:

- $2N + 1$ equations from the balance of moments
- $3N$ equations from the balance of forces
- 3 kinematic conditions of the spherical joint B_i position

Finally, 4 of the 6 equations of the static balance of the mobile platform are added to equal unknowns and equations:

- 3 are from the balance of moments, where the moment τ that the platform must apply to the patient and the moment of the bars forces are considered
- 1 is from the balance of bars forces in Z direction, where a condition of null vertical force on the patient is imposed

This means that the sum of the forces applied by the bars on the mobile platform in X and Y will be generally unbalanced. It is the user who wears it the one who balances them. They depend on the position and the material and size of the bars. The manipulator in isolation will not be able to remain stable in every position, but, depending on the tilt of the platform, it is possible to find positions where the balance happens.

Regarding the condition of force in Z direction over the patient, that force could counteract the own weight of the platform and the patient, which is an interesting feature of many rehabilitation devices. However, that force is limited by the strength of the bars.

4 Results and Discussion

As an example, Fig. 5 shows the simulation of the manipulator motion during Exercise A, rotation around X, considering $\alpha = 0$ and $\alpha = 1$. In Fig. 5a and 5b it can be seen the difference in the deflection of the bars 1 and 2 for both situations. In Fig. 5a, they work in a buckling mode to be able to apply the positive moment on the mobile platform while in Fig. 5b they are in bending mode. The torque that the motors apply on the bars and their position is seen in Fig. 5c and 5d. The influence of α is mainly seen in the behaviour of bars 1 and 2. As mentioned, during the exercises the manipulator reaches positions due to its compliance that it wouldn't take in isolation, at the expense of applying a force on the patient that compensates the unbalance of forces in the mobile platform in X and Y. In Fig. 6, those forces are shown for $\alpha = 0$ and $\alpha = 1$ in Exercise A, B and C. In Fig. 6a, for the rotation around X, it is seen that only appears an unbalance in Y direction, and the force reaches a value of −35 N on the patient at −20° tilting when $\alpha = 0$. That same force remains positive when $\alpha = 1$, reaching a maximum value of 13 N.

In Fig. 6b, for the rotation around Y, it is seen how, for the selected bars, the manipulator cannot tilt more than 16°. The length of the bars should be increased to reach those positions. In this case, the forces on the patient would remain below 10 N. In Fig. 6c, for the rotation around Z, the forces on the patient also remain below 10 N. Although there are reports applying more than 200 N on healthy people [9], it is currently being consulted which load could be acceptable on a real patient.

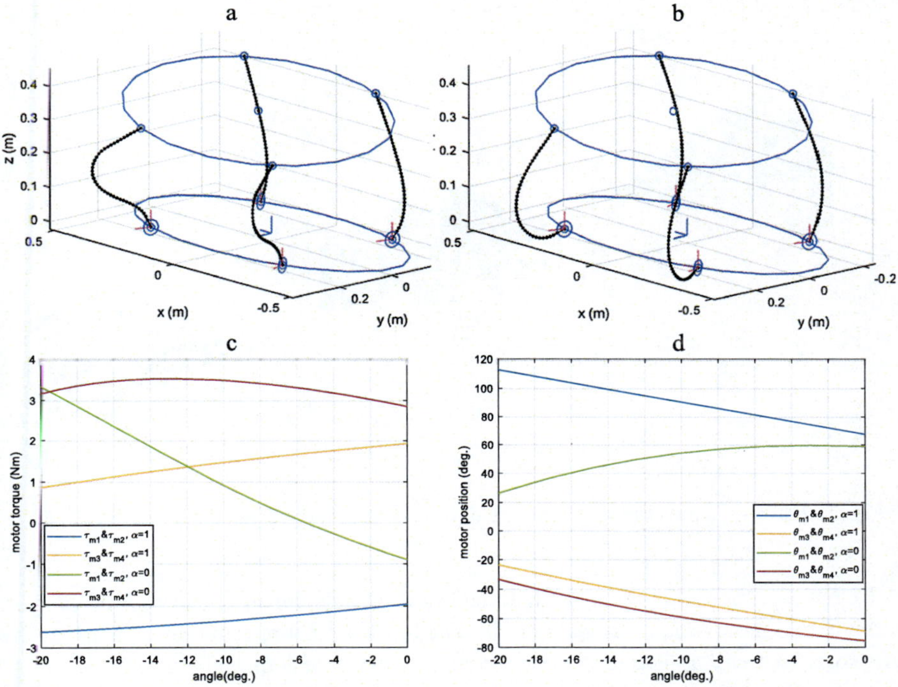

Fig. 5. Simulation of Exercise A: a) Manipulator at 20° for $\alpha = 0$. b) Manipulator at 20° for $\alpha = 1$. c) Torque applied by the motors. d) Motors position.

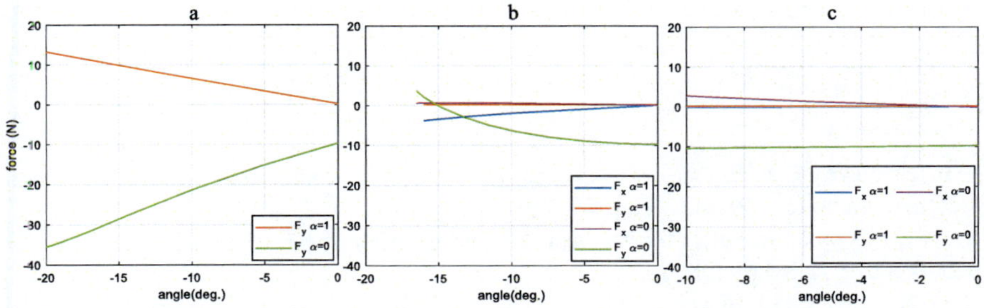

Fig. 6. Force applied on the patient: a) Exercise A. b) Exercise B. c) Exercise C.

To illustrate how the unbalanced force varies in this manipulator, Fig. 7 shows the value of the unbalanced force in Y direction when a translation in Y direction is applied to the mobile platform at several rotation angles around X axis, depicted in the isolines. The center of the mobile platform remains at x = 0 and z = 0,84 m coordinates. To consider the moment that the manipulator applies, −10 Nm, 0 Nm and 10 Nm have been simulated. Only when the lines displayed take a zero value the manipulator has the capability of being statically balanced alone. In the other positions, an external force to counteract the one displayed is necessary to balance the platform. The graphs also give an idea of all the possible positions that the manipulator can reach, and how they change when the manipulator must apply a moment.

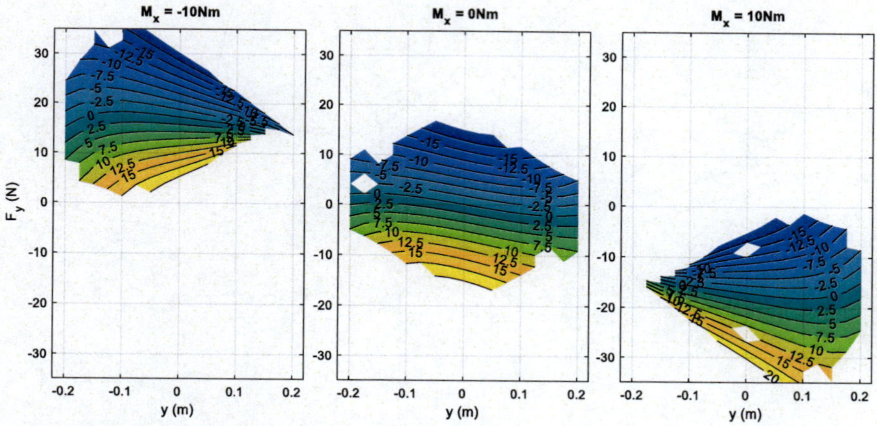

Fig. 7. Force in Y direction as a function of the moment applied and the translation in Y.

To sum up, these diagrams can help taking decisions during the design of the bars for the manipulator when it is going to interact with a person, providing a map for the static force that will appear in the contact and the range of achievable positions.

Acknowledgements. Authors want to acknowledge the support given by the Spanish Government (Ministerio de Economía y Competitividad y Fondo Europeo de Desarrollo Regional, FEDER, DPI2015-64450-R), and the Basque Government for the project IT949-16.

References

1. Reinkensmeyer, D.J., Emken, J.L., Cramer, S.C.: Robotics, motor learning, and neurologic recovery. Annu. Rev. Biomed. Eng. **6**(1), 497–525 (2004)
2. Xie, S.: Advanced Robotics for Medical Rehabilitation. Springer, Cham (2016). https://doi.org/10.1007/978-3-319-19896-5
3. Nam, K.Y., Kim, H.J., Kwon, B.S., Park, J.W., Lee, H.J., Yoo, A.: Robot-assisted gait training (Lokomat) improves walking function and activity in people with spinal cord injury: a systematic review. J. Neuroeng. Rehabil. **14**(1), 1–13 (2017)
4. Bryson, C.E., Rucker, D.C.: Toward parallel continuum manipulators. In: 2014 IEEE International Conference on Robotics and Automation, Hong Kong, pp. 778–785 (2014)

5. Burgner-Kahrs, J., Rucker, D.C., Choset, H.: Continuum robots for medical applications: a survey. IEEE Trans. Rob. **31**(6), 1261–1280 (2015)
6. Diaz-Caneja, D., Campa, F.J., Altuzarra, O.: Design and modelling of a parallel continuum manipulator for trunk motion rehabilitation. J. Med. Devices **15**(1), 011109 (2021)
7. Winter, D.A.: Biomechanics and Motor Control of Human Movement. Wiley (2009)
8. McGill, S., Seguin, J., Bennett, G.: Passive stiffness of the lumbar torso in flexion, extension, lateral bending, and axial rotation. Effect of belt wearing and breath holding. Spine **19**(6), 696–704 (1994)
9. Mack, G., Gardner-Morse, I., Stokes, A.F.: Trunk stiffness increases with steady-state effort. J. Biomech. **34**(4), 457–463 (2001)

Design and Analysis of a Compliant Parallel Robot with Cardan Joints for a Cryogenic Working Environment

Philipp Jahn(✉), Jakob Hentschel, and Annika Raatz

Institute of Assembly Technology, Gottfried Wilhelm Leibniz University Hanover, An der Universität 2, 30823 Garbsen, Germany
jahn@match.uni-hannover.de

Abstract. Automation is ubiquitous and is finding its way into more and highly specialized applications - including cryo-conservation. However, manual handling of biological or toxic samples is still the norm in research and commercial storage facilities. In biobanks, samples are stored, retrieved or moved by hand wearing rigid, protective clothing due to the significant risk of injury to personnel from cold burns. In addition, sample integrity is compromised by temperature changes or contamination.

In this paper, we present an approach for the automation of handling processes in cryobanks for cryopreservation at temperatures below $-130\,°C$. The automation system is based on a parallel robot, as its structure allows the positioning of the drives outside the cooled workspace. The robot's joints and manipulators are located inside the biobank, cooled with liquid nitrogen. The passive joints of the robot are implemented as Cardan joints, which are composed of multiple flexure hinges. In the following, we describe the design of the kinematic chains of the parallel robot's structure. The elaboration of these is based on the results of FE-modeling of individual components under cryogenic conditions. We then subject the overall structure to static mechanical simulations, both in its initial position and under deflection of the actuator and compare the occurring deviations with the calculations based on the inverse solution.

Keywords: Compliant parallel robot · Cryopreservation · Flexure hinges

1 Introduction

In today's research and production facilities, cryo-conservation of biological materials meets many challenges during the handling of samples. On the one hand, the manual handling of biological or toxic samples poses a significant danger for the workers (due to cold burns or exposure to hazardous materials) and the risk of compromising the samples' integrity and stability (due to temperature changes or contamination). Technical advances, on the other hand, permit storage capacities of more than a million test samples per container. These advances make the sample storage more economical, but the enormously increasing number of sample tubes can no longer be handled efficiently and

O. Altuzarra and A. Kecskémthy (Eds.): ARK 2022, SPAR 24, pp. 220–227, 2022.
https://doi.org/10.1007/978-3-031-08140-8_24

safely by solely manual operations [1]. Different approaches have been made to automate parts of the handling process using gas locks and serial robot structures [2]. However, a completely automated handling system inside a cryogenic environment is still missing. This paper presents a new approach for a full automation of handling operations inside a biobank, using a compliant parallel robot with passive Cardan joints.

1.1 Related Research

Compliant mechanisms and flexure hinges are primarily applied in the field of robot-supported micro-operations. YU [3] proposed a 3-DOF parallel robot with compliant joints. They showed the validity of compliant joints by replacing the kinematic pairs implemented usually in parallel robots. However, the tests were only conducted for a minimal range of motion (0,009 mm^2 working space), too small for the application in a cryobank. YUN [4] presented a 6-DOF compliant dual redundant parallel robot with wide-range flexure hinges. This robot can achieve a wide range of motion (40.000 mm^2 working space, see Sect. 2.2). Still, the accuracy is limited to a rough positioning due to the geometric errors the flexure hinges introduce into the structure. CHEN [5] proposed a parallel robot with SCARA motions (selective compliance assembly robot arm). The robot's accuracy was raised by implementing an error model and a kinematic calibration. However, the proposed structure is not suitable for the application in a cryobank because the radial movement of the drives makes it impossible to place these drives outside of a cryogenic storage tank, a prerequisite to prevent any heat carryover into the system. Finally, KOZUKA [6] researched the use of compliant joints in a parallel robot structure for precision assembly applications. A serially-layered flat compliant joint was developed in his work, allowing large bending angles up to 90°. The overall stiffness of the structure is extremely high, which results in increased operational accuracy. Yet, the proposed robot structure itself is only suited for a very limited range of motion.

All of the mentioned works are aimed at a work environment at room temperature. An approach for a fully automated, robot-supported handling process under cryogenic conditions does not yet exist and requires extensive research.

1.2 Previous Work

The concept of the parallel robot we present in this paper is subject to several constraints of geometric and material-scientific nature due to the extreme work environment it is intended for. In previous works, we investigated parallel kinematic structures using evolutionary algorithms like particle swarm optimization and similar strategies [7]. We determined a delta structure with vertical drives in the form of electro spindles with universal joints as passive joint geometries as the most suitable. With the prior conducted simulations and optimizations, we determined that an individual maximum bending angle ψ_{max} of at least 30° (in each direction of bending) is necessary to implement flexure hinges into the robot structure. The challenge here was the low mobility of the flexure hinges compared to conventional swivel joints and the previously unexplored influence of low temperatures on the deformation behavior of flexure hinges. On the one hand, a base material suitable for the extreme operating conditions had to be identified for manufacturing the joints. On the other hand, a joint geometry adapted to the task was necessary to achieve the desired bending angles under the cryogenic conditions while providing a sufficiently rigid behavior along and around the parasitic axes (all axes that are not identical with the primary bending axis are considered parasitic). In [8], we were able to show with simulations and test scenarios that flexure hinges made from the titanium alloy TiAl6V4 can achieve the necessary bending angles under cryogenic conditions without permanent deformation.

These flexure hinges, however, are only one part of the approach for the automation of handling processes in cryobanks. In Sect. 2, we present a design of flexure hinges for the passive Cardan joints and the kinematic chains, and the overall structure of the envisaged parallel robot in more detail. In Sect. 3, we simulate the deformation of the robot structure in its initial position (due to its weight) and during deflection (end effector moved to the perimeter of the working area). We then compare the deviation of the robot's structure with the calculated analytical solution we get solving the inverse kinematics to determine the anticipated positioning accuracy of the envisaged robot. Finally, in Sect. 4, we give an overview on the current conducted work and further research.

2 Constructive Implementation

Based on the general joint structure researched by KOZUKA and ARATA [6], we developed an adapted joint structure (shown in Fig. 1) suited for the cryogenic work environment to meet the requirements of this unique application (e.g. a sufficiently high stiffness along the parasitic axes). The flexure hinges with a total thickness of T_J consist of serially layered flexures separated by rigid shuttles. To generate and evaluate several design candidates, we varied the length l_{SP} and thickness t_{SP} of the flexures and the distance s between them. The bending angle of a single flexure can be approximately calculated based on Eq. (1). The total bending angle can be found by multiplying this angle by the number of flexures in the joint.

$$\psi_{max} = \arcsin\left(\frac{s}{2 \cdot l_{RP} - 2 \cdot l_S + l_{SP}}\right) \tag{1}$$

In simulations we were able to show that joints with 27 flexures could reach bending angles up to $\psi_{max} = 40°$, while still providing only minimal deviations along the parasitic axes (see Sect. 3.2).

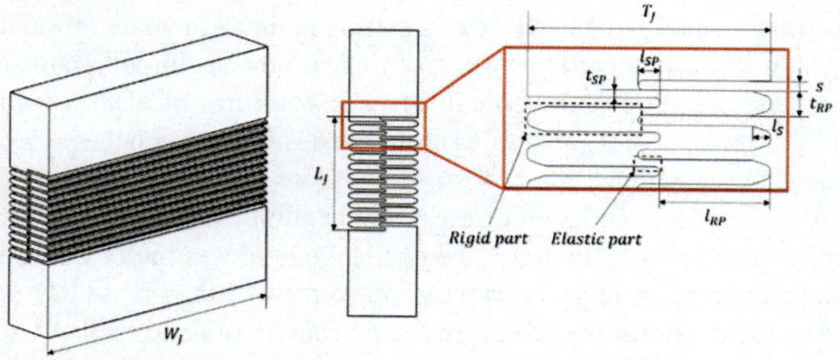

Fig. 1. Serially layered flat compliant joint [6]

2.1 Structure of a Single Kinematic Chain

The structure of one of the three identical kinematic chains of the parallel robot design is shown in Fig. 2; the single kinematic chain consists of an upper (Fig. 2a) and a lower (Fig. 2b) compliant universal joint, which are connected by two parallel guide rods (1). The upper universal joint is connected to the piston rod of the linear actuator via a coupling rod (2). The lower universal joint is bolted to the end effector platform (3) via an angle (4). In addition, the end effector platform is located within the working space below the universal joints in all positions. The joints of the first bending axis (5, red) and the second bending axis (6, green) are interlocked by the angle $\alpha = 31.23°$. This ensures that the joints of the first axis are not deflected in the initial position of the end effector and that the overall construction is thus largely stress-free. The angle ensures that the joint of the first axis can be deflected to the maximum extent without causing collisions within the structure. In addition, only in this way does the joint have a symmetrical bidirectional bending angle from the initial position of the end effector.

2.2 Overall Construction of the Robot

The overall structure of the parallel robot (Fig. 3) consists of a total of three identical kinematic chains. According to [9], these are each arranged offset at an angle of 120°. The kinematic chains are connected to the end effector platform at the bottom. At the upper end, the respective coupling rod is bolted to the associated piston rod of the linear actuator. These are supported by plain bearings in the lid (7) of the cryobank. The three linear drives (8) generate the translatory drive movement. They are screwed onto the lid of the cryobank via a motor mount and are additionally secured at the top by a frame (9). This ensures that they remain vertically aligned during operation. The gripper (10) is inserted with its cylindrical carrier from below into the end effector platform and then screwed in place from above. Two Micronic 96-3 sample racks (11) and a QR code scanner (12) for sample identification are arranged at the bottom of the cryobank, according to Fig. 3a. This arrangement defines the working space of the robot. The overall robot structure has a moving mass of about 1800 g, which is very light. The slim

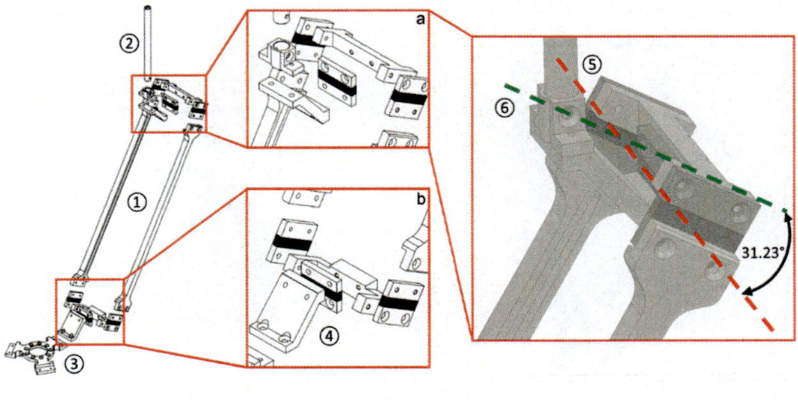

Fig. 2. Kinematic chain and Universal joints

structure of the kinematic chains and the end effector thus ensure significantly higher dynamics of the system. Also shown is the parallel robot structure in its initial position. The tip of the end effector is centered and elevated above the storage racks loaded with sample tubes. The working area (13) is a square of 200 × 200 mm.

3 Simulation of Static Deviations

In this chapter, we carry out the numerical calculations for validating the design using ANSYS and the associated solver ANSYS-Mechanical. Our goal is to verify the expected deviations from the calculated solution in our simulated environment. We generate structural, mechanical models of the individual flexure hinges and the overall robot structure. In our investigations, we assume a stationary and homogeneous temperature field with a temperature of $-196\ ^{\circ}C$ (which equals the boiling point of the cooling medium Nitrogen) with regard to the thermal boundary conditions. Due to the large cross-sections of most structural components, these are perfectly rigid. We expect the only significant deformation within the joint structures.

3.1 Preconditions of the Simulations

Structural, Mechanical Consideration of the Individual Compliant Joints. In the conducted simulations, we want to verify the in [8] analytically calculated maximum bending angle $\psi_{max} = 40^{\circ}$ by FE-modeling a single joint. In addition, we carry out strain and stress investigations on the flexure hinge to ensure that at no point during deflection, the joints are submitted to permanent deformation.

Structural, Mechanical Consideration of the Overall Robot Structure. The design of the kinematic chains imposes a pivot point around the axes of the individual joints. It is, therefore, necessary to investigate whether and to what extent the bending behavior of the joints within the parallel structure deviates from the individual investigations.

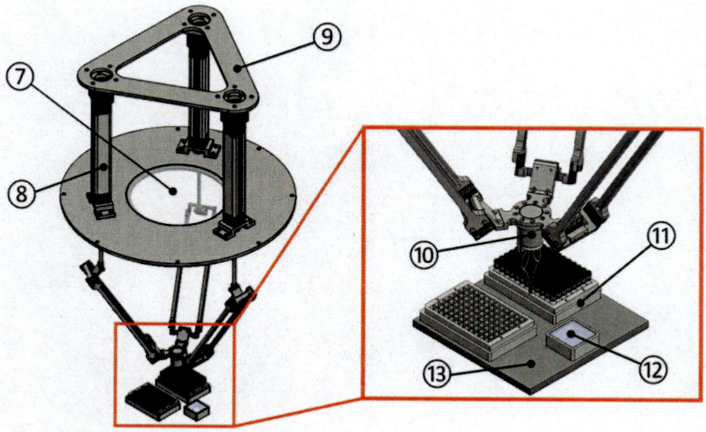

Fig. 3. Overall construction of the robotic structure

Based on the inverse kinematics, we calculated the maximum bending angle of a single flexure hinge under maximum deflection to be $\psi_{max} = 37,01°$. We simulated two specific configurations: On the one hand, we looked at the overall deviation of the robot in its initial position (see Sect. 3.2). On the other hand, we moved the end effector to the perimeter of the working space regarding the flexure hinge under maximum deflection.

3.2 Results of the Simulations

The rotational stiffness of the individual joint while bending around the main bending axis is 0.1669 $^{Nmm}/_{\circ}$. The maximum equivalent stress occurring from a rotational deflection of $\psi_{max} = 40°$ is $\sigma_{max} = 830.17\,\text{MPa}$ and thus is in the linear-elastic range of the material $\left(R_p = 1250\,\text{MPa}\right)$.

The stresses and deformations occurring in the flexures due to the robot's weight in the initial position can be neglected ($\sigma_{weight} = 0,0114\,\text{MPa}$). The maximum stress within the flexure hinges is $\sigma_{max} = 486,55\,\text{MPa}$, which is still in the linear-elastic range of the material and confirms our assumption that no permanent deformations occur during the deflection. Figure 4 shows a single flexure hinge and the complete deflected robot structure.

For the calculation of the positioning accuracy, we compare the results of the FE modelling and the computation of the inverse kinematics. For the inverse kinematics, we calculate the position of the end effector ${}^{0}\overrightarrow{r_{0,E,ideal}}$ and the associated drive coordinates $\theta_{i,ideal}$ for the deflection of the end effector. We apply these coordinates to the structure in the simulations and determine the occurring position ${}^{0}\overrightarrow{r_{0,E,sim}}$. To obtain a measurement for the positioning accuracy of the entire robot structure, we computehe deviation between the ideal and the simulated displacement utilizing Eq. (2).

$$\Phi = \left|{}^{0}\overrightarrow{r_{\Delta}}\right| = \left|{}^{0}\overrightarrow{r_{0,E,sim}} - {}^{0}\overrightarrow{r_{0,E,ideal}}\right| = \left|\begin{pmatrix} X_{E,sim} - X_{E,ideal} \\ Y_{E,sim} - Y_{E,ideal} \\ Z_{E,sim} - Z_{E,ideal} \end{pmatrix}\right| \tag{2}$$

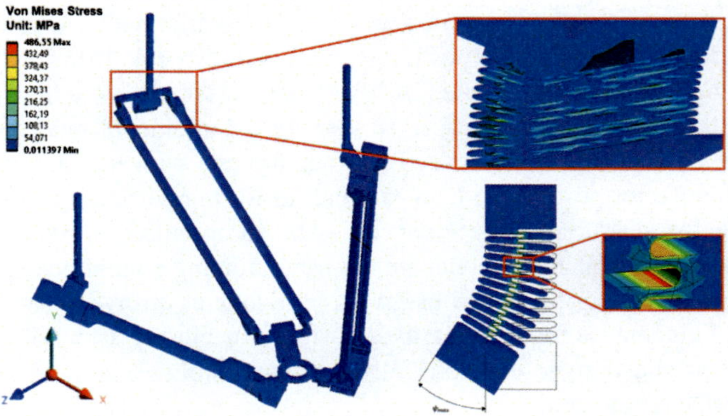

Fig. 4. Single flexure and complete robot structure

The deflection over the working space is increased equally in X and Z direction in 5 mm increments. The resulting overall deviation from the ideal position is shown in Fig. 5. The plot suggests a linear dependency of Φ in relation to the deflection in X and Z direction, which would make it possible to compensate the deviation with a simple feed-forward control (See Sect. 4). However, ANSYS is prone to calculation errors when the simulated objects are submitted to extensive deformations: The objects tend to inflate and contort unrealistically due to mesh deformations even if nonlinear adaptivity algorithms (NLAD) are implemented [10]. For this reason, the positioning accuracy in this work is initially determined only for moderate deflections of the end effector platform where the errors due to large deformations are within acceptable limits. At a deflection of 30 mm in X and Z direction, the total deviation from the ideal end effector position is 0.44 mm. Despite the potential errors, the results still give a seasonable estimate of the order of magnitude of the positioning accuracy. We assume that the deviation continues to increase slightly with further deflection. This assumption must still be verified in further investigations, which we describe in Sect. 4.

4 Summary and Outlook

In this paper, we presented a robot structure for the automated handling processes in cryobanks. We described the single flexure hinges used as passive joints, the kinematic chains of the envisaged parallel robot and the overall robot structure itself. We based these elaborations on the FE-modeling of individual components and the whole robot under cryogenic conditions. We subjected the overall structure to static mechanical simulations, both in its initial position and under deflection of the actuator and compared the occurring deviations with the calculations based on the inverse kinematics. The results we presented showed that the used flexure hinges were able to bend $\psi_{max} = 40°$ without permanent deformation, thus guaranteeing a long service life. We calculated the deviation of the simulated end effector position from the ideal position (derived from the inverse kinematics) under deflection. We introduced the factor Φ as a measurement for the position accuracy of the end effector. The results, however, are influenced by errors occurring in simulations with large deformations. A reliable model can only be achieved with extensive measurements in a physical technology demonstrator. The final goal of our work is to develop a parametric kinematic model for the robot and perform the kinematic calibration, position control, and dynamic feed-forward control of the technology demonstrator created. A key objective is to achieve a high positioning accuracy in all areas of the workspace. To identify the additional model parameters, we will trace suitable trajectories (e.g. Lissajous figures) with the demonstrator and record the end effector position with an external measuring device (e.g. laser tracker). With the collected measurement data, we can set up an optimization problem in which the parameters of the kinematic model are adjusted utilizing a numerical optimization algorithm (e.g. Levenberg-Marquard based on least-squares error) so that the solution of the forward kinematics fits the measured positions as closely as possible. This will be the basis that allows us to realize a fully automated, reliable storage solution in a cryogenic work environment.

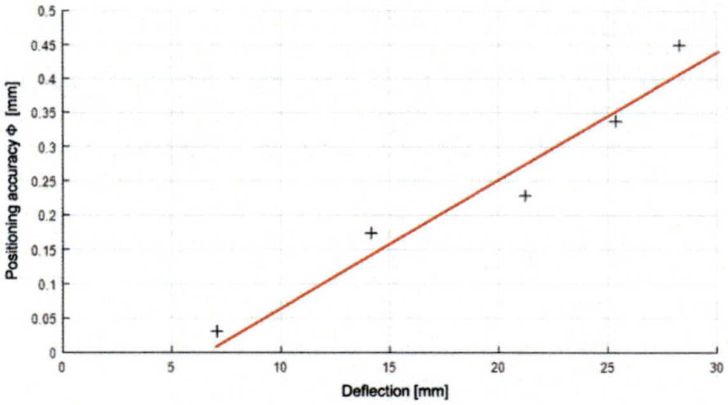

Fig. 5. Positioning accuracy Φ depending on the deflection

References

1. Alikani, M., Parmegiani, L.: Human reproductive cell cryopreservation, storage, handling, and transport: risks and risk management. Semin. Reprod. Med. **36**, 265–272 (2018). https://doi.org/10.1055/s-0038-1676849
2. Kaindl, J., Meiser, I., Majer, J., et al.: Zooming in on cryopreservation of hiPSCs and neural derivatives: a dual-center study using adherent vitrification. Stem Cells Transl. Med. **8**, 247–259 (2019). https://doi.org/10.1002/sctm.18-0121
3. Yu, G., Wang, L., Wu, J., et al.: Stiffness modeling approach for a 3-DOF parallel manipulator with consideration of nonlinear joint stiffness. Mech. Mach. Theory **123**, 137–152 (2018). https://doi.org/10.1016/j.mechmachtheory.2018.01.005
4. Yun, Y., Li, Y.: Design and analysis of a novel 6-DOF redundant actuated parallel robot with compliant hinges for high precision positioning. Nonlinear Dyn. **61**, 829–845 (2010). https://doi.org/10.1007/s11071-010-9690-x
5. Chen, Y., Xie, F., Liu, X., et al.: Error modeling and sensitivity analysis of a parallel robot with SCARA (selective compliance assembly robot arm) motions. Chin. J. Mech. Eng. **27**, 693–702 (2014). https://doi.org/10.3901/CJME.2014.0423.082
6. Kozuka, H., Arata, J., Ichikawa, S., et al.: Multi-DOF compliant-parallel mechanism for high precision machine in optical industry. J. Robot. Soc. Jpn. **32**, 481–489 (2014)
7. Schappler, M., Jahn, P., Raatz, A., et al.: Combined Structural and Dimensional Synthesis of a Parallel Robot for Cryogenic Handling Tasks. In: Annals of Scientific Society for Assembly, Handling and Industrial Robotics 2021, vol. 125, pp. 65–77. Springer, Cham (2022)
8. Jahn, P., Ihmig, F., Raatz, A.: Design of a parallel robot with additively manufactured flexure hinges for a cryogenic work environment. Procedia CIRP **103**, 280–285 (2021). https://doi.org/10.1016/j.procir.2021.10.045
9. Jahn, P., Raatz, A.: Numerical simulation and statistical analysis of a cascaded flexure hinge for use in a cryogenic working environment. In: Schüppstuhl, T., Tracht, K., Henrich, D. (eds.) Annals of Scientific Society for Assembly, Handling and Industrial Robotics, pp. 81–94. Springer, Heidelberg (2020). https://doi.org/10.1007/978-3-662-61755-7_8
10. Ansys Inc.: Adapt large deflection simulations confidently - nonlinear adaptivity. White Paper, Canonsburg (2021)

Design and Inverse Kinematics of a Novel Tendon-Driven Continuum Manipulator Capable of Twisting Motion

Yuhang Lei(✉), Yusuke Sugahara, and Yukio Takeda

Tokyo Institute of Technology, Meguro-ku, Tokyo, Japan
lei.y.ab@m.titech.ac.jp

Abstract. Tendon-driven Continuum Manipulator (TCM) is a kind of flexible manipulator with outstanding performance on dexterity, safety and intrinsic force sensing ability. These features make TCM prospective for the application in human contact scenarios such as Robot-assisted Minimally Invasive Surgery (RMIS). As an improvable point, this paper proposed a novel design of TCM with capability of one additional rotational DOF (twisting motion) compared with the previous design. Firstly, the basic composition and working principle of the proposed design are introduced. Then the corresponding inverse kinematic model as well as the verification experiment are introduced.

Keywords: Tendon-driven Continuum Manipulator · Twisting motion · Spatial curve shape · Inverse kinematics

1 Introduction

Tendon-driven Continuum Manipulator (TCM) is a kind of flexible manipulator, with a continuous elastic structure in the middle as a backbone. The actuators are placed at the base and the motion of the manipulator is controlled via tendons. An illustration of a conventional TCM is given in Fig. 1.

The spacer discs are fixed to the elastic backbone and three holes on each of them are used to guide the tendons. As shown in Fig. 1, conventional TCM can realize 2 DOFs bending motion in two directions, however, the twisting of the manipulator along the backbone is not controllable for current existing TCMs [1]. TCM has certain advantages on dexterity, safety for human contact and the intrinsic ability of sensing the external force applied to the end disc. These features make it suitable for human contact (collaborative) scenarios.

One probable application of TCM is the Robot-assisted Minimally Invasive Surgery (RMIS) field. RMIS has been gradually applied to many classes of surgeries as a response to the increasing needs from the patients for less pain and smaller incision caused during the operation, represented by da Vinci surgical system [2]. Currently, one remaining drawback in the RMIS systems is the limitation of the movement due to traditional rigid link structure. Although current RMIS manipulator can realize most surgical operation

O. Altuzarra and A. Kecskémthy (Eds.): ARK 2022, SPAR 24, pp. 228–236, 2022.
https://doi.org/10.1007/978-3-031-08140-8_25

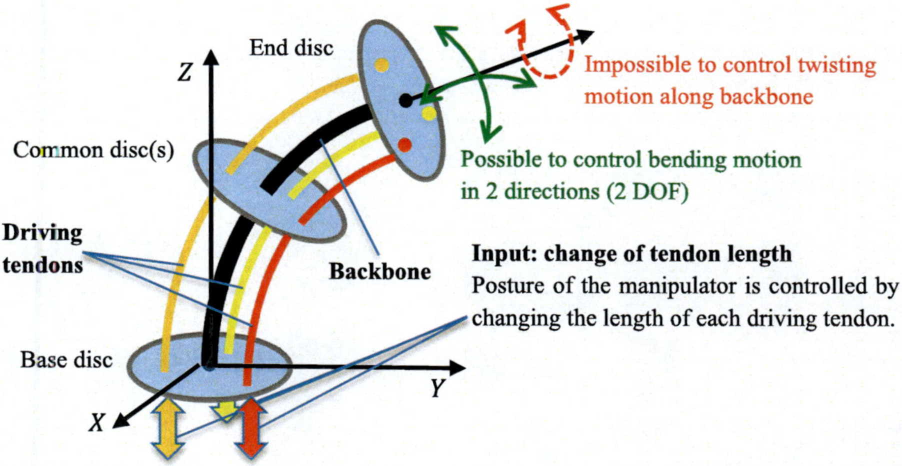

Fig. 1. Composition and working principle of a conventional TCM

with enough accuracy, they can still face difficulties when trying to pass through narrow paths inside patients' body, and the twisting of the manipulator along its shaft axis usually takes much effort which is a commonly used motion such as in suturing operation [3]. Therefore, it will be beneficial if the TCM can realize twisting motion along the backbone.

As an attempt to increase such dexterity of TCM, this research proposed a novel TCM which is capable of controllable twisting motion. This paper focuses on the introduction of the new design and the corresponding kinematic analysis along with the verification experiment.

2 Design Concept and Proof-of-Concept

2.1 Remodeling of Conventional TCM

As same in conventional TCM, the backbone length 'L' and the distance between driving tendons and backbones 'r' are set constant. Meanwhile, the discs are always perpendicular to the backbone. Besides these similarities, some changes are made to realize the target function of this design as follows:

Tendon Material: Polyoxymethylene (POM) which is a kind of elastic material is utilized for both backbone and driving tendons in this research. In order to clarify the property that it can be both pulled and pushed, the tendons will be referred to as driving rods in this paper. In most conventional TCMs which only have 2 DOFs, the material of the driving tendons are usually soft wires which can only provide pulling force. However, in order to realize 3 DOF movement with 3 actuators, these tendons are required to provide both pull and push force as proved in [4]. The utilization of elastic rods instead of soft wires also have other advantages such as increased buckling stability, enhanced down-scalability, backlash elimination and increased load capacity as introduced in [4].

Backbone Fixation: In conventional TCM design, the backbone is totally fixed to the discs and the movable range of twisting motion is strongly limited due to the resisting torque of the backbone. Under this consideration, the backbone is connected to the end and common discs with bearings in order to release the twisting constraints.

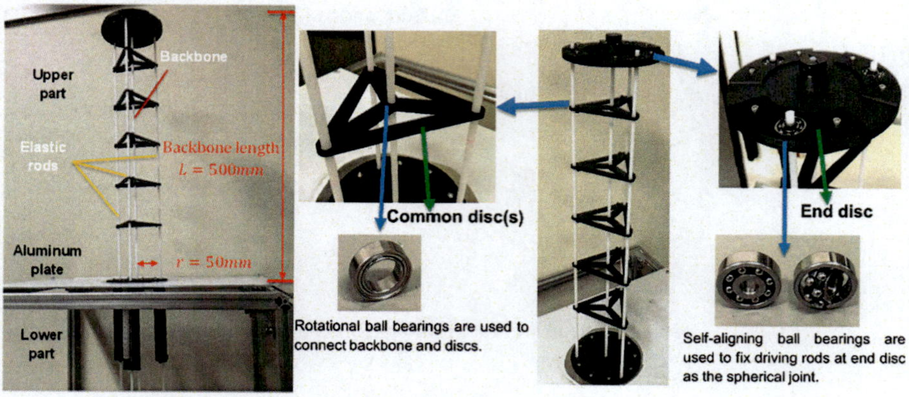

Fig. 2. Overview of prototype and fixation method

Driving Rods Fixation: In order to realize flexible twisting, the driving rods are also freely fixed at the end disc with a 3 DOF spherical joint and at common discs, the pinholes having very thin edges are considered as 3 DOF spherical joints as well. Due to such design, theoretically no twisting torque will occur in the driving rods.

2.2 Proof-of-Concept

Based on the afore-mentioned considerations, a prototype was made as shown in Fig. 2 for the purpose of validation experiment of the kinematic model in next chapter. In this prototype, actuators are not implemented, and the prototype is manually driven.

An Aluminum plate is used as the base disc with four pinholes for three driving rods and backbone. In the upper part, end and common discs are made through 3D printing. As introduced earlier, ball bearings are used for the fixation of backbone with each disc and self-aligning ball bearings are used to fix driving rods at the end disc (shaft retaining rings are used for axial fixation). For the lower part, base part is 3D printed including linear guides for the manual actuation of the driving rods.

3 Inverse Kinematics

3.1 Nomenclature

The nomenclature used in the following analysis is developed based on Kai Xu and Nabil Simaan's work [5, 6]. An assumption is also preserved from their work for the calculation which is that the shape of the backbone is always a planar circular arc. The validation experiment of this assumption has been done and the result has proved the validity of it, though some experiment-based corrections need to be applied to the kinematic model for the concern of calculation accuracy [5].

All the important parameters are shown in Fig. 3 where only the 1st elastic rod is shown, other two driving rods can be described similarly.

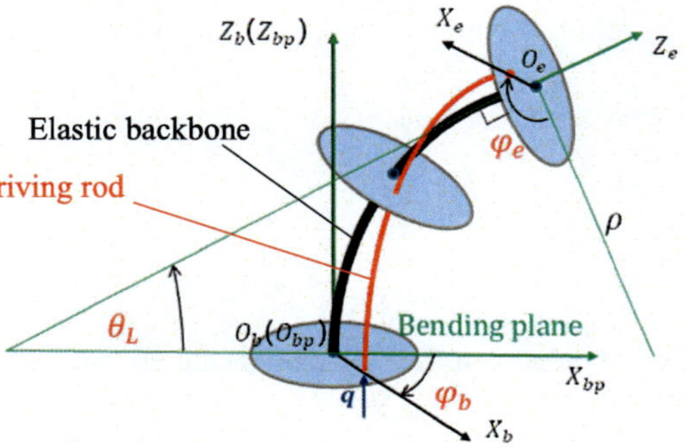

Fig. 3. Nomenclature of the kinematic model

Following is the definition of each parameter and explanation of some important terms:

Base Disc and End Disc Coordinate Systems: $\{X_b, Y_b, Z_b\}$ and $\{X_e, Y_e, Z_e\}$ are the coordinate systems of base disc and end disc, respectively. The origin points of these two coordinate systems O_b and O_e are placed at the center of the base and end discs, respectively. Z_b, Z_e axes are tangent to the curve of the backbone and X_b, X_e axes point to the intersection of the 1st driving rod and base disc and end disc, respectively.

Bending Plane Coordinate System: $\left\{X_{bp}, Y_{bp}, Z_{bp}\right\}$ is called the bending plane coordinate system. Bending plane is defined as the plane where the backbone lies in. The bending plane always exists since the shape of the backbone is assumed to be a circular arc. The bending plane is always perpendicular to the base disc and the bending plane coordinate system is defined such that $X_{bp}O_{bp}Z_{bp}$ becomes the bending plane. The green lines in Fig. 3 lie in the bending plane.

Orientation Parameters: In order to describe the orientation of the end disc, the orientation parameters of the manipulator are defined as

$$\boldsymbol{\psi} = \begin{bmatrix} \theta_L \\ \varphi_b \\ \varphi_e \end{bmatrix}. \tag{1}$$

θ_L: The angle between the base plane and the tangent of the backbone at the end tip. It describes the **bending motion** (how deep the manipulator is bent). $\theta_L = \frac{\pi}{2}$ means the backbone is a straight line perpendicular to the base disc and no force or moment exists in the backbone under this condition.

φ_b: The angle between the direction pointing from the center to the 1st rod and bending plane at the base disc. φ_b describes the **rotation motion** which refers to the rotation along the vertical direction. More specifically, φ_b describes which direction the manipulator is bent to.

φ_e: The angle between the direction pointing from the center to the 1st rod and bending plane at the end disc. The difference between φ_e and φ_b describes the **twisting motion** (along the backbone) which means $\varphi_e - \varphi_b$ is how much the manipulator is twisting.

ρ: The radius of curvature of the backbone. It changes w.r.t. θ_L. The relation is given by

$$\rho(\theta_L) = \frac{L}{\frac{\pi}{2} - \theta_L}. \tag{2}$$

$\boldsymbol{q}$: The input of this mechanism. It is defined as the change of length (displacement) of each rod, expressed as

$$\boldsymbol{q} = \begin{bmatrix} q_1 \\ q_2 \\ q_3 \end{bmatrix}. \tag{3}$$

3.2 Inverse Kinematics

The following inverse kinematic analysis is based on the piecewise constant curvature approach and the key subject to be solved is to find the relation between the driving rod length and orientation parameters. The difficult point in this procedure lies in the spatial curve shape of the driving rod. In this section, the approach for solving such issue is introduced by taking the 1st driving rod for example. The approach used for expressing the shape of the target curve is to derive the parametric equation of it by describing a target vector $\boldsymbol{c}$ whose start point fixed to the origin and end point changes along the target curve w.r.t. parameter. Vector $\boldsymbol{c}$ is further divided into the combination of two vectors, $\boldsymbol{b}$ and $\boldsymbol{r}$ as shown in Fig. 4, which will be explained later.

Consider an arbitrary posture where the orientation parameters $\boldsymbol{\psi} = [\theta_L \varphi_b \varphi_e]^T$ are known. Then, the arc length element of the backbone is set to be s, of which the integration range is [0, L] (L is the constant value of the backbone length). For an arbitrary s, two more intermediate parameters are defined as θ_s and φ_s as shown in Fig. 4(a). θ_s is used to describe the end point of $\boldsymbol{b}$ and φ_s is used to describe the end point of $\boldsymbol{r}$ w.r.t. s, respectively. With these two intermediate parameters, vectors $\boldsymbol{b}$ and $\boldsymbol{r}$ can be described with s.

The end point of vector $\boldsymbol{b}$ changes along the backbone w.r.t. s while its start point being fixed at the origin point O_b. Then vector $\boldsymbol{r}$ is defined to start from the end point of vector $\boldsymbol{b}$, while the angle between vector $\boldsymbol{r}$ and the bending plane $Y_{bp}O_{bp}Z_{bp}$ changes w.r.t. s and its norm always equals to r (the distance between backbone and driving rod),

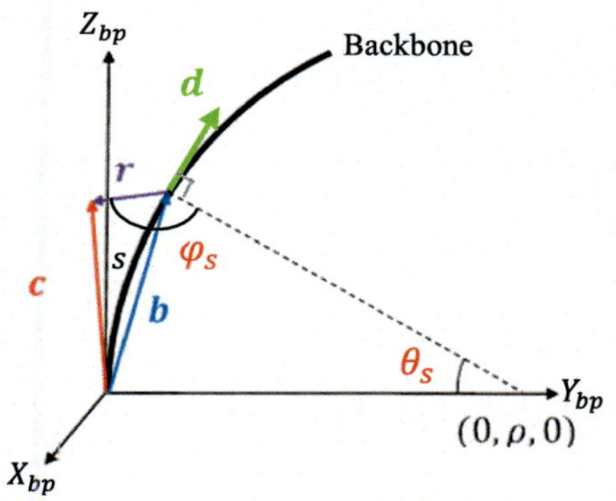

(a) For arbitrary s parameter

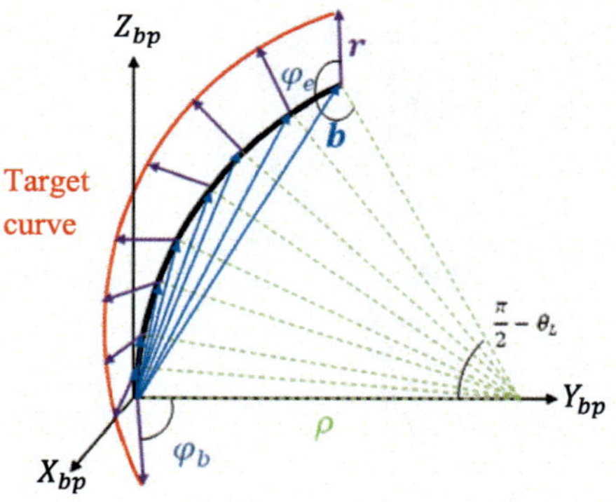

(b) For all s along backbone

Fig. 4. Illustration of target curve derivation (of 1st driving rod)

representing the twisting motion of the manipulator. In this process, vector $\boldsymbol{r}$ is set to be always perpendicular to the backbone curve. With above-defined vectors $\boldsymbol{b}$ and $\boldsymbol{r}$, the target vector $\boldsymbol{c}$ can be expressed and the shape of the driving rod at this moment can be described by connecting the end points of $\boldsymbol{c}$ for all parameter s along the backbone as shown in Fig. 4(b).

Next step is to find the parametric equations of $\boldsymbol{b}$ and $\boldsymbol{r}$, respectively. First consider the parametric expression of $\boldsymbol{b}$, since the shape of the backbone is assumed to be a planar circular arc, then the relation between s, θ_s and $\boldsymbol{b}$ can be easily obtained as

$$\theta_s = \frac{s}{L}\left(\frac{\pi}{2} - \theta_L\right), \tag{4}$$

$$\boldsymbol{b} = (0, \rho(1 - \cos\theta_s), \rho \sin\theta_s). \tag{5}$$

Then in order to express φ_s with s, the twisting of the rod along the backbone is assumed to be expressed by a planar curve in a side view coordinate system Y_iOZ_i. This side view is obtained by watching from the direction of $\boldsymbol{r}$ along the backbone, as shown in Fig. 5(a). The obtained side view curve of the driving rod is assumed to be a parabola in coordinate system Y_iOZ_i as shown in Fig. 5(b), Then, by calculating the equation of this curve, the relation between s, φ_s and $\boldsymbol{r}$ can be obtained as:

$$\varphi_s = \frac{\varphi_e - \varphi_b}{L^2} s^2 + \varphi_b, \tag{6}$$

$$\boldsymbol{r} = \boldsymbol{R}\boldsymbol{r}'. \tag{7}$$

Here, $\boldsymbol{r}$ is considered to be rotated from $\boldsymbol{r}'$ with rotation matrix $\boldsymbol{R}$. $\boldsymbol{r}'$ is a vector having the same start point and norm as $\boldsymbol{r}$, with its direction towards the center of the backbone arc. This means that it lies in the bending plane and perpendicular to the backbone. $\boldsymbol{R}$ can be obtained with Rodrigues' rotation formula knowing the rotation angle φ_s and the rotation axis vector $\boldsymbol{d} = (0, \sin\theta_s, \cos\theta_s)$. Therefore, $\boldsymbol{r}$ can be expressed as a function of θ_s

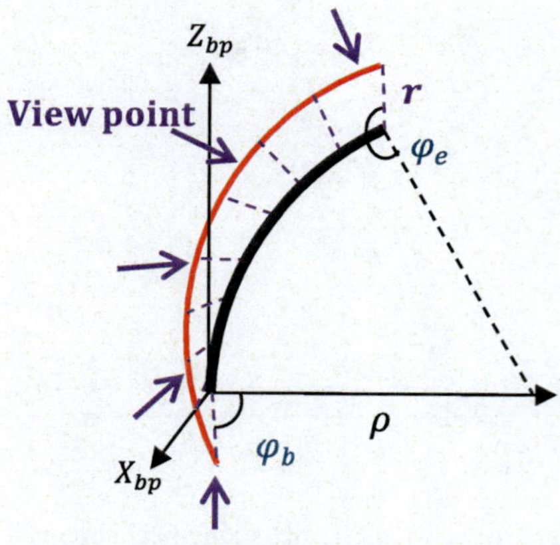

(a) Illustration of view point

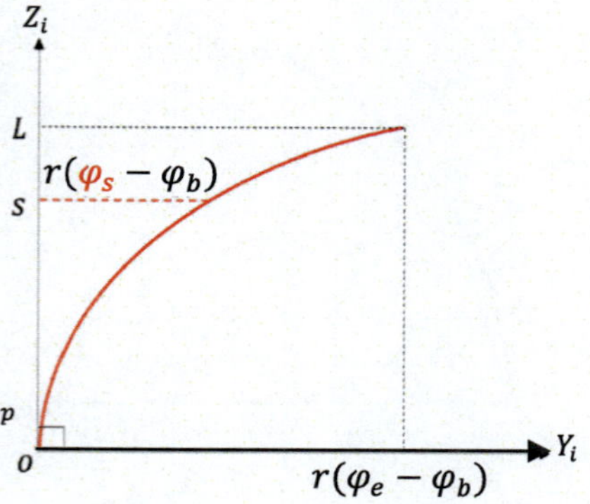

(b) Relation between s and φ_s can be expressed on the side view curve

Fig. 5. Imaginary side view of driving rod

and φ_s. Then, the parametric equation of the target curve can be obtained as $\boldsymbol{c}(s) = \boldsymbol{b} + \boldsymbol{r}$. Finally, the length of each driving rod and the inverse kinematic model can be obtained as ($i = 1, 2, 3$ is number of driving rods):

$$L_i = \int_0^L \left| \frac{d\boldsymbol{c}(s)}{ds} \right| ds = L_i(\theta_L, \varphi_b, \varphi_e), \tag{8}$$

$$\boldsymbol{q} = \boldsymbol{L}(\theta_L, \varphi_b, \varphi_e) - L = \boldsymbol{q}(\theta_L, \varphi_b, \varphi_e). \tag{9}$$

3.3 Validation Experiment of the Inverse Kinematic Model

The verification experiment of the derived kinematic model was conducted by comparing the measured data and the simulation result. The motion of the manipulator was divided into 10 discrete parts for manual control. The output orientation parameters were calculated from the posture information of the end disc which was obtained through 3D motion capture system. Figure 6 shows the result and illustration of one of the trials where three orientation parameters change together. Table 1 shows the mean error of all four trials of experiment.

The result shows that this new TCM is capable of 3 DOF rotational motion including one twisting motion. The error of θ_L and φ_b can reach within 5° of theoretical value. However, the error of φ_e (including twisting angle $\varphi_e - \varphi_b$) is around 20°. The probable source is due to the limitation of the assumption that the backbone shape would be circular arc where the experiment-based corrections to the kinematic model have not been applied to this research yet. Besides, the reason of the irregular performance at position 3 in Fig. 6 is also a future subject to be solved.

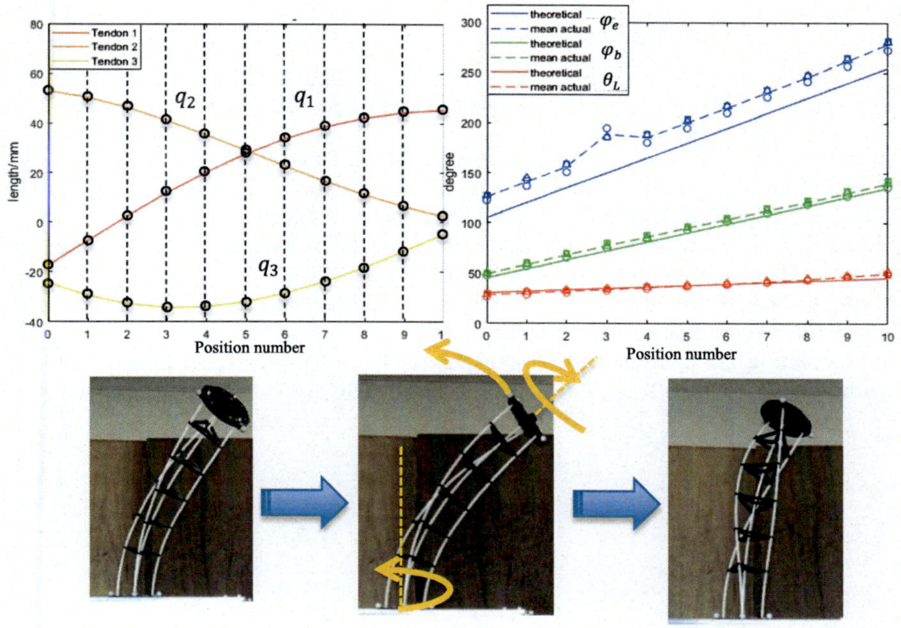

Fig. 6. Kinematic experiment result

Table 1. Mean error of kinematic experiment (unit: degree)

Motion target	Trial 1: Bending θ_L changes 15°	Trial 2: Rotating φ_b changes 90°	Trial 3: Twisting $\varphi_e - \varphi_b$ changes 60°	Trial 4: General Trial 1,2,3 together
θ_L	−3.89	−2.64	−1.76	0.25
φ_b	4.12	3.06	5.85	4.56
φ_e	17.87	21.67	34.09	22.76
$\varphi_e - \varphi_b$	13.75	18.60	18.25	18.20

4 Conclusion

Based on the conventional Tendon-driven Continuum Manipulator with 2 DOF, this research proposed a novel Tendon-driven Continuum Manipulator with 3 DOF where the capability of twisting motion is added. In this paper, the assumption that the shape of backbone would always be circular arc is preserved as in Kai Xu and Nabil Simaan's work [5, 6]. Then, a new kinematic model corresponding to the 3 DOF motion is derived and the validity of such structure to achieve 3 DOF motion is proved through the validation experiments. The accuracy of such motion is within 20° and the probable error sources will be analyzed in further research.

References

1. Li, Z., Wu, L., Ren, H., Yu, H.: Kinematic comparison of surgical tendon-driven manipulators and concentric tube manipulators. Mech. Mach. Theory **107**, 148–165 (2016). https://doi.org/10.1016/j.mechmachtheory.2016.09.018
2. https://www.intuitive.com/en-us/products-and-services/da-vinci/systems
3. Kang, H.: Robotic assisted suturing in minimally invasive surgery. J. Min. Access Surg. **5**, 1 (2009). https://doi.org/10.4103/0972-9941.51313, Source: PubMed
4. Simaan, N.: Snake-like units using flexible backbones and actuation redundancy for enhanced miniaturization. 0-7803-8914-X/05/$20.00 ©2005. IEEE (2005)
5. Xu, K., Simaan, N.: An investigation of the intrinsic force sensing capabilities of continuum robots. IEEE Trans. Robot. **24**(3), 576–587 (2008)
6. Xu, K., Simaan, N.: Intrinsic wrench estimation and its performance index for multisegment continuum robots. IEEE Trans. Robot. **26**(3), 555–561 (2010)

Geometric Insights into Kinematically-Singular Configurations of Planar Continuum Robots

Neel Shihora and Nabil Simaan(✉)

Department of Mechanical Engineering, Vanderbilt University,
Nashville, TN 37212, USA
{neel.shihora,nabil.simaan}@vanderbilt.edu

Abstract. Continuum robots present designers with the challenges of interpreting the geometric conditions that lead to singularity. Unlike serial robots, where it is easy to relate Plücker line coordinates to physical locations fixed in each link-attached frame, such geometric understanding is hard to visualize for continuum robots. In this paper, we explore the conditions for the singularity of continuum robots comprised of two and three segments. We derive the conditions for singularity and visualize these singular configurations with their corresponding conditions of singularity. We start the analysis assuming circular curvature, and then we discuss how the same analysis can be extended to non-circular curvature cases. We also define safety regions for singularity that depend on an assumed modal shape variation due to the elastic deflections of these robots. These safety zones can be used to produce designs and path plans that guarantee singularity-free performance despite norm-bounded deflections in configuration space.

Keywords: Continuum robots · Singularity · Instantaneous screw axes · Instantaneous centers of motion

1 Introduction

Continuum robots (CRs) have garnered increased interest for surgical applications because of their ability to circumvent obstacles and due to their inherent safety as a result of their compliance. Unlike rigid-link serial robots, where the locations of the instantaneous screws of motion are easy to relate to local frames of the links, CRs exhibit non-trivial loci of their screw axes. These screw axes depend on the equilibrium bending shapes of the CRs. As a result, the geometric insight into which configuration constitutes a kinematic singularity is harder to obtain. This paper aims to clarify these geometric conditions for multi-segment CRs performing planar motions. We show that closed-form expressions for the loci of the motion screws is attainable for circular bending CRs. We also use a general approach based on end effector twist to facilitate generalized treatment of non-circular bending CRs.

O. Altuzarra and A. Kecskémthy (Eds.): ARK 2022, SPAR 24, pp. 237–247, 2022.
https://doi.org/10.1007/978-3-031-08140-8_26

While it is possible to use kinematic regularization in the neighbourhood of singularity (e.g. using the singularity-robust Jacobian [10]), such methods do not present users with the necessary geometric understanding and only support local-redundancy resolution that would require real-time shape sensing if one wishes to account for deflections/shape uncertainty. If one wishes to apply a global optimization and path planning approach, having knowledge of singularity zones within the robot configuration space would be helpful. Our goal is therefore to derive the singularity zones in configuration space *a-priori*, which can in the future be considered for design optimization, path planning, and control.

In this paper, we limit our discussion to CRs comprised from serially-stacked continuum segments that can independently control their bending. We therefore exclude discussions of configurations where elastic instabilities can occur (e.g. as in [2,4]) and focus primarily on kinematic singularity. Thus, relevant works on singularity and kinematic conditioning of CRs include [9] who considered the singularity conditions of articulated snake-like robots and [5] who defined the kinematic manipulability ellipsoids for CRs. Subsequently, [6] considered the singular configuration in which a continuum segment (CS) is straight and addressed the numerical instability that occurs near this configuration. Recently [7] presented a sufficient condition for kinematic singularities for CRs and also considered workspace reachability conditions.

Despite previous works, the geometric insight regarding the conditions leading to kinematic singularity of CRs remain lacking. The contribution in this paper takes inspiration from past works on singularity analysis using line geometry as in [8,12]. We therefore consider the use of axodes of motion and their spatial relationship to determine easy-to-understand geometric conditions that explain the nature of singularity. Furthermore, since CRs have large model-parameter uncertainty (e.g. material and geometric uncertainties), we address the question of how to guarantee that the robot does not encounter a kinematic singularity despite a maximal allowable shape deflection or shape uncertainty.

2 Kinematic Modeling for Continuum Robots: Preliminaries

Figure 1 shows an example multi-segment continuum robot with three segments. The bending shape of a segment depends on the minimal energy of its structure. Under certain design conditions and with minimal loading, these robots mostly bend in circular arcs [13]. The CS may deviate from the circular bending shape due to assembly errors, friction, and external loading. The analysis below uses axodes of motion of each segment, we present the instantaneous kinematics next. We first explain the nomenclature and kinematics for a single segment and then explain the kinematics for a multi-segment design.

2.1 Subsegment Kinematics

In the following analysis, we assume minimal/negligible torsional deflections about the backbone. This assumption holds for torsionally stiff designs and when small loads are applied on the robot as demonstrated in [16]. With these assumptions, the configuration of the kth CS ($k \in [1, n]$) is determined by its configuration space vector $\boldsymbol{\psi}_k = [\theta_k(L_k), \delta_k]^{\mathrm{T}}$, where L_k is the length of the kth segment. The configuration variable $\theta_k(L_k)$ is the bending angle at the segment end-effector ($s = L_k$) and δ_k characterizes the plane in which the segment bends, Fig. 1). Here s designates the arc-length parameter measured from the segment's base disk, Also, we assume that $\theta_k(0) = \pi/2$, which we denote as θ_0.

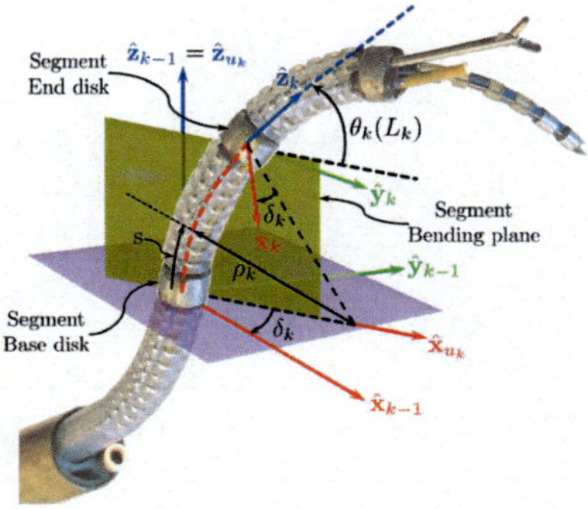

Fig. 1. A multi-segment continuum robot as in [11] with the configuration variables $\theta_k(L_k)$ and δ_k shown for the kth ($k = 2$) continuum segment.

If the robot suffers from friction, it may deviate from circular bending. Following the approach in [14], we capture a non-circular bending shape, using a family of shapes prescribed by a Barycentric interpolation between two generator curves prescribed by their respective curvature functions $\kappa_{a_k}(s)$ and $\kappa_{b_k}(s)$.

$$\kappa(s, t_k) = t_k \kappa_{b_k}(s) + (1 - t_k)\kappa_{a_k}(s), \quad t_k \in [0, 1] \tag{1}$$

where t_k is the shape interpolation parameter. Inspired by [3], the curvature functions of the generator curves are assumed to use the following modal representation:

$$\kappa_{a_k}(s) = \mathbf{a}_k^{\mathrm{T}} \boldsymbol{\eta}(s), \quad \kappa_{b_k}(s) = \mathbf{b}_k^{\mathrm{T}} \boldsymbol{\eta}(s), \qquad \mathbf{a}_k, \mathbf{b}_k, \boldsymbol{\eta} \in \mathrm{IR}^{m+1} \tag{2}$$

For practical designs m is small (generally less than $m = 3$) [15], therefore a monomial modal basis remains acceptably well conditioned, i.e. $\boldsymbol{\eta}(s) = \left[s^0, s^1, s^2, \cdots, s^m\right]^{\mathrm{T}}$ and $\mathbf{a}_k = \left[a_{k0}, a_{k1}, \cdots, a_{km}\right]^{\mathrm{T}}$ and $\mathbf{b}_k = \left[b_{k0}, b_{k1}, \cdots, b_{km}\right]^{\mathrm{T}}$. With the above curvature distribution, the local tangent angle at arc-length s is given by:

$$\theta_k(s, t_k) = \theta_0 - \int_0^s \kappa(s, t_k) ds \tag{3}$$

This relationship between t_k and $\theta_k(s, t_k)$ allows us to redefine the configuration space vector as $\boldsymbol{\psi}_k = \left[t_k \ \delta_k\right]^{\mathrm{T}}$ for segment bending with variable curvature. Now, given this parameterizations for modeling the variable curvature bending of the continuum robot, the segment curve defined within a frame $\{U\}$ as shown[1] in Fig. 1:

[1] We use ${}^{a}\mathbf{x}$ to designate the representation of vector $\mathbf{x}$ in a frame {A} with axes $\hat{\mathbf{x}}_a$, $\hat{\mathbf{y}}_a$, $\hat{\mathbf{z}}_a$.

$$^{u_k}\mathbf{x}(s,t_k) = \left[\int_0^s \cos(\theta_k(\eta,t_k))d\eta, \quad 0, \quad \int_0^s \sin(\theta_k(\eta,t_k))d\eta\right]^{\mathrm{T}} \tag{4}$$

The position of the segment end disk relative to its base is given in frame $\{k-1\}$ by using $s = L_k$ in the above equation and then rotating back to frame $\{k-1\}$:

$$^{k-1}\mathbf{p}_k = {}^{k-1}\mathbf{R}_{u_k}\ \mathbf{x}(L_k,t_k), \quad {}^{k-1}\mathbf{R}_{u_k} = \left(e^{\delta_k[e_3]^\wedge}\right) \tag{5}$$

where $^{k-1}\mathbf{R}_{u_k}$ is a simple rotation about the z-axis of frame $\{k-1\}$. It is expressed above using the matrix exponential with e_3 designating the canonic representation of the z-axis and $[\cdot]^\wedge$ denoting the cross-product matrix form (wedge operator).

The orientation of the end-disk frame relative to the base disk frame is given by a moving frame sequence as:

$$^{k-1}\mathbf{R}_k = e^{\delta_k[e_3]^\wedge}\ e^{(\theta_0-\theta_k(L_k))[e_2]^\wedge}\ e^{-\delta_k[e_3]^\wedge} \tag{6}$$

where e_i designate the canonic basis for IR^3, $i = 1,2,3$. The twist of the end disk relative to the base disk is defined as $\boldsymbol{\xi}_{k/k-1} \triangleq [\mathbf{v}_{k/k-1}^{\mathrm{T}}, \boldsymbol{\omega}_{k/k-1}^{\mathrm{T}}]^{\mathrm{T}}$, where $\mathbf{v} = \dot{\mathbf{p}}$. The $(\dot{\ })$ operator designates time derivative and $\boldsymbol{\omega}$ is the spatial angular velocity of the end disk relative to the base disk in frame $\{k-1\}$. Using this twist definition, the instantaneous direct kinematics is given by $\boldsymbol{\xi}_{k/k-1} = J_{\xi\psi_k}\dot{\psi}_k$and the Jacobian is:

$$\mathbf{J}_{\xi\psi_k} = \begin{bmatrix} \frac{\partial}{\partial t_k}\left({}^{k-1}\mathbf{p}_k\right) & \frac{\partial}{\partial \delta_k}\left({}^{k-1}\mathbf{p}_k\right) \\ \left[\left(\frac{\partial}{\partial t_k}{}^{k-1}\mathbf{R}_k\right){}^{k-1}\mathbf{R}_k^{\mathrm{T}}\right]^\vee & \left[\left(\frac{\partial}{\partial \delta_k}{}^{k-1}\mathbf{R}_k\right){}^{k-1}\mathbf{R}_k^{\mathrm{T}}\right]^\vee \end{bmatrix} \tag{7}$$

The special case of circular bending can be described by the constant curvature

$$\kappa(s) = 1/\rho_k = \left(\frac{\theta_0 - \theta_k(L_k)}{L_k}\right) \tag{8}$$

where, ρ_k is the constant radius of curvature of the kth CS as shown in Fig. 1. The direct kinematics and instantaneous kinematics of the continuum robot bending with constant curvature are also defined by Eqs. (5), (6), and (7). After simplifying (7) for segment bending with constant curvature, the Jacobian matrix $\mathbf{J}_{\xi\psi_k} = [(\mathbf{J}_{\xi\psi_k})_v^{\mathrm{T}}, (\mathbf{J}_{\xi\psi_k})_\omega^{\mathrm{T}}]^{\mathrm{T}}$ is given by:

$$(\mathbf{J}_{\xi\psi_k})_v = L_k \begin{bmatrix} c_{\delta_k}\left(\frac{(\theta_{L_k}-\pi/2)c_{\theta_{L_k}} - s_{\theta_{L_k}} + 1}{(\theta_{L_k}-\pi/2)^2}\right) & s_{\delta_k}\left(\frac{1-s_{\theta_{L_k}}}{\theta_{L_k}-\pi/2}\right) \\ -s_{\delta_k}\left(\frac{(\theta_{L_k}-\pi/2)c_{\theta_{L_k}} - s_{\theta_{L_k}} + 1}{(\theta_{L_k}-\pi/2)^2}\right) & c_{\delta_k}\left(\frac{1-s_{\theta_{L_k}}}{\theta_{L_k}-\pi/2}\right) \\ \left(\frac{(\theta_{L_k}-\pi/2)s_{\theta_{L_k}} + c_{\theta_{L_k}}}{(\theta_{L_k}-\pi/2)^2}\right) & 0 \end{bmatrix} \tag{9}$$

$$(\mathbf{J}_{\xi\psi_k})_\omega = \begin{bmatrix} -s_{\delta_k} & c_{\delta_k} c_{\theta_{L_k}} \\ -c_{\delta_k} & -s_{\delta_k} c_{\theta_{L_k}} \\ 0 & s_{\theta_{L_k}} - 1 \end{bmatrix} \tag{10}$$

where we have used the shorthand notation $c_{\delta_k} = \cos(\delta_k), s_{\delta_k} = \sin(\delta_k), c_{\theta_{L_k}} = \cos(\theta_k(L_k)), s_{\theta_{L_k}} = \sin(\theta_k(L_k))$, and $\theta_{L_k} = \theta_k(L_k)$.

2.2 Multi-segment Kinematics

To capture the kinematics of a multi-segment robot having n segments, we usually express it in terms of the kinematics of its constituent segments given by $\boldsymbol{\xi}_{k/k-1} = \mathbf{J}_{\xi\psi_k}\dot{\boldsymbol{\psi}}_k$. To do so, we define an augmented configuration vector as:

$$\breve{\boldsymbol{\psi}} = \left[\boldsymbol{\psi}_1^{\mathrm{T}}, \ldots, \boldsymbol{\psi}_n^{\mathrm{T}}\right]^{\mathrm{T}} \tag{11}$$

Since the end-effector twist is the composition of the induced end effector twists due to motion of each segment, we can write $\boldsymbol{\xi} = \mathbf{J}_{\xi\breve{\psi}}\dot{\breve{\boldsymbol{\psi}}}$. Expressed explicitly, we write:

$$\boldsymbol{\xi} = \sum_{k=1}^{n} \mathbf{S}_k \mathbf{J}_{\xi\psi_k}\dot{\boldsymbol{\psi}}_k = \underbrace{\left[\mathbf{S}_1\mathbf{J}_{\xi\psi_1}, \ \cdots, \mathbf{S}_n\mathbf{J}_{\xi\psi_n}\right]}_{\mathbf{J}_{\xi\breve{\psi}}} \dot{\breve{\boldsymbol{\psi}}} \tag{12}$$

where the matrices $\mathbf{S}_k$, $k = 1, 2, \ldots, n$ are the twist transformation matrices expressing the contribution of the kth segment to the absolute twist of the end disk.

$$\mathbf{S}_k = \begin{bmatrix} {}^0\mathbf{R}_{k-1} & [({}^0\mathbf{p}_k - {}^0\mathbf{p}_{n_s})^\wedge] \\ \mathbf{0} & {}^0\mathbf{R}_{k-1} \end{bmatrix} \tag{13}$$

where ${}^0\mathbf{p}_k$ is the position of the end disk of the kth segment and ${}^0\mathbf{R}_k = \prod_{i=1}^{k} {}^{i-1}\mathbf{R}_i$.

3 Instantaneous Screws of Motion

Before discussing the proposed singularity condition, in this section we present the necessary preliminaries required to define that singularity condition. We use the instantaneous screws of motion, which reduces to instantaneous centers of motion (ICMs) for planar case, for defining singularity condition. Given the configuration space velocity $\dot{\boldsymbol{\psi}}_k$, the motion of the segment end-effector with respect to the segment base can be fully described by the axis-order Plücker coordinates of the screw axis and the screw pitch:

$$\$_k = \begin{bmatrix} \mathbf{s}_{0k} \times \hat{\mathbf{s}}_k + \lambda_k \hat{\mathbf{s}}_k \\ \hat{\mathbf{s}}_k \end{bmatrix} \tag{14}$$

where $\hat{\mathbf{s}}_k$ represents the direction of the line, $\mathbf{s}_{0k} \times \hat{\mathbf{s}}_k$ is the moment of the line about the origin and λ_k is the pitch of the screws which is equal to zero as the discussion in this subsection is limited to the planar case with $\dot{\delta}_k = 0$.

Given the end-effector twist, $\boldsymbol{\xi} = [\mathbf{v}_n^{\mathrm{T}}, \boldsymbol{\omega}_n^{\mathrm{T}}]^{\mathrm{T}}$, one can find the constituent elements of the Plücker coordinates describing its ICM as:

$$\hat{\mathbf{s}}_n = \frac{\boldsymbol{\omega}_n}{\|\boldsymbol{\omega}_n\|}, \qquad \mathbf{s}_{0n} = \frac{\boldsymbol{\omega}_n \times \mathbf{v}_n}{\|\boldsymbol{\omega}_n\|^2} \tag{15}$$

The equation above may also be used with the relative twist for each segment to find the location of its ICM in its base frame.

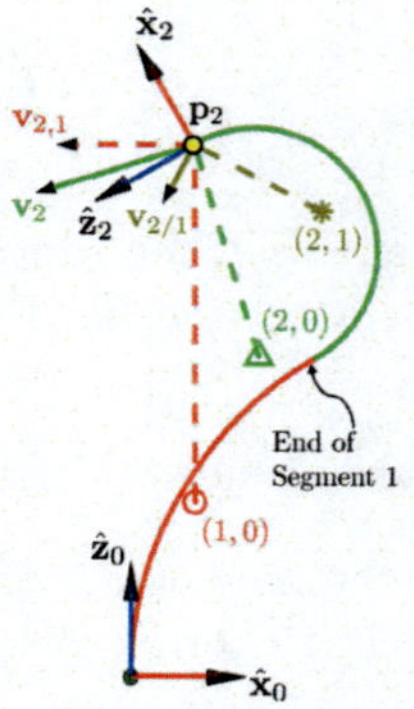

Fig. 2. Key linear velocity vectors of this planar 2-segment CR whose associated screws of motion are perpendicular to the page and passes through their respective ICMs (1, 0), (2, 1), and (2, 0).

Although Eq. (14) is expressed in general terms of twist components, it is possible to manipulate it to yield closed-form expression for the location of the ICM in a frame attached at the base of a CS. For example, if a CS bends in circular fashion, the location of the ICM describing relative motion of the segment end frame relative to its base is denoted $\mathbf{r}$ and is given by:

$$\mathbf{r} = L\left[\frac{\tilde{\theta} - sin(\tilde{\theta})}{\tilde{\theta}^2}, \frac{1 - cos(\tilde{\theta})}{\tilde{\theta}^2}, 0\right]^{\mathrm{T}}, \tag{16}$$

$$\tilde{\theta} = \theta_0 - \theta_L$$

Equation (16) expresses the ICM for each segment in a local frame. It can be easily used with the direct kinematics equations to express these ICMs in world frame. The ICM of the end-effector motion relative to the world is harder to obtain in closed-form and is derived using the twist notation in Eq. (15).

4 Singularity Condition

Referring back to Fig. 2 we define $\mathbf{v}_{2,1}$ as the induced end-effector velocity due to the motion of segment 1 while segment 2 is locked. We also define $\mathbf{v}_{2/1}$ as the velocity of the end effector due to motion of segment 2 while segment 1 is locked. The vector $\mathbf{v}_{2/1}$ can be obtained directly using the kinematics for segment 2 only. The velocity $\mathbf{v}_{2,1}$ can be found using the overall robot kinematics while assuming $\dot{\boldsymbol{\psi}}_1 = 0$. The end-effector velocity, $\mathbf{v}_2$ is the vectorial sum $\mathbf{v}_2 = \mathbf{v}_{2,1} + \mathbf{v}_{2/1}$ and, in a non-singular configuration, can be arbitrarily controlled within the pencil of lines (flat pencil) defined by $\mathbf{v}_{2,1}$ and $\mathbf{v}_{2/1}$. Singularity associated with a loss of a degree-of-freedom is marked by the degeneracy of this flat pencil to a line. This occurs when $\mathbf{v}_{2,1}$ and $\mathbf{v}_{2/1}$ point in the same direction.

Using $\mathbf{v}_{2,1}$ and $\mathbf{v}_{2/1}$ one can calculate the locus of the ICMs (1,0), (1,2) and (2,0) where the notation (i,k) designates the ICM of frame i with respect to k. These ICMs are depicted in Fig. 2 for a non-singular) configuration and they are denoted by $\circ$, $\triangle$ and $*$ markers. By definition, $\mathbf{v}_{2/1} \perp (\mathbf{p}_2 - (2,1))$ and $||\mathbf{v}_{2/1}|| = \dot{\theta}_2 \, ||\mathbf{p}_2 - (2,1)||$. Similarly,$\mathbf{v}_{2,1} \perp (\mathbf{p}_2 - (1,0))$ and $||\mathbf{v}_{2,1}|| = \dot{\theta}_1 \, ||\mathbf{p}_2 - (1,0)||$. We also note that (2,0) is collinear with (1,0) and (2,1) as predicted by Aronhold-Kennedy's theorem [1].

Since we identified the singularity condition as $\mathbf{v}_{2,1} \times \mathbf{v}_{2/1} = 0$, we also conclude that this occurs whenever the end-effector is collinear with (1,0) and (2,0) (or (2,1)). Note that this is also consistent with the singularity of a 2R serial kinematic chain. We therefore can use the area of the triangle defined by points $\mathbf{p_2}$, (1,0), (2,1) shown in Fig. 3a) as an easy visualization that conveys proximity to singularity. Using this method, one can solve for the possible singular configurations for the 2-segment continuum robot where this triangle vanishes. This is illustrated in Fig. 3b) where we consider the first segment locked at a given configuration and ask what poses of the second segment correspond with singularity. This process leads to three singular configurations as shown in the Fig. 3b).

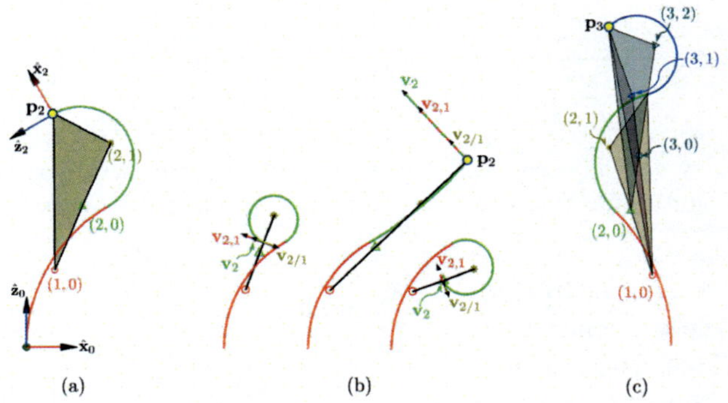

Fig. 3. (a) An imaginary triangle (shaded polygon) whose vertices are $(1,0)$, $(2,1)$, and the end-effector (EE). The area of this triangle designates closeness to singularity. (b) An illustration of the only three singular configurations for $\theta_{L_1} = \pi/6$ for a planar circular-bending continuum robot with equal segment lengths. (c) a planar 3-segment circular bending CR with equal segment lengths. Vanishing of the areas of both or one of the imaginary triangles (shaded polygons) designate singularity.

If one has a continuum robot with three segments, one can find the singularities by considering the conditions for the vanishing of the triangles corresponding with the motion of any two-segment pairs as depicted in Fig. 3c.

5 Simulation Study

To illustrate the methodology of this paper, we created a simulation code for a multi-segment robot with online computation of its ICMs and kinematics. For our simulations we considered a two-segment continuum robot with equal segment lengths. For simulations assuming circular bending, we considered admissible values of the bending angles as $\theta_1(L_1), \theta_2(L_2) \in [-\frac{3\pi}{2}, \frac{5\pi}{2}]$ spanning the whole configuration space of the robot. For simulations considering non-circular bending, we chose the vectors of modal factors defining the shape generator curves as $\mathbf{a}_k = [0.25, -15/L_k]^{\mathrm{T}}$ and $\mathbf{b}_k = [0.25, 15/L_k]^{\mathrm{T}},\ k = 1, 2$.

Figure 4 shows the singularity trajectories of robot end-effector position. For the clarity in presentation, we chose to plot the robot in only five distinct singular configurations. It is clear from these plots that unlike rigid-link serial kinematic chain, it is not trivial to discern whether the configurations presented in Fig. 4 are singular by simple visual inspection of the segment shapes. The values of the configuration space variables at these configurations are listed in Table 1.

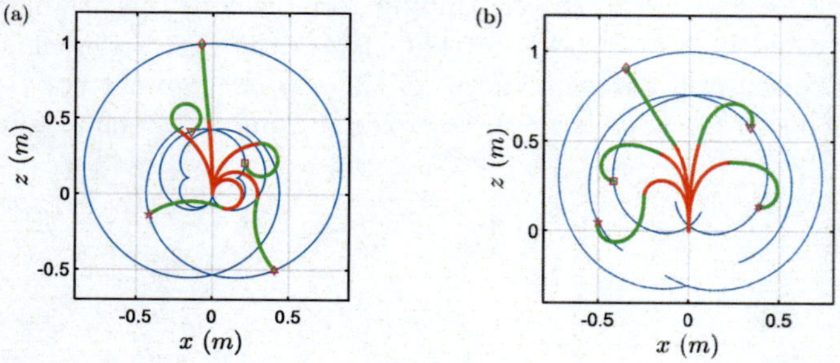

Fig. 4. Singularity end effector paths and sample singular configurations for a 2-segment planar continuum robot. (a) circular bending case, (b) non-circular bending.

While operating, the CRs may deflect under the influence of an external load. This deflection may go undetected due to their inherent structural compliance. We considered a constant deflection of $\pm 5°$ in bending angle at the end-effector allowing us to designate the singularity safety zones (shaded curves/regions in Fig. 5 in the configuration space of the robot. These zones are defined to ensure that despite shape uncertainty, the robot should not encounter a singularity. The Fig. 6 also shows a visual representation of the boundaries of these singularity zones by overlaying the maximal uncertainty configurations on top of the nominal singular configurations marked by $\triangledown$ symbol in Fig. 4 of the robot.

Table 1. Example singular configurations

Marker	θ_{L_1} (rad)	θ_{L_2} (rad)	t_1	t_2
✡	−1.8064	2.5220	0.6941	0.9302
□	0.0873	−3.0369	0.3240	0.0378
◇	1.6842	1.5315	0.4049	0.4844
▽	2.5569	−3.7350	0.5205	0.9302
☆	−3.5517	2.5962	0.0691	0.9302

Figure 6 also numerically verifies that the using area of the triangles or their minimal height correspond with singularity indicated by the minimal singular value or inverse condition number of the translational portion of the Jacobian matrix $\mathbf{J}_{\xi\breve{\psi}}$. It is clear from Fig. 6 that for given configuration of segment 1,

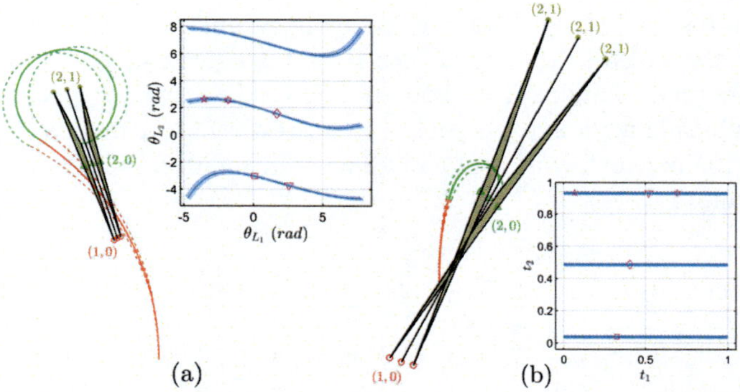

Fig. 5. Configuration space of a 2-segment planar continuum robot (a) circular bending (b) non-circular bending. The markers corresponds to their respective configurations in Fig. 4. The two nominal singular configurations (marked by ▽ in Fig. 4 with solid line style backbones are shown with shape deflections (dashed backbones) of ± 5° in θ_{L_1} and θ_{L_2} for the circular bending case and ± 5°(≈ ± 0.0115 *units* in t_k) in t_1 and t_2 for the non-circular bending case.

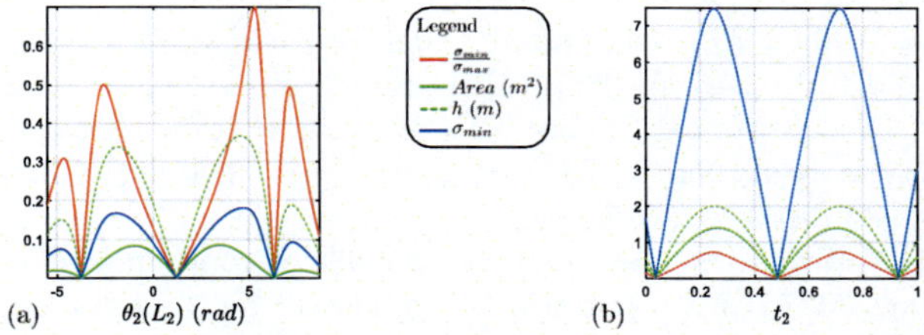

Fig. 6. Plots of inverse condition number (ICN), σ_{min}, area of the triangle, and height of the triangle (a) *vs* $\theta_2(L_2)$ for given $\theta_1(L_1) = 2.5569$, and (b) *vs* t_2 for given $t_1 = 0.5205$

there are only three values of $\theta_2(L_2) \in [-\frac{3\pi}{2}, \frac{5\pi}{2}]$ and $t_2 \in [0, 1]$ at which robot is in singularity.

6 Conclusion

This contribution presented two preliminary, yet useful, concepts for continuum robot singularity analysis. While it is possible to determine singularity from examination of the minimal singular values, it is hard to obtain insights into the nature of the singularity and since continuum robots suffer from uncertainty in their bending shape there is a need for establishing some safety zones around nominal singularities. The paper addressed these two needs by presenting the instantaneous kinematics and making use of the instantaneous centers of motion to show that singularity can be geometrically determined for circular-bending continuum robots based on the condition of collinearity of the instantaneous centers of motion of segment pairs with the end effector point. We showed that the approach also holds for non-circular-bending continuum robots and demonstrated that the areas of the respective ICM triangles can give useful insights to finding singular configurations and solving for the singular paths of the end effector. While this work is preliminary, it suggests that future extension of this approach to 3D is worthwhile and can allow additional insights for robot design and path planning.

References

1. Bottema, O., Roth, B.: Theoretical Kinematics, vol. 24. Courier Corporation, North Chelmsford (1990)
2. Briot, S., Goldsztejn, A.: Singularity conditions for continuum parallel robots. IEEE Trans. Robot. **38**, 1–19 (2021). https://doi.org/10.1109/TRO.2021.3076830
3. Chirikjian, G., Burdick, J.: A modal approach to hyper-redundant manipulator kinematics. IEEE Trans. Robot. Autom. **10**(3), 343–354 (1994). https://doi.org/10.1109/70.294209
4. Gilbert, H.B., Hendrick, R.J., Webster, R.J., III.: Elastic stability of concentric tube robots: a stability measure and design test. IEEE Trans. Robot. **32**(1), 20–35 (2016). https://doi.org/10.1109/TRO.2015.2500422
5. Gravagne, I., Walker, I.: Manipulability and force ellipsoids for continuum robot manipulators. In: Proceedings 2001 IEEE/RSJ International Conference on Intelligent Robots and Systems. Expanding the Societal Role of Robotics in the the Next Millennium (Cat. No. 01CH37180), vol. 1, pp. 304–311 (2001). https://doi.org/10.1109/IROS.2001.973375
6. Jones, B.A., Walker, I.D.: Limiting-case analysis of continuum trunk kinematics. In: Proceedings 2007 IEEE International Conference on Robotics and Automation, pp. 1363–1368 (2007). https://doi.org/10.1109/ROBOT.2007.363174
7. Mayer, A., Sawodny, O.: Singularity and workspace analysis for modular continuum robots. In: 2018 IEEE Conference on Control Technology and Applications (CCTA), pp. 280–285 (2018). https://doi.org/10.1109/CCTA.2018.8511408
8. Merlet, J.P.: Singular configurations of parallel manipulators and Grassmann geometry. Int. J. Robot. Res. **8**(5), 45–56 (1989)
9. Mochiyama, H., Kobayashi, H.: The shape Jacobian of a manipulator with hyper degrees of freedom. In: Proceedings 1999 IEEE International Conference on Robotics and Automation (Cat. No. 99CH36288C), vol. 4, pp. 2837–2842 (1999). https://doi.org/10.1109/ROBOT.1999.774027
10. Nakamura, Y., Hanafusa, H.: Inverse kinematic solutions with singularity robustness for robot manipulator control. J. Dyn. Syst. Measure. Control **108**(3), 163–171 (1986). https://doi.org/10.1115/1.3143764
11. Sarli, N., Giudice, G.D., De, S., Dietrich, M.S., Herrell, S.D., Simaan, N.: TURBot: a system for robot-assisted transurethral bladder tumor resection. IEEE/ASME Trans. Mechatron. **24**(4), 1452–1463 (2019). https://doi.org/10.1109/TMECH.2019.2918137
12. Simaan, N., Shoham, M.: Singularity analysis of a class of composite serial in-parallel robots. IEEE Trans. Robot. Autom. **17**(3), 301–311 (2001). https://doi.org/10.1109/70.938387
13. Simaan, N., Taylor, R., Flint, P.: A dexterous system for laryngeal surgery. In: IEEE International Conference on Robotics and Automation, Proceedings, ICRA 2004, vol. 1, pp. 351–357 (2004). https://doi.org/10.1109/ROBOT.2004.1307175

14. Wang, L., Simaan, N.: Investigation of Error Propagation in Multi-backbone Continuum Robots. In: Lenarčič, J., Khatib, O. (eds.) Advances in Robot Kinematics, pp. 385–394. Springer, Cham (2014). https://doi.org/10.1007/978-3-319-06698-1_40
15. Wang, L., Simaan, N.: Geometric calibration of continuum robots: joint space and equilibrium shape deviations. IEEE Trans. Robot. **35**(2), 387–402 (2019). https://doi.org/10.1109/TRO.2018.2881049
16. Xu, K., Simaan, N.: Analytic formulation for kinematics, statics, and shape restoration of multibackbone continuum robots via elliptic integrals. J. Mech. Robot. **2**(1), 011006 (2009). https://doi.org/10.1115/1.4000519

A Gazebo Simulator for Continuum Parallel Robots

Andrea Gotelli[1,2(✉)], Federico Zaccaria[1,2,3], Olivier Kermorgant[1,2], and Sébastien Briot[1,4]

[1] Laboratoire des Sciences du Numérique de Nantes (LS2N), Nantes, France
{Andrea.Gotelli,Olivier.Kermorgant,Sebastien.Briot}@ls2n.fr
[2] École Centrale de Nantes, Nantes, France
[3] DIN, University of Bologna, Bologna, Italy
federico.zaccaria3@unibo.it
[4] Centre National de la Recherche Scientifique (CNRS), Paris, France

Abstract. Continuum Parallel Robots (CPRs) combine properties of continuum and parallel rigid-link robots. They inherit simplicity, compliance, and cost-effectiveness from the first as payload capacity and stiffness from the latter. In this paper, we propose to use Gazebo and ROS to provide a generalized simulator for CPRs, in terms of their joints and geometry, while we use the Cosserat rod theory to model their deformable bodies. We exploit our simulator to solve the direct and inverse geometrico-static models of CPRs and to provide a useful base for simulations.

1 Introduction

Continuum Parallel Robots (CPRs) are manipulators with flexible elastic links, arranged in parallel. Like rigid links parallel robots, they have a good payload capacity and stiffness, but they also show the simplicity, compliance, and cost-effectiveness of continuum robots. Early research on these robots started with [7] and, nowadays, CPRs have a wide range of applications [3,8].

In this work, we consider robots composed by a single distal plate, a single base and n deformable links, or limbs, ($\{n \in \mathbb{N}, n > 1\}$), as in [8]. A generic representation of such a robot is given in Fig. 1b. In what follows, we consider three possible types of actuated base joints: prismatic, revolute and an actuation that changes the length of the rod, like the one in [8], namely extensible limb. With this layout, the structure of these robots is simple, inexpensive, lightweight, compliant, and easily scalable, making it possible to realize CPRs of a few millimeters [3].

The combination of rigid and deformable bodies complicates the identification of the CPRs configuration. It depends on the robot geometry and the efforts among its bodies; this problem is called geometrico-static. In general, as it does not admit an analytical solution, simulations can be used to find a solution numerically.

However, there are no generalized simulators available in the literature. For every case study, researchers develop their own code [3,8]. These specific simulations can be severely optimized to achieve real-time computation as in [8],

O. Altuzarra and A. Kecskémthy (Eds.): ARK 2022, SPAR 24, pp. 248–256, 2022.
https://doi.org/10.1007/978-3-031-08140-8_27

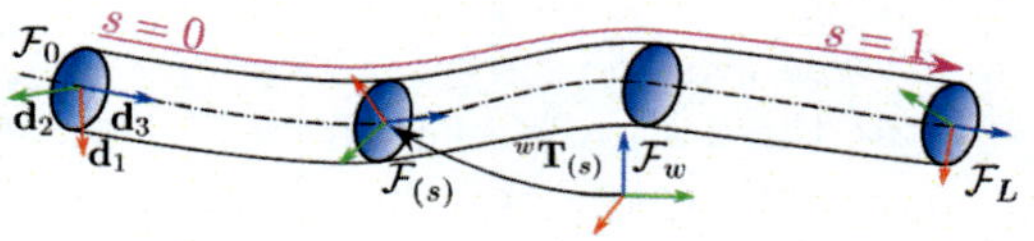

(a) Representation of a Cosserat rod.

(b) On the right, an example of a generic assembly of a CPR with a graphical representation of the numerical integration.

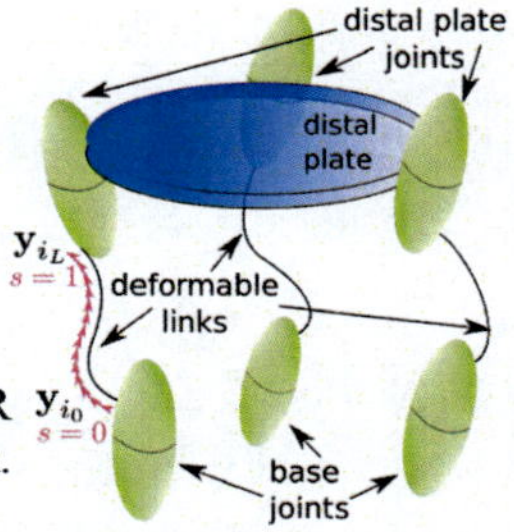

Fig. 1. Representation of Cosserat rod (on the left) and of a generic CPR (on the right).

but they are not prone to modification of the robot geometry or assembly. Any modification requires modifying the code or, in the worst case, refactoring the whole simulator.

In this paper, we propose a generic simulation for CPRs, based on Gazebo and able to identify the robot assembly automatically. We choose to use Gazebo and ROS as nowadays they are standard in many robotics applications. In particular, Gazebo allows integrating new capabilities, in terms of custom physics and rendering [6].

The paper is structured as follows: in Sect. 2 we discuss modeling for the deformable links. In Sect. 3, we present the generalized modeling strategy for CPRs. In Sect. 4, we briefly introduce the architecture of the simulator. In Sect. 5, we present some case studies to show the capabilities of our simulator. Finally we discuss our conclusion in Sect. 6.

2 Cosserat Rod Theory

The Cosserat rod theory is widely used and discussed in the literature to model the geometrico-static properties of continuum robots[7], as it has been proven to be reliable and precise [2]. A rod can be represented as a sequence of cross-sections stacked on top of each other along the rod centerline, as shown in Fig. 1a. This line is parameterized by $s \in [0\,1]$ being a normalized curvilinear abscissa, defined by $s = p/\ell$, with $p \in [0\,\ell]$ the non-normalized curvilinear abscissa along the beam, and ℓ the beam length at rest. Following the assumptions discussed in [8], we attach a frame $\mathcal{F}_{(s)}$ to every rod cross-section, defined with the principal axis of inertia of the cross-section at s, namely $\mathbf{d}_1$, $\mathbf{d}_2$ and $\mathbf{d}_3 = \mathbf{d}_1 \times \mathbf{d}_2$. The frame configuration is defined by an homogeneous transformation ${}^{w}\mathbf{T}_{(s)} \in SE(3)$ that describes its pose in terms of the position of its origin ${}^{w}\mathbf{p}_{(s)} \in \mathbb{R}^3$ and orientation ${}^{w}\mathbf{R}_{(s)} \in SO(3)$ of its frame axes with respect to a reference frame $\mathcal{F}_w$:

$$
{}^{w}\mathbf{T}_{(s)} = \langle {}^{w}\mathbf{p}_{(s)}, {}^{w}\mathbf{R}_{(s)} \rangle = \begin{bmatrix} {}^{w}\mathbf{R}_{(s)} & {}^{w}\mathbf{p}_{(s)} \\ \mathbf{0}_{[1\times 3]} & 1 \end{bmatrix} \tag{1}
$$

In this paper, we define a derivative with respect to the arc length $\frac{d}{ds}$ with the symbol $(\cdot)'$. In order to establish the evolution of the continuum body with respect to the arc length, we compute the derivative of position and orientation respectively as ${}^{w}\mathbf{p}'_{(s)}$ and ${}^{w}\mathbf{R}'_{(s)}$. Their counterparts, in the local coordinate of the cross-section, are: ${}^{s}\mathbf{v}_{(s)} = {}^{w}\mathbf{R}_{(s)}{}^{T\,w}\mathbf{p}'_{(s)}$ and ${}^{s}\mathbf{u}_{(s)} = \left[{}^{w}\mathbf{R}_{(s)}{}^{T\,w}\mathbf{R}'_{(s)}\right]^{\vee}$, where we used the map $(\cdot)^{\vee} : \mathbb{R}^{3\times3} \rightarrow \mathbb{R}^{3}$ such that: $\mathbf{a} \in \mathbb{R}^{3}, (\hat{\mathbf{a}})^{\vee} = \mathbf{a}$, $\hat{\mathbf{a}}$ being the cross-product matrix associated with the vector $\mathbf{a}$. In the following, we drop the prefix ${}^{w}(\cdot)$ for the quantities expressed with respect to $\mathcal{F}_w$.

In our work, we implement the system of ordinary differential equations (ODEs) for a Cosserat rod [3] for every i^{th} limb of a CPR, with $i = 1, \ldots, n$.

$$\begin{cases} \mathbf{p}_{i}{}'_{(s)} = \ell_i \left(\mathbf{R}_{i_{(s)}}{}^{s}\mathbf{v}_{i_{(s)}}\right) \\ \mathbf{R}_{i}{}'_{(s)} = \ell_i \left(\mathbf{R}_{i_{(s)}}{}^{s}\mathbf{u}_{i_{(s)}}\right) \\ \mathbf{n}_{i}{}'_{(s)} = -\ell_i \bar{\mathbf{f}}_{i_{(s)}} \\ \mathbf{m}_{i}{}'_{(s)} = -\ell_i \left(\mathbf{p}_{i}{}'_{(s)} \times \mathbf{n}_{i_{(s)}} + \bar{\mathbf{l}}_{i_{(s)}}\right) \end{cases} \tag{2}$$

$$\mathbf{n}_{i_{(s)}} = \mathbf{R}_{i_{(s)}} \mathbf{K}_{SE_i} \left[{}^{s}\mathbf{v}_{i_{(s)}} - {}^{s}\mathbf{v}^{0}_{i_{(s)}}\right] \tag{3}$$

$$\mathbf{m}_{i_{(s)}} = \mathbf{R}_{i_{(s)}} \mathbf{K}_{BT_i} \left[{}^{s}\mathbf{u}_{i_{(s)}} - {}^{s}\mathbf{u}^{0}_{i_{(s)}}\right] \tag{4}$$

where $\bar{\mathbf{f}}_{i_{(s)}}$ and $\bar{\mathbf{l}}_{i_{(s)}}$ are respectively some external distributed forces and moments along the centerline of the ith limb, while $\mathbf{n}_{i(s)}$ and $\mathbf{m}_{i(s)}$ are respectively the internal forces and moments that act from a cross-section s to the next one $s + ds$ having ds as an infinitesimal increment of s. The ODEs (2) are paired with Equations (3) and (4) linking the deformation with the internal stresses of the rod (Hooke's law). The deformation is a difference between the current rod curvatures and the ones in a load free state: ${}^{s}\mathbf{u}^{0}_{(s)}$ and ${}^{s}\mathbf{v}^{0}_{(s)}$. The two diagonal matrices $\mathbf{K}_{SE} = diag\,(GA, GA, EA)$ and $\mathbf{K}_{BT} = diag\,(EI_{xx}, EI_{yy}, E\,(I_{xx} + I_{yy}))$ account for geometrical and physical properties of the rod: E and G are the rod Young and shear modules, A is the cross-section area while I_{xx} and I_{yy} the principal moment of inertia.

We define the state of a rod as $\mathbf{y}_{i_{(s)}} = \left(\mathbf{p}_{i_{(s)}}, \mathbf{R}_{i_{(s)}}, \mathbf{n}_{i_{(s)}}, \mathbf{m}_{i_{(s)}}\right)$, where the operator $(\cdot, \ldots, \cdot)$ is a vertical concatenation of the vector listed between parentheses, in the case of a matrix, it vertically concatenates each of its columns. For every i^{th} limb, starting from the state at the rod base: $\mathbf{y}_{i(s=0)} = \mathbf{y}_{i_0}$, the numerical integration of (2) gives the state at the rod tip $\mathbf{y}_{(s=1)} = \mathbf{y}_{L}$. This final state contains the pose of the last cross section $\mathbf{T}_{i(s=1)} = \mathbf{T}_{i_L}$, and the wrench of the internal stresses $\mathbf{W}_{i(s=1)} = \mathbf{W}_{i_L}$.

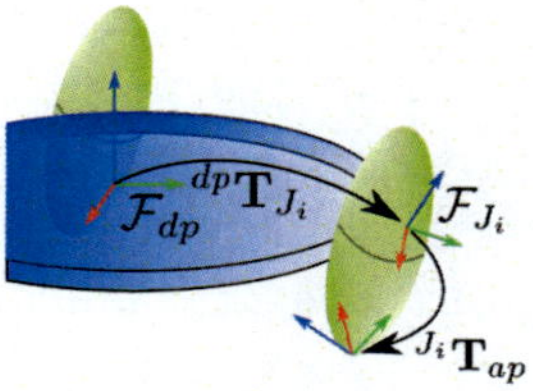

(a) Distal plate representation.

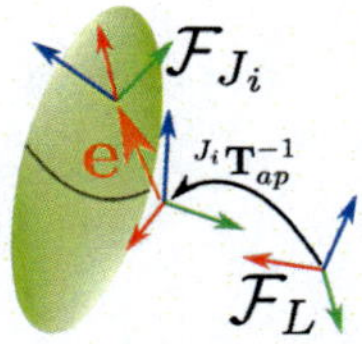

(b) Representation of the geometrical constrains.

Fig. 2. Representation of the distal plate joint. On the left its definition, on the right considerations about the geometrical constrains.

3 Generic Modeling of CPRs

We assume that base joints are responsible for the motion of the deformable links and that they define the initial state of the rod: $\mathbf{y}_{i_0}$, or the rod length ℓ_i. At the distal plate, every rod tip state $\mathbf{y}_{iL}$ must comply with the assembly constraints coming from the coupling between the tip of the rod and the joint. Moreover, the static balance of the distal plate must be satisfied. We detail these computations in what follows.

3.1 Assembly Constrains

The distal plate joints are attached to the platform body, each one with its own frame $\mathcal{F}_{J_i}$, as represented in Fig. 2a. These frames are described with a rigid body transformation ${}^{dp}\mathbf{T}_{J_i}$ with respect to the distal plate frame $\mathcal{F}_{dp}$. For every joint, ${}^{J_i}\mathbf{T}_{ap}$ describes how the rod tip should connect with the joint body. If the rod tip does not reach the attach point, we define an error as shown in Fig. 2b. The projection of ${}^{J_i}\mathbf{T}_{ap}^{-1}$ through $\mathbf{T}_L$ gives a pseudo joint origin $\mathcal{F}_{\bar{J}_i} = \langle \mathbf{p}_{\bar{J}_i}, \mathbf{R}_{\bar{J}_i} \rangle$. We then define: $\mathbf{e}_i = \left(\mathbf{e}_{ipos}, \mathbf{e}_{iori}\right)$, $\mathbf{e}_i \in \mathbb{R}^6$; the geometrical error in $SE(3)$ between the rod tip and the joint, with $\mathbf{e}_{ipos} = \mathbf{p}_{\bar{J}_i} - \mathbf{p}_{J_i}$ and $\mathbf{e}_{iori} = \left(\mathbf{R}_{J_i}^T \mathbf{R}_{\bar{J}_i} - \mathbf{R}_{J_i} \mathbf{R}_{\bar{J}_i} T\right)^\vee$. These formulations are detailed in [8], where all joints are assumed ideal and body-less. Note that $\mathbf{e}_i$ can be computed in different ways. However, we believe that our formulation is more computationally efficient and allows to account for the joint geometry and DoFs generically. If the joint has some DoFs, the corresponding part of $\mathbf{e}_i$ is neglected, as there are no constraints on that motion. We introduce the vector ${}^{J_i}\mathbf{a} = (a_x, a_y, a_z) \in \mathbb{R}^3$, for which each component can have two possible values: 1 if the joint is free along the corresponding axis and 0 otherwise. It defines a matrix $\tilde{I} = I - diag\left({}^{J_i}\mathbf{a}\right)$, with I the identity matrix, that we use to compose a block-diagonal matrix ${}^{J_i}\boldsymbol{I}_{dof} \in \mathbb{R}^{6\times6}$. This last matrix cancels the component of $\mathbf{e}_i$ along the joint DoFs. As an example, we detail the calculation for a spherical joint. As it can rotate in all directions, we have to consider only the translation part of the error. It follows that ${}^{J_i}\mathbf{a} = (1, 1, 1)$ and $\tilde{I} = \mathbf{0}_{3\times3}$, and the error is computed as follows:

$$\mathbf{e}_{p_i}^{jnt} = \mathcal{R}_{J_i} {}^{J_i}\mathcal{I}_{dof} \mathcal{R}_{J_i}^T \mathbf{e}_i \tag{5}$$

$$\begin{aligned}\mathbf{e}_{p_i}^{jnt} &= \mathcal{R}_{J_i} \begin{bmatrix} \mathbb{1}_{3\times3} & \mathbf{0}_{3\times3} \\ \mathbf{0}_{3\times3} & \tilde{I} \end{bmatrix} \mathcal{R}_{J_i}^T \begin{bmatrix} \mathbf{e}_{ipos} \\ \mathbf{e}_{iori} \end{bmatrix} \\ &= \mathcal{R}_{J_i} \begin{bmatrix} \mathbb{1}_{3\times3} & \mathbf{0}_{3\times3} \\ \mathbf{0}_{3\times3} & \mathbf{0}_{3\times3} \end{bmatrix} \begin{bmatrix} {}^{J_i}\mathbf{e}_{ipos} \\ {}^{J_i}\mathbf{e}_{iori} \end{bmatrix} \\ &= \mathcal{R}_{J_i} \begin{bmatrix} {}^{J_i}\mathbf{e}_{ipos} \\ \mathbf{0}_{3\times1} \end{bmatrix} = \begin{bmatrix} \mathbf{e}_{ipos} \\ \mathbf{0}_{3\times1} \end{bmatrix}\end{aligned} \tag{6}$$

Table 1. Definition of $\mathcal{I}_{dof}$ based on joint type

Type	Axis	$\mathcal{I}_{dof}$	${}^{J_i}\mathbf{a}^T$
Revolute	x	$diag\left(I, \tilde{I}\right)$	(1, 0, 0)
	y	$diag\left(I, \tilde{I}\right)$	(0, 1, 0)
	z	$diag\left(I, \tilde{I}\right)$	(0, 0, 1)
Prismatic	x	$diag\left(\tilde{I}, I\right)$	(1, 0, 0)
	y	$diag\left(\tilde{I}, I\right)$	(0, 1, 0)
	z	$diag\left(\tilde{I}, I\right)$	(0, 0, 1)
Fixed		$diag\,(I, I)$	(0, 0, 0)
Spherical		$diag\left(I, \tilde{I}\right)$	(1, 1, 1)

where $\mathcal{R}_{J_i} = diag\left(\mathbf{R}_{J_i}, \mathbf{R}_{J_i}\right)$ is a block-diagonal matrix to project an error in $SE(3)$ from the joint frame to the world frame. In Eq. (5), we provide the general computation of the actual error while, in Eq. (6), there is the formula applied to a spherical joint. For other joints, the Table 1 refers for the various cases.

Dual considerations apply for the static equilibrium of the distal plate. The wrench from the rod tip transfers on the joint origin: ${}^{J_i}\mathbf{W}_{J_i} = -\mathbf{Ad}_{\mathbf{T}_{rJ_i}}\mathbf{W}_L$, where $\mathbf{Ad}_{\mathbf{T}_{rJ_i}}$ is the adjoint transformation from the rod tip frame and the joint frame defined in [5]. It can be dived in two parts: ${}^{j}\mathbf{W}_{J_i}^{r} = {}^{J_i}\mathcal{I}_{dof}{}^{J_i}\mathbf{W}_{J_i}$ that is reciprocal to the joint twist and ${}^{J_i}\mathbf{W}_{J_i}^{p} = \left[I - {}^{J_i}\mathcal{I}_{dof}\right]{}^{J_i}\mathbf{W}_{J_i}$ that develops a non zero power.

The contribution of the ith limb on the distal plate balance is obtained again with the shifting law: $\mathbf{W}_{idp} = \mathcal{R}_{dp}\,\mathbf{Ad}_{\mathbf{T}_{J_i dp}}{}^{J_i}\mathbf{W}_{J_i}^{r}$.

3.2 Unique Solver

In order to find the configuration of a CPR, the geometrico-static problem is divided into two sub problems: the direct and inverse geometrico-static problem (DGM and IGM, respectively). The DGM consists in finding the pose of the distal plate knowing the wrench applied on it and the base joint actuations. In the IGM, knowing the distal plate pose and wrench, the joints coordinates are found.

In the literature, the CPRs geometrico-static problem is typically solved with the shooting method as in [8]. The shooting method consists in manipulating two vectors: the guess and the residual vector. The former, denoted as $\mathbf{x}_0$, is a vector containing all the variables in the geometrico-static problem while the second, denoted as $\mathbf{r}$ represents a cost function which module must be null. They are defined as:

$$\mathbf{x}_0 = \left(\mathbf{W}_{1(0)}, \ldots, \mathbf{W}_{n(0)}, \ldots, \mathbf{q}, \mathbf{p}_{dp}, \mathbf{ea}_{dp}\right) \tag{7}$$

$$\mathbf{r} = \left(\mathbf{e}_{p1}{}^{jnt}, \mathbf{e}_{W_1}^{pow}, \ldots, \mathbf{e}_{pn}^{jnt}, \mathbf{e}_{W_n}^{pow}, \ldots, \mathbf{e}_{W}^{bal}, (\mathbf{r}_{DGM}|\mathbf{r}_{IGM})\right) \tag{8}$$

where $\mathbf{q}$ is the vector of joints values, $\mathbf{W}_{i(0)}$ the wrench at the ith rods base and $\mathbf{p}_{dp}$ and $\mathbf{ea}_{dp}$ describes the platform position and Euler angles orientation respectively. The residual vector $\mathbf{r}$ contains the constrains $\mathbf{e}_{p_i}^{jnt}$ and $\mathbf{e}_{W_i}^{pow} = \mathbf{W}_i^p$ and the static equilibrium of the distal plate $\mathbf{e}_W^{bal}$, which satisfies the following equation.

$$\mathbf{e}_W^{bal} = \mathbf{W}_{dp}^{\mathbf{g}} + \sum_{i=0}^{n} \mathbf{W}_{idp} + \mathbf{W}_{dp}^{\mathbf{ext}} = \mathbf{0} \tag{9}$$

where $\mathbf{W}_{dp}^{\mathbf{g}}$ and $\mathbf{W}_{dp}^{\mathbf{ext}}$ are gravity and external wrenches acting on the distal plate.

Finally, the section $(\mathbf{r}_{DGM}|\mathbf{r}_{IGM})$ allows to handle both DGM and IGM with the same numerical framework [9]. It constrains either $\mathbf{q}$ or $\langle \mathbf{p}_{dp}, \mathbf{ea}_{dp} \rangle$ depending on the problem type. Using $\cdot^*$ for the desired values, we can describe $\mathbf{r}_{DGM}|\mathbf{r}_{IGM}$ as follows: $\mathbf{r}_{DGM} = (\mathbf{q}^* - \mathbf{q})$ and $\mathbf{r}_{IGM} = \left(\mathbf{p}_{dp}^* - \mathbf{p}_{dp}, \left[\mathbf{R}_{dp}^{*T}\mathbf{R}_{dp} - \mathbf{R}_{dp}^{*}\mathbf{R}_{dp}^{T}\right]^{\vee}\right)$.

With this method, the variables in the vector $\mathbf{x}_0$ are computed in order to cancel the norm of $\mathbf{r}$. This problem requires a nonlinear solver: we used here the Levenberg-Marquardt algorithm implemented in the Ceres library [1].

4 The Gazebo Simulator

Gazebo is a widely used simulator for the dynamic of rigid bodies [4]. It is open-source and based on the Open Dynamic Engine or Bullet physics. It hosts many robotics applications and, with our work, we want to bring the CPRs in this worldwide framework. Even if it does not support deformations, it provides extensions for its physics and rendering through plugins. Plugins allow users to control the Gazebo environment and physics. We developed our own plugin which accounts for the physics of the deformable links, applies the effects on the rigid bodies and displays the rod shapes using Gazebo rendering capabilities.

In order to simulate a CPR we need to specify its properties: deformable links are described in a dedicated *.yaml* file while the rigid bodies are described in a *.sdf* (simulation description format) file. This file contains all the physical, geometrical and visualization properties of the bodies.

We developed three main routines in our plugin. The first one is activated when all the parts of the CPR are loaded in Gazebo; our plugin starts retrieving their needed properties in order to construct the robot model: joints poses, geometries and axis ${}^{J_i}\mathbf{a}$ for all $i = 1, \ldots, n$. A second routine initializes the simulation finding a solution for the robot model. The shooting method requires an

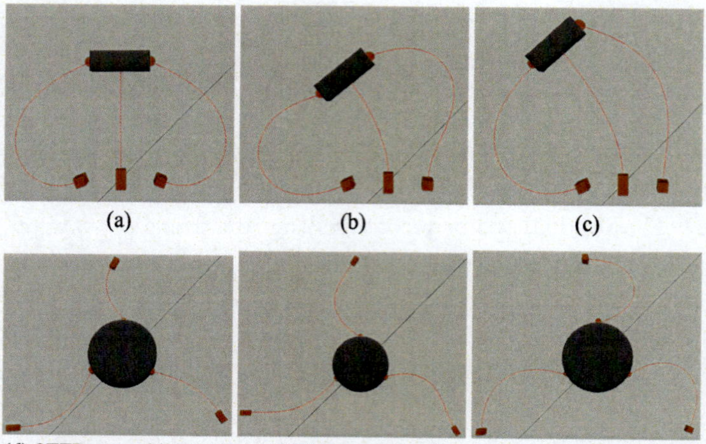

(a) (b) (c)

(d) 3EFR moved in IGM. (e) 3PFR moved in IGM. (f) 3RFR moved in IGM.

Fig. 3. Simulation of four planar CPRs with different actuations. In Figs. 3a, 3b and 3c, we show a robot with two actuated revolute joints and an extensible limb which is then fixed on the distal plate. In Fig. 3d, we present a robot with three extensible limbs (3EFR). In Fig. 3e, we depict a robot with prismatic joints at the base (3PFR). A robot with revolute joints at the base (3RFR) in Fig. 3f.

initial guess for begin the iterative process. This guess can be given by the user or can be found automatically. In the last case, the solver will start assuming all the forces and actuation values as zero. If no solution is found, then bounded random values for wrench and joints actuation are used to find a suitable solution. If no solution is found after a fixed number of tentative, the simulator exits with error. Once the robot is initialized, the third routine allows the user to move the robot. For this purpose, we created a Graphical User Interface (GUI) to ease the usage of the simulator, which is available online in our GitHub repository[1]. A video demonstration can be found on YouTube[2].

5 Simulations Results

In this section, we want to present the simulation of some CPRs starting with a version of the simulator for planar CPRs for which loads, deformations and displacements happen in the plane only. As a result, their model is simpler compared to the one of a spatial CPR such as [8]. Illustrations of DGM and IGM simulations are represented respectively in Fig. 3. Another version of the simulator is dedicated to the general case of spatial CPRs. Figure 4 shows the examples of three Stewart-Gough like platforms with different types of joints at the distal plate.

[1] https://github.com/aGotelli/A_Gazebo_Simulator_For_Continuum_Parallel_Robots.git.

[2] https://youtu.be/6k5aZPOQjQ8.

We used the simulator developed in [8] and carried out an identical robot model to be loaded in Gazebo and simulated with our simulator. We compared the length obtained for the extensible limb for different configurations of the robot. The relative error is negligible as its order is $\propto 10^{-4}$ m for rods length of $\propto 10^{0}$ m, which shows the accuracy of our simulator.

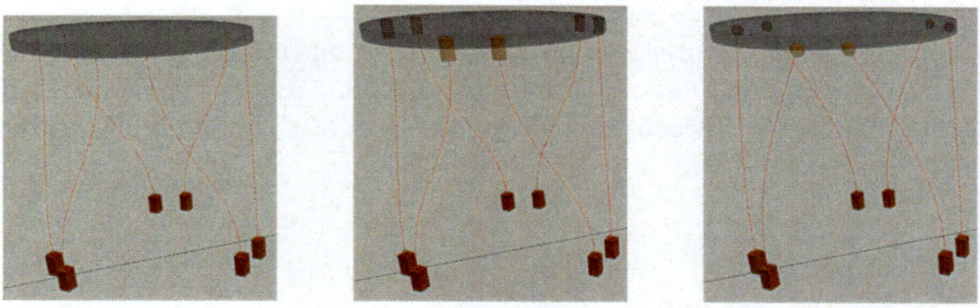

(a) Fixed joints. (b) Revolute joints. (c) Spherical joints.

Fig. 4. Simulation of three Stewart-Gough like platforms with different joint types.

6 Conclusions

In conclusion, we developed a simulator for CPRs, with ROS interfaces and designed for non-programmer users. Nonetheless, it can be used, extended, or modified in order to simulate specific CPRs. There are no other generic simulators in the literature and thus every researcher had to develop their own simulator in their own language, starting from scratch. We generalized the shooting method in order to deal with a wide range of CPRs, accounting for the geometry and dimensions of the joints.

Future works will target the implementation of a theory different from the one we used. We aim to accomplish real-time performances in Gazebo simulation in to order efficiently simulate CPRs, while preserving our generalized approach.

References

1. Agarwal, S., Mierle, K., et al.: Ceres solver. http://ceres-solver.org (online)
2. Antman, S.: Nonlinear Problems of Elasticity. Applied Mathematical Sciences, Springer New York (2006). https://books.google.fr/books?id=_MagNyCXNqMC
3. Black, C.B., Till, J., Rucker, D.C.: Parallel continuum robots: modeling, analysis, and actuation-based force sensing. IEEE Trans. Rob. **34**(1), 29–47 (2018)
4. Koenig, N., Howard, A.: Design and use paradigms for gazebo, an open-source multi-robot simulator. In: 2004 IEEE/RSJ International Conference on Intelligent Robots and Systems (IROS) (IEEE Cat. No. 04CH37566), vol. 3, pp. 2149–2154 (2004)
5. Murray, R.M., Li, Z., Sastry, S.S.: A Mathematical Introduction to Robotic Manipulation. CRC Press, Boca Raton (2017)
6. Okoli, F., Lang, Y., Kermorgant, O., Caro, S.: Cable-driven parallel robot simulation using Gazebo and ROS. Robot Design, Dynamics and Control. CISM International Centre for Mechanical Sciences, pp. 288–295 (2019)

7. Rucker, D.C., Jones, B.A., Webster, R.J., III.: A geometrically exact model for externally loaded concentric-tube continuum robots. IEEE Trans. Rob. **26**(5), 769–780 (2010)
8. Till, J., Bryson, C.E., Chung, S., Orekhov, A., Rucker, D.C.: Efficient computation of multiple coupled Cosserat rod models for real-time simulation and control of parallel continuum manipulators. In: International Conference on Robotics and Automation (ICRA), pp. 5067–5074 (2015)
9. Zaccaria, F., Briot, S., Chikhaoui, M.T., Idá, E., Carricato, M.: An analytical formulation for the geometrico-static problem of continuum planar parallel robots. In: ROMANSY 23 - Robot Design, Dynamics and Control, ROMANSY 2020. CISM International Centre for Mechanical Sciences, vol. 601, pp. 512–520 (2020, 2021). https://doi.org/10.1007/978-3-030-58380-4_61

Exploring Electric Field Perturbations as the Actuator for Nanorobots and Nanomachines

Caitlyn Mundrane[1], Meysam Chorsi[2], Olga Vinogradova[3], Horea Ilies[2], and Kazem Kazerounian[1,2](✉)

[1] Department of Biomedical Engineering, University of Connecticut, Storrs, CT 06269, USA
kazem.kazerounian@uconn.edu
[2] Department of Mechanical Engineering, University of Connecticut, Storrs, CT 06269, USA
[3] Department of Pharmaceutical Sciences, University of Connecticut, Storrs, CT 06269, USA

Abstract. Nanomachines and nanorobots are nanoscale mechanisms that transform an input motion or force into some desirable output. These machines offer significant applications in nanotechnology, biotechnology, and medicine. To perform a specific task, nanomachines must have a controllable kinematic structure. Our team in past has proposed amino acids as the links for these kinematic chains (peptide chains). The kinematic formulation of these chains are now well developed. One noteworthy outcome has been design and development of 7-R closed loop one-degree-of-freedom spatial mechanisms (cyclic peptide). Unlike in macro-scale machines and robots, traditional kinematic pairs such as prismatic and revolute actuators cannot be employed in nano kinematic chains. However, a one-degree-of-freedom nano-scale kinematic chain can be a viable alternative to traditional actuators and can be used as the driving module in more complex nano systems. We have designed, simulated, fabricated, and tested several 7-R closed loop systems. Due to the one-degree-of-freedom nature of 7-R kinematic chains, their geometry as a function of one input parameter is completely predictable and repeatable. However, controlling this input parameter in real applications remains an open question. In this work, we explore perturbation of the electric force field as a control mechanism to manipulate the motions of these kinematic nano robots and machines. Two different external electric field models, uniform and non-uniform, are developed. Protofold II, an in-house developed protein folding software, is used to simulate the range of motion for closed-loop cyclic 7-R peptides with electric field perturbations. The results show that one-degree-of-freedom closed loop systems have the potential to be precisely controlled by external physical stimuli. This discovery presents significant potential in design and implementation of nanorobots and nano machines for applications in biotechnology.

Keywords: Electric field · Nanomachines · Peptide · Kinematic closure

O. Altuzarra and A. Kecskémthy (Eds.): ARK 2022, SPAR 24, pp. 257–265, 2022.
https://doi.org/10.1007/978-3-031-08140-8_28

1 Introduction

Utilizing proteins molecules as kinematic chains of mini-rigid bodies connected by revolute joints based on the vast body of knowledge available on their macro-scale siblings has been well documented in the last two decades [5,6,11,12,14]. These kinematic chains in nature constitute a broad range of molecular machinery responsible for carrying out essentially all critical functions in biological systems. However the design and fabrication of artificial nanomachines and nano robots and their manipulation of matter pose monumental scientific and technological challenges. The level of complexity associated with the natural protein machinery makes it extremely difficult to mimic the structure and control strategies observed in nature. Nevertheless, an effective way to tackle these challenges is to use established and tested methodologies in kinematics and mechanisms. For example it is well known that the one-degree-of-freedom (DOF) machines are robust, and widely used designs because of their completely predictable, deterministic, and repeatable motion, which is fully determined by a single input parameter. In fact, the vast majority of macroscale mechanisms that have been engineered throughout human history are either fundamentally one-DOF mechanisms or use them as building blocks. But unlike in macro-scale machines and robots, traditional kinematic pairs such as prismatic and revolute actuators cannot be employed in nano kinematic chains. Instead, a one-degree-of-freedom nano-scale closed loop kinematic chain can be a viable alternative to traditional controllable actuators and can be used as the driving module in more complex nano systems. Our team has developed novel 7-R closed loop one-degree-of-freedom spatial mechanisms (cyclic peptide) [1].

In this paper, we build on the theoretical and computational framework of systematically designing 1-DOF closed-loop linkages nano kinematic chains using peptide molecules. Furthermore, we explore control mechanisms to manipulate the motions of these machines via external physical stimuli, specifically via the application of perturbations in external electric fields.

2 Kinematic Model

Proteins are composed of amino acids (N_i-C_{α_i}-C_i). Amino acids are consecutively linked by a covalent chemical bond known as a peptide bond ($C_i - N_{i+1}$). Due to the secondary resonance stabilization [7], peptide bonds reside in a single plane called the peptide plane. Therefore, peptide planes are considered to be rigid groups of atoms linked in the chain by covalent bonds at the alpha carbons (C_α) and they can only rotate about the $N - C_\alpha$ and $C_\alpha - C$ bonds known as torsional angles ϕ and ψ, respectively. Proline is an exception to this rule with only one dihedral angle, ψ. Therefore, the polypeptide chain CGPG (C for cysteine, G for glycine, and P for proline) has 7 degrees of freedom (cysteine and glycine have 2 DOF, and proline has 1 DOF). Consequently, the closed form of the peptide, will have 1 DOF. This extra degree of freedom is the redundancy. Thus, the combination of the joint angles can be freely changed independently from the placement of the end-effector.

To set a reference for the angle measurements, the zero reference position description (ZRPD) method is used [2,4]. In the zero position, a suitable configuration of the manipulator is designated as the zero reference position where all of the joint variables are set to zero. The unit vectors (u), body vectors (b), end-effector position (P) and end-effector orientation unit vectors (u_a, axial and u_t, transverse) completely define the kinematic structure of the manipulator. In zero position, the CGPG peptide is modeled as a kinematic linkage as presented in Fig. 1A (0 index shows the zero position). The end-effector position in zero configuration is defined as:

$$[P_{0H}] = \sum_{i=1}^{7} [b_{0i}] \tag{1}$$

The details about the formulation and indexing scheme is presented in [13]. In general, the loop closure equations yield a nonlinear relation between the position variables [1]. The closed form solution of these equations is not straight forward. For the CGPG nanomachine, 12 dependent nonlinear equations need to be solved for a total of 6 unknowns. Here we use unconstrained minimization method to solve the system of nonlinear equations since this method eliminates the need to calculate the Hessian matrix. Figure 1B demonstrates the cyclic structure of CGPG after one of the kinematic closure solutions were applied to the open structure.

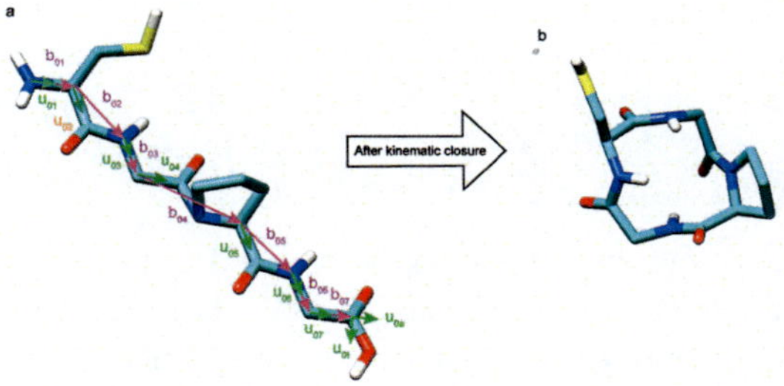

Fig. 1. The CGPG peptide is modeled as a kinematic linkage. a, Open chain CGPG peptide at zero reference position and the corresponding body and unit vectors. b, Closed chain form of CGPG peptide after kinematic loop closure

3 Defining Cyclic Peptide Energies

To understand what control mechanisms can be used to alter cyclic peptide motion we must understand how these nanomachines can be energetically defined. A common way to represent the energy profile of a polypeptide or even larger protein structures is by way of the force field equation used in molecular dynamics and molecular mechanics [3,9,18]. This equation is used to model the series of forces, internal and external, that energetically define the structure and motion of molecules. Therefore, the force field equation serves as a predictive tool for searching molecular structure under different environments. In general, the force field equation is represented by:

$$G^{tot}(\mathbb{A}) = G^{elec}(\mathbb{A}) + G^{vdw}(\mathbb{A}) + G^{bonds}(\mathbb{A}) + G^{angles}(\mathbb{A}) + G^{dihedrals}(\mathbb{A}) \quad (2)$$

Using this model, geometry dictates the total amount of energy a molecule has for a given configuration. While different versions exist depending on application, force field models are built on the same general framework, that the potential energy of a molecule depends upon structurally linked geometries and spatial proximity geometries or electrostatic interactions. The structurally linked parameters include bond lengths, bond angle, dihedral angles, and electrostatic interactions include van der Waal's interactions experienced between all atoms in proximity and proximal interactions between charged particles.

Since the molecule for our purposes is being represented by a kinematic model, we are not concerned in the internal energies of the structure. These are instead kinematically determined based on the kinematic model created for the arrangement of links and joints for a polypeptide chain. Therefore, the modified force field equation for our purposes eliminates these structurally linked energy contributions and is:

$$\begin{aligned} G^{total}(\mathbb{A}) = & \sum_{\alpha_i \epsilon \mathbb{A}} \sum_{\alpha_j \epsilon \mathbb{A}-\{a_i\}} \frac{1}{4\pi\varepsilon_{i,j}} \frac{q_i q_j}{\|r_i - r_j\|} + \\ & \sum_{\alpha_i \epsilon \mathbb{A}} \sum_{\alpha_j \epsilon \mathbb{A}-\{a_i\}} \epsilon_{i,j} \left[\left(\frac{D_{i,j}}{d_{i,j}} \right)^{12} - 2 \left(\frac{D_{i,j}}{d_{i,j}} \right)^{6} \right] \end{aligned} \quad (3)$$

Here the interactions between electrically charged particles are calculated by Coulomb's law (first term) [19], and van der Waals interactions are calculated using the Lennard-Jones potential function (second term) [19].

In previous work, our in-house developed software package Protofold II was designed for calculating energies and running protein simulations [13]. We used the software package for the running of our force field calculations on cyclic peptide CGPG. Originally, Protofold II was designed for open-loop polypeptide chains, so a modified approach was taken to utilize its calculation capabilities. Figure 2A provides a schematic of the approach, whereby solving the nonlinear equations for a closed-loop linkage provide the mobility for a given cyclic peptide structure. This mobility analysis allows for the generation of a series of PDBs, files that represent polypeptide chain structures with all the atom coordinate information. A series of PDBs can then be fed into Protofold II over which energy calculations can be run to generate a full image of the energy curves for a cyclic peptide. Figure 2B shows the full energy curve for CGPG using this process. The curves represent only two solutions of a specific linkage configuration of CGPG representing motion along two seperate branches. In theory, there are other potential energy curves that would exist if specific criteria for the linkage arrangement are altered thereby changing the overall shape of the energy curve generated. Based on solution 1, there exists a local minima where the configuration of CGPG statistically will most likely exist. Often in energy search simulations such as molecular dynamics, the search algorithm gets stuck in local minima and will struggle to search for other potential lower local minima. But in this case, because we have the full mobility for CGPG, we can generate the full energy space of the structure.

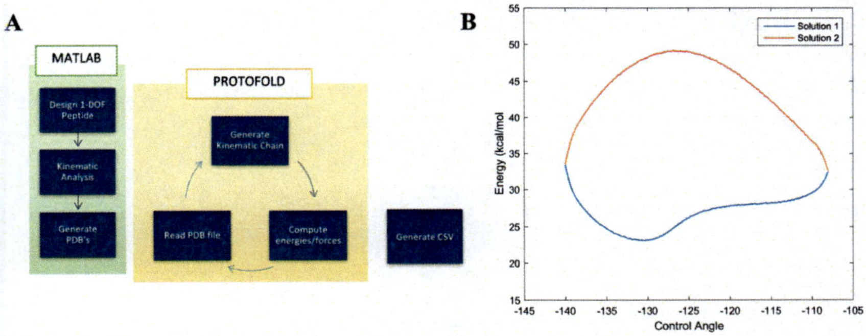

Fig. 2. A, Schematic of calculating energy curves, forces, and torques for cyclic peptide in Protofold II **B**, Two solution energy curves for CGPG as calculated using Protofold II. Control angle is a dihedral angle along CGPG as the chosen single parameter.

4 Force Field Perturbation Methods

Ultimately, the goal is to eventually control the actuation of 1-DOF kinematic chain cyclic peptides, though currently there is no known mechanism to provide concentrated forces and torques at the nanoscale. However, there are forces that do govern the motion of the peptide represented by the aforementioned force field equation. This a good starting point for finding modes of perturbing the motion of these cyclic peptide structures. We also know that peptides have the capacity to respond to external stimuli as well, whether that be through the introduction of other molecules coming into interaction, the pH of the solution, the temperature of the surrounding environment, or if there is a presence of an electromagnetic field. These external stimuli also have the potential to influence the spatial configuration of these molecules and are therefore important in understanding the behavior of a peptide.

One of the most intriguing control mechanisms of the physical field is to place proteins an electric field. We proposed several different electric field models implemented in Protofold II to examine how external fields can potentially impact cyclic peptide structure. Equations 4 and 5 represent a potential uniform electric field model and a potential non-uniform electric field model respectively.

$$G_E^{total}F(\mathbb{A}) = G_0^{total}(\mathbb{A}) - \left[\mu_i E^i + \frac{1}{2}\alpha_{ij}E^iE^j + \frac{1}{3}\Theta_{ij}E^{ij}\right] \tag{4}$$

$$G_E^{total}F(\mathbb{A}) = G_0^{total}(\mathbb{A}) + \frac{1}{4\pi\varepsilon_{i,e}}\frac{q_iq_e}{\|r_i - r_e\|} \tag{5}$$

For the uniform electric field, we consider the molecular response to an electric field as a multipole power series expansion where the applied field is a function of a series of multipole moments to represent and describe how the charge is distributed in the molecule [8,10,16]. This model examines the response as change in the original energy with contributions from its dipole moment, μ_i, dipole polarizability, α_{ij}, and the quadrupole moment, Θ_{ij} as interacting with an external uniform electric field. Here the dipole moment and quadrupole moment are simply calculated as Eqs. 6 and 7 respectively and the atomic polarizability is approximated using a dipole-interaction model [15,17]:

$$\mu_i = \sum_{n=1}^{N} q_n r_{n,i} \tag{6}$$

$$\Theta_{ij} = \sum_{n=1}^{N} q_n r_{n,i} r_{n,j} \tag{7}$$

To evaluate this model, we applied varying uniform electric field strengths of 0.3, 0.5, 0.7, and 1.0 V/$\AA$ along the dipole moment of CGPG and then plotted the energy curves as seen in Fig. 3B.

A non-uniform electric field model defined via a charged particle in space allows one, at least in principle, to tailor the strength of the field to a specific protein or to a specific direction of motion along the curves defined by the appropriate kinematic solutions such as those shown in Fig. 2B for CGPG. Since all charged particles generate electric field lines, a charged particle in close proximity to a molecule in space can generate a non-uniform external electric field that interacts with the molecule. We return to Coulomb's Law for charged particles where instead of a molecular pairwise interaction between two atoms within the structure, we calculate an additional electrostatic term between all atoms in the molecule and the external charged particle with position r_i and charge q_e.

To evaluate this non-uniform model, we set an external charged particle a fixed distance z away from the molecule and then ranged the position of the charged particle in the xy-plane to observe changes in the energy curve CGPG. Figure 3A shows one example of this approach where an externally charged particle 5×10^3 C is shifted in the xy-plane with a fixed distance of 15 $\AA$ away from CGPG.

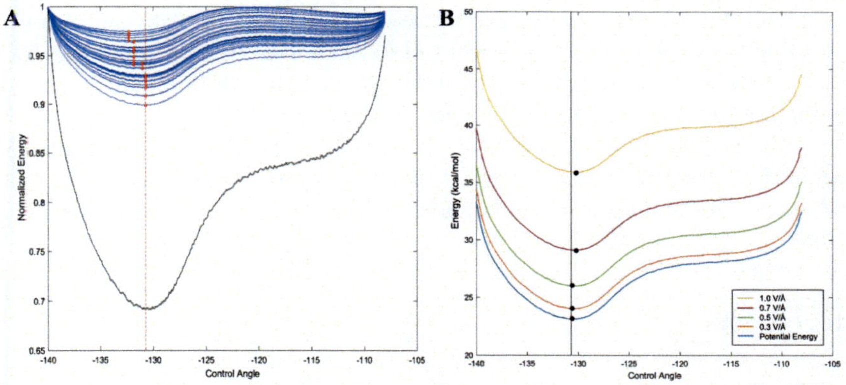

Fig. 3. **A**, Varying energy curves for CGPG (in blue) in a non-uniform electric field with externally charged particle of 5×10^3 C being shifted in the xy-plane a fixed distance of 15 $\AA$ from CGPG. Black curve shows original energy curve. Largest shift in control angle is approximately 1.6° **B**, Energy curves for CGPG for varying uniform electric fields placed along dipole moment of molecule using multipole power series expansion model with applied uniform fields of 0.3, 0.5, 0.7, and 1.0 V/$\AA$. Largest shift in control angle is approximately 0.5°.

The results of these two models are not to definitively conclude that one model is better or provides more statistically significant results, but rather to show how force field perturbations can be implemented in Protofold II. For a designed cyclic peptide like CGPG which has a much narrow range of mobility, we would expect shifts to be very small but for flexible and larger designed cyclic peptides, these models might potentially show larger shifts in global minimum energy. By having the full mobility analysis for a closed peptide linkage, in this case CGPG, we can calculate the energy curves for this structure and then modify the force field equations in Protofold II to observe how the energy curves for a given structure might change such that the global minimum energy of the curve has adapted a new conformation.

5 Conclusion

In this paper we presented a theoretical and computational framework to systematically design controllable, nanomachines as 1-DOF closed-loop linkages using peptide structure as a basis for nanomachine building blocks. Here we presented a designed 4 amino acid unit cyclic peptide, CGPG, whose mobility analysis was calculated by solving its closed-loop nonlinear equations using unconstrained minimization method. By solving these equations, we could then use the mobility analysis for CGPG to compute its energies at any given instance in its motion. Using Protofold II software package, modified to account for a closed-loop linkage condition, we calculated the energy of CGPG under two different external electric models, uniform and non-uniform, to observe if the minimum energy of the curve shifted in response to external stimuli. The results show small shifts in input angle for CGPG under different external conditions and provide a framework under which additional models could be potentially implemented to continue further observing peptide behavior. The future of this work serves to show how cyclic peptides may be designed kinematically and then from a molecular standpoint, energetically defined so as to further perturb and explore the functional space for the design of nanomachines that have prescribed function and motion.

References

1. Chorsi, M.T., Tavousi, P., Mundrane, C., Gorbatyuk, V., Ilies, H., Kazerounian, K.: One degree of freedom 7-R closed loop linkage as a building block of nanorobots. In: Lenarčič, J., Siciliano, B. (eds.) Advances in Robot Kinematics 2020, ARK 2020. Springer Proceedings in Advanced Robotics, vol. 15, pp. 41–48. Springer, Cham (2020). https://doi.org/10.1007/978-3-030-50975-0_6
2. Gupta, K.C.: Kinematic analysis of manipulators using the zero reference position description. Int. J. Robot. Res. **5**(2), 5–13 (1986)
3. Halgren, T.A.: Merck molecular force field. i. Basis, form, scope, parameterization, and performance of MMFF94. J. Comput. Chem. **17**(5-6), 490–519 (1996)
4. Kazerounian, K., Gupta, K.: Manipulator dynamics using the extended zero reference position description. IEEE J. Robot. Autom. **2**(4), 221–224 (1986)
5. Kazerounian, K., Latif, K., Alvarado, C.: Protofold: a successive kinetostatic compliance method for protein conformation prediction. J. Mech. Des. **127**, 712–717 (2005)
6. Kazerounian, K., Latif, K., Rodriguez, K., Alvarado, C.: Nano-kinematics for analysis of protein molecules. J. Mech. Des. **127**(4), 699–711 (2005)
7. Lehninger, A.L., Nelson, D.L., Cox, M.M., Cox, M.M., et al.: Lehninger Principles Of Biochemistry. Macmillan, New York (2005)
8. Morris, M., Jordan, M.J.: Modeling molecular response in uniform and non-uniform electric fields. J. Chem. Phys. **138**(5), 054,111 (2013)
9. Rappé, A.K., Casewit, C.J., Colwell, K., Goddard III, W.A., Skiff, W.M.: UFF, a full periodic table force field for molecular mechanics and molecular dynamics simulations. J. Am. Chem. Soc. **114**(25), 10024–10035 (1992)
10. Stone, A.: The Theory of Intermolecular Forces. OUP, Oxford (2013)
11. Subramanian, R., Kazerounian, K.: Improved molecular model of a peptide unit for proteins. In: International Design Engineering Technical Conferences and Computers and Information in Engineering Conference, vol. 42568, pp. 601–613 (2006)
12. Subramanian, R., Kazerounian, K.: Kinematic mobility analysis of peptide based nano-linkages. Mech. Mach. Theory **42**(8), 903–918 (2007)
13. Tavousi, P., Behandish, M., Ilieş, H.T., Kazerounian, K.: Protofold ii: enhanced model and implementation for kinetostatic protein folding. J. Nanotechnol. Eng. Med. **6**(3), 034,601 (2015)
14. Tavousi, P., Kazerounian, K., Ilies, H.: Synthesizing functional mechanisms from a link soup. J. Mech. Des. **138**(6), 062303 (2016)
15. Van Duijnen, P.T., Swart, M.: Molecular and atomic polarizabilities: Thole's model revisited. J. Phys. Chem. A **102**(14), 2399–2407 (1998)

16. Vegiri, A., Schevkunov, S.V.: A molecular dynamics study of structural transitions in small water clusters in the presence of an external electric field. J. Chem. Phys. **115**(9), 4175–4185 (2001)
17. Wang, J., Cieplak, P., Li, J., Hou, T., Luo, R., Duan, Y.: Development of polarizable models for molecular mechanical calculations i: parameterization of atomic polarizability. J. Phys. Chem. B **115**(12), 3091–3099 (2011)
18. Wang, J., Wolf, R.M., Caldwell, J.W., Kollman, P.A., Case, D.A.: Development and testing of a general amber force field. J. Comput. Chem. **25**(9), 1157–1174 (2004). J. Comput. Chem. **26**(1), 114–114 (2005)
19. Weiner, S.J., et al.: A new force field for molecular mechanical simulation of nucleic acids and proteins. J. Am. Chem. Soc. **106**(3), 765–784 (1984)

Mechanism Synthesis

General Procedure for Path Generation Synthesis Based on Kinematic Constraints

Aitor Muñoyerro[1(✉)], Alfonso Hernández[2], Mónica Urizar[2], and Oscar Altuzarra[2]

[1] SENER Aeroespacial, Avda. de Zugazarte 56, 48992 Getxo, Spain
aitor.munoyerro@aeroespacial.sener

[2] Faculty of Engineering in Bilbao, University of the Basque Country (UPV/EHU), Plaza Ingeniero Torres Quevedo 1, 48013 Bilbao, Spain

Abstract. This paper presents a general formulation intended for path generation synthesis based on the derivation of kinematic constraint equations in their matrix form and their subsequent assembly in a system of linear equations. The procedure enables obtaining in a simple way the partial derivatives of the nodal coordinates of a mechanism with respect to the dimensional or input parameters. The main utility of this type of calculation is its application to gradient calculation within optimal dimensional synthesis, being valid for any planar mechanism composed of any number of elements connected by revolute joints (R), prismatic (P) or the combination R+P. The proposed approach has been implemented in general-purpose GUI software to solve path generation problems.

1 Introduction

Dimensional synthesis of linkages has been studied long before computational optimization methods became available. In late 1800s Burmester [1] presented a geometrical theory to design planar mechanisms whose coupler link goes through a limited number of positions (e.g. maximum of 5 positions for four bar linkage). Classical textbooks on mechanism theory, e.g. Erdman & Sandor [2], deal with graphical and analytical methods for exact synthesis. In contrast, optimization methods allow to obtain approximate solutions for a larger number of prescribed positions and for any type of objective function.

Calculating partial derivatives is one of the main problems in gradient-based optimal synthesis methods. A simple way to calculate such terms is to use approximate numerical methods (i.e., finite differences), but these methods have disadvantages, such as increasing the number of function evaluations, and thus, the computational cost, as well as inaccuracies that appear especially in highly nonlinear problems, reducing the efficiency of the algorithm and the quality of the solutions obtained. Therefore, it is always preferable to calculate the gradient analytically [3, 4].

The classical approach within the optimal synthesis problems consists of solving the nonlinear system resulting from the null gradient of the error function, usually posed as the squared sum of differences between the prescribed and generated characteristics. The vast majority of papers focus on particular mechanisms, mainly a four bar linkage [5,

O. Altuzarra and A. Kecskémethy (Eds.): ARK 2022, SPAR 24, pp. 269–276, 2022.
https://doi.org/10.1007/978-3-031-08140-8_29

6]. Among the most noteworthy general purpose developments, in [7] an optimization method based on minimum distance error function is presented. Additionally, in [4] an effective solution to the exact gradient calculation is provided by developing a method that is general and systematic, albeit not automatic. The novelty of the present paper compared to [4] is the substitution of the loop equations by kinematic constraint equations and natural coordinates to describe the position of multibody systems using only Cartesian coordinates of different points [8, 9]. This avoids identification of loop closure equations, and enables solving the problem in a fully automatic way by means of a linear system obtained by assembly of equations derived from kinematic constraints. A novel approach to solving both the exact gradient problem and optimal dimensional synthesis is proposed, with advantages of simplicity of formulation, suitability for implementation in general purpose programs, and low computational cost.

The baseline algorithm described in this paper does not use global methods, but they could be implemented in a simple way by using the same error function proposed for the gradient method. Indeed, metaheuristic methods are currently of great interest in the field of dimensional synthesis, and have been successfully applied to different types of mechanisms, e.g., genetic algorithms [10].

2 Analytical Formulation of Partial Derivatives

Gradient based methods require to compute the gradient of the error function, which is composed of derivatives of the type $\frac{\partial \boldsymbol{v}_i}{\partial s_k} \forall i = 1 \ldots np$, where $\boldsymbol{v}_i$ is a vector containing the variables of the synthesis (e.g. coordinates of a coupling point, orientation of an element, etc.) for an i-th synthesis position, np is the total number of synthesis position, and s_k is a term that belongs to the vector $\boldsymbol{s}$ where all the optimization parameters are collected. Two types of synthesis parameters can be distinguished:

The *dimensional parameters*, which appear in the *geometric constraint* equations are generically denoted by a_j. They can be either bar lengths or coordinates defining the position in the plane of the fixed nodes.

In the case of *unprescribed timing* synthesis, the values assigned to the input variable of the mechanism at each *i-th* synthesis position, commonly referred to as *input parameters*, are also considered as parameters to be optimized and denoted by φ_i.

Therefore, the vector $\boldsymbol{s}$ is expressed as

$$\boldsymbol{s} = \left[a_1 \; a_2 \; \ldots \; a_{nd} \; \varphi_1 \; \varphi_2 \; \ldots \; \varphi_{np} \right]^T \tag{1}$$

where nd is the number of dimensional parameters, and np is the number of synthesis positions.

Two types of kinematic constraints are formulated. On the one hand, *geometric constraints*, are those where independent variable does not appear explicitly. They are expressed as $\boldsymbol{\Phi}^G(\boldsymbol{x}(\varphi)) = 0$, where $\boldsymbol{x}$ is the coordinates vector and φ is the independent variable. Since this paper deals with single DoF mechanisms, for the sake of clarity the input parameter will be used as independent variable instead of time. On the other hand, *driving constraints* are those where φ appear explicitly, formulated as $\boldsymbol{\Phi}^D(\boldsymbol{x}(\varphi), \varphi) = 0$. Note that, according to these definitions, *geometric constraints* are of holonomic-scleronomic kind while *driving constraints* are holonomic-rheonomic.

The combination of all geometric and driving constraints is expressed by the following nonlinear system of equations that allows the position problem to be solved numerically.

$$\boldsymbol{\Phi}(\boldsymbol{x}(\varphi), \varphi) = \begin{bmatrix} \Phi_1(\boldsymbol{x}(\varphi)) \\ \Phi_2(\boldsymbol{x}(\varphi)) \\ \cdots \\ \cdots \\ \Phi_R(\boldsymbol{x}(\varphi), \varphi) \end{bmatrix} = 0 \tag{2}$$

2.1 Kinematic Constraints Formulation

Different constraints that can appear in a general case of planar mechanism can be defined, considering joints of type R, P, and R + P for both linear and rotary actuators. The reader may consult a detailed development of kinematic constraint equations in [6, 7]. After posing of these kinematic constraints, the proposed procedure obtains their derivatives with respect to the optimization parameters in order to perform the optimal synthesis, as it will be explained in Sect. 2.2.

As an example, the constraint equation for rotary actuation driving is given by expression (3), as shown in Table 1.

Table 1. Example of rotary actuation driving constraint

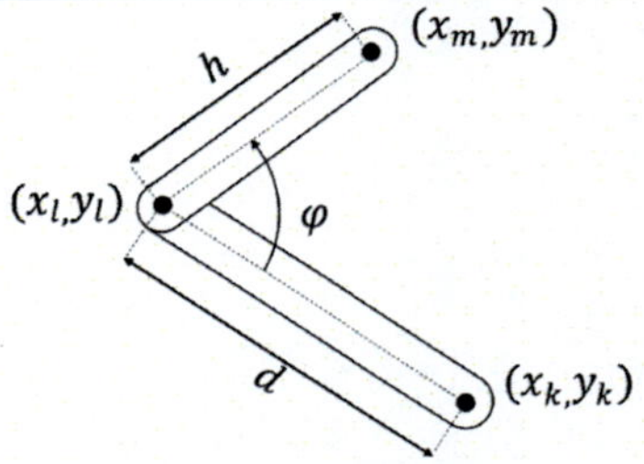

$$q = \boldsymbol{x}_{klm}{}^T \cdot \boldsymbol{N} \cdot \boldsymbol{x}_{klm} \tag{3}$$

Where q is the cross product of vectors from node "l" to node "k", and from "l" to "m":

$$q = 2 \cdot d \cdot h \cdot \sin(\varphi) \tag{4}$$

$$\boldsymbol{x}_{klm} = [x_k, y_k, x_l, y_l, x_m, y_m]^T \tag{5}$$

d, h: length of bars between which the relative angle is imposed.
φ: kinematic degree of freedom (angle) imposed by the actuator.

where:

$$\boldsymbol{N} = \begin{bmatrix} 0 & 0 & 0 & 1 & 0 & -1 \\ 0 & 0 & -1 & 0 & 1 & 0 \\ 0 & -1 & 0 & 0 & 0 & 1 \\ 1 & 0 & 0 & 0 & -1 & 0 \\ 0 & 1 & 0 & -1 & 0 & 0 \\ -1 & 0 & 1 & 0 & 0 & 0 \end{bmatrix} \tag{6}$$

2.2 Derivation of Kinematic Constraints

After formulating kinematic constraint equations (e.g. Eq. (3)), they shall be derived with respect to a generic design variable, or with respect to the input variable, and squared. Subsequently, the resulting equations are organized in such a way that they all have a common structure, as shown in expression (7):

$$0 = \left(\frac{\partial \boldsymbol{x}}{\partial s_k}\right)^T \cdot \boldsymbol{C_r} \cdot \left(\frac{\partial \boldsymbol{x}}{\partial s_k}\right) - 2 \cdot \left(\frac{\partial \boldsymbol{x}}{\partial s_k}\right)^T \cdot \boldsymbol{p_r} + k_r \tag{7}$$

This common structure facilitates the above-mentioned matrix assembly operation and guarantees the correct matrix expression of the function. In this way, new equations are obtained where the unknowns are the partial derivatives of the nodal coordinates.

For example, in rotary actuator driving constraint the following terms will be the ones to substitute in Eq. (7):

$$\boldsymbol{p_{klm}} = \boldsymbol{F_{klm}} \cdot \boldsymbol{x} \tag{8}$$

$$\boldsymbol{J_{klm}} = \boldsymbol{F_{klm}} \cdot \boldsymbol{x} \cdot \boldsymbol{x}^T \cdot \boldsymbol{F_{klm}} \tag{9}$$

$$\boldsymbol{C_r} = \boldsymbol{J_{klm}} \tag{10}$$

$$\boldsymbol{p_r} = \frac{1}{2} \cdot \boldsymbol{p_{klm}} \cdot \left(\frac{\partial q}{\partial s_k}\right) \tag{11}$$

$$k_r = \frac{1}{4}\left(\frac{\partial q}{\partial s_k}\right)^2 \tag{12}$$

$$\boldsymbol{F} = \begin{bmatrix} 0 & 0 & 0 & 1 & 0 & -1 \\ 0 & 0 & -1 & 0 & 1 & 0 \\ 0 & -1 & 0 & 0 & 0 & 1 \\ 1 & 0 & 0 & 0 & -1 & 0 \\ 0 & 1 & 0 & -1 & 0 & 0 \\ -1 & 0 & 1 & 0 & 0 & 0 \end{bmatrix} \tag{13}$$

$\boldsymbol{F_{klm}}$ is obtained by expanding $\boldsymbol{F}$ to the positions of nodes k, l, m to a $(2nx2n)$ matrix, where n is the total number of nodes

$$\begin{array}{ll} \frac{\partial q}{\partial s_k} = 2 \cdot d \cdot sin(\varphi) & \text{If } s_k = h \\ \frac{\partial q}{\partial s_k} = 2 \cdot h \cdot sin(\varphi) & \text{If } s_k = d \\ \frac{\partial q}{\partial s_k} = 2 \cdot d \cdot h \cdot cos(\varphi) & \text{If } s_k = \varphi \\ \frac{\partial q}{\partial s_k} = 0 & \text{If } s_k \neq d \wedge s_k \neq h \wedge s_k \neq \varphi \end{array} \tag{14}$$

2.3 Automatic Calculation of Partial Derivatives

In a general case, mechanisms that give rise to over-constrained systems of equations can occur frequently. The most appropriate solution is to approach the calculation by means of a least squares problem.

Deriving Eq. (2) with respect to a dimensional or input parameter, denoted s_k, where $k = 1, 2, \ldots, n_d + n_p$, and then posing the quadratic expression, one obtains Eq. (15). Left hand side will be equal to zero if $\frac{\partial x}{\partial s_k}$ is compatible with all kinematic constraints, and greater than zero if it is not.

$$\left(\frac{\partial \boldsymbol{x}}{\partial s_k}\right)^T \cdot \left(\frac{\partial \boldsymbol{\Phi}}{\partial \boldsymbol{x}}\right)^T \cdot \frac{\partial \boldsymbol{\Phi}}{\partial \boldsymbol{x}} \cdot \frac{\partial \boldsymbol{x}}{\partial s_k} + 2 \cdot \frac{\partial \boldsymbol{\Phi}}{\partial \boldsymbol{x}} \cdot \frac{\partial \boldsymbol{x}}{\partial s_k} \cdot \frac{\partial \boldsymbol{\Phi}}{\partial s_k} + \left(\frac{\partial \boldsymbol{\Phi}}{\partial s_k}\right)^T \cdot \frac{\partial \boldsymbol{\Phi}}{\partial s_k} = 0 \tag{15}$$

Therefore, the following function $\psi\left(\frac{\partial \boldsymbol{x}}{\partial s_k}\right)$ can be posed, and one way to solve Eq. (15) is to minimize $\psi\left(\frac{\partial \boldsymbol{x}}{\partial s_k}\right)$.

$$\psi\left(\frac{\partial \boldsymbol{x}}{\partial s_k}\right) = \left(\frac{\partial \boldsymbol{x}}{\partial s_k}\right)^T \cdot (\boldsymbol{\Phi}_x)^T \cdot \boldsymbol{\Phi}_x \cdot \frac{\partial \boldsymbol{x}}{\partial s_k} + 2 \cdot \boldsymbol{\Phi}_x \frac{\partial \boldsymbol{x}}{\partial s_k} \cdot \boldsymbol{\Phi}_s + (\boldsymbol{\Phi}_s)^T \cdot \boldsymbol{\Phi}_s \tag{16}$$

Each of the equations obtained when deriving an r-th constraint from Eq. (2) with respect to a dimensional or input parameter will be expressed in matrix form as shown in Eq. (17), and identified as $\psi_r\left(\frac{\partial \boldsymbol{x}}{\partial s_k}\right) = 0$.

$$\psi_r\left(\frac{\partial \boldsymbol{x}}{\partial s_k}\right) = \left(\frac{\partial \boldsymbol{x}}{\partial s_k}\right)^T \cdot \boldsymbol{C_r} \cdot \left(\frac{\partial \boldsymbol{x}}{\partial s_k}\right) - 2\cdot\left(\frac{\partial \boldsymbol{x}}{\partial s_k}\right)^T \cdot \boldsymbol{p_r} + k_r for r = 1, 2, \ldots, R \tag{17}$$

The function $\psi\left(\frac{\partial \boldsymbol{x}}{\partial s_k}\right)$ in Eq. (16) can be expressed as follows:

$$\sum_{r=1}^{R} \psi_r(\partial \boldsymbol{x}/\partial s_k) = \left(\frac{\partial \boldsymbol{x}}{\partial s}\right)^T \cdot \sum_{r=1}^{R} \boldsymbol{C_r} \cdot \left(\frac{\partial \boldsymbol{x}}{\partial s_k}\right) - 2\cdot\left(\frac{\partial \boldsymbol{x}}{\partial s_k}\right)^T \cdot \sum_{r=1}^{R} \boldsymbol{p_r} + \sum_{r=1}^{R} k_r \tag{18}$$

where the following terms $\sum_{r=1}^{R} \boldsymbol{C_r}, \sum_{r=1}^{R} \boldsymbol{p_r}$ and $\sum_{r=1}^{R} k_r$, called global terms, are obtained by expansion and assembly of local terms $\boldsymbol{C_r}, \boldsymbol{p_r}$ and k_r formulated for each r-th constraint, as it is done in finite element method.

Finally, by posing the minimization of the function $\psi(\partial \boldsymbol{x}/\partial s_k)$, the system of linear Eq. (19) is obtained, which allows us to directly calculate the partial derivatives $\left(\frac{\partial \boldsymbol{x}}{\partial s_k}\right)$:

$$\boldsymbol{C}\left(\frac{\partial \boldsymbol{x}}{\partial s_k}\right) = \boldsymbol{p} \tag{19}$$

From system (19), the exact derivatives that are used in Sect. 3 to carry out the optimization process can be extracted. It is worth mentioning that the terms $\boldsymbol{C}$ and $\boldsymbol{p}$ are defined from analytical expressions.

3 Optimization Method

In path generation problems, error vector $\boldsymbol{E}$ can be posed as:

$$\boldsymbol{E} = \left[\left(x_1 - x_1^d\right) \ldots \left(x_N - x_N^d\right) \left(y_1 - y_1^d\right) \ldots \left(y_N - y_N^d\right) \right]^T \tag{20}$$

where x_i, y_i are the positions generated by the target node on each synthesis position $i = 1, 2 \ldots N$, and x_i^d, y_i^d are the objective values to achieve in optimization problem. Least squares optimization will be performed, with function to be minimized denoted by F and expressed as:

$$F = \sum_{i=1}^{2N} E_i^2 \tag{21}$$

Expanding the Taylor series error function F and posing the zero gradient condition, we obtain the following system of equations to be solved iteratively:

$$\left(\left[J^k\right]^T \left[J^k\right]\right)\left\{\Delta s^k\right\} = -\left[J^k\right]^T \cdot \boldsymbol{E}^k \tag{22}$$

where $[J]^k$ is the jacobian matrix of $\boldsymbol{E}$ evaluated according to design on k^{th} iteration. It is built by using nodal coordinates derivatives calculated on each iteration according to analytical method described in Sect. 2. On the other hand, $\boldsymbol{E}^k$ is the error vector evaluated according to design on k^{th} iteration, and Δs^k is the unknown of the system, which is the parameter correction vector obtained on k^{th} iteration, to be applied according to expression (23):

$$\Delta s^{k+1} = s^k + \Delta s^k \tag{23}$$

4 Illustrative Example

Below is an example of path generation consisting of 21 points shown in red in Fig. 2, describing two orthogonal straight lines and a semicircle.

The kinematic structure used is composed of 7 hinged bars. The example is solved with a self-developed software GIMSYNT programmed in MATLAB. In the GUI (Guided User Interface) shown in Fig. 1 the problem is pre-processed. The user introduces an initial design of the mechanism by choosing the joints and the positions of the linkage nodes in any configuration of the circuit, so that dimensional parameters are defined. The user also enters the objective path. The program automatically calculates the motion of the mechanism along the chosen circuit. Finally, the user assigns a correspondence between the prescribed points and those of the trajectory generated by the initial mechanism, so that the input parameters for the first iteration of the optimization are automatically defined.

Once the preprocessing is complete, the program automatically executes the optimization and generates files with results in graphical and numerical format. Figure 2(a) shows the initial design used in this example, with error E = 1010.38, and Fig. 2(b) the optimal design to which it converges, with E = 0.052. This example of path generation is solved in 152 iterations and a total time of 18 s.

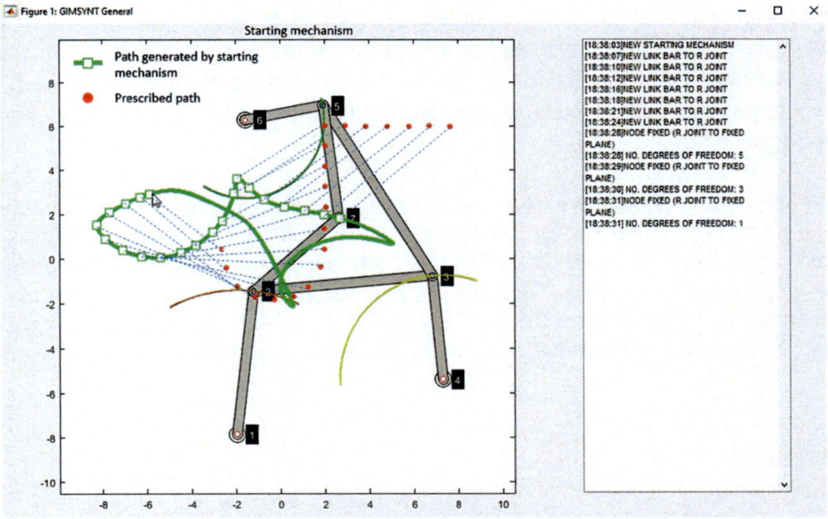

Fig. 1. Guided User Interface of path generation in GIMSYNT

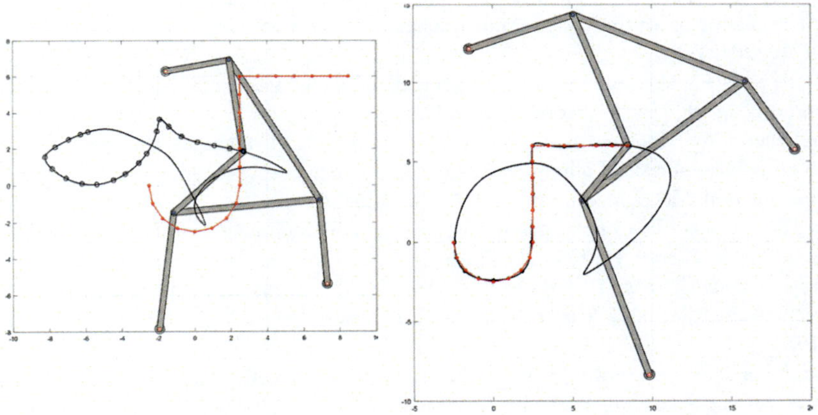

Fig. 2. Path generation example (a) starting design E = 1010.38 (b) optimal design E = 0.052

5 Conclusions

In this paper, an automatic method for the dimensional synthesis of planar mechanisms of any level of complexity has been proposed. Any planar mechanism with R and P joints or their combination, and with or without redundant constraints, can be synthesized. The partial derivatives needed for the optimum dimensional synthesis approach based on the null-gradient condition are automatically obtained, thanks to a formulation based on nodal coordinates and kinematic constraints. With this formulation we can obtain in a systematic way a function from differentiating the constraint equations. The minimization of this function results in exact analytical expressions for the calculation of partial derivatives. Moreover, the proposed procedure permits the Jacobian matrix to be computed automatically for any general case, without the need to identify the loop equations manually. An illustrative example shows the good performance of the proposed procedure.

Acknowledgments. The authors wish to acknowledge the financial support received from the Spanish government, *Ministerio de Ciencia e Innovación* (Project PID2020-116176GB-I00) financed by MCIN/AEI/10.13039/501100011033, the support for the research group, Project IT949–16, provided by the *Dept. Educación, Política Lingüística y Cultura* from the regional Basque Government, and Program BIKAINTEK 2020 (Ref. 012-B2/ 2020) provided by the *Dept. Desarrollo Económico, Sostenibilidad y Medio Ambiente* from the regional Basque Government.

References

1. Burmester, L.: Lehrbuch der Kinematic. A. Felix, Leipzig (1888)
2. Erdman, A.G., Sandor, G.N.: Mechanism Design: Analysis and Synthesis, vol. 1. Prentice Hall (1984)
3. Sancibrián, R., et al.: Optimal Dimensional Synthesis of Linkages Using Exact Jacobian Determination in the SQP Algorithm. Mechanics Based Design of Structures and Machines, vol. 40 (2012)
4. Mariappan, J., Krishnamurty, S.: A generalized exact gradient method for mechanism synthesis. Mech. Mach. Theory **31**, 413–421 (1996)
5. Hernández, A., Muñoyerro, A., Urízar, M., Amezua, E.: Comprehensive approach for the dimensional synthesis of a four-bar linkage based on path assessment and reformulating the error function. Mech. Mach. Theory **156**, 104–126 (2021)
6. Bai, S., Angeles, J.: Coupler-curve synthesis of four-bar linkages via a novel formulation. Mech. Mach. Theory. **94**, 177–187 (2015)
7. Fernandez de Bustos, I., Urkullu, G., García, V., Ansola, R.: Optimization of Planar Mechanisms by using a Minimum Distance Function. Mech. Mach. Theory **138**, 149–168 (2019)
8. García de Jalon, J., Serna, M.A., Avilés, R.: Computer method for kinematic analysis of lower-pair mechanisms—I velocities and accelerations. Mech. Mach. Theory **16**(5), 543–556 (1981)
9. García de Jalon, J., Serna, M.A., Avilés, R.: Computer method for kinematic analysis of lower-pair mechanisms—II position problems. Mech. Mach. Theory **16**(5), 557–566 (1981)
10. Shao, Y., Xiang, Z., Liu, H., Li, L.: Conceptual design and dimensional synthesis of cam-linkage mechanisms for gait rehabilitation. Mech. Mach. Theory **104**, 31–42 (2016)

Finding Straight Line Generators Through the Approximate Synthesis of Symmetric Four-Bar Coupler Curves

Aravind Baskar(✉), Mark Plecnik, and Jonathan Hauenstein

University of Notre Dame, Notre Dame, IN 46556, USA
abaskar@nd.edu

Abstract. In this paper, equations for the approximate synthesis of symmetric four-bar coupler curves are formulated. Our approach specifies a number of desired trace points, and finds a number of four-bar linkages with a coupler trace that approximately passes through these points. The computed linkages correspond to all the minima of the posed objective. The objective posed simultaneously enforces kinematic accuracy, loop closure, and leads to polynomial first order necessary conditions with a monomial structure that remains the same for any number of specified desired trace points. This last characteristic makes our result more general. To simplify computations, ground pivot locations are set as chosen parameters, and a root count analysis is conducted that shows our objective has a maximum of 73 critical points. The theoretical work is applied to the computational design of straight line coupler paths. To perform this exercise, the choice of ground pivots was varied, and a parameter homotopy for each choice (504 in total) was executed. These computations found the expected linkages (Watt, Evans, Roberts, Chebyshev) and other linkages resembling them but with sizable variations on their dimensions. The t-SNE algorithm was employed to organize the computed straight line generators into a visual atlas.

1 Introduction

The synthesis of a point path by a four-bar linkage has been addressed in [1] for the exact case, and in [2] for the approximate case. Here we address a subcase, that is the synthesis of symmetric coupler curves. We are motivated to study symmetric curves as we note that many of the special straight line generators found over time produce symmetric curves, e.g. the Watt linkage, the Evans linkage, the Roberts linkage, the Chebyshev linkage, and the Chebyshev lambda linkage [3]. In search of more such interesting geometries, symmetry constraints are installed. This reduces the well known nine dimensional design space of four-bar linkages down to seven dimensions. In addition, to aid in computational tractability, the positions of ground pivots were set, reducing the design space to three dimensions. The relevant kinematic constraints were formulated into an

O. Altuzarra and A. Kecskémthy (Eds.): ARK 2022, SPAR 24, pp. 277–285, 2022.
https://doi.org/10.1007/978-3-031-08140-8_30

optimization problem which was solved completely for all minima using polynomial homotopy continuation. The result is used to search for straight line generators by systematically varying ground pivot locations and computing several parameter homotopies. Our computational search found the well known straight line generators as well as several variants of their geometries. The resulting linkage designs are organized into an atlas using the t-SNE unsupervised machine learning algorithm.

2 Mathematical Formulation of Four-Bars

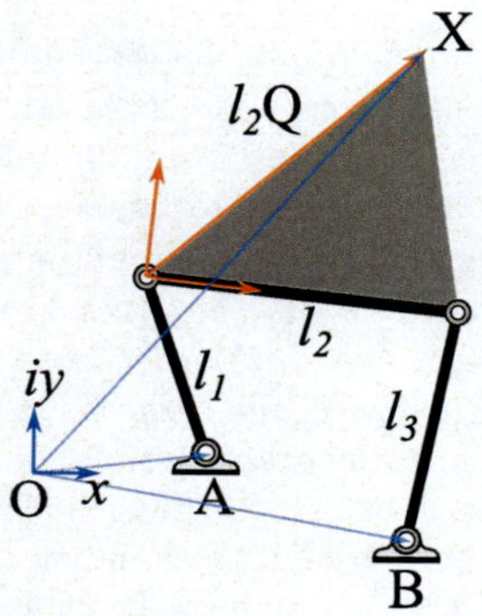

Fig. 1. Schematic of a four-bar linkage in the complex plane

Consider a planar four-bar linkage as shown in Fig. 1 in the complex plane. Let A and B represent the two fixed pivots, respectively. For representing vector variables such as the fixed pivots, isotropic coordinates [4] are used here. Hence, additional variables A^* and B^* denoting the conjugate variables of A and B, respectively, are introduced. This is an alternative approach to the Cartesian framework in order to gain certain advantages [4] during the mathematical formulation stage as well as in the implementation of numerical continuation solution technique that follows. Let l_1, l_2, and l_3 denote the lengths of the three moving links as shown. The coupler trace point (normalized by the coupler base length l_2) is represented in the local frame of the coupler as Q and its conjugate counterpart Q^*. Thus, the design variables of the four-bar linkage are summarized as $\boldsymbol{d} = \{A, A^*, B, B^*, l_1, l_2, l_3, Q, Q^*\}$. If X and its conjugate X^* denote the locus of the trace point of interest in the global frame, then it satisfies the equation $f(\boldsymbol{d}, X, X^*) = 0$ given by:

$$\begin{vmatrix} Q^*(A-X) & g(X,X^*) & l_2Q(A^*-X^*) & 0 \\ 0 & l_2Q^*(A-X) & g(X,X^*) & Q(A^*-X^*) \\ (-1+Q^*)(B-X) & h(X,X^*) & l_2(-1+Q)(B^*-X^*) & 0 \\ 0 & l_2(-1+Q^*)(B-X) & h(X,X^*) & (-1+Q)(B^*-X^*) \end{vmatrix} = 0, \quad (1)$$

where

$$g(X,X^*) = -l_1^2 + l_2^2QQ^* + (A-X)(A^*-X^*) \qquad \text{and}$$
$$h(X,X^*) = -l_3^2 + l_2^2(-1+Q)(-1+Q^*) + (B-X)(B^*-X^*).$$

As is well known for four-bar linkages, Eq. 1 is a sextic equation with circularity 3. It comprises of 16 distinct monomial terms in X, X^*, namely,

$$\left\{ \begin{matrix} X^3X^{*3}, X^3X^{*2}, X^3X^*, X^3, X^2X^{*3}, X^2X^{*2}, \\ X^2X^*, X^2, XX^{*3}, XX^{*2}, XX^*, X, X^{*3}, X^{*2}, X^*, 1 \end{matrix} \right\}$$

in which the coefficient of the leading term X^3X^{*3} is equal to 1. Four-bar linkages that share an identical coupler locus occur as *Roberts cognate triplets* in the four-bar design space (see pp. 168–176 of [5]). For a design $\boldsymbol{d}_1 = \{A, A^*, B, B^*, l_1, l_2, l_3, Q, Q^*\}$, its other two cognates can be expressed as:

$$\boldsymbol{d}_2 = \Big\{B, B^*, A + Q(B-A), A^* + Q^*(B^*-A^*), l_2\sqrt{(1-Q)(1-Q^*)},$$
$$l_3\sqrt{(1-Q)(1-Q^*)}, l_1\sqrt{(1-Q)(1-Q^*)}, \frac{1}{1-Q}, \frac{1}{1-Q^*}\Big\},$$
$$\boldsymbol{d}_3 = \Big\{A + Q(B-A), A^* + Q^*(B^*-A^*), A, A^*, l_3\sqrt{QQ^*},$$
$$l_1\sqrt{QQ^*}, l_2\sqrt{QQ^*}, \frac{Q-1}{Q}, \frac{Q^*-1}{Q^*}\Big\}. \quad (2)$$

In our experiment, we restrict the model to four-bars that generate symmetric coupler curves. We do this for two reasons. First, much of the straight line linkages reported in the literature [3] such as Watt, Evans, Roberts, and Chebyshev linkages generate symmetric coupler curves about some axis of symmetry in the plane. And second, the inclusion of additional conditions on the design variables to this effect simplifies the model significantly and enables faster computations.

2.1 Symmetric Coupler Curves

The following derives the necessary and sufficient conditions for a four-bar linkage to generate symmetric coupler curves. While some of these conditions can be found in the literature, we present a direct proof here via analytical geometry and subsequent algebraic analysis.

Following the isotropic coordinates convention, points (P, P^*) on a generic line in the complex plane satisfy:

$$L(P, P^*) = K^*P + KP^* + c = 0, \quad (3)$$

where K, K^* $(\neq 0)$ are isotropic parameters and c is a real parameter. If (X, X^*) is any point in the plane, then its symmetric reflection about the axis given by Eq. 3 is

$$(X_m, X_m^*) = \left(-\frac{c + KX^*}{K^*}, -\frac{c + K^*X}{K}\right). \tag{4}$$

It follows that, for a four-bar coupler curve to be symmetric about an axis $L(P, P^*) = 0$, (X_m, X_m^*) given by Eq. 4 must also satisfy Eq. 1, that is, $f(\boldsymbol{d}, X_m, X_m^*) = 0$. Since the equation and its reflection must be identical, the coefficients of the 16 monomial terms in X, X^* can be equated element-wise to arrive at 15 conditions (disregarding the unit leading term of the monomial X^3X^{*3}) on the design variables $\boldsymbol{d}$ and the axis parameters K, K^*, c. As the symmetric behavior is unaffected by scaling, rotation, and translation, the fixed pivots can be plugged in as $A = A^* = 0$ and $B = B^* = 1$ which further simplifies the conditions. Note that this choice of fixed pivots is made only for enabling the derivation of the conditions of symmetry and is not a global choice for the latter sections. The conditions corresponding to the monomials X^3X^{*2}, X^3X^*, X^3 are, respectively, as follows:

$$3c + K + K^* + K^*Q + KQ^* = 0, \tag{5}$$

$$-3c^2 - 2cK^* - 2cK^*Q - K^{*2}Q + K^2Q^* = 0, \tag{6}$$

$$c(c + K)(c + KQ^*) = 0. \tag{7}$$

The conjugate of these conditions also occur for the monomials $X^2X^{*3}, XX^{*3}, X^{*3}$.

Equation 7 shows that either $c = 0$, $c = -K$, or $c = -KQ^*$. Each of these three conditions can be analyzed separately in conjunction with Eqs. 5 and 6, and then with the other 12 coefficient conditions (not all independent). The algebra is not included for brevity and we present only the results:

Four-bar linkages with design variables $\boldsymbol{d} = \{A, A^*, B, B^*, l_1, l_2, l_3, Q, Q^*\}$ that generate coupler curves symmetric about an axis $K^*P + KP^* + c = 0$ can be of the following two classes:

Class A $c = 0, K = -K^*, Q = Q^*$. These correspond to four-bars whose trace point lies along the line that connects the two floating pivots. For these, the reflection of the linkage about its ground link in any given configuration is also part of its configuration space, thus enabling the occurrence of symmetric coupler curves. The cognates of such four-bars also meet these conditions with the ground-pivots of all three cognates lying along the axis of symmetry.

Class B This class can be split into three types which themselves form a Roberts cognate triplet.

1. $c = 0, Q = -\frac{K}{K^*} = \frac{1}{Q^*}, l_1 = l_2$
2. $c = -K = -K^*, Q + Q^* = 1, l_1 = l_3$
3. $c = -KQ^* = -K^*Q, Q^* = \frac{Q}{Q-1}, l_2 = l_3$

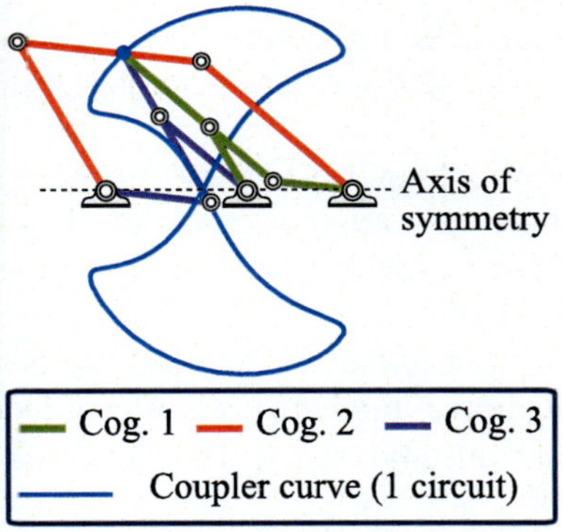

(a) Class A

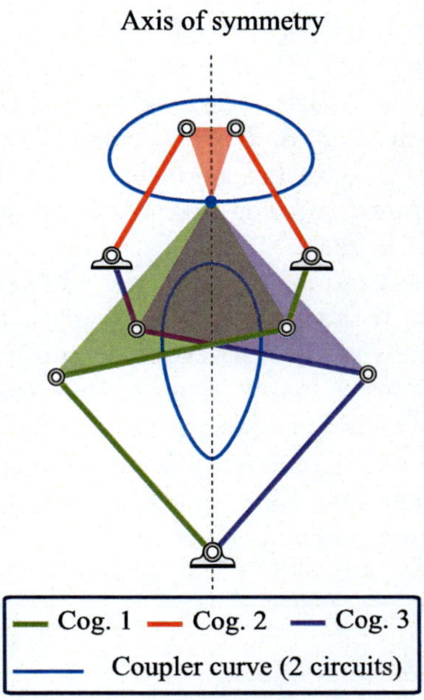

(b) Class B

Fig. 2. Two classes of four-bar linkages which generate symmetric coupler curves

In Fig. 2, examples of the two classes of four-bars that generate symmetric coupler curves are shown as cognate triplets. Arguably, four-bars of Class B are more interesting because, unlike Class A, the symmetric curves generated by them are not simply reflections about the ground link. The two classes of four-bars overlap in the design space in some cases, notably the Chebyshev and Watt straight line linkages. Another well-known symmetric straight line linkage, the Roberts linkage, is of Class B. The axis of symmetry in the four-bars of Class B is the perpendicular bisector of the fixed link corresponding to the cognate #2, while passing through the ground pivot shared between the cognates #1 and #3. For more geometric description of the four-bars of Class B and their conditions, refer to [6].

In this work, we limit the following design procedure to the four-bars of Class B based on the reasoning above. In particular, we solve for cognate #2 of Class B and compute cognates #1 and #3 based on the transformations presented in Eq. 2.

3 Optimization Model for Approximating Straight Lines

For four-bar linkages of Class B and cognate #2, a generic design is represented by $\boldsymbol{d} = \{A, A^*, B, B^*, l, l_2, l, Q, 1 - Q\}$. Note that $l_1 = l_3 = l$ and $Q^* = 1 - Q$ based on the conditions derived earlier. This simplifies the coupler equation in terms of $\{A, A^*, B, B^*, l, l_2, Q\}$. As the variables l and l_2 occur only in the form of squares, $l_{2s} = l_2^2$ and $l_s = l^2 - l_2^2 Q(1 - Q)$ are introduced to simplify the equation further and to reduce the total degree. At this stage, a decision is made to treat A, A^*, B, B^* as specified design parameters instead of treating them as variables. This brings down the number of variables to 3, namely, l_s, l_{2s} and Q, as opposed to being 7 which would be a much harder problem outside the scope of this work.

As mentioned earlier, the coupler curve of a four-bar linkage is degree six. Hence, if the exact synthesis approach is taken, a maximum of only six design positions along a straight line can be specified. Approximate synthesis process allows for as many design specifications as desired. The optimization problem is one of minimizing the error residue of the coupler equation over all the design positions. We chose the $\boldsymbol{L}^2$-norm to retain the polynomial nature of the objective function, thus allowing the use of a numerical continuation approach to solve any resulting polynomial system.

The objective of the optimization problem is a sum of squares of the residue of the coupler equation over all the design positions, $j = 1, 2, ..., N$:

$$\frac{1}{2}\sum_{j=1}^{N} \eta_j^2, \tag{8}$$

where $\eta_j = f(A, A^*, B, B^*, l_s, l_{2s}, Q, X_j, X_j^*)$. The design variables are l_s, l_{2s}, Q, while A, A^*, B, B^* are the design parameters and X_j, X_j^* are the design positions. The first-order necessary conditions of optimality are then derived symbolically as:

$$\sum_{j=1}^{N} \eta_j \begin{pmatrix} \frac{\partial \eta_j}{\partial l_s} \\ \frac{\partial \eta_j}{\partial l_{2s}} \\ \frac{\partial \eta_j}{\partial Q} \end{pmatrix} = 0. \tag{9}$$

This system of 3 equations in 3 unknowns has a monomial structure that is invariant to the number of design positions N. This allows us to specify more design positions without increasing the complexity of the system. In particular, the total degree of this polynomial system is 648, which forms a trivial upper bound of the number of critical points of the objective function. One can computer tighter bounds such as a 2-homogeneous Bézout bound [4] of 186 and the BKK bound [4] of 73. This is confirmed by explicitly solving a randomly chosen *ab initio* system using the numerical continuation solver `Bertini` [7,8] via a 2-homogeneous homotopy of 186 startpoints. Such a start system is usually constructed by forming a polynomial system respecting the same multi-homogeneous structure using linear expressions which are easily solved to form the startpoints [1,9]. Then, using a predictor-corrector numerical path tracking, the startpoints are deformed continuously to the target points of the ab initio system. Solving this *ab initio* system yielded 73 solutions matching the BKK bound, while the rest diverged off to infinity as expected. Thus, one can use a parameter homotopy [8] and track 73 solution paths to solve any other system with the same monomial structure.

4 Design of Experiments

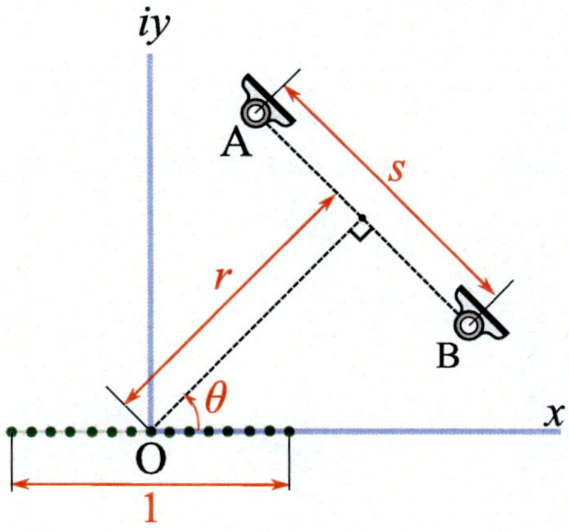

Fig. 3. Design specification for approximate straight line generating four-bar linkages

Parameter homotopy runs are carried out for the design of approximate straight line generating four-bar linkages with cognate #2 symmetric four-bars of Class B being the primary focus. The design specification is chosen to be discrete points of equal step-length along the x-axis in the range $[-0.5, 0.5]$. Specifying a high number of design positions ($N = 100$ in this work) reduces the possibility of undesirable coupler curve behavior between the desired positions. The ground link is described by four parameters, two for each fixed pivot. We add a constraint that restricts the ground link such that the axis of symmetry passing through the mid-point of the design specification as shown in Fig. 3 resulting in three parameters r, θ, and s as illustrated. The parameter θ can be restricted to be within $[0°, 90°]$ as the other values are topologically equivalent. We sample the space by employing a discretization scheme as follows:

$$r \in \{0.25i\}_{i=0}^{8} \qquad \theta \in \{15° j\}_{j=0}^{6} \qquad s \in \{0.25k\}_{k=1}^{8}, \tag{10}$$

which yields a total of $9 \cdot 7 \cdot 8 = 504$ distinct problems. The computation time required for solving a single parameter homotopy run of 73 paths is about 15s and only the solutions that correspond to physical linkages are investigated. Moreover, since this computation yields all critical points, only the local minima for each computation are retained while all saddle points are rejected. This yields 2461 linkages which are then further refined based on an allowable structural error tolerance of $\frac{1}{100}$ of unity in the y direction of the desired segment and a maximum link length constraint of 2. This results in 59 linkages of which cognates #1 and #3 are computed based on Eq. 2.

For exhibiting these $59 \cdot 3 = 177$ linkages, we used the machine learning technique t-SNE [10], a nonlinear dimensional reduction tool to allow us to visualize

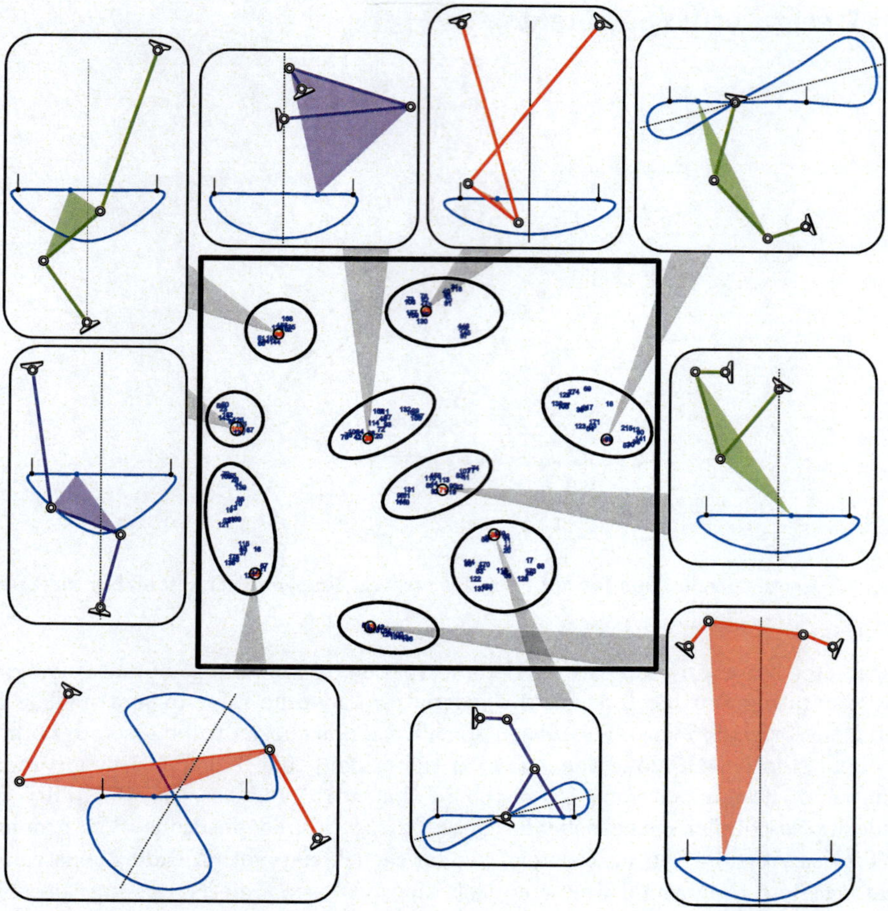

Fig. 4. An atlas of four-bar linkages that generate an approximate straight line segment visualized using t-SNE. Of these, cognate #2 symmetric four-bars (shown in red) are found via numerical continuation experiments while the other two Roberts cognates are readily computed using the appropriate transformations.

data in 2D. Using the link dimensions to represent each four-bar linkage and setting the hyper-parameter of t-SNE, namely, *perplexity*, at 5, Fig. 4 is produced. It shows bunches of linkages that qualitatively resemble classical straight line generators as observed in the representative set of nine four-bar linkages displayed to scale. This computational approach produced many that serve as a useful atlas for designers.

5 Conclusion

In this paper, the synthesis equations were formulated, characterized, and solved for a four-bar linkage with ground pivots specified to produce a desired symmetric coupler curve approximately. The solution is applied to search for four-bar approximate straight line generators by solving an optimization problem. The mathematical model is restricted to one particular class of four-bars which generate symmetric coupler curves to reduce the computational challenges associated with solving a generic four-bar linkage system. The validity of our approach is affirmed by rediscovering the classical approximate straight line generators by solving an easier system. In addition, we found more approximate straight line generators, each of which seems to be a variant of the classical linkages, but with substantially different dimensions. Using the t-SNE algorithm, our results are organized into a 2D atlas, which could be a useful reference for mechanical designers in need of more straight line options. A future direction related to this paper would be to more thoroughly investigate what we term as Class A linkages as well as extending to generic four-bar linkages.

Acknowledgements. This material is based upon work supported by the National Science Foundation under Grant Nos. CMMI-2041789 and CMMI-2144732. In addition, the authors thank Parker Edwards for discussions related to t-SNE, and Caroline Hills for discussions about the work in general.

References

1. Wampler, C.W., Morgan, A.P., Sommese, A.J.: Complete solution of the nine-point path synthesis problem for four-bar linkages. J. Mech. Des. **114**(1), 153–159 (1992)
2. Sánchez Marín, F.T., Pérez González, A.: Global optimization in path synthesis based on design space reduction. Mech. Mach. Theory **38**(6), 579–594 (2003)
3. Nolle, H.: Linkage coupler curve synthesis: a historical review-I. Developments up to 1875. Mech. Mach. Theory **9**(2), 147–168 (1974)
4. Wampler, C.W.: Isotropic coordinates, circularity, and Bézout numbers: planar kinematics from a new perspective. In: Proceedings of the ASME 1996 Design Engineering Technical Conferences and Computers in Engineering Conference. Volume 2A: 24th Biennial Mechanisms Conference, Irvine, California, USA, 18-22 August 1996 (1996)
5. Hartenberg, R.S., Denavit, J.: Kinematic Synthesis of Linkages. McGraw-Hill Book Company, New York (1964)
6. Natesan, A.K.: Kinematic analysis and synthesis of four-bar mechanisms for straight line coupler curves. Thesis. Rochester Institute of Technology (1994). https://scholarworks.rit.edu/theses/4658
7. Bates, D.J., Hauenstein, J.D., Sommese, A.J., Wampler, C.W.: Bertini: software for Numerical Algebraic Geometry. bertini.nd.edu
8. Bates, D.J., Hauenstein, J.D., Sommese, A.J., Wampler, C.W.: Numerically Solving Polynomial Systems with Bertini. SIAM, Philadelphia (2013)
9. Morgan, A., Sommese, A.: A homotopy for solving general polynomial systems that respects m-homogeneous structures. Appl. Math. Comput. **24**(2), 101–113 (1987)
10. Van der Maaten, L., Hinton, G.: Visualizing data using t-SNE. J. Mach. Learn. Res. **9**(11) (2008)

A Method for the Complete Set of Solutions of the Finite-Position Synthesis Problem

Alba Perez Gracia(✉)

Universitat Politecnica de Catalunya (UPC), Barcelona, Spain
alba.perez.gracia@upc.edu

Abstract. The equations that appear in the formulation of finite-position synthesis problems tend to increase their complexity and become quite large when solving for multiple-joint systems. In this work, we create a new set of equations that are suited to find all solutions for very large synthesis problems, targeting interval branch-and-cut algorithms. We obtain a system of equations in which all the variables are naturally bounded. Here we present the formulation and preliminary results, while its application to interval analysis is still a work in progress.

1 Introduction

The formulation of the finite-position synthesis of spatial chains problem [6] leads to complex equations that, except in the simplest cases, need further simplification. Regardless of the simplification or elimination method, the obtained set of equations increases its complexity quickly when adding joints to the chain.

The interest for finding faster and more complete methods for solving the complex equations that are obtained in kinematics problems, let those be synthesis, forward or inverse kinematics, has been continuous. Algebraic methods based on resultants were first developed, however they are limited to relatively simple systems. Similarly, methods based on algebraic geometry yield good insights [7] but are for medium-size problems at most. Continuation methods [9] increased the size of the problems that can be solved, and have been refined in multiple ways. Among the most interesting ones are those that can solve systems with Bezout numbers in the hundreds of milions, thanks to finite-root generation [5] or cyclic coefficient-parameter continuation [1], which avoid the multiple degenerate roots to compute the finite subsets of solutions.

Interval algorithms [8] may be the best option in cases such as the synthesis of the 5R chain, with an initial Bezout bound of more than 10^{100} using our formulation. However, they have the limitation of having to define initial intervals for the variables [3], which are also important in the performance of the method. In the case of synthesis problems, it is not possible to know the length of those intervals if they are to contain all solutions. This has limited the results to a certain range of values.

O. Altuzarra and A. Kecskémthy (Eds.): ARK 2022, SPAR 24, pp. 286–293, 2022.
https://doi.org/10.1007/978-3-031-08140-8_31

In this work we follow our previous formulation of the synthesis problem [4], but manipulate the equations differently to obtain a set of equations that are a function of bound variables. It is then possible to use an interval solver then to find all solutions to the synthesis problem. The aim is the complete set of solutions of the nR serial chains, with up to five joints.

2 Methodology

Given a serial chain formed by joint axes $S_j = s_j + \epsilon s_{0j}$ of joint variables θ_j, write the relative forward kinematics as the product of exponentials of the joint axes. Equate to a relative position P^{0i},

$$e^{S_1\theta_1^i} \dots e^{S_{k-1}\theta_{k-1}^i} e^{S_k\theta_k^i} = P^{0i}, \tag{1}$$

and solve for both the joint axes and the joint variables to solve the function-to-form problem. This is the well-known parameterized set of equations presented by Mavroidis [2] to generalize Burmester's theory. For our purposes, it is convenient to split the serial chain at the previous-to-last joint,

$$e^{S_1\theta_1^i} \dots e^{S_{k-1}\theta_{k-1}^i} = P^{0i} e^{-S_k\theta_k^i}, \tag{2}$$

and use these as our design equations.

Notice that, while the directions of the joint axes are bound by the unit condition $s_j \cdot s_j = 1$, the components of the moment of the joint axes, s_{0j}, can take any value.

In order to obtain a final set of equations as a function of bound variables, we use standard elimination of the structural parameters of each S_j. We obtain a system of linear equations,

$$\left[M(P^{0i}, \theta_j^i)\right] \begin{Bmatrix} S_1 \\ \vdots \\ S_j \\ S_{12} \\ \vdots \end{Bmatrix} = \left\{V(P^{0i}, \theta_j^i)\right\}, \quad j = 1 \dots k, \tag{3}$$

where $S_{12} = S_1 S_2$ is the result of the Clifford product of the axes. This arrangement will be later used, in a modified way, to linearly obtain the structural parameters.

The design equations are obtained by imposing the Plucker conditions on the obtained expressions of the joint axes. These are a function of the joint variables θ_j^i, which are always bound. The final set of equations F is

$$F(P^{0i}, \theta_j^i) = \begin{cases} s_j \cdot s_j - 1 \\ s_j \cdot s_{0j} \\ s_{12} - s_1 \times s_2 \\ s_{12}^0 - s_1 \times s_{20} + s_{10} \times s_2 \\ \vdots \end{cases}, \quad j = 1 \dots k. \tag{4}$$

The degree of these equations on the joint variables θ_j^i depends on the number of joints of the serial chain. Here we present the development from the simplest case, the 2R serial chain, to the most complex one, the serial 5R chain.

3 The 2R Chain

Let the two joint axes of the chain be S_1 and S_2, with rotation angles θ_1 and θ_2. Define two relative displacements of the chain, P^{01} and P^{02}. The forward kinematics equations must reach each of those positions, and expressing them as in Eq. (2),

$$e^{S_1\theta_1^i} = P^{0i}e^{-S_2\theta_2^i}, \quad i = 1, 2, \tag{5}$$

they can be used to linearly solve for the structural variables as a function of the joint variables. In the case of the RR chain, we obtain the following linear system,

$$\left[M(P^{01}, P^{02}, \theta)\right] \begin{Bmatrix} s_{1x} \\ s_{1y} \\ \vdots \\ s_{2z0} \end{Bmatrix} = \left\{V(P^{01}, P^{02}, \theta)\right\}. \tag{6}$$

Here $\left[M(P^{01}, P^{02}, \theta)\right]$ is a 12×16 matrix with coefficients shown in Eq. (9), $\left\{V(P^{01}, P^{02}, \theta)\right\}$ is a 16×1 vector shown in Eq. (10), and $\theta = (\theta_1^1, \theta_2^1, \theta_1^2, \theta_2^2)$ are the joint variables.

The system in Eq. (6) has two dependent equations, from the two conditions that ensure the unit dual quaternion. These are redundant with the expression of each joint as a rotation about a line, up to a sign. Because of this, we can eliminate two equations and obtain a square 12×12 matrix $[M_{ind}]$ (shown in Eq. (9)) and a 12×1 vector V_{ind} (Eq. (10)) that can be solved for the joint axes S_1 and S_2,

$$\begin{Bmatrix} S_1 \\ S_2 \end{Bmatrix} = \left[M_{ind}(P^{01}, P^{02}, \theta)\right]^{-1} V_{ind}(P^{01}, P^{02}, \theta). \tag{7}$$

To these solutions for the joint axes, which are parameterized by the joint variables, we need to impose the line conditions in order for the system to maintain the structure of a displacement. These conditions form the design equations for the synthesis, which are quadratic in structure. However we need to keep in mind that each of the joint axis is expressed as a sixth-degree function of the joint variables. The equations to minimize are

$$F(P^{01}, P^{02}, \theta) = \begin{cases} s_1 \cdot s_1 = 1, \\ s_2 \cdot s_2 = 1, \\ s_1 \cdot s_{10} = 0, \\ s_2 \cdot s_{20} = 0. \end{cases} \tag{8}$$

The equations in (8) define a variety of dimension zero, as we have four independent equations in the four unknowns $\theta_1^1, \theta_2^1, \theta_1^2, \theta_1^2$. Before further simplification, the equations have degree twelve in both numerator and denominator.

$$\left[M_{ind}(P^{01}, P^{02}, \theta)\right] = \tag{9}$$

$$\begin{bmatrix}
s_{\frac{\theta_1^1}{2}} & 0 & 0 & 0 & 0 & 0 & p_w s_{\frac{\theta_2^1}{2}} & -p_z s_{\frac{\theta_2^1}{2}} & p_y s_{\frac{\theta_2^1}{2}} & 0 & 0 & 0 \\
0 & s_{\frac{\theta_1^1}{2}} & 0 & 0 & 0 & 0 & p_z s_{\frac{\theta_2^1}{2}} & p_w s_{\frac{\theta_2^1}{2}} & -p_x s_{\frac{\theta_2^1}{2}} & 0 & 0 & 0 \\
0 & 0 & s_{\frac{\theta_1^1}{2}} & 0 & 0 & 0 & -p_y s_{\frac{\theta_2^1}{2}} & p_x s_{\frac{\theta_2^1}{2}} & p_w s_{\frac{\theta_2^1}{2}} & 0 & 0 & 0 \\
0 & 0 & 0 & 0 & 0 & 0 & -p_x s_{\frac{\theta_2^1}{2}} & -p_y s_{\frac{\theta_2^1}{2}} & -p_z s_{\frac{\theta_2^1}{2}} & 0 & 0 & 0 \\
0 & 0 & 0 & s_{\frac{\theta_1^1}{2}} & 0 & 0 & p_{w0} s_{\frac{\theta_2^1}{2}} & -p_{z0} s_{\frac{\theta_2^1}{2}} & p_{y0} s_{\frac{\theta_2^1}{2}} & p_w s_{\frac{\theta_2^1}{2}} & -p_z s_{\frac{\theta_2^1}{2}} & p_y s_{\frac{\theta_2^1}{2}} \\
0 & 0 & 0 & 0 & s_{\frac{\theta_1^1}{2}} & 0 & p_{z0} s_{\frac{\theta_2^1}{2}} & p_{w0} s_{\frac{\theta_2^1}{2}} & -p_{x0} s_{\frac{\theta_2^1}{2}} & p_z s_{\frac{\theta_2^1}{2}} & p_w s_{\frac{\theta_2^1}{2}} & -p_x s_{\frac{\theta_2^1}{2}} \\
0 & 0 & 0 & 0 & 0 & s_{\frac{\theta_1^1}{2}} & -p_{y0} s_{\frac{\theta_2^1}{2}} & p_{x0} s_{\frac{\theta_2^1}{2}} & p_{w0} s_{\frac{\theta_2^1}{2}} & -p_y s_{\frac{\theta_2^1}{2}} & p_x s_{\frac{\theta_2^1}{2}} & p_w s_{\frac{\theta_2^1}{2}} \\
0 & 0 & 0 & 0 & 0 & 0 & -p_{x0} s_{\frac{\theta_2^1}{2}} & -p_{y0} s_{\frac{\theta_2^1}{2}} & -p_{z0} s_{\frac{\theta_2^1}{2}} & -p_x s_{\frac{\theta_2^1}{2}} & -p_y s_{\frac{\theta_2^1}{2}} & -p_z s_{\frac{\theta_2^1}{2}} \\
s_{\frac{\theta_1^2}{2}} & 0 & 0 & 0 & 0 & 0 & q_w s_{\frac{\theta_2^2}{2}} & -q_z s_{\frac{\theta_2^2}{2}} & q_y s_{\frac{\theta_2^2}{2}} & 0 & 0 & 0 \\
0 & s_{\frac{\theta_1^2}{2}} & 0 & 0 & 0 & 0 & q_z s_{\frac{\theta_2^2}{2}} & q_w s_{\frac{\theta_2^2}{2}} & -q_x s_{\frac{\theta_2^2}{2}} & 0 & 0 & 0 \\
0 & 0 & s_{\frac{\theta_1^2}{2}} & 0 & 0 & 0 & -q_y s_{\frac{\theta_2^2}{2}} & q_x s_{\frac{\theta_2^2}{2}} & q_w s_{\frac{\theta_2^2}{2}} & 0 & 0 & 0 \\
0 & 0 & 0 & 0 & 0 & 0 & -q_x s_{\frac{\theta_2^2}{2}} & -q_y s_{\frac{\theta_2^2}{2}} & -q_z s_{\frac{\theta_2^2}{2}} & 0 & 0 & 0 \\
0 & 0 & 0 & s_{\frac{\theta_1^2}{2}} & 0 & 0 & q_{w0} s_{\frac{\theta_2^2}{2}} & -q_{z0} s_{\frac{\theta_2^2}{2}} & q_{y0} s_{\frac{\theta_2^2}{2}} & q_w s_{\frac{\theta_2^2}{2}} & -q_z s_{\frac{\theta_2^2}{2}} & q_y s_{\frac{\theta_2^2}{2}} \\
0 & 0 & 0 & 0 & s_{\frac{\theta_1^2}{2}} & 0 & q_{z0} s_{\frac{\theta_2^2}{2}} & q_{w0} s_{\frac{\theta_2^2}{2}} & -q_{x0} s_{\frac{\theta_2^2}{2}} & q_z s_{\frac{\theta_2^2}{2}} & q_w s_{\frac{\theta_2^2}{2}} & -q_x s_{\frac{\theta_2^2}{2}} \\
0 & 0 & 0 & 0 & 0 & s_{\frac{\theta_1^2}{2}} & -q_{y0} s_{\frac{\theta_2^2}{2}} & q_{x0} s_{\frac{\theta_2^2}{2}} & q_{w0} s_{\frac{\theta_2^2}{2}} & -q_y s_{\frac{\theta_2^2}{2}} & q_x s_{\frac{\theta_2^2}{2}} & q_w s_{\frac{\theta_2^2}{2}} \\
0 & 0 & 0 & 0 & 0 & 0 & -q_{x0} s_{\frac{\theta_2^2}{2}} & -q_{y0} s_{\frac{\theta_2^2}{2}} & -q_{z0} s_{\frac{\theta_2^2}{2}} & -q_x s_{\frac{\theta_2^2}{2}} & -q_y s_{\frac{\theta_2^2}{2}} & -q_z s_{\frac{\theta_2^2}{2}}
\end{bmatrix}$$

$$V_{ind}(P^{01}, P^{02}, \theta) = \begin{Bmatrix} p_x c_{\frac{\theta_2^1}{2}} \\ p_y c_{\frac{\theta_2^1}{2}} \\ \vdots \\ q_{w0} c_{\frac{\theta_2^2}{2}} \end{Bmatrix}. \tag{10}$$

4 The 3R Chain

The synthesis of the spatial 3R chain is a more complex problem with a higher number of solutions. Let its three joint axes be S_i, with rotation angles θ_i, $i = 1, \ldots 3$. It is well known that 4 relative displacements P^{0j}, $j = 1, \ldots 4$, define the workspace for the 3R chain.

The forward kinematics equations must reach those positions,

$$e^{S_1 \theta_1^i} e^{S_2 \theta_1^i} = P^{0i} e^{-S_3 \theta_3^i}, \quad i = 1, \ldots, 4, \tag{11}$$

and are used to linearly solve for the structural variables as a function of the joint variables. In the case of the 3R chain, we obtain the following linear system,

$$\left[M(P, \theta)\right] \begin{Bmatrix} S_1 \\ S_2 \\ S_{12} \\ S_3 \end{Bmatrix} = \left\{V(P, \theta)\right\}, \quad i = 1, 4. \tag{12}$$

Here $[M(P, \theta)]$ is a 24×24 matrix and $\{V(P, \theta)\}$ is a 24×1 vector, $\theta = (\theta_1^1, \ldots, \theta_3^4)$ are the joint variables, and $P = (P^{01}, \ldots, P^{04})$ are the target positions. We define the dual vector $S_{12} = S_1 \times S_2$.

The $[M(P, \theta)]$ matrix becomes too big to include it here. Equation (13) shows elements m_{11} and $m_{24,24}$ of this matrix; the rest of them have a similar structure.

$$m_{11} = \sin \frac{\theta_{11}}{2} \cos \frac{\theta_{21}}{2}, \qquad m_{24,24} = p_{4w} \sin \frac{\theta_{34}}{2}. \tag{13}$$

We can solve linearly for the 24 structural variables defined in Eq. (12). The resulting expressions are functions of the joint variables θ_i^j.

The design equations are defined by the Plucker conditions of the axes plus the fact that $S_{12} = S_1 \times S_2$; a total of 24 quadratic equations (Eq. 14). We check the validity and convergence of the equations with a numerical solver and random positions, obtaining a single RRR chain and inverse kinematics solution for each run.

$$F(P^{0i}, \theta_j^i) = \begin{cases} s_j \cdot s_j = 1, \\ s_j \cdot s_{j0} = 0, \quad j = 1, 2, 3, \\ s_{12} = s_1 \times s_2, \\ s_{12}^0 = s_1 \times s_{20} + s_{10} \times s_2. \end{cases} \tag{14}$$

5 The 4R and 5R Chain

The complexity increases when we are dealing with the 4R and especially the 5R chains. For both cases, let the axes of the chain be $S_i, i = 1, \ldots n$, with rotation angles θ_i. It is well known that $k = 8$ relative displacements for the case of the 4R chain and $k = 20$ relative displacements for the 5R chain define the workspace for each chain. We denote those as P^{0j}, $j = 1, \ldots k$.

We use the following expression for the forward kinematics equations,

$$e^{S_1 \theta_1^i} \ldots e^{S_{n-1} \theta_{n-1}^i} = P^{0i} e^{-S_n \theta_n^i}, \quad i = 1, \ldots, k, \tag{15}$$

which are used to linearly solve for the combinations of structural variables as a function of the joint variables. In the case of the 4R chain, the system of Eq. (3) consists of a 48×48 matrix and 48×1 vector, and in the case of the 5R, the linear system yields a 210×210 matrix and a 210×1 vector, with the structure shown in Eq. (3), joint variables $\theta = (\theta_1^1, \ldots, \theta_n^k)$ and target positions $P = (P^{01}, \ldots, P^{0k})$.

These linear systems have considerable dimension but their coefficients are still relatively simple, so that the calculations are fast. A couple of elements of the matrix for the 5R chain are shown in Eq. (16),

$$m_{11} = \sin \frac{\theta_{11}}{2} \cos \frac{\theta_{21}}{2} \cos \frac{\theta_{31}}{2} \cos \frac{\theta_{41}}{2}, \qquad m_{210,210} = p_{20,w} \sin \frac{\theta_{5,20}}{2} \tag{16}$$

The set of Eq. (4) consists of 32 quadratic equations for the spatial 4R chain, and 76 quadratic equations for the case of the spatial 5R chain.

6 Implementation and Results

To check the validity of the method, the equations have been tested for the 2R, 3R and 4R cases with a numerical solver, following the scheme shown below,

Solver

`Initialization` $P = (P^{01}, \ldots, P^{0j}), \qquad \Theta = (\theta_1^1, \ldots, \theta_k^j) = \Theta_0$

`Minimization`

$$\begin{Bmatrix} S_1 \\ \vdots \end{Bmatrix} = \Big[M_{ind}(P, \Theta)\Big]^{-1} V_{ind}(P, \Theta) \tag{17}$$

minimize: $\quad f = \sqrt{F(P, \Theta)^2} = \sqrt{\sum_{i=1,2} (s_i \cdot s_i - 1)^2 + (s_i \cdot s_{i0})^2 + \ldots}$

`until f <` ϵ

The equations are implemented in *Matlab*'s numerical solver *lsqnonlin* with the Levenberg-Marquardt algorithm. For the 2R, the unique two solutions are recovered easily. For the 3R, a set of 1000 runs finds 8 different solutions consistently. These are shown reaching the first position in Fig. 1 with values in Table 1, illustrating the fact that it is complicated to set some a-priory bounds to the robot dimensions. For all of them the solver arrives to a solution in less than one second.

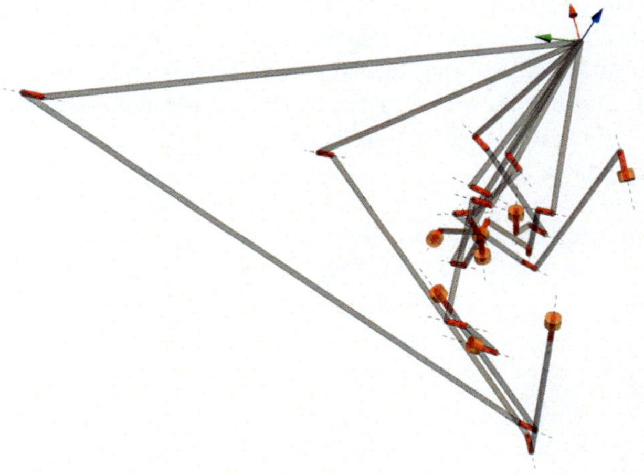

Fig. 1. Eigth solutions for the 3R chain at the first goal position, joints depicted in red and links in grey. It is hard to have a correct initial guess on the joint parameters of the potential solutions.

Table 1. The four relative positions and two extreme solutions for the 3R problem

Positions	$P_{01} = 0.85 + 0.26i + 0.22j + 0.38k + \epsilon(-1.41i - 0.75j - 1.36k + 1.23)$
	$P_{02} = -0.20 - 0.90i + 0.22j + 0.30k + \epsilon(-0.03i - 0.90j - 1.68k - 3.37)$
	$P_{03} = -0.84 - 0.32i + 0.01j + 0.43k + \epsilon(-2.58i + 0.41j + 3.07k + 2.56)$
	$P_{04} = 0.42 + 0.57i - 0.53j - 0.46k + \epsilon(-1.07i - 0.52j - 1.52k - 0.91)$
Solution 1	$S_1 = (0.18, 0.24, -0.95) + \epsilon(13.33, -7.87, 0.48)$
	$S_2 = (0.01, 0.23, 0.97) + \epsilon(-2.28, 3.79, -0.85)$
	$S_3 = (-0.87, 0.16, 0.46) + \epsilon(6.76, -2.36, 13.82)$
Solution 2	$S_1 = (0.94, 0.27, 0.19) + \epsilon(0.93, -2.95, -0.44)$
	$S_2 = (0.60, 0.37, -0.71) + \epsilon(-2.28, 3.79, -0.85)$
	$S_3 = (-0.20, -0.97, 0.14) + \epsilon(0.60 - 0.43, -2.21)$

However solving the synthesis equations numerically is nothing new. These equations have been designed so that they can be applied to an interval solver such as [8], because of the fact that the variables are bound. The next step is to implement them in a branch-and-bound interval algorithm in order to obtain the complete set of solutions. We expect the equations to behave well in the solver, but it is still a work in progress.

7 Conclusions

A new formulation for the finite-position synthesis of serial chains is presented, that allows bounding the unknowns in order to find the complete set of solutions. In addition, this formulation allows including joint angle limits in a very natural way, as well as adding geometric relations between joint axes. The equations behave well in numerical solvers, and the next step is to implement them in an interval branch-and-cut solver in order to obtain the complete set of solutions for the nR spatial chains, with $n \leq 5$. In this sense, it is a work in progress that we expect to complete soon.

Acknowledgements. This work has been partially supported by Agencia Estatal de Investigación under project PID2020-117509GB-I00/AEI/10.13039/501100011033.

References

1. Baskar, A., Bandyopadhyay, S.: An algorithm to compute the finite roots of large systems of polynomial equations arising in kinematic synthesis. Mech. Mach. Theory **133**, 493–513 (2019). https://doi.org/10.1016/j.mechmachtheory.2018.12.004
2. Lee, E., Mavroidis, C.: Geometric design of 3R robot manipulators for reaching four end-effector spatial poses. Int. J. Robot. Res. **23**(3), 247–254 (2004)
3. Lee, E., Mavroidis, C., Merlet, J.P.: Five precision point synthesis of spatial RRR manipulators using interval analysis. J. Mech. Des. **126**, 842–849 (2004)

4. Perez-Gracia, A., McCarthy, J.: Kinematic synthesis of spatial serial chains using Clifford algebra exponentials. Proc. Inst. Mech. Eng. C J. Mech. Eng. Sci. **220**(7), 953–968 (2006). https://doi.org/10.1243/09544062JMES166
5. Plecnik, M.M., Fearing, R.S.: Finding only finite roots to large kinematic synthesis systems. J. Mech. Robot. **9**(2) (2017). https://doi.org/10.1115/1.4035967
6. Roth, B.: Finite-position theory applied to mechanism synthesis. J. Appl. Mech. **34**(3), 599–605 (1967). https://doi.org/10.1115/1.3607749
7. Schrocker, H., Pfurner, M., Schadlbauer, J.: Constraint varieties in mechanism science. In: St. John, A., Sitharam, M., Sidman, J. (eds.) Handbook of Geometric Constraint Systems Principles. CRC Press (2018)
8. Shabani, A., Sarabandi, S., Porta, J., Thomas, F.: A fast branch-and-prune algorithm for the position analysis of spherical mechanisms. In: IFToMM WC 2019: Advances in Mechanism and Machine Science, pp. 549–558 (2019)
9. Wampler, C.W., Morgan, A.P., Sommese, A.J.: Numerical continuation methods for solving polynomial systems arising in kinematics. J. Mech. Des. **112**(1), 59–68 (1990). https://doi.org/10.1115/1.2912579

Classification of Mobilities - New Insights on an Old Topic

Thiago Hoeltgebaum(✉), Luan Meneghini, Vinicius N. Artmann, and Daniel Martins

Universidade Federal de Santa Catarina, R. Eng. Agronômico Andrei Cristian Ferreira s/n - Trindade, Florianópolis, SC 88040-900, Brazil
thiagohbaum@gmail.com

Abstract. The topic mobility of mechanisms has been studied for more than 150 years and there is still space to enhance its understanding. Davies' Method integrates graph and screw theory to generate matrices that represent the mechanism statics and/or kinematics. These tools help create new definitions. This paper brings a literature review highlighting the relevance of the subject with respect to mechanism science. Davies' Method is presented to set the basic knowledge for the developments proposed by this paper, focusing on the structures that define the kinematics of mechanisms and providing an alternative view for the mobility topic. Thus, it is possible to define, in a proper form, distinct types of mobility: Actuated, Passive, and Adjustment.

1 Introduction

The mobility of mechanisms has been an academic topic since the second half of 19th Century. Chebyshev [1], Gruebler [2], Kutzbach [3], and Malyshev [4] had their names perpetuated into equations that are still in use nowadays.

With the evolution of mechanism representation methods and the advent of computer science, other contributions emerged using kinematic constraint equations and rank evaluations to determine mobility (*e.g.* Davies [5]).

Consequently, one might think that *Mobility of Mechanisms* is a topic of the past, that there is no room for new propositions. This paper shows exactly the opposite. By integrating advanced analysis techniques (*i.e.* graph and screw theory) it is possible to enhance the understanding and create novel perspectives about mobility[1].

[1] In this paper, mobility and degree-of-freedom are used as synonyms and all the developments are valid for parallel mechanisms.

O. Altuzarra and A. Kecskéméthy (Eds.): ARK 2022, SPAR 24, pp. 294–302, 2022.
https://doi.org/10.1007/978-3-031-08140-8_32

A survey was performed within the archives of one of the most prominent journals in the field, the Mechanism and Machine Theory Journal (MMT), in order to investigate the number and contents related to the mobility of mechanisms.

Gogu [6] presented a chronological and critical mobility review comprising the whole mechanism science history until 2004. Thus, this present paper does not intend to focus on history, but rather to investigate the novelty of mobility publications. Hence, this survey was directed to the subsequent period of 2004 to 2021.

Papers in mechanism theory, use the mobility in either one of the three possibilities:

- **Analysis:** Properties such as workspace, number of links and joints are taken into account and the mobility is the output of the process;
- **Synthesis:** A mechanism is designed using the degrees of freedom and further characteristics as an input;
- **New Equations/Methods:** The authors propose novel methods that can evaluate the mobility of mechanisms, which is the central aspect of the research. The results of the present survey belong to this specific item.

By examining the MMT archives, there are at least 18 publications presenting novel methods to compute the mechanism mobility. Most of the developments are related to the solution of mobility for parallel mechanisms: [7–15].

Other researchers focused on the over-constrained mechanisms [16–19], deployable structures [20], compliant mechanisms [21], and even for a double loop single DOF kinematic chains [22].

This initial survey shows that researches about mobility are still relevant to mechanism science and therefore, new points of view, perspectives and methods are welcome to increase the understanding about the subject.

2 Davies' Method

By using Graph and Screw Theory, Davies adapted Kirchhoff's circulation and cutset laws to multibody systems enabling the mechanism representation through matrices and the evaluation of kinematics and statics [23]. Davies' Method became an important tool for the analysis and synthesis of mechanisms. More details about the method can also be found at: [24–26].

For this paper, it is sufficient to understand the method output: the M_N matrix. The properties and characteristics of such matrix will be explained through the example of a scooter suspension (Fig. 1).

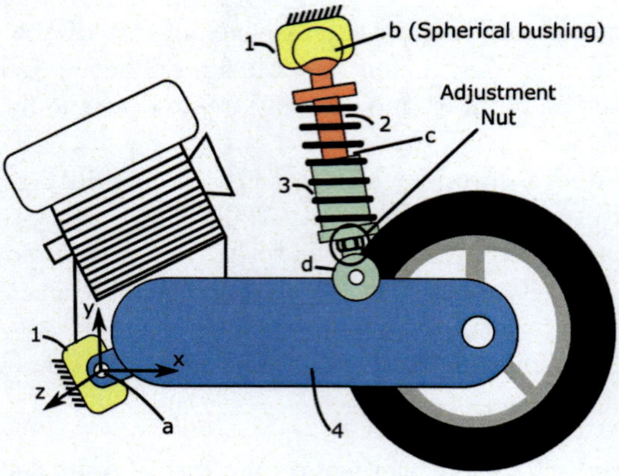

Fig. 1. Rear suspension of the Suzuki Burgman AN 125 scooter. This four-bar linkage is composed of four joints: a and d are revolute, b is spherical bushing, and c is a cylindrical pair. The adjustment nut is used to tune the suspension ride-height.

The dimension of a M_N matrix is $\lambda\nu \times F$, where λ is the screw system order, ν is the number of closed loops, and F is the total number of freedoms present in the mechanism. Thus, every column of Matrix M_N is a motion screw, $\$_i$, associated to a mechanism freedom.

It is important to remind the difference between *freedom* and *mobility. Freedom* is considered at joint level and it is analyzed how many movements a specific joint allows. *Mobility*, on the other hand, is considered at the mechanism level and is analyzed how many independent coordinates are needed to be defined so the mechanisms can properly work.

The suspension system in Fig. 1 has the following freedoms:

- Joint a is revolute, it only allows one rotation around the z-axis;
- Joint b is spherical, allowing three rotations around x, y, z axes;
- Joint c is cylindrical, allowing one rotation and one translation in x-axis;
- Joint d is revolute, it only allows one rotation around the z-axis;

Therefore, the M_N matrix in Eq. 1[2] has seven columns, accordingly labeled.

$$[M_N]_{6\times 7} = \begin{array}{c} \begin{matrix} \$_{at} & \$_{br} & \$_{bs} & \$_{bt} & \$_{cs} & \$_{cu} & \$_{dt} \end{matrix} \\ \left[\begin{array}{c|ccc|cc|c} 0 & 1 & 0 & 0 & 0 & 0 & 0 \\ 0 & 0 & 1 & 0 & 1 & 0 & 0 \\ 1 & 0 & 0 & 1 & 0 & 0 & 1 \\ 0 & 0 & z_b & -y_b & z_c & 0 & -y_d \\ -z_a & -z_b & 0 & x_b & 0 & 1 & x_d \\ y_a & y_b & -x_b & 0 & -x_c & 0 & 0 \end{array}\right] \end{array} \quad (1)$$

[2] Subscripts r, s, t denote rotations around x, y, z axes and u, v, w denote translations along x, y, z axes, respectively [27].

Every mechanism can be represented by a M_N matrix, being possible to simulate its kinematics, investigate singularities, evaluate the redundant constraints and analyze different aspects of the mechanism mobility. In Sect. 3, The scooter suspension mechanism will also be used to exemplify how the M_N is useful to evaluate the mobilities of a mechanism.

3 Classification of Mobilities

According to the IFToMM[3], the mobility or degree of freedom (DoF) is the number of independent coordinates to define the configuration of a kinematic chain or mechanism [29].

The mobility is often related to the number of required actuations for the mechanism operation; for instance levers, motors, pistons are widely used in mechanisms and machines as actuators. However, this paper brings a different and more strict interpretation of mobility by the point of view of Davies' Method: Mobility is the number of **linear dependent** motion screws present in the M_N matrix.

Take the example of a race-car rear wing adjustment system in Fig 2. The rank of its M_N matrix (Eq. 2) is $r = 3$, which means that there are **three linear independent columns**, and therefore, **one linear dependent motion screw**. It is known that such mechanism has one degree of freedom which agrees with the rank evaluation.

Usually, rank evaluations through M_N matrices result in infinitesimal or local mobilities, since the method uses specific joint coordinates. However, it is possible to use the *mixed matrix* and *generic rank* defined by Murota [28] to calculate the finite/global mobility. A mixed matrix is composed by two types of numbers: the accurate numbers (normally 0 or 1), and the inaccurate numbers, or system parameters, such as the joint coordinates in the M_N matrix.

The generic rank is defined as the maximum rank in mixed matrix. This definition enables the evaluation of the global mobility since excludes the possibilities of singularities.

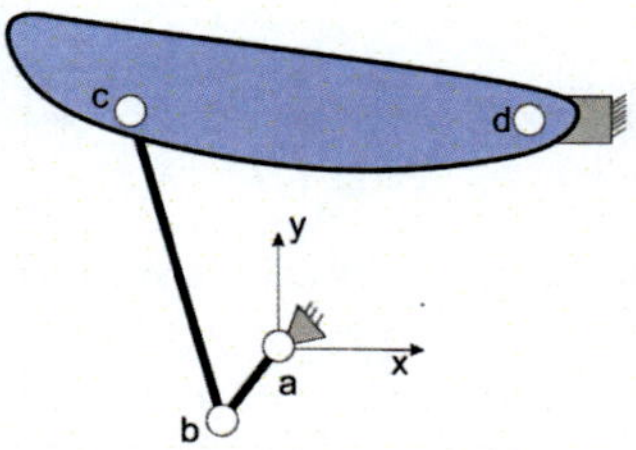

Fig. 2. Schematic mechanism of a race-car rear wing adjustment. This four-bar linkage is composed of four revolute joints: a, b, c, and d.

[3] IFTOMM stands for International Federation for the Promotion of Mechanism and Machine Science.

$$[M_N]_{6\times4} = \begin{array}{c} \begin{array}{cccc} \$_{at} & \$_{bt} & \$_{ct} & \$_{dt} \end{array} \\ \left[\begin{array}{c|c|c|c} 0 & 0 & 0 & 0 \\ 0 & 0 & 0 & 0 \\ 1 & 1 & 1 & 1 \\ -y_a & -y_b & -y_c & -y_d \\ x_a & x_b & x_c & x_d \\ 0 & 0 & 0 & 0 \end{array}\right] \end{array} \quad (2)$$

This interpretation allows the definition of different types of mobility. Take for example the scooter suspension mechanism in Fig 1. The system is expected to have one degree-of-freedom. However, the rank of M_N matrix (Eq. 1) is $r = 5$, thus containing two linear dependent motion screws. The identification of such extra DOF led to the following classification of mobilities:

- *Actuated Mobility*;
- *Passive Mobility*;
- *Adjustment Mobility.*

The *actuated mobility* is an input of power, force, or movement used to guarantee the working principle of a mechanism. Electric motors and hydraulic cylinders are examples of components used to actuate a kinematic pair, as shows Fig. 3.

An actuated mobility is found also in Fig. 1. The spring actuates the joint c leveling the suspension and absorbing the road impacts. The motion screw, $\$_{cu}$ is linear dependent to the following motion screws: $\$_{at}$, $\$_{bt}$. $\$_{dt}$

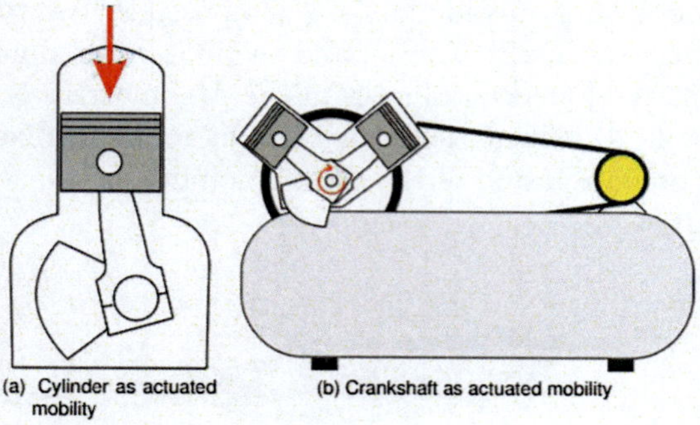

Fig. 3. Examples of actuated mobility.

Passive mobility is present while the mechanism is functioning, but it does not require actuation nor interferes with the output or the kinematics. Even tough the definition is new, passive mobilities widely occur in linkages, sometimes without the knowledge or intention of the project engineer.

Such mobility might enhance self-aligning characteristics but can also induce redundant constraints. For instance, when two spherical joints are connected in series or when a spherical and a cylindrical pair are adjacent, a passive mobility occurs. The latter is present in the suspension mechanism from Fig. 1.

The M_N matrix (Eq. 2) contains two linear dependent motion screws and, therefore, two degrees-of-freedoms. One is related to the mechanism actuation, as previously discussed.

The other degree of freedom is a passive mobility. Joint b is spherical bushing and joint c is cylindrical. It is possible to verify that the motion screw $\$_{bs}$ is linearly dependent to $\$_{cs}$ since $z_b = z_c$ and $xz_b = x_c$. Thus, it is possible to rotate link 2 around its own axis. This characteristic does not change the mechanism main kinematics nor requires actuation.

The *adjustment mobility* usually exists to guarantee the geometric position of a given link, but it does not interfere with the mechanism's functioning. The input of the adjustment mobility is not constantly required .

The issue with this type of mobility is that it is not always easy to notice in a mechanism. Figure 4 brings an example of the adjustment mobility in a steering system with the following characteristics:

- Link 220 is the steering rack, which is fixed to the chassis and promotes linear movement of link 230 culminating in wheel steering.
- Link 320 is the tie-rod. It connects the steering rack to the wheel pivot (120).
- Joints 303 and 125 are spherical joints.

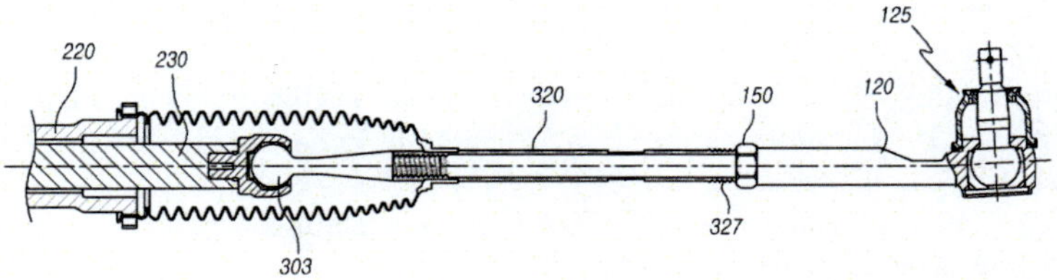

Fig. 4. Adjustment mobility present at a steering tie-rod. Adapted from [30]

Between the tie-rod (320) and the pivot (120) there is a threaded connection (327) with a nut (150). Normally, the toe-in set-up is made by adjusting the distance of spherical joints (303) and (125) using this thread (327). When the system is mounted, the nut (150) locks this adjustment mobility.

An adjustment mobility is also present in Fig. 1. Depending on the weight to be carried by the scooter it is possible to adjust the suspension length and pre-load of the spring. The adjustment mobility, $\$_{adj}$ can be interpreted as a motion screw on hold. When, needed the screw integrates into the M_N matrix and a mobility appears, as shown in Eq. 3.

$$
{}_{6\times 7} = \begin{array}{c} \begin{array}{ccccccc} \$_{at} & \$_{br} & \$_{bs} & \$_{bt} & \$_{cs} & \$_{cu} & \$_{dt} \end{array} \\ \left[\begin{array}{c|ccc|cc|c} 0 & 1 & 0 & 0 & 0 & 0 & 0 \\ 0 & 0 & 1 & 0 & 1 & 0 & 0 \\ 1 & 0 & 0 & 1 & 0 & 0 & 1 \\ 0 & 0 & z_b & -y_b & z_c & 0 & -y_d \\ -z_a & -z_b & 0 & x_b & 0 & 1 & x_d \\ y_a & y_b & -x_b & 0 & -x_c & 0 & 0 \end{array}\right] \end{array} \qquad \$_{adj} = \begin{bmatrix} 0 \\ 0 \\ 0 \\ 0 \\ 1 \\ 0 \end{bmatrix} \tag{3}
$$

4 Conclusions and Future Works

The Mobility is still a relevant research topic in mechanism theory. This paper used the Davies' Method and proposed a different point of view and proposed the classification of different types of mobilities. Due to the limited number of pages, it was not possible to show the mathematical proof of all the types of mobilities, but an extended version of this paper is expected to be submitted to a journal. The research continues with the objective of creating tools that enhance the design capabilities of mechanism and machine engineers.

Acknowledgements. This study was financed in part by the Coordenação de Aperfeiçoamento de Pessoal de Nível Superior - Brasil (CAPES) and by the Conselho Nacional de Desenvolvimento Científico e Tecnológico (CNPq).

References

1. Chebychev, P.A.: Théorie des mécanismes connus sous le nom de parallélogrammes, 2ème partie. In: Mémoires présentésa l'Académie impériale des sciences de Saint-Pétersbourg par divers savants (1869)
2. Grúbler, M.: Allgemeine Eigenschaften der zwangsläufigen ebenen kinematischen Ketten. Der Civilingenieur, Leipzig (1883)
3. Kutzbach, K.: Mechanische leitungsverzweigung, ihre gesetze und anwendungen. Maschinenbau **8**(21), 710–716 (1929)
4. Malishev, A.P.: Analysis and synthesis of mechanisms with a viewpoint of their structure. Izvestiya Tomskogo Technological Institute (1923). (in Russian)
5. Davies, T.H.: Mechanical networks - II: formulae for degrees of mobility and redundancy. Mech. Mach. Theory **18**, 95–101 (1983). https://doi.org/10.1016/0094-114X(83)90101-5
6. Gogu, G.: Mobility of mechanisms: a critical review. Mech. Mach. Theory **40**, 1068–1097 (2005). https://doi.org/10.1016/j.mechmachtheory.2004.12.014
7. Zhao, J.S., Zhou, K., Feng, Z.J.: Theory of degrees of freedom for mechanisms. Mech. Mach. Theory **39**(6), 621–643 (2004). https://doi.org/10.1016/j.mechmachtheory.2003.12.005
8. Guest, S.D., Fowler, P.W.: A symmetry-extended mobility rule. Mech. Mach. Theory **40**(9), 1002–1014 (2005). https://doi.org/10.1016/j.mechmachtheory.2004.12.017
9. Yang, D.C., Xiong, J., Yang, X.D.: A simple method to calculate mobility with Jacobian. Mech. Mach. Theory **43**(9), 1175–1185 (2008). https://doi.org/10.1016/j.mechmachtheory.2007.08.001

10. Wampler, C.W., Hauenstein, J.D., Sommese, A.J.: Mechanism mobility and a local dimension test. Mech. Mach. Theory **46**(9), 1193–1206 (2011). https://doi.org/10.1016/j.mechmachtheory.2011.04.011
11. Chen, C.: The order of local mobility of mechanisms. Mech. Mach. Theory **46**(9), 1251–1264 (2011). https://doi.org/10.1016/j.mechmachtheory.2011.04.007
12. Muller, A., Shai, O.: Constraint graphs for combinatorial mobility determination. Mech. Mach. Theory **108**, 260–275 (2017). https://doi.org/10.1016/j.mechmachtheory.2016.10.012
13. Huo, X., Sun, T., Song, Y.: A geometric algebra approach to determine motion/constraint, mobility and singularity of parallel mechanism. Mech. Mach. Theory **116**, 273–293 (2017). https://doi.org/10.1016/j.mechmachtheory.2017.06.005
14. Li, C., et al.: Cell division method for mobility analysis of multi-loop mechanisms. Mech. Mach. Theory **141**, 67–94 (2019). https://doi.org/10.1016/j.mechmachtheory.2019.07.002
15. Zhu, X., Shen, H., Wu, C., Chablat, D., Yang, T.: Computer-aided mobility analysis of parallel mechanisms. Mech. Mach. Theory **148**, 103810 (2020). https://doi.org/10.1016/j.mechmachtheory.2020.103810
16. Liu, R., Serré, P., Rameau, J.F.: A tool to check mobility under parameter variations in over-constrained mechanisms. Mech. Mach. Theory **69**, 44–61 (2013). https://doi.org/10.1016/j.mechmachtheory.2013.05.005
17. Rameau, J.F., Serré, P.: Computing mobility condition using Groebner basis. Mech. Mach. Theory **91**, 31–38 (2015). https://doi.org/10.1016/j.mechmachtheory.2015.04.003
18. Bartkowiak, R., Woernle, C.: Necessary and sufficient mobility conditions for single-loop overconstrained nH mechanisms. Mech. Mach. Theory **103**, 65–84 (2016). https://doi.org/10.1016/j.mechmachtheory.2016.03.023
19. Chen, Y., Feng, J., Liu, Y.: A group-theoretic approach to the mobility and kinematic of symmetric over-constrained structures. Mech. Mach. Theory **105**, 91–107 (2016). https://doi.org/10.1016/j.mechmachtheory.2016.06.004
20. Chen, Y., Feng, J.: Mobility of symmetric deployable structures subjected to external loads. Mech. Mach. Theory **93**, 98–111 (2015). https://doi.org/10.1016/j.mechmachtheory.2015.07.007
21. Hao, G., Kong, X.: A normalization-based approach to the mobility analysis of spatial compliant multi-beam modules. Mech. Mach. Theory **59**, 1–19 (2013). https://doi.org/10.1016/j.mechmachtheory.2012.08.013
22. Wang, J., Ting, K.L., Xue, C.: Discriminant method for the mobility identification of single degree-of-freedom double-loop linkages. Mech. Mach. Theory **45**, 740–755 (2010). https://doi.org/10.1016/j.mechmachtheory.2009.12.004
23. Davies, T.H.: Kirchhoff's circulation law applied to multi-loop kinematic chains. Mech. Mach. Theory **16**, 171–183 (1981). https://doi.org/10.1016/0094-114X(81)90033-1
24. Mejia, L., Simas, H., Martins, D.: Force capability in general 3 DoF planar mechanisms. Mech. Mach. Theory **91**, 120–134 (2015). https://doi.org/10.1016/j.mechmachtheory.2015.04.013
25. Mejia, L., Simas, H., Martins, D.: Wrench capability in redundant planar parallel manipulators with net degree of constraint equal to four, five or six. Mech. Mach. Theory **105**, 58–79 (2016). https://doi.org/10.1016/j.mechmachtheory.2016.06.020
26. Laus, L.P., Simas, H., Martins, D.: Machine efficiency determined using graph and screw theories with application in robotics. Mech. Mach. Theory **148**, 103748 (2020). https://doi.org/10.1016/j.mechmachtheory.2019.103748

27. Hunt, K.H.: Kinematic Geometry of Mechanisms. The Oxford Engineering Science Series, vol. 7. Clarendon, Oxford (1978)
28. Murota, K.: Matrices and Matroids for System Analysis. Algorithms and Combinatorics Series. Springer, Heidelberg (2009)
29. Ionescu, T.G., Antonescu, P., Biro, I., Bögelsack, G., Klein Breteler, A.J.: Terminology for the mechanism and machine science, Chapter 0-13. Mech. Mach. Theory **38**, 767-901 (2003)
30. Hwan, K.K.: Korean patent: tie rod assembly of steering apparatus for vehicle (2012). https://patents.google.com/patent/KR20140037352A/en

First and Second Order Centrodes of Slider-Crank Mechanisms by Using Instantaneous Invariants

Chiara Lanni(✉), Giorgio Figliolini, and Luciano Tomassi

DICEM, University of Cassino and Southern Lazio, Via G. Di Biasio 43, 03043 Cassino, FR, Italy
{lanni,figliolini,luciano.tomassi}@unicas.it

Abstract. A general algorithm to trace the first and second order centrodes of slider-crank mechanisms is proposed by using the instantaneous geometric and kinematic invariants. Bresse's circles can be also traced in order to validate the instantaneous positions of the velocity and acceleration poles. In particular, the second order centrodes give kinematic properties of the coupler motion and, thus, they are computed and traced for a constant angular velocity of the driving crank. Significant examples are included in the paper to validate the proposed algorithm.

Keywords: Instantaneous geometric and kinematic invariants · Slider-crank mechanisms · First and second order centrodes · Coupler motion · Bresse's circles

1 Introduction

Linkages and mainly those derived by four-bar kinematic chains through different kinematic inversions, find several applications in different fields, where they can play the role of kinematic structure of industrial and non-industrial robots, as walking machines, and/or the role of mechanisms to move the fingers of grippers and robotic hands, as described in [1–4]. In particular, the kinematic analysis and synthesis of planar mechanisms can be developed with the aid of suitable geometric loci, as the fixed and moving centrodes and Bresse's circles, as shown in [5–9]. Interesting applications can be also found in spherical and spatial mechanisms, in terms of pitch cones [10] and axodes [11], along with the spherical equivalent of Bresse's circles [12]. Moreover, the instantaneous geometric and kinematic invariants that are directly related to the rigid body motion, can be very useful for the kinematic analysis and synthesis of mechanisms and to determine the main geometric loci of the coupler motion. In fact, these loci take a convenient algebraic form, when they are referred to the canonical frame, as shown in [13–25].

This paper deals with the formulation of a general algorithm to trace the first and second order centrodes of both types of centered and offset slider-crank mechanisms, by using the instantaneous geometric and kinematic invariants. Bresse's circles can be also traced in order to validate the instantaneous positions of the velocity and acceleration

O. Altuzarra and A. Kecskémthy (Eds.): ARK 2022, SPAR 24, pp. 303–310, 2022.
https://doi.org/10.1007/978-3-031-08140-8_33

poles. In particular, the second order centrodes are computed and traced for a constant angular velocity of the driving crank. Significant results are shown in order to validate the proposed algorithm.

2 Instantaneous Invariants

The instantaneous invariants have been introduced in order to evaluate coefficients, constants and parameters that can characterize the motion of a planar mechanisms by geometrical and kinematical points of view.

The instantaneous geometric invariants are related to the rigid motion and the instantaneous kinematic invariants are time-dependent. In particular, the instantaneous geometric invariants are defined invariants among any pairs of fixed and moving reference frames or coordinate systems, since related to the motion characteristics of the coupler link and thus, they are independent by the particular choice of the reference frames.

Referring to the offset slider-crank mechanism of Fig. 1, the pairs of fixed $\mathcal{F}$ (O, X, Y) and moving f (Ω, x, y) reference frames, were chosen along with the corresponding fixed and moving canonical reference frames $\tilde{\mathcal{F}}$ $(P_1, \tilde{X}, \tilde{Y})$ and $\tilde{f}$ $(P_1, \tilde{x}, \tilde{y})$, which origin coincides with the instantaneous center of rotation P_1 of the coupler link AB. The $\tilde{Y}$-axis is orthogonal at P_1 point to the fixed centrode π and oriented toward the moving centrode that is not shown in Fig. 1. Thus, the $\tilde{X}$-axis is tangent to both centrodes at P_1 point and oriented clockwise with respect to the $\tilde{Y}$-axis, while the moving canonical reference frame $\tilde{f}$ is assumed as coincident with $\tilde{\mathcal{F}}$ at the referring configuration. These canonical frames are very important for the kinematic analysis and synthesis of planar mechanisms, because the geometric loci, which are of kinematic interest, take a simple mathematical form, when expressed with respect to them.

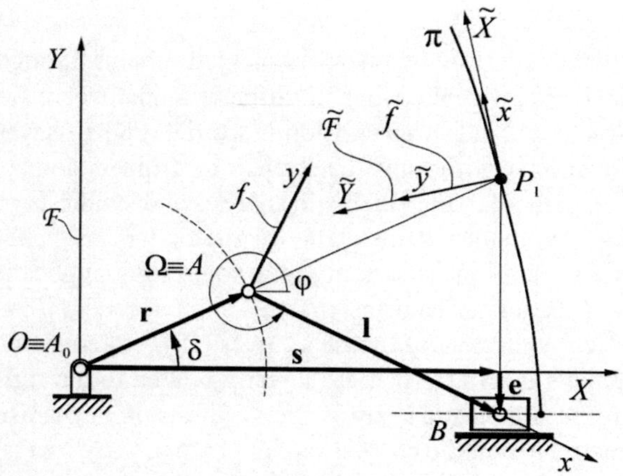

Fig. 1. Original ($\mathcal{F}$ and f) and canonical ($\tilde{\mathcal{F}}$ and $\tilde{f}$) reference frames for a general offset slider-crank mechanism.

The position vector $\mathbf{r}_\Omega$ of the origin Ω of f (Ω, x, y) can be expressed as

$$\mathbf{r}_\Omega = r\,[\cos\delta \;\; \sin\delta\,]^T \tag{1}$$

where r and δ are respectively, the A_0A crank length and the oriented counter-clockwise angle of A_0A with respect to the X-axis. Thus, during the mechanism motion, $\mathcal{F}$ and $\tilde{\mathcal{F}}$ remain fixed to the frame, while f and $\tilde{f}$ move as attached to the coupler link AB of the slider-crank mechanism. The same rigid body motion can be also obtained by rolling the moving on the fixed centrode by giving the successive positions of P_1 as the tangent point between them.

The instantaneous geometric invariants a_n and b_n are the n-order (where n is a natural number) derivatives of the Cartesian-coordinates $\tilde{X}_{P1}$ and $\tilde{Y}_{P1}$ of P_1 with respect to the oriented angle φ of $\tilde{f}$ with respect to $\tilde{\mathcal{F}}$ during the coupler motion

$$a_n = \frac{\mathrm{d}^n \tilde{X}_{P1}}{\mathrm{d}\varphi^n} \quad b_n = \frac{\mathrm{d}^n \tilde{Y}_{P1}}{\mathrm{d}\varphi^n} \tag{2}$$

For the convenient starting configuration of the mechanism, where δ and φ are equal to zero, points P_1 and B are coincident and both canonical frames coincide each other, the instantaneous geometric invariants up to the second order are given by the following expressions

$$a_0 = b_0 = a_1 = b_1 = a_2 = 0 \tag{3}$$

$$b_2 = \sqrt{\left(\frac{\mathrm{d}^2 X_\Omega}{\mathrm{d}\varphi^2} + \frac{\mathrm{d}Y_\Omega}{\mathrm{d}\varphi}\right)^2 + \left(\frac{\mathrm{d}^2 Y_\Omega}{\mathrm{d}\varphi^2} - \frac{\mathrm{d}X_\Omega}{\mathrm{d}\varphi}\right)^2} \tag{4}$$

Referring to Eq. (1), the first and second derivatives with respect to the oriented angle φ of the Cartesian coordinates X_Ω and Y_Ω that represent the components of the position vector $\mathbf{r}_\Omega$, are given by

$$\frac{\mathrm{d}X_\Omega}{\mathrm{d}\varphi} = -r \sin\delta \frac{\mathrm{d}\delta}{\mathrm{d}\varphi} \quad \frac{\mathrm{d}Y_\Omega}{\mathrm{d}\varphi} = r\cos\delta \frac{\mathrm{d}\delta}{\mathrm{d}\varphi} \tag{5}$$

$$\frac{\mathrm{d}^2 X_\Omega}{\mathrm{d}\varphi^2} = -r\left[\cos\delta\left(\frac{\mathrm{d}\delta}{\mathrm{d}\varphi}\right)^2 + \sin\delta\frac{\mathrm{d}^2\delta}{\mathrm{d}\varphi^2}\right] \quad \frac{\mathrm{d}^2 Y_\Omega}{\mathrm{d}\varphi^2} = -r\left[\sin\delta\left(\frac{\mathrm{d}\delta}{\mathrm{d}\varphi}\right)^2 - \cos\delta\frac{\mathrm{d}^2\delta}{\mathrm{d}\varphi^2}\right] \tag{6}$$

Similarly, the first and second derivatives of the crank angle δ with respect to φ can be expressed in sequence, as follows

$$\sin\delta = \frac{e - l\sin\varphi}{r}, \quad \frac{\mathrm{d}\delta}{\mathrm{d}\varphi} = -\frac{l\cos\varphi}{r\cos\delta}, \quad \frac{\mathrm{d}^2\delta}{\mathrm{d}\varphi^2} = \frac{l\left(\sin\varphi\cos\delta - \cos\varphi\sin\delta\frac{\mathrm{d}\delta}{\mathrm{d}\varphi}\right)}{r\cos^2\delta} \tag{7}$$

The proposed formulation allows the computations of the instantaneous geometric invariants a_n and b_n for $n = 0, 1, 2$ which are very useful to express in a canonical algebraic form, the most significant geometric loci with respect to $\tilde{f}(P_1, \tilde{x}, \tilde{y})$.

This computation can be very complex when referring to the canonical frames directly, from which the convenience to make use of a different pair of frames, as f and $\mathcal{F}$, which are closer to the mechanism motion than the canonical frames. The first and second derivatives of the crank angle φ with respect to δ can be expressed in sequence, as follows:

$$\varphi = \sin^{-1}\left(\frac{e - r\sin\delta}{l}\right) \quad \dot{\varphi} = \frac{d\varphi}{d\delta} = -\frac{r\cos\delta}{l\cos\varphi}\dot{\delta} \tag{8}$$

$$\ddot{\varphi} = \frac{d^2\varphi}{d\delta^2} = \frac{\dot{\delta}^2 r\sin\delta\,\cos\varphi - \ddot{\delta}\,r\cos\delta\,\cos\varphi + \frac{\dot{\delta}^2 r^2\cos^2\delta\,\sin\varphi}{l\cos\varphi}}{l\cos^2\varphi} \tag{9}$$

where $\dot{\delta}$ and $\ddot{\delta}$ are the input angular velocity and acceleration of A_0A crank, respectively.

3 First and Second Order Centrodes

The fixed and moving centrodes of first and second order are traced by the instantaneous center of rotation P_1 and the acceleration pole P_2, with respect to the fixed $\mathcal{F}$ and moving f reference frames.

The position of a generic coupler point M can be expressed as

$$X_M = r_{\Omega x} + x_M\cos\varphi - y_M\sin\varphi \quad Y_M = r_{\Omega y} + x_M\sin\varphi + y_M\cos\varphi \tag{10}$$

In particular, when M coincides with P_1, Eqs. (10) give

$$X_{P1} = r_{\Omega x} + x_{P1}\cos\varphi - y_{P1}\sin\varphi \quad Y_{P1} = r_{\Omega y} + x_{P1}\sin\varphi + y_{P1}\cos\varphi \tag{11}$$

and since P_1 is the velocity pole, one has $\frac{dX_{P1}}{dt} = \frac{dY_{P1}}{dt} = 0$, and in turn

$$\begin{aligned} \frac{\mathrm{d}X_\Omega}{\mathrm{d}\varphi}\frac{\mathrm{d}\varphi}{\mathrm{d}t} - x_{P1}\frac{\mathrm{d}\varphi}{\mathrm{d}t}\sin\varphi - y_{P1}\frac{\mathrm{d}\varphi}{\mathrm{d}t}\cos\varphi = 0 \\ \frac{\mathrm{d}Y_\Omega}{\mathrm{d}\varphi}\frac{\mathrm{d}\varphi}{\mathrm{d}t} + x_{P1}\frac{\mathrm{d}\varphi}{\mathrm{d}t}\cos\varphi - y_{P1}\frac{\mathrm{d}\varphi}{\mathrm{d}t}\sin\varphi = 0 \end{aligned} \tag{12}$$

Thus, Eqs. (12) give

$$x_{P1} = \frac{\mathrm{d}X_\Omega}{\mathrm{d}\varphi}\sin\varphi - \frac{\mathrm{d}Y_\Omega}{\mathrm{d}\varphi}\cos\varphi \quad y_{P1} = \frac{\mathrm{d}X_\Omega}{\mathrm{d}\varphi}\cos\varphi + \frac{\mathrm{d}Y_\Omega}{\mathrm{d}\varphi}\sin\varphi \tag{13}$$

and considering the Eqs. (5), one has

$$x_{P1} = -r\frac{\mathrm{d}\delta}{\mathrm{d}\varphi}(\sin\delta\sin\varphi + \cos\delta\cos\varphi) \quad y_{P1} = r\frac{\mathrm{d}\delta}{\mathrm{d}\varphi}(\sin\delta\cos\varphi + \cos\delta\sin\varphi) \tag{14}$$

Substituting the second of Eqs. (7) in Eqs. (14), the following parametric equations of the first order moving centrode l_1 in f are obtained

$$x_{P1} = l\cos\varphi(\tan\delta\sin\varphi + \cos\varphi) \quad y_{P1} = l\cos\varphi(\tan\delta\cos\varphi - \sin\varphi) \tag{15}$$

The parametric equations of the first order fixed centrode λ_1 in $\mathcal{F}$ take the following form

$$X_{P1} = r\cos\delta\left(1 + \frac{l\cos\varphi}{r\cos\delta}\right) \quad Y_{P1} = r\sin\delta\left(1 + \frac{l\cos\varphi}{r\cos\delta}\right) \tag{16}$$

Likewise, when point M coincides with P_2, the second order fixed centrode λ_2 can be expressed as

$$X_{P2} = r_{\Omega x} + x_{P2}\cos\varphi - y_{P2}\sin\varphi \quad Y_{P2} = r_{\Omega y} + x_{P2}\sin\varphi + y_{P2}\cos\varphi \tag{17}$$

and since P_2 is the acceleration pole, one has $\frac{dX^2_{P2}}{dt^2} = \frac{dY^2_{P2}}{dt^2} = 0$, and in turn

$$\begin{aligned}
&\frac{\mathrm{d}^2X_\Omega}{\mathrm{d}\varphi^2}\left(\frac{\mathrm{d}\varphi}{\mathrm{d}t}\right)^2 + \frac{\mathrm{d}X_\Omega}{\mathrm{d}\varphi}\frac{\mathrm{d}^2\varphi}{\mathrm{d}t^2} - x_{P2}\left[\cos\varphi\left(\frac{\mathrm{d}\varphi}{\mathrm{d}t}\right)^2 + \sin\varphi\frac{\mathrm{d}^2\varphi}{\mathrm{d}t^2}\right] + y_{P2}\left[\sin\varphi\left(\frac{\mathrm{d}\varphi}{\mathrm{d}t}\right)^2 - \cos\varphi\frac{\mathrm{d}^2\varphi}{\mathrm{d}t^2}\right] = 0\\
&\frac{\mathrm{d}^2Y_\Omega}{\mathrm{d}\varphi^2}\left(\frac{\mathrm{d}\varphi}{\mathrm{d}t}\right)^2 + \frac{\mathrm{d}Y_\Omega}{\mathrm{d}\varphi}\frac{\mathrm{d}^2\varphi}{\mathrm{d}t^2} - x_{P2}\left[\sin\varphi\left(\frac{\mathrm{d}\varphi}{\mathrm{d}t}\right)^2 - \cos\varphi\frac{\mathrm{d}^2\varphi}{\mathrm{d}t^2}\right] - y_{P2}\left[\cos\varphi\left(\frac{\mathrm{d}\varphi}{\mathrm{d}t}\right)^2 + \sin\varphi\frac{\mathrm{d}^2\varphi}{\mathrm{d}t^2}\right] = 0
\end{aligned} \tag{18}$$

Thus, the parametric equations of the second order moving centrode l_2 in f take the following form

$$\begin{aligned}
x_{P2} =& \frac{1}{\dot\varphi^4 + \ddot\varphi^2}\Bigg[\left(\frac{\mathrm{d}^2X_\Omega}{\mathrm{d}\varphi^2}\cos\varphi + \frac{\mathrm{d}^2Y_\Omega}{\mathrm{d}\varphi^2}\sin\varphi\right)\dot\varphi^4 + \left(\frac{\mathrm{d}X_\Omega}{\mathrm{d}\varphi}\cos\varphi + \frac{\mathrm{d}^2X_\Omega}{\mathrm{d}\varphi^2}\sin\varphi\right.\\
&+ \left.\frac{\mathrm{d}Y_\Omega}{\mathrm{d}\varphi}\sin\varphi - \frac{\mathrm{d}^2Y_\Omega}{\mathrm{d}\varphi^2}\cos\varphi\right)\dot\varphi^2\ddot\varphi + \left(\frac{\mathrm{d}X_\Omega}{\mathrm{d}\varphi}\sin\varphi - \frac{\mathrm{d}Y_\Omega}{\mathrm{d}\varphi}\cos\varphi\right)\ddot\varphi^2\Bigg]
\end{aligned} \tag{19}$$

$$\begin{aligned}
y_{P2} =& \frac{1}{\dot\varphi^4 + \ddot\varphi^2}\Bigg[\left(-\frac{\mathrm{d}^2X_\Omega}{\mathrm{d}\varphi^2}\sin\varphi + \frac{\mathrm{d}^2Y_\Omega}{\mathrm{d}\varphi^2}\cos\varphi\right)\dot\varphi^4 + \left(-\frac{\mathrm{d}X_\Omega}{\mathrm{d}\varphi}\sin\varphi + \frac{\mathrm{d}^2X_\Omega}{\mathrm{d}\varphi^2}\cos\varphi\right.\\
&+ \left.\frac{\mathrm{d}Y_\Omega}{\mathrm{d}\varphi}\cos\varphi + \frac{\mathrm{d}^2Y_\Omega}{\mathrm{d}\varphi^2}\sin\varphi\right)\dot\varphi^2\ddot\varphi + \left(\frac{\mathrm{d}X_\Omega}{\mathrm{d}\varphi}\cos\varphi + \frac{\mathrm{d}Y_\Omega}{\mathrm{d}\varphi}\sin\varphi\right)\ddot\varphi^2\Bigg]
\end{aligned} \tag{20}$$

It is well known, that the fixed and moving centrodes of first order are always tangent each other at the instantaneous center of rotation and that the rigid body motion can be reproduced by the pure rolling of the moving centrode on the fixed one. Instead, the fixed and moving centrodes of second order intersect each other in more than one point, for which the position of the acceleration pole is not straightforward to determine. However, one way to find the acceleration pole is to use Bresse's circles because they intersect each other in both poles P_1 and P_2.

Thus, the inflection and stationary circles are determined by still using the instantaneous invariants and even for validation purposes, because the acceleration pole must be one of the intersection points between the second order centrodes, but also one of the two intersection between Bresse's circles.

In particular, the inflection circle $\mathcal{I}$ is the geometric locus of the coupler points, which show an inflection point in their paths and is always tangent to both centrodes at the instantaneous center of rotation P_1. Referring to $\tilde{f}$, one has

$$\tilde{x}^2 + \tilde{y}^2 - b_2\tilde{y} = 0 \tag{21}$$

where the diameter b_2 of $\mathcal{I}$ is obtained by the Eq. (4).

Likewise, the stationary circle $\mathcal{S}$ that is the geometric locus of the coupler points, which show a pure normal acceleration, is given by

$$\ddot\varphi\left(\tilde{x}^2 + \tilde{y}^2\right) + b_2\dot\varphi^2\tilde{x} = 0 \tag{22}$$

where $\dot{\varphi}$ and $\ddot{\varphi}$ are the angular velocity and acceleration, respectively, which are expressed by Eqs. (8) and (9). Equations (19), (20) and (22) depend by the kinematic properties by means of $\dot{\varphi}$ and $\ddot{\varphi}$, while the inflection circle is related to the geometry of the coupler motion.

The proposed algorithm has been implemented in Matlab and validated through several significant examples dealing with centered and offset slider-crank mechanisms. In particular, Figs. 2 and 3 have been obtained for $r = 10$ cm, $l = 2r$, $\dot{\delta} = 1$ r/s, $\ddot{\delta} = 0$ and e (offset) $= -50$ cm. When $\delta = 0$ deg, point B is at top dead center and thus, it coincides with P_1, as shown in Fig. 2b) and for $\delta = 90$ deg, both Bresse's circles degenerates in two orthogonal straight lines, Fig. 3a).

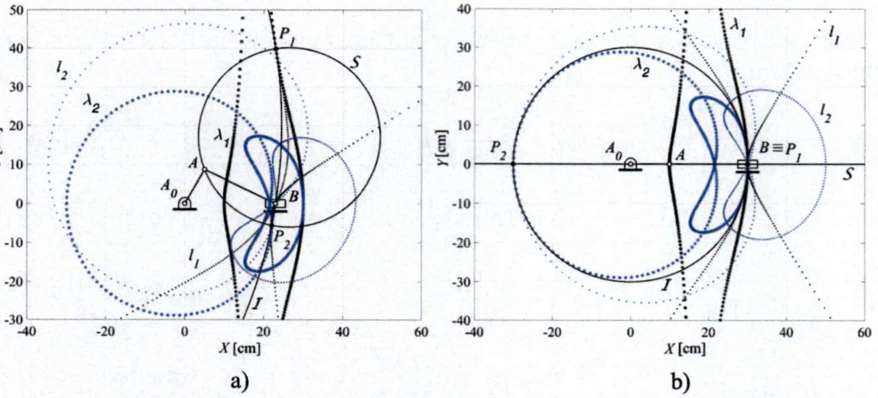

Fig. 2. Fixed and moving centrodes of first and second order, along with the inflection and stationary circles, for a centered slider-crank mechanism: a) $\delta = 60$ deg; b) $\delta = 0$ deg.

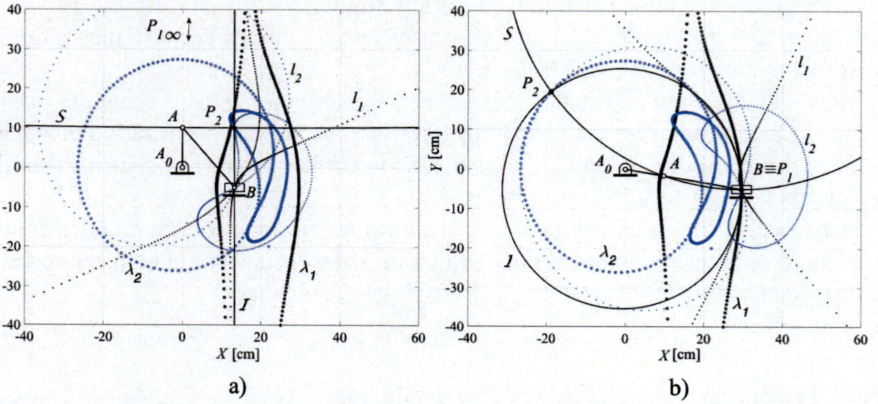

Fig. 3. Fixed and moving centrodes of first and second order, along with the inflection and stationary circles of an offset slider-crank mechanism: a) $\delta = 90$ deg; b) Top dead center.

4 Conclusions

A general algorithm to trace the first and second order centrodes of centered and offset slider-crank mechanisms, along with Bresse's circles, has been formulated by using the instantaneous geometric and kinematic invariants. In particular, the second order centrodes have been computed and traced for a constant angular velocity of the driving crank, otherwise the fixed and moving second order centrodes would change accordingly, by losing their practical meaning. The proposed algorithm was implemented in a Matlab program and significant examples have been obtained and discussed in order to validate the proposed algorithm.

References

1. Di Benedetto, A., Pennestrì, E.: Introduction to the Kinematics of Mechanisms, vol. 1, 2, 3, Casa Editrice Ambrosiana, Milan (1991). (in Italian)
2. Krause, M.: Analysis der ebenen Bewegung, Vereinigung Wissenschaftlicher Verleger, Berlin (1920)
3. Figliolini, G., Rea, P., Conte, M.: Mechanical design of a novel biped climbing and walking robot. In: Parenti-Castelli, V., Schielen, W. (eds.) ROMANSY 18 - Robot, Design, Dynamics and Control, pp. 199–206. Springer, Vienna (2010).https://doi.org/10.1007/978-3-7091-0277-0_23
4. Figliolini, G., Lanni, C., Kaur, R.: Kinematic synthesis of spherical four-bar linkages for five-poses rigid body guidance. In: Uhl, T. (ed.) Proceedings of the 15th IFToMM World Congress on Mechanism and Machine Science. MMS. Advances in Mechanism and Machine Science, vol. 73, pp. 639–648. Springer, Cham (2019).https://doi.org/10.1007/978-3-030-20131-9_64
5. Figliolini, G., Lanni, C.: Jerk and jounce relevance for the kinematic performance of long-dwell mechanisms. In: Uhl, T. (ed.) Proceedings of the 15th IFToMM World Congress on Mechanism and Machine Science, MMS, vol. 73, pp. 219–228, Springer, Cham (2019). https://doi.org/10.1007/978-3-030-20131-9_22
6. Figliolini, G., Lanni, C.: Geometric loci for the kinematic analysis of planar mechanisms via the instantaneous geometric invariants. In: Gasparetto, A., Ceccarelli, M. (eds.) MEDER 2018. MMS, vol. 66, pp. 184–192. Springer, Cham (2019). https://doi.org/10.1007/978-3-030-00365-4_22
7. Figliolini, G., Lanni, C., Tomassi, L.: Kinematic analysis of slider – crank mechanisms via the Bresse and Jerk's circles. In: Carcaterra, A., Paolone, A., Graziani, G. (eds.) AIMETA 2019. LNME, pp. 278–284. Springer, Cham (2020). https://doi.org/10.1007/978-3-030-41057-5_23
8. Figliolini, G., Lanni, C., Angeles, J.: Kinematic analysis of the planar motion of vehicles when traveling along tractrix curves. ASME. J. Mech. Robot. **12**(5), 054502 (2020)
9. Pennestrì, E., Belfiore, N.P.: On the numerical computation of generalized Burmester points. Meccanica **30**(2), 147–153 (1995)
10. Figliolini, G., Angeles, J.: Synthesis of the pitch cones of N-lobed elliptical bevel gears. ASME J. Mech. Des. **133**(3), 031002–031011 (2011)
11. Figliolini, G., Rea, P., Angeles, J.: The synthesis of the axodes of RCCC linkages. ASME J. Mech. Robot. **8**(2), 021011 (2015)
12. Figliolini, G., Angeles, J.: The spherical equivalent of Bresse's circles: The case of crossed double-crank linkages. ASME J. Mech. Robot. **9**(1), 011014 (2017)
13. Bottema, O.: On instantaneous invariants. In: Proceedings of the International Conference for Teachers of Mechanism, pp. pp. 159–164. Yale University, New Haven, CT (1961)

14. Veldkamp, G.R.: Curvature Theory in Plane Kinematics. T.H. Delft, Groningen (1963)
15. Kamphuis, H.J.: Application of spherical instantaneous kinematics to the spherical slider-crank mechanism. J. Mech. **4**(1), 43–56 (1969)
16. Roth, B., Yang, A.T.: Application of instantaneous invariants to the analysis and synthesis of mechanisms. ASME J. Eng. Industr. **99**(1), 97–103 (1977)
17. Bottema, O., Roth, B.: Theoretical Kinematics. Dover, New York, NY (1990)
18. Yang, A.T., Pennock, G.R., Hsia, L.M.: Instantaneous invariants and curvature analysis of a planar four-link mechanism. ASME J. Mech. Des. **116**(4), 1173–1176 (1994)
19. Roth, B.: On the advantages of instantaneous invariants and geometric kinematics. Mech. Mach. Theor. **89**, 5–13 (2015)
20. Inalcik, A., Ersoy, S., Stachel, H.: On instantaneous invariants of hyperbolic planes. Math. Mech. Solids **22**(5), 1047–1057 (2017)
21. McDonald, E.W.: Determination of the reducible cases of the fixed centrode of three-bar motion. Am. Math. Mon. **33**(2), 90–93 (1926)
22. Carter, W.J.: Second Acceleration in four-bar mechanisms as related to rotopole motion. J Appl. Mech. Brief Notes **25**, 293–294 (1958)
23. Schiller, R.W.: Method for determining the instantaneous center of acceleration. J. Appl. Mech. **32**(1), 217–218 (1965)
24. Hwang, W.-M., Fan, Y.-S.: Coordination of two crank angles and two acceleration poles for a slider crank mechanism using parametric equations. Proc. Inst. Mech. Eng. Part C J. Mech. Eng. Sci. **222**(3), 483–491 (2008)
25. Sun, H., Liu, T.: Analysis of instantaneous center of zero acceleration of rigid body. Mod. Appl. Sci. **3**(4), 191–195 (2009)

Spherical 4R Linkage Algebraic v_i-v_j Input-Output Equations

M. John D. Hayes(✉), Mirja Rotzoll, and Zachary Copeland

Department of Mechanical and Aerospace Engineering,
Carleton University, Ottawa, ON, Canada
john.hayes@carleton.ca

Abstract. In this paper the algebraic polynomial equations relating the relative orientations between the six distinct pairs of rigid links in an arbitrary spherical 4R mechanism are derived. First, the forward kinematics transformation matrix of an arbitrary spherical open 4R kinematic chain is computed in terms of its Denavit-Hartenberg parameters, where all angles are converted to their tangent half-angle parameters. This transformation matrix is mapped to its corresponding four non-zero Study soma coordinates. The serial kinematic chain is conceptually closed by equating the forward kinematics transformation to the identity matrix. Gröbner bases and resultants are then used to eliminate the two intermediate joint angle parameters leaving an algebraic polynomial in terms of the selected input and output (IO) joint angle parameters and the four twist angle or link length parameters. This yields six independent algebraic IO Equations. Their utility is demonstrated with two function generator continuous approximate synthesis examples.

Keywords: Spherical four-bar linkage · v_i-v_j algebraic input-output equations · Continuous approximate synthesis

1 Introduction

Relative motion between mechanically constrained rigid bodies on the surface of a sphere has fascinated philosophers, mathematicians, and engineers for millennia [3]. The design of predictable motion of a four-bar spherical mechanism appears to have its origins in the development of *universal joints* based on gimbals, which have been investigated since antiquity [10]. Arguably the most significant modern spherical four-bar mechanism is the *Agile Eye* [5], which is used as a camera pointing system. While there is a substantial volume of archival literature regarding spherical 4R mechanisms, this type of mechanical system still excites the imagination, see [8] for a recent example. Hence, we believe there is sufficient justification to present the work on the derivation of the six v_i-v_j algebraic input-output (IO) equations reported in this paper.

O. Altuzarra and A. Kecskémethy (Eds.): ARK 2022, SPAR 24, pp. 311–319, 2022.
https://doi.org/10.1007/978-3-031-08140-8_34

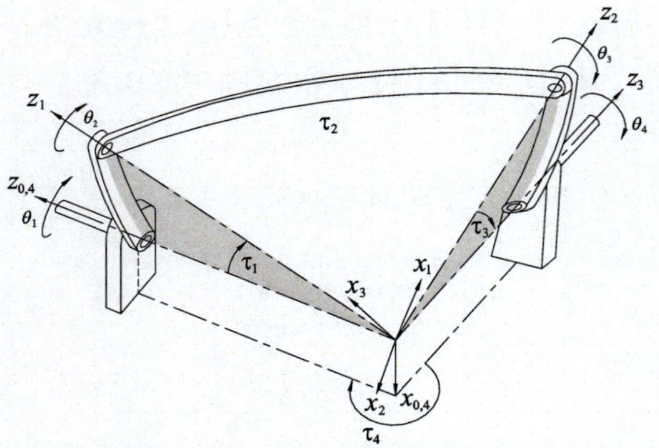

Fig. 1. Spherical 4R DH reference frames and parameters.

Consider the arbitrary spherical 4R linkage illustrated in Fig. 1. The IO equation expresses the implicit functional relationship between the input and output angles, θ_i and θ_j in terms of the constant arc lengths between the four R-pair centres, τ_i. The derivation of the algebraic form of the spherical IO equation [9] makes use of the original Denavit-Hartenberg (DH) parametrisation of the kinematic geometry [6]. It requires that all measures of angle be converted to algebraic parameters using the tangent half-angle substitutions:

$$v_i = \tan\frac{\theta_i}{2}, \quad \cos\theta_i = \frac{1-v_i^2}{1+v_i^2}, \quad \sin\theta_i = \frac{2v_i}{1+v_i^2},$$
$$\alpha_i = \tan\frac{\tau_i}{2}, \quad \cos\tau_i = \frac{1-\alpha_i^2}{1+\alpha_i^2}, \quad \sin\tau_i = \frac{2\alpha_i}{1+\alpha_i^2}.$$

The forward kinematics of an arbitrary kinematic chain is obtained as a linear transformation matrix in terms of the DH parameters. This linear transformation can then be mapped to the corresponding eight Study soma coordinates [9]. For spherical kinematic chains there are only four homogeneous soma coordinates. The ideal generated by the three soma that equate to zero are used to derive the algebraic IO equations relating the six distinct edges of an arbitrary spherical quadrangle.

Making these substitutions, the algebraic form of the IO equation is derived as [9]

$$Av_1^2v_4^2 + Bv_1^2 + Cv_4^2 + 8\alpha_1\alpha_3\left(\alpha_4^2+1\right)\left(\alpha_2^2+1\right)v_1v_4 + D = 0, \tag{1}$$

where

$$\begin{aligned}
A &= A_1A_2 = (\alpha_1\alpha_2\alpha_3 - \alpha_1\alpha_2\alpha_4 + \alpha_1\alpha_3\alpha_4 - \alpha_2\alpha_3\alpha_4 + \alpha_1 - \alpha_2 + \alpha_3 - \alpha_4) \\
&\qquad (\alpha_1\alpha_2\alpha_3 - \alpha_1\alpha_2\alpha_4 - \alpha_1\alpha_3\alpha_4 - \alpha_2\alpha_3\alpha_4 - \alpha_1 - \alpha_2 - \alpha_3 + \alpha_4)\,, \\
B &= B_1B_2 = (\alpha_1\alpha_2\alpha_3 + \alpha_1\alpha_2\alpha_4 - \alpha_1\alpha_3\alpha_4 - \alpha_2\alpha_3\alpha_4 + \alpha_1 + \alpha_2 - \alpha_3 - \alpha_4) \\
&\qquad (\alpha_1\alpha_2\alpha_3 + \alpha_1\alpha_2\alpha_4 + \alpha_1\alpha_3\alpha_4 - \alpha_2\alpha_3\alpha_4 - \alpha_1 + \alpha_2 + \alpha_3 + \alpha_4)\,, \\
C &= C_1C_2 = (\alpha_1\alpha_2\alpha_3 - \alpha_1\alpha_2\alpha_4 - \alpha_1\alpha_3\alpha_4 + \alpha_2\alpha_3\alpha_4 - \alpha_1 + \alpha_2 + \alpha_3 - \alpha_4) \\
&\qquad (\alpha_1\alpha_2\alpha_3 - \alpha_1\alpha_2\alpha_4 + \alpha_1\alpha_3\alpha_4 + \alpha_2\alpha_3\alpha_4 + \alpha_1 + \alpha_2 - \alpha_3 + \alpha_4)\,, \\
D &= D_1D_2 = (\alpha_1\alpha_2\alpha_3 + \alpha_1\alpha_2\alpha_4 + \alpha_1\alpha_3\alpha_4 + \alpha_2\alpha_3\alpha_4 - \alpha_1 - \alpha_2 - \alpha_3 - \alpha_4) \\
&\qquad (\alpha_1\alpha_2\alpha_3 + \alpha_1\alpha_2\alpha_4 - \alpha_1\alpha_3\alpha_4 + \alpha_2\alpha_3\alpha_4 + \alpha_1 - \alpha_2 + \alpha_3 + \alpha_4)\,.
\end{aligned}$$

While the derivation of this algebraised v_1-v_4 IO equation is novel and far from intuitive, the algebraic form of this fourth degree polynomial in the v_1-v_4 IO angle parameters is not. The earliest derivations of similar equations representing manipulatable octahedra, identical in form, is due to R. Bricard in 1897 [2].

2 Derivation of the Six Spherical v_i-v_j IO Equations

Using the eight bi-cubic coefficient definitions from Eq. (1), the remaining five v_i-v_j equations contain all eight of the bi-cubic coefficients, but in different permutations:

$$A_1B_2v_1^2v_2^2 + A_2B_1v_1^2 + C_1D_2v_2^2 + 8\alpha_2\alpha_4\left(\alpha_1^2+1\right)\left(\alpha_3^2+1\right)v_1v_2 + C_2D_1 = 0; \tag{2}$$

$$A_1B_1v_1^2v_3^2 + A_2B_2v_1^2 + C_2D_2v_3^2 + C_1D_1 = 0; \tag{3}$$

$$A_1D_2v_2^2v_3^2 + B_2C_1v_2^2 + B_1C_2v_3^2 - 8\alpha_1\alpha_3\left(\alpha_2^2+1\right)\left(\alpha_4^2+1\right)v_2v_3 + A_2D_1 = 0; \tag{4}$$

$$A_1C_1v_2^2v_4^2 + B_2D_2v_2^2 + A_2C_2v_4^2 + B_1D_1 = 0; \tag{5}$$

$$A_1C_2v_3^2v_4^2 + B_1D_2v_3^2 + A_2C_1v_4^2 + 8\alpha_2\alpha_4\left(\alpha_1^2+1\right)\left(\alpha_3^2+1\right)v_3v_4 + B_2D_1 = 0. \tag{6}$$

The v_1-v_4 IO Equation. To obtain this IO equation from the ideal generated by the three soma coordinates that equate to zero, both v_2 and v_3 are eliminated by first computing the Gröbner bases using the Maple 2021 "`tdeg`" monomial ordering with the list sequence (v_3, v_2, v_4, v_1). This is *graded reverse lexicographic order*, also known as *degrevlex* in the literature [1], with indeterminate ordering $v_3 > v_2 > v_4 > v_1$. This monomial ordering sorts the terms by total degree before breaking ties between terms with identical degree by comparing the smallest indeterminate first and considering a higher degree as smaller in the term ordering. In this case, 12 bases are computed, all functions of all four v_i. We eliminate v_2 and v_3 by computing the bases of these 12 with the reverse monomial ordering by using "`plex`", which is the *pure lexicographic order*, also known as *lex*. This results in 10 new bases, with one that is a function of only

v_1 and v_4 and the four α_i, which represents the IO equation we are looking for. This polynomial splits into three factors. The first two are $(1+v_1^2)(1+v_4^2)$, a product that is always greater than zero, and can be safely factored out, leaving us with Eq. (1). This, and some of the other IO equations are computable in one application of the elimination monomial ordering called "`lexdeg`", but the computation time is about 3500 s s compared to 120 s for the sequential application of "`tdeg`" and "`plex`" on an Intel Core i7-7700 CPU @ 3.60 GHz.

It is important to note that we are using the standard Denavit-Hartenberg [6] relative joint angle parameters, which are each a measure of the angle a link makes with the previous link in the kinematic chain. This fact enables us to derive the remaining five IO equations such that the same eight bi-cubic coefficient factors characterise all six IO equations. This is generally not the case when vector loop methods are used together with trigonometry, see [7] for a detailed example.

The v_1-v_2 IO Equation. The derivation steps are precisely the same as for the v_1-v_4 IO equation. Eliminating v_3 and v_4 from the same three soma coordinates, the resulting v_1-v_2 IO equation splits into three similar factors. The first two, $(1+v_1^2)(1+v_2^2)$, can be safely factored out, leaving us with Eq. (2).

The v_1-v_3 IO Equation. The derivation steps are precisely the same as for the previous two IO equations. But, after the elimination of v_2 and v_4 from the same three soma coordinates, the resulting v_1-v_3 IO equation splits into five factors. The first two are $(1+v_1^2)(1+v_3^2)$, and can be safely factored out. The next two are

$$(\alpha_2^2\alpha_3^2+2\alpha_2\alpha_3+1)v_3^2+\alpha_2^2\alpha_3^2-2\alpha_2\alpha_3+1, \tag{7}$$

$$(\alpha_2^2-2\alpha_2\alpha_3+\alpha_3^2)v_3^2+\alpha_2^2+2\alpha_2\alpha_3+\alpha_3^2. \tag{8}$$

In order for either, or both, of Eqs. (7) or (8) to be identically zero the arc length parameters α_2 and α_3 must be complex, meaning these two factors may also eliminated, leaving us with Eq. (3).

The v_2-v_3 IO Equation. To derive this IO equation using elimination methods on the three soma coordinates we have been using requires a very different approach. We were successful by first applying grevlex to the three soma coordinates using the list sequence (v_1, v_4, v_2, v_3), then applying *graded lexicographic order* using "`grlex`" to the bases identified with grevlex. After each computation we obtain 12 bases, all in terms of the four α_i and the four v_i, with the exception of one in the graded lexicographic order set of bases, which is in terms of the four α_i, but only v_1, v_2, and v_3, and is used in the elimination steps. Next, resultants are used to eliminate v_4 first, then v_1. We obtain a v_2-v_3 IO equation that splits into nine factors.

The first five of these factors are simple to divide out since they are trivially non-zero: the first is -1; the other four are the squares of a single α_i added to a positive integer. The next three factors are functions of v_2 and v_3, but only α_1, α_2, and α_3:

$$\begin{aligned}(\alpha_1\alpha_2 - \alpha_1\alpha_3 + \alpha_2\alpha_3 + 1)^2 v_2^2 v_3^2 + (\alpha_1\alpha_2 + \alpha_1\alpha_3 - \alpha_2\alpha_3 + 1)^2 v_2^2 + \\ 8\alpha_1\alpha_3(\alpha_2^2+1)v_2v_3 + (\alpha_1\alpha_2 - \alpha_1\alpha_3 - \alpha_2\alpha_3 - 1)^2 v_3^2 + (\alpha_1\alpha_2 + \alpha_1\alpha_3 + \alpha_2\alpha_3 - 1)^2;\end{aligned} \tag{9}$$

$$\begin{aligned}(\alpha_1\alpha_2\alpha_3 + \alpha_1 - \alpha_2 + \alpha_3)^2 v_2^2 v_3^2 + (\alpha_1\alpha_2\alpha_3 - \alpha_1 + \alpha_2 + \alpha_3)^2 v_2^2 - \\ 8\alpha_1\alpha_3(\alpha_2^2+1)v_2v_3 + (\alpha_1\alpha_2\alpha_3 + \alpha_1 + \alpha_2 - \alpha_3)^2 v_3^2 + (\alpha_1\alpha_2\alpha_3 - \alpha_1 - \alpha_2 - \alpha_3)^2;\end{aligned} \tag{10}$$

$$\begin{aligned}\alpha_3(\alpha_1\alpha_2+1)(\alpha_1-\alpha_2)v_2^2 + 2\alpha_1\alpha_3(\alpha_2^2+1)v_2v_3 - \alpha_1(\alpha_2\alpha_3+1)(\alpha_2-\alpha_3)v_3^2 + \\ \alpha_2(\alpha_1+\alpha_3)(\alpha_1\alpha_3-1).\end{aligned} \tag{11}$$

In order for Eqs. (9), (10), and/or (11) to be identically zero the arc length parameters α_1, α_2, and/or α_3 must be complex numbers, so we may safely divide these three factors out, leaving only Eq. (4) as the desired IO equation.

The v_2-v_4 IO Equation. The derivation steps for the v_2-v_4 IO equation are the same as those for the v_1-v_3 IO equation. The the second set of Gröbner bases computed using the pure lexicographic order with list sequence (v_3, v_1, v_2, v_4) lead to an IO equation that splits into five factors, the first two are trivial. The next two are

$$(\alpha_1^2\alpha_2^2 + 2\alpha_1\alpha_2 + 1)v_2^2 + \alpha_1^2\alpha_2^2 - 2\alpha_1\alpha_2 + 1, \tag{12}$$

$$(\alpha_1^2 - 2\alpha_1\alpha_2 + \alpha_2^2)v_2^2 + \alpha_1^2 + 2\alpha_1\alpha_2 + \alpha_2^2. \tag{13}$$

For either, or both of Eqs. (12) and (13) to equate to zero, it requires both α_1 and α_2 to be complex. We can therefore factor both of these out, leaving only the desired v_2-v_4 IO, Eq. (5).

The v_3-v_4 IO Equation. Finally, the derivation steps for the v_3-v_4 IO equation are precisely the same as for the v_1-v_4 and v_1-v_2 IO equations. After the elimination of v_1 and v_2 from the same three soma coordinates, the resulting v_3-v_4 IO equation splits into three factors. The first two are safely divided out, leaving us with Eq. (6).

3 Application

The utility of these six spherical 4R algebraic IO equations is nicely demonstrated by the following two continuous approximate dimensional synthesis examples for function generation.

$v_3 = f(v_1)$ Function Generator. For this example, we will apply a modified form of the continuous approximate syntheses method [4] to function generation for $v_3 = f(v_1)$ using Eq. (3). The prescribed function is

$$v_3 = 2 + \tan\left(\frac{v_1^2}{v_1^2+1}\right), \tag{14}$$

over the range $-2 \leq v_1 \leq 2$.

The v_1-v_3 IO equation relates the input angle parameter to a measure of the transmission angle. We begin by squaring Eq. (3) to eliminate the residual error values that are equal in magnitude yet opposite in sense. We partition the result into a 9×1 array of angle parameters and a 9×1 array of associated link arc length parameter coefficients. The array $\mathbf{s}_{v_1,v_3}$ of angle parameters is used to generate the synthesis equation

$$\mathbf{s}_{v_1,v_3} = [v_1^4 v_3^4, v_1^4 v_3^2, v_1^4, v_1^2 v_3^4, v_1^2 v_3^2, v_1^2, v_3^4, v_3^2, 1]^T, \tag{15}$$

while the corresponding 9×1 array of link arc length parameters $\boldsymbol{\alpha}_{v_1,v_3}$ are the coefficients scaled by the v_1 and v_3 variable elements in $\mathbf{s}_{v_1,v_3}$. Once the v_3 parameters are replaced by the prescribed function $v_3 = f(v_1)$, the array is integrated between the desired bounds of $v_{1_{\min}}$ and $v_{1_{\max}}$, the only synthesis equation required is revealed with the numerical minimisation of an Euclidean inner product, which can be generalised for all six $v_j = f(v_i)$ function generator possibilities:

$$\min_{(\alpha_1,\alpha_2,\alpha_3,\alpha_4)\in\mathbb{R}} \left(\boldsymbol{\alpha}_{v_i,v_j} \cdot \int_{v_{i_{\min}}}^{v_{i_{\max}}} \mathbf{s}_{v_i,f(v_i)} dv_i\right). \tag{16}$$

The numerical integrator and optimiser used in Maple 2021 require initial guesses for the link arc length parameters. Three reasonable initial guesses are those that satisfy the exact synthesis problem after setting $\alpha_4 = 1$ to normalise the equation. For the three precision pairs of $(v_1, v_3) = (-2, 3.0296), (0, 2), (2, 3.0296)$ the standard Maple solver could not identify any real, or non-trivial solutions. We decided instead to use the optimiser, which also requires initial guesses. We arbitrarily selected $(\alpha_1, \alpha_2, \alpha_3) = (3/10, 7/12, 11/10)$: these Maple algorithms are more reliable when rational numbers, ideally with prime numerators, are used for the initial guesses rather than floating point decimals. For the exact synthesis problem, the Maple optimiser returned the floating point numbers listed in Table 1, with $\alpha_4 = 1$. For the integrator initial guesses we used the same as those for the exact synthesis optimiser and the continuous approximate synthesis yielded the results listed in Table 1.

The *structural error* (S.E.) of a function generator is considered to be a useful performance indicator for the utility of the identified linkage. It is defined to be the difference between the prescribed linkage output value and the actual generated output value for a given input value [6]. Using Eq. (3) and the prescribed function equation itself, we can visualise the S.E. by examining the plots of the corresponding v_1-v_3 curves, see Fig. 2a, and by computing the area between the two curves. Since we have algebraic expressions for the three curves generated by the exact and continuous approximate synthesis and the prescribed function, it is a simple matter to re-define the S.E. as the difference of the integrals of the prescribed and generated functions. Comparing two different linkages designed to generate the same function over the same range, the one with the smallest structural error generates the prescribed function with the greatest precision. For the synthesis results listed in Table 1 of the $v_3 = f(v_1)$ function generator, we observe that the linkage identified with our modified continuous approximate synthesis method generates the prescribed function with a S.E. that is an order of magnitude smaller than the S.E. of the linkage identified with the standard exact precision point method.

$v_4 = f(v_1)$ Function Generator. The prescribed $v_4 = f(v_1)$ function for this example is

$$v_4 = 2 + \tan\left(\frac{v_1}{v_1^2+1}\right), \tag{17}$$

over the same range as the previous example, $-2 \leq v_1 \leq 2$. We will again use the modified continuous approximate synthesis using Eq. (1) after making the substitution $v_4 = f(v_1)$ and integrating the square of the resulting equation over the specified bounds. In this case the 13×1 array $\mathbf{s}_{v_1,v_4}$ of angle parameters is

$$\mathbf{s}_{v_1,v_4} = [v_1^4v_4^4, v_1^4v_4^2, v_1^4, v_1^3v_4^3, v_1^3v_4, v_1^2v_4^4, v_1^2v_4^2, v_1^2, v_1v_4^3, v_1v_4, v_4^4, v_4^2, 1]^T, \tag{18}$$

while the corresponding 13×1 array of link arc length parameters $\boldsymbol{\alpha}_{v_1,v_4}$ are the coefficients scaled by the v_1 and v_4 elements in $\mathbf{s}_{v_1,v_4}$. Proceeding as in the previous example, we use as our initial guesses for the integrator the α_i that satisfy the exact synthesis problem by solving the three equations generated by the three IO precision pairs of $(v_1, v_4) = (-2, 1.5772), (0, 2), (2, 2.4228)$. It is important to note that the exact synthesis using the spherical 4R algebraic IO equation can, according to Bezout's theorem, lead to as many as 216 solutions for α_1, α_2, and α_3, all in terms of α_4. In this example there are 32 solutions. We used the first, $\alpha_1 = -0.1083$, $\alpha_2 = 0.5183$, and $\alpha_3 = 1.0432$ for an initial guess in the optimiser. The resulting continuous approximate synthesis results are listed in Table 1. Again, the S.E. for linkage identified with the continuous approach is an order of magnitude smaller than that of the linkage identified with exact synthesis, as illustrated in Fig. 2b.

Table 1. Exact and continuous approximate synthesis results.

Function	Synthesis	α_1	α_2	α_3	α_4	S.E
$v_3 = 2 + \tan\left(\frac{v_1^2}{v_1^2+1}\right)$	Exact	0.0226	0.2023	1.3459	1	0.0052
	Continuous	0.0372	0.3460	1.3244	0.7998	0.0007
$v_4 = 2 + \tan\left(\frac{v_1}{v_1^2+1}\right)$	Exact	−0.1083	0.5183	1.0432	1	0.1010
	Continuous	−0.1030	0.4920	0.7512	0.6199	0.0165

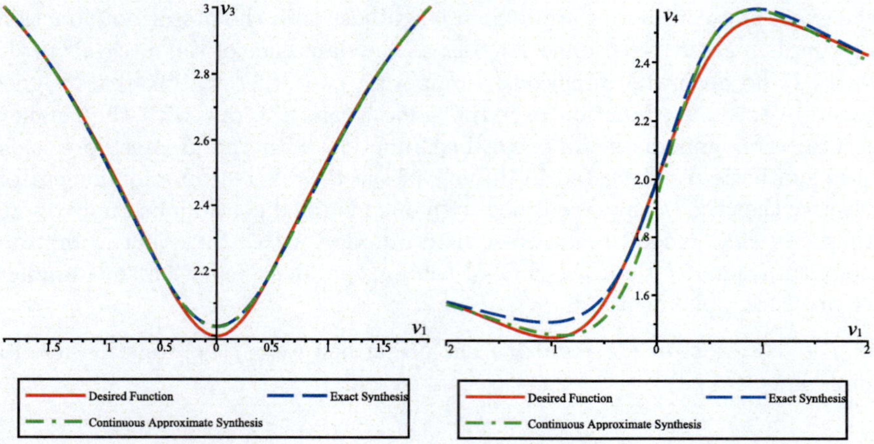

(a) Synthesis results for $v_3 = f(v_1)$. (b) Synthesis results for $v_4 = f(v_1)$.

Fig. 2. Synthesis results.

4 Conclusions

In this paper we have derived the six possible spherical 4R algebraic IO equations that describe the relative input and output angles between different pairs of edges in a spherical quadrangle. The equations were derived using Study's soma coordinates that represent the displacement space of all spherical 4R kinematic chains, and elimination methods to reveal the desired algebraic IO equation. We also illustrated the utility of the equations to function generator synthesis using a novel continuous approximate synthesis approach to implicitly minimise the structural error.

References

1. Adams, W., Loustaunau, P.: An Introduction to Gröbner Bases, vol. 3. American Mathematical Society, Graduate Studies in Mathematics (1994)
2. Bricard, R.: Mémoire sur la théorie de l'octaèdre articulé. J. Math. Pures Appl. **3**, 113–148 (1897)
3. Ceccarelli, M. (ed.): Distinguished Figures in Mechanism and Machine Science, Their Contributions and Legacies Part 1. Springer, New York (2007). https://doi.org/10.1007/978-1-4020-6366-4
4. Copeland, Z., Rotzoll, M., Hayes, M.J.D.: Concurrent type and dimensional continuous approximate function generator synthesis for all planar four-bar mechanisms. In: 11th CCToMM Symposium on Mechanisms, Machines, and Mechatronics, Ontario Tech University, Oshawa, ON, Canada (2021)
5. Gosselin, C.M., Hamel, J.F.: The agile eye: a high-performance three-degree-of-freedom camera-orienting device. In: Proceedings of the 1994 IEEE International Conference on Robotics and Automation (1994)

6. Hartenberg, R., Denavit, J.: Kinematic Synthesis of Linkages. McGraw-Hill, Book Co., New York (1964)
7. Hayes, M.J.D., Rotzoll, M., Iraei, A., Nichol, A., Bucciol, Q.: Algebraic differential kinematics of planar 4R linkages. In: 20th International Conference on Advanced Robotics, ICAR 2021, Ljubljana, Slovenia (2021)
8. Moazami, S., Zargarzadeh, H., Palanki, S.: Kinematics of spherical robots rolling over 3D terrains. Complexity **2019**. (2019) https://doi.org/10.1155/2019/7543,969
9. Rotzoll, M., Hayes, M.J.D., Husty, M.L., Pfurner, M.: A general method for determining algebraic input-output equations for planar and spherical 4R linkages. In: Lenarčič, J., Siciliano, B. (eds.) ARK 2020. SPAR, vol. 15, pp. 90–97. Springer, Cham (2021). https://doi.org/10.1007/978-3-030-50975-0_12
10. Willis, R.: Principles of Mechanism, 2nd edn. Cambridge University Press, Cambridge (1841)

Spherical Mechanisms

Kinematics of a Gear-Based Spherical Mechanism

Federico Thomas(✉)

Institut de Robòtica i Informàtica Industrial (CSIC-UPC), ETSEIB, Diagonal 647, Pavelló E, 08028 Barcelona, Spain
f.thomas@csic.es

Abstract. Kazuki Abe and his collaborators have recently presented an actuated gear-based spherical mechanism called ABENICS. It has received a lot of attention, not only because of its eye-catching motions during operation, but also and mostly, because it can successfully be used when large motion ranges and a high stiffness are required. Nevertheless, the main disadvantage of Abe et al.'s design is that it is an over-actuated mechanism: it requires four instead of only three actuators. In this paper, we propose a variation on this mechanism which requires three actuators, thus simplifying its control and its potential cost. The kinematics of this new mechanism is studied in detail, including its forward and inverse kinematics, as well as its singularities.

1 Introduction

Many different parallel spherical mechanisms have been proposed in the past. The mechanism proposed in [3] is probably the most famous one within this category. For more recent alternative designs, see [4] and the references therein. These mechanisms provide high positioning accuracy and excellent dynamic characteristics. Nevertheless, achieving both a large motion range and a high stiffness is a challenging goal for them [2]. A gear-based mechanism might be a good alternative to alleviate these limitations, however it is not obvious how to design a spherical gear to attain arbitrary spherical motions. Although the idea of engraving tooth patterns on a sphere has long been proposed as an improvement for universal joints (see, for example, [5]), it seems that the use of a cross gear engraved over the full surface of a sphere to accomplish general spherical motions has only been recently proposed in [1] under the name of ABENICS. This design consists of a cross spherical gear and two monopole gears. The main drawback of the ABENICS design is that it is an over-actuated parallel mechanism: it uses four actuators to control the three degrees of freedom of the moving

O. Altuzarra and A. Kecskémthy (Eds.): ARK 2022, SPAR 24, pp. 323–331, 2022.
https://doi.org/10.1007/978-3-031-08140-8_35

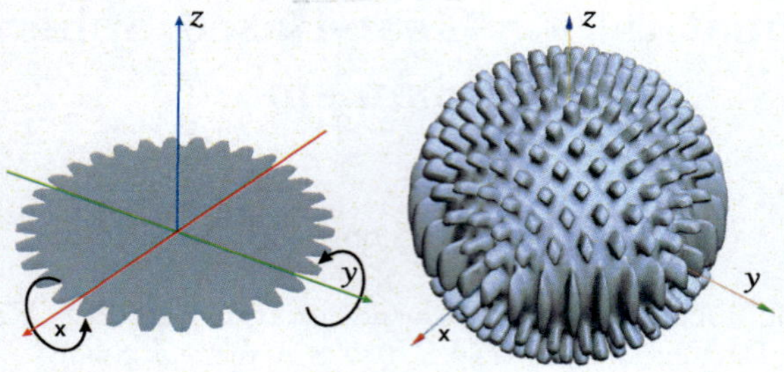

Fig. 1. The cross spherical gear is made by cutting two axisymmetric teeth patterns on a sphere. In this case, the $x-$ and the $y-$ axes.

sphere. The two orientation angles of each monopole are actuated, thus leading to a dependency between them which increases their control complexity. In this paper, we explore the possibility of using three monopoles in such a way that only one orientation angle of each monopole is actuated and the order is left free.

The cross spherical gear is made by cutting two axisymmetric teeth patterns on a sphere. According to Fig. 1, the sphere is first cut around the x-axis following the profile of an ordinary involute gear, and then this operation is repeated around the y-axis. The resulting cross spherical gear has four poles located at $\mathbf{p}_1 = (R, 0, 0)$, $\mathbf{p}_2 = (0, R, 0)$, $\mathbf{p}_3 = (-R, 0, 0)$, and $\mathbf{p}_4 = (0, -R, 0)$.

The monopole gears have a simpler teeth pattern which can mesh with the cross spherical gear. Their radii are half that of the sphere. As shown in Fig. 2, when their $z-$ and $x-$ axes are aligned with the rotational axis and the pole, respectively, their cross-sections with the $xy-$plane also have the typical involute gear profile. The monopoles have only one pole, and hence their name. In their local reference frames, this pole is located, according to Fig. 2, at $\mathbf{q} = (R/2, 0, 0)$.

When meshing a monopole with the spherical gear, we have to previously match its pole with one of the four poles of the sphere. Nevertheless, due to the symmetry of the sphere, observe that it is equivalent to match a monopole with $\mathbf{p}_1$ or $\mathbf{p}_2$, or with $\mathbf{p}_3$ or $\mathbf{p}_4$. Thus, it can simply be said that the monopoles have to be matched with either the $x-$ or the $y-$ axis.

The ABENICS mechanism consists of a cross spherical gear and two monopoles located at a right angle on a maximum circle of the sphere. The orientation of each monopoles is controlled by two actuators. This arrangement of the monopoles, and the fact that their locations are fully controlled, make the kinematics analysis of the mechanism rather trivial. Apparently, there was no simple way to avoid the over-actuation and to keep the kinematics of the mechanism simple at the same time. This is the problem essentially treated in this paper.

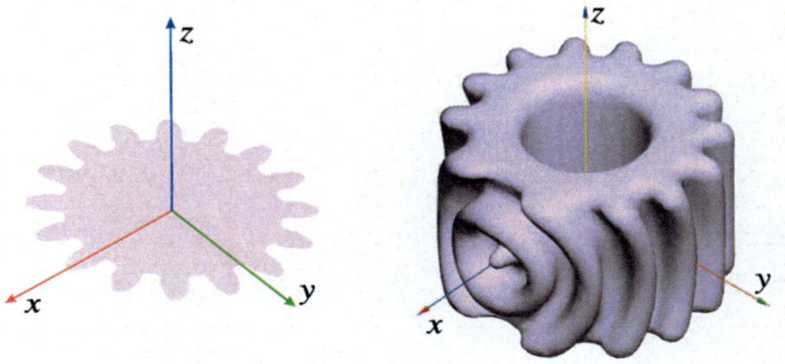

Fig. 2. The monopoles are made by cutting a teeth pattern on a cylinder that can mesh with the cross spherical gear.

This paper is structured as follows. Section 2 presents the proposed variation on the ABENICS mechanism. Section 3 deals with its inverse kinematics. The main challenge imposed by the new design is the resolution of its forward kinematics for which we found a particularly elegant closed-form formula presented in Sect. 4. The mechanism singularities are deduced in Sect. 5. The correctness of the presented formulas is verified for an example in Sect. 6. This paper is concluded in Sect. 7 with some final considerations.

2 The Proposed Mechanism

In the proposed mechanism, three monopoles are regularly distributed on the xy-plane of the sphere. According to the reference frames in Fig. 1, Fig. 2, and Fig. 3, their local reference frames in the mechanism will be given by the displacements

$$\mathbf{D}_1 = \mathbf{T}(R + R/2, 0, 0)\,\mathbf{R}_z(\pi)\,\mathbf{R}_x(\pi/2), \tag{1}$$

$$\mathbf{D}_2 = \mathbf{R}_z(2\pi/3)\,\mathbf{D}_1, \tag{2}$$

$$\mathbf{D}_3 = \mathbf{R}_z(-2\pi/3)\,\mathbf{D}_1. \tag{3}$$

where the term $\mathbf{R}_x(\pi/2)$ is added to simplify the inverse kinematics as we will show later. The orientation of the sphere will be given by the proper orthogonal matrix $\mathbf{R} = (r_{ij})_{1\leq i,j\leq 3}$.

Without loss of generality, it has been decided to match monopole 1 with the sphere x-axis, and the other two monopoles with the y−axis. Matching all the monopoles with the same axis would make any rotation of the sphere about this axis uncontrollable, as it will become clear in the following section. In the proposed design, the angles θ_1, θ_2, and θ_3 around each monopole's z−axis are actuated, while ϕ_1, ϕ_2, and ϕ_3 are left as passive angles (see Fig. 3).

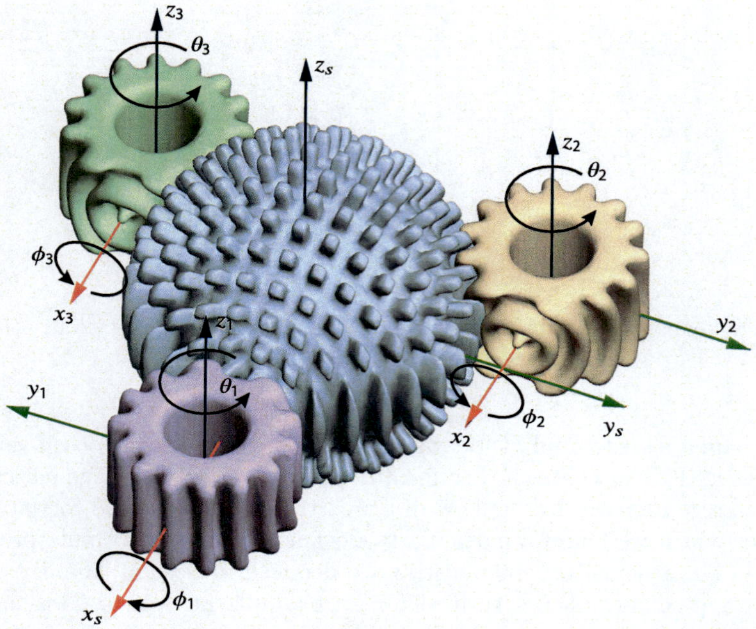

Fig. 3. Arrangement of a cross spherical gear and three monopoles in the proposed new mechanism. The definition of their local reference frames and associated monopoles orientation angles are also shown. While θ_1, θ_2, and θ_3 are actuated, ϕ_1, ϕ_2, and ϕ_3 are passive.

3 Inverse Kinematics

Observe that, if a monopole is matched with the sphere x-axis (y–axis), its orientation is invariant with respect to any rotation of the sphere about this axis. Indeed, such a rotation induces a lateral sliding of the monopole over the sphere which does not alter its orientation. Thus, only the orientation of the x–axis (y–axis) is relevant in the computation of the monopole's orientation.

Since the orientation of the sphere is given by the rotation matrix $\mathbf{R}$, we have that the orientation angles of monopole 1 are given by (see Fig. 4)

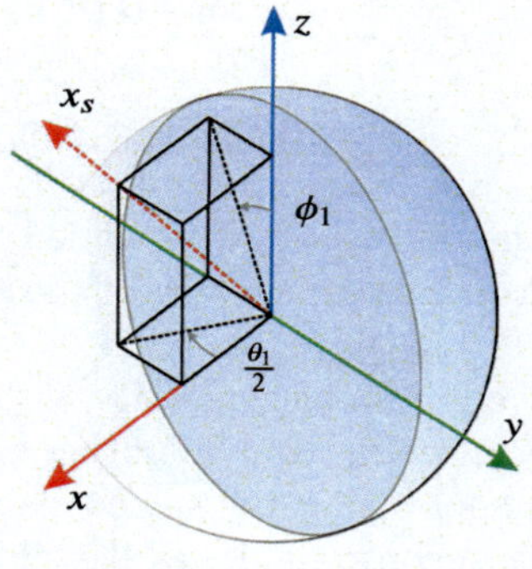

Fig. 4. Computation of the inverse kinematics for monopole 1 which is matched with the sphere x-axis, whose orientation is given by $\mathbf{x}_s = \mathbf{R}\,\mathbf{x} = (r_{1,1}, r_{2,1}, r_{3,1})^T$.

$$\phi_1 = \mathrm{atan2}(r_{2,1}, r_{3,1}) \text{ and } \theta_1 = 2\arccos(r_{1,1}). \tag{4}$$

In other words, the location of monopole 1 can be expressed in the world reference frame as

$$\mathbf{M}_1 = \mathbf{D}_1\,\mathbf{R}_x(\phi_1)\,\mathbf{R}_z(\theta_1). \tag{5}$$

Now, observe that, due to the invariance of the spherical gear with respect to rotations of $\pm\pi/2$ about its z-axis, the orientation angles of monopoles 2 and 3 can still be obtained using (4)—despite they are matched to the sphere y-axis—by simply performing the substitutions

$$\mathbf{R} \leftarrow \mathbf{R}_z(-2\pi/3)\,\mathbf{R}\,\mathbf{R}_z(\pi/2), \quad \text{and} \quad \mathbf{R} \leftarrow \mathbf{R}_z(2\pi/3)\,\mathbf{R}\,\mathbf{R}_z(\pi/2), \tag{6}$$

respectively. By performing these substitutions, it can be checked that the orientations of monopoles 2 and 3 are given by the angles

$$\begin{aligned} \phi_2 &= \operatorname{atan2}\left(-\frac{1}{2}r_{2,2} - \frac{\sqrt{3}}{2}r_{1,2}, r_{3,2}\right), & \theta_2 &= 2\arccos\left(\frac{\sqrt{3}}{2}r_{2,2} - \frac{1}{2}r_{1,2}\right), \\ \phi_3 &= \operatorname{atan2}\left(\frac{1}{2}r_{1,2} - \frac{\sqrt{3}}{2}r_{2,2}, r_{3,2}\right), & \theta_3 &= 2\arccos\left(-\frac{1}{2}r_{1,2} - \frac{\sqrt{3}}{2}r_{2,2}\right). \end{aligned} \tag{7}$$

4 Forward Kinematics

The set of equations in (4) and (7) can be rewritten as

$$r_{11} = \cos\frac{\theta_1}{2}, \tag{8}$$

$$r_{21} = k_1 \sin\phi_1, \tag{9}$$

$$r_{31} = k_1 \cos\phi_1, \tag{10}$$

$$r_{12} = \frac{1}{\sqrt{3}}(k_3 \sin\phi_3 - k_2 \sin\phi_4) = -\cos\frac{\theta_2}{2} - \cos\frac{\theta_3}{2}, \tag{11}$$

$$r_{22} = \frac{1}{\sqrt{3}}\left(\cos\frac{\theta_2}{2} - \cos\frac{\theta_3}{2}\right) = -k_2 \sin\phi_2 - k_3 \sin\phi_3, \tag{12}$$

$$r_{32} = k_2 \cos\phi_2 = k_3 \cos\phi_3. \tag{13}$$

where k_1, k_2, and k_3 are constants that cancel when computing the arctangent in the inverse kinematics computations.

Now, observe that using equations (8), (11), and (12), the rotation matrix representing the orientation of the sphere—in terms of the actuated angles, θ_1, θ_2, and θ_3—can be expressed as

$$\mathbf{R} = \begin{pmatrix} c_1 & -c_2 - c_3 & \boxed{r_{13}} \\ \boxed{r_{21}} & \frac{1}{\sqrt{3}}(c_2 - c_3) & \boxed{r_{23}} \\ \boxed{r_{31}} & \boxed{r_{32}} & \boxed{r_{33}} \end{pmatrix}, \tag{14}$$

where the boxed entries are unknown, and $c_i = \cos\frac{\theta_i}{2}$. Thus, the forward kinematics problem reduces to find the different ways in which this matrix can be

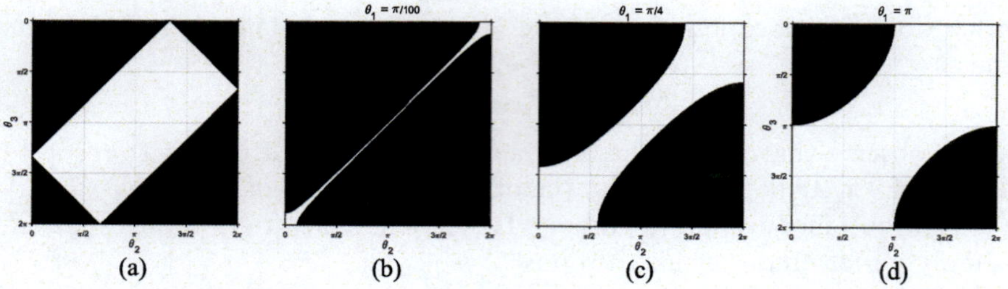

Fig. 5. The feasible actuation region for θ_2 and θ_3 is obtained by intersecting the region in (a) with a region that depends on θ_1. This latter region is shown for $\theta_1 = \pi/100$ (b), $\theta_1 = \pi/4$ (c), and $\theta_1 = \pi$ (d). In all cases, the light areas represent the feasible regions.

completed. Since it must be proper orthogonal, the problem can be reduced to solve the system of equations resulting from imposing the algebraic conditions $\mathbf{R}\mathbf{R}^T = \mathbf{I}$ and $\det(\mathbf{R}) = 1$. Nevertheless, using this approach the problem becomes quite involved. Fortunately, a simpler and neater solution is possible by relying on Euler angles.

An arbitrary rotation matrix can be expressed in terms of XZY Euler angles as

$$\mathbf{R} = \mathbf{R}_x(\alpha_1)\mathbf{R}_z(\alpha_2)\mathbf{R}_y(\alpha_3) = \begin{pmatrix} C_2C_3 & -S_2 & C_2S_3 \\ \boxed{S_1S_3 + C_1C_3S_2} & C_1C_2 & \boxed{C_1S_2S_3 - C_3S_1} \\ \boxed{C_3S_1S_2 - C_1S_3} & C_2S_1 & \boxed{C_1C_3 + S_1S_2S_3} \end{pmatrix}, \tag{15}$$

where $S_i = \sin\alpha_i$ and $C_i = \cos\alpha_i$, and

$$\alpha_1 = \operatorname{atan2}(r_{32}, r_{22}),\ \alpha_2 = \arcsin(-r_{12}),\ \text{and}\ \alpha_3 = \operatorname{atan2}(r_{13}, r_{11}). \tag{16}$$

Since the calculation of α_1, α_2, and α_3 does not require the values of the boxed entries in (15), we can readily conclude from (14) that

$$\begin{aligned}\mathbf{R} = \mathbf{R}_x &\left(\operatorname{atan2}\left(\pm\sqrt{3 - 3(c_2 + c_3)^2 - (c_2 - c_3)^2}, c_2 - c_3\right)\right)\\ &\mathbf{R}_z\left(\arcsin\left(c_2 + c_3\right)\right)\\ &\mathbf{R}_y\left(\operatorname{atan2}\left(\pm\sqrt{1 - c_1^2 - (c_2 + c_3)^2}, c_1\right)\right).\end{aligned} \tag{17}$$

Therefore, given the values of θ_1, θ_2, and θ_3, four solutions for $\mathbf{R}$ are obtained, one for each combination of signs of the two square roots appearing in (17).

To finish this section, it is worth remembering that, if $(\alpha_1, \alpha_2, \alpha_3)$ is a valid set of XZY Euler angles, $(\pi + \alpha_1, \pi - \alpha_2, \pi + \alpha_3)$ is as well a valid set. Nevertheless, this fact is irrelevant in our case because both sets lead to the same completion of $\mathbf{R}$.

5 Singularities

When the pole of monopole $i = 1, 2, 3$ is in contact with the sphere, ϕ_i is undefined. That is, when the orientation of the sphere is

$$\mathbf{R} = \mathbf{R}_z(n\pi)\mathbf{R}_x(\omega),\ \mathbf{R} = \mathbf{R}_z\left(\frac{2\pi}{3} + n\pi\right)\mathbf{R}_y(\omega),\ \text{or}\ \mathbf{R} = \mathbf{R}_z\left(-\frac{2\pi}{3} + n\pi\right)\mathbf{R}_y(\omega), \tag{18}$$

with $n \in \mathbb{Z}$ and $\omega \in \mathbb{R}$, the mechanism is in a singularity of the inverse kinematics. The set of these singularities can algebraically identified as those cases in with the two arguments of the atan2 functions in (4) or (7) are simultaneously zero.

Besides these rather obvious singularities, we also have the singularities of the forward kinematics. There are values of the actuated angles for which there is a change in the number of forward kinematics solutions. In these cases, at least one of the squared roots arguments in (17) are zero. Moreover, observe that the forward kinematics problem can be solved provided that the following conditions are satisfied

$$(c_2 + c_3)^2 + \frac{1}{3}(c_2 - c_3)^2 \geq 1 \quad \text{and} \quad c_1^2 + (c_2 + c_3)^2 \geq 1. \tag{19}$$

These inequalities define a feasible region in the actuation space whose boundary is the singular set, the set where the number of solutions of the forward kinematics drops from 4 to 2, or 1. While the region defined by the first inequality in (19) is independent of θ_1 [see Fig. 5 (a)], the second one depends on it [see the resulting region in the $\theta_1\theta_2$ plane for different values of θ_1 in Fig. 5 (a), (b), and (c)]. The feasible region obviously results from intersecting both regions.

6 Example

Let us consider the particular case in which the actuated angles are arbitrarily chosen as $\theta_1 = 2.4093$, $\theta_2 = 4.4438$, and $\theta_3 = 3.4215$. Given these values, the forward kinematics of the mechanism can be solved using equation (17). The four obtained results appear in Table 1. To verify the correctness of these results, it is possible to compute the inverse kinematics for each solution using (4) and (7) to recover the value of the actuated angles. Solving the inverse kinematics for each case also gives us the corresponding values of ϕ_1, ϕ_2, and ϕ_3, that is, the passive joints angles. They are also given in Table 1. The graphical representation of the four assembly modes appears in Fig. 6.

Table 1. Forward kinematics solutions for $\theta_1 = 2.4093$, $\theta_2 = 4.4438$, and $\theta_3 = 3.4215$. The corresponding values for the passive joints angles are also included.

Assembly mode	**R**	ϕ_1	ϕ_2	ϕ_3
1	$\begin{pmatrix} 0.35802 & 0.74558 & 0.56208 \\ 0.93337 & -0.26938 & -0.23719 \\ -0.025432 & 0.60954 & -0.79234 \end{pmatrix}$	1.5981	−0.69768	0.9077
2	$\begin{pmatrix} 0.35802 & 0.74558 & 0.56208 \\ -0.60955 & -0.26938 & 0.74558 \\ 0.7073 & -0.60955 & 0.35802 \end{pmatrix}$	−0.7113	−2.4439	2.2339
3	$\begin{pmatrix} 0.35802 & 0.74558 & -0.56208 \\ -0.60955 & -0.26938 & -0.74558 \\ -0.7073 & 0.60954 & 0.35802 \end{pmatrix}$	−2.4303	−0.69768	0.9077
4	$\begin{pmatrix} 0.35802 & 0.74558 & -0.56208 \\ 0.93337 & -0.26938 & 0.23719 \\ 0.025432 & -0.60954 & -0.79234 \end{pmatrix}$	1.5436	−2.4439	2.2339

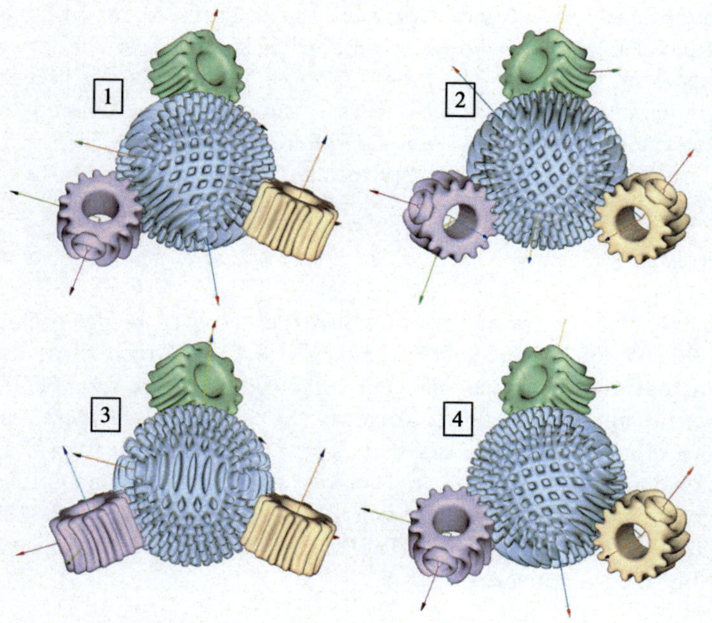

Fig. 6. The four assembly modes given in Table 1.

7 Conclusion

We have presented a variation on the recently proposed ABENICS mechanism that simplifies its actuation. The main challenge associated with this variation was the resolution of its forward kinematics and the characterization of its singularities. Curiously enough, however, the use of Euler angles have been shown to be very convenient to complete rotation matrices in a way that allowed us to derive elegant close-form solution formulas for these two problems.

Acknowledgements. We gratefully acknowledge the financial support of the Spanish Ministry of Economy and Competitiveness through the project PID2020-117509GB-I00.

References

1. Abe, K., Tadakuma, K., Tadakuma, R.: ABENICS: active ball joint mechanism with three-DoF based on spherical gear meshings. IEEE Trans. Robot. **37**(5), 1806–1825 (2021)
2. Bajaj, N.M., Spiers, A.J., Dollar, A.M.: State of the art in artificial wrists: a review of prosthetic and robotic wrist design. IEEE Trans. Robot. **35**(1), 261–277 (2019)
3. Gosselin, C.M., St Pierre, E., Gagne, M.: On the development of the agile eye. IEEE Robot. Autom. Mag., **3**(4), 29–37 (1996)
4. Kumar, S., Bongardt, B., Simnofske, M., et al.: Design and kinematic analysis of the novel almost spherical parallel mechanism active ankle. J. Intell. Robot. Syst. **94**, 303–325 (2019)
5. Ting, L., Cunyun, P.: On grinding manufacture technique and tooth contact and stress analysis of ring-involute spherical gears. Mech. Mach. Theory **44**(10), 1807–1825 (2009)

Modeling of a Remote Center of Motion Spherical Parallel Tensegrity Mechanism for Percutaneous Interventions

H. El Jjouaoui[1](✉), G. Cruz-Martinez[2], J.-C. Avila Vilchis[2], A. Vilchis González[2], S. Abdelaziz[1], and P. Poignet[1]

[1] LIRMM, Univ Montpellier, CNRS, Montpellier, France
hamza.el-jjouaoui@lirmm.fr

[2] Autonomous University of Mexico State, Toluca, Mexico

Abstract. The present paper deals with the mathematical modeling of a new 2 DOF remote center of motion spherical parallel tensegrity mechanism, dedicated to percutaneous needle interventions. Analytical inverse kinematic and numerical direct kinematic models are developed. Trilateration approach is considered in order to determine the coordinates of the joints that constitute the system. A 3D prototype of the mechanism has been developed for future evaluations. This work constitutes a first step towards the control of the mechanism.

Keywords: Spherical RCM parallel structure · Tensegrity system · Modeling

1 Introduction

In interventional radiology, needle puncture is widely used for cancer diagnosis and treatment, such as biopsy and ablation [1,2]. To perform these gestures, it is necessary to manually adjust the needle position and orientation. Using a robotic assistant instead of the radiologist's hand to position the needle is of interest since it increases the needle position accuracy [3,4].

Using a tensegrity architecture to design a robotic assistant is of great interest, particularly when stiffness variation is required [5]. Tensegrity structures were introduced by Richard B. Fuller [6]. They can be defined as structures composed of rigid compressed elements (bars) forming a self-equilibrium that preserve its stable state using the forces produced by the tension of flexible elements (springs, cables) that are linked to the rigid parts [7]. Designing a robot based on tensegrity allows to produce efficient structures [8] with variable stiffness, high precision as well as high volume-to-mass ratio and stiffness-to-mass ratio [9].

O. Altuzarra and A. Kecskémthy (Eds.): ARK 2022, SPAR 24, pp. 332–339, 2022.
https://doi.org/10.1007/978-3-031-08140-8_36

The main challenge for such robotized medical interventions is to design a remote center of motion (RCM) mechanism that allows a rotational movement around a fixed point, which is in our case the needle insertion point. There are several mechanism architectures in the literature that guarantee a rotation around an RCM [10]. Our approach is based on the use of a spherical parallel RCM [11] that is redesigned to incorporate the concept of tensegrity. Pantographs, constituted by rigid curved bars, are introduced to the mechanism. These bars are connected to each other using revolute joints. The joints axis are directed towards the RCM [12]. The pantographs have the form of spherical parallelograms that allow the incorporation of cables and springs to define the system as a spherical RCM tensegrity mechanism. It is a 2 DOF mechanism driven by 4 cables. The 2 degrees of redundancy are used to vary the stiffness of the mechanism.

This work deals with the kinematic modeling of such a system. In Sect. 2, the system description is introduced. In Sect. 3, the inverse and the direct kinematic models as well as the workspace estimation are derived. The trilateration approach that allows the computation of the joints coordinates is also introduced. Finally, conclusions and perspectives are discussed in Sect. 4.

2 System Description

The system, as illustrated in Fig. 1, is a spherical RCM mechanism. The needle guide is the end effector of the mechanism. It is manipulated using two spherical pantographs. The first pantograph is located in a sphere surface of radius R_1. It is connected to the base at the joint A_1. Similarly, the second pantograph is located in a sphere surface of radius R_2 and is connected to the base at the joint A_2. The manipulation of each pantograph is obtained by manipulating a pair of two cables. The first cable is attached to the joint B_1, passes through a pulley located at D_1 and some other pulleys before being winded on a first actuated pulley. The second cable is attached to the joint C_1, passes through a pulley located at A_1 and some other pulleys before being winded on the second actuated pulley. Similarly, the second pantograph is manipulated using two actuated cables 3 and 4.

3 System Modeling

The origin of the reference frame $\mathcal{R}_0 = (O, \mathbf{x}_0, \mathbf{y}_0, \mathbf{z}_0)$ is defined as the RCM of the mechanism (Fig. 1). The end effector orientation is defined by $\mathbf{x} = [\eta\ \mu]^T$ (Fig. 2, right). The coordinates of the joint T are expressed in the reference frame $\mathcal{R}_0$ as $\mathbf{T} = [T_x\ T_y\ T_z]^T$. The joint variables are defined by $\boldsymbol{\beta} = [\beta_1\ \beta_2]^T$. The variable β_i represents the angle between $(A_iB_i^{'})$ and $(A_iD_i^{'})$. The points $B_i^{'}$ and $D_i^{'}$ represent respectively the projection of the points B_i and D_i on the plane $(P1)$ that is considered as a tangential plane (Fig. 2, left). These joint variables are measured using optical encoders located at the joints A_1 and A_2 (Fig. 1).

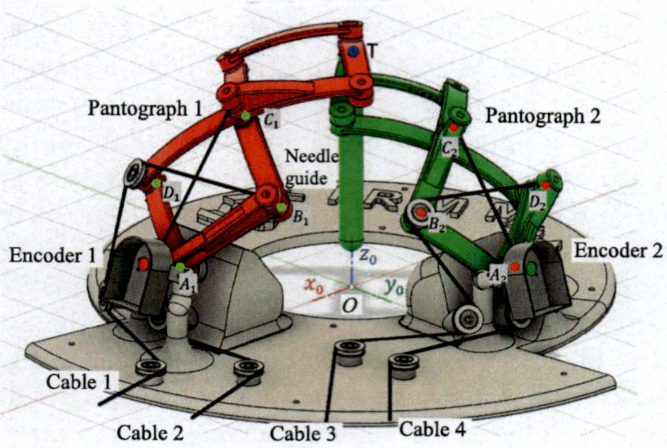

Fig. 1. System overview.

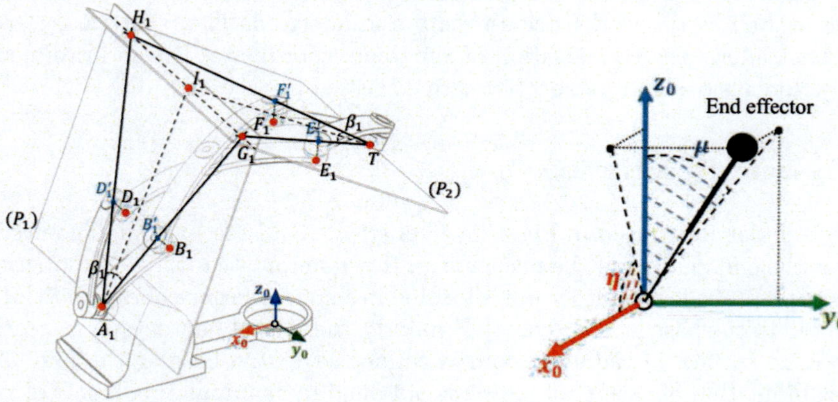

Fig. 2. Left, joint variables definition. Right, end effector orientation.

3.1 Inverse Kinematic Model

The inverse kinematic (IK) model allows to express the joint variables $\boldsymbol{\beta} = [\beta_1\ \beta_2]^T$ according to the orientation $\mathbf{x} = [\eta\ \mu]^T$ of the needle guide. This orientation is supposed to be known and can be expressed using the coordinates of $\mathbf{T} = [T_x\ T_y\ T_z]^T$. The IK model is determined in two steps. First, a relationship between β_i and θ_i is established. θ_i represents the angle between (OT) and (OA_i) (Fig. 3). In the second step, a relationship between the angle θ_i and the end effector position $\mathbf{T}$ is derived.

From Fig. 2, left, one can notice that $\|A_1 I_1\| = f_1 \cos(\beta_1/2)$ where $f_1 = \|A_1 G_1\|$. The distance f_1 is fixed whereas $\|A_1 I_1\|$ is variable. This latter can be used to compute the distance $\|A_1 K_1\| = \|A_1 I_1\| \cos(\theta_1/2)$, as it can be observed from Fig. 3. Besides, $\|A_1 K_1\| = R_1 \sin(\theta_1/2)$ where $R_1 = \|OA_1\|$. Based on these 3 equations, one can express the relationship between β_1 and θ_1, and similarly the relationship between β_2 and θ_2 for the second pantograph:

$$\beta_i = 2 * \cos^{-1}(R_i/f_i * \tan(\theta_i/2)) \tag{1}$$

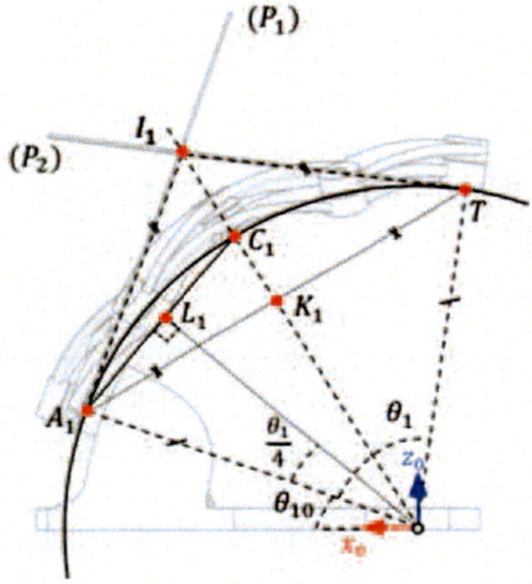

Fig. 3. Definition of the relationship between θ_1 and β_1

Knowing the coordinates of the vectors $\mathbf{T}$, $\mathbf{A}_1$ and $\mathbf{A}_2$, it is possible to compute the angle θ_1 between $\mathbf{T}$ and $\mathbf{A}_1$, and the angle θ_2 between $\mathbf{T}$ and $\mathbf{A}_2$:

$$\begin{cases} \theta_1 = \pm\cos^{-1}(\dfrac{\mathbf{A}_1^T\mathbf{T}}{\|\mathbf{A}_1\|\|\mathbf{T}\|}) & = \pm\cos^{-1}(\dfrac{T_x \cos(\theta_{10}) + T_z \sin(\theta_{10})}{R_1}) \\ \theta_2 = \pm\cos^{-1}(\dfrac{\mathbf{A}_2^T\mathbf{T}}{\|\mathbf{A}_2\|\|\mathbf{T}\|}) & = \pm\cos^{-1}(\dfrac{T_y \cos(\theta_{20}) + T_z \sin(\theta_{20})}{R_1}) \end{cases} \tag{2}$$

where θ_{10} is the angle between (OA_1) and $\mathbf{x}_0$ axis and θ_{20} is the angle between (OA_2) and $\mathbf{y}_0$ axis. Only positive solutions are considered, since the robot can evolve only in the upper hemisphere.

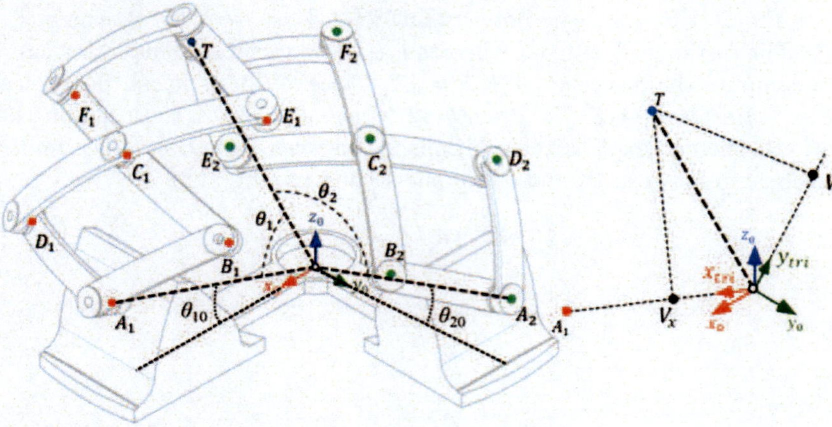

Fig. 4. Robot parametrization.

3.2 Direct Kinematic Model

The direct kinematic model allows expressing the orientation of the needle guide $\mathbf{x} = [\eta\ \mu]^T$ according to the joint variables $\boldsymbol{\beta} = [\beta_1\ \beta_2]^T$. This is performed also in two steps. First, the relationship between θ_i and β_i is established by inverting (1):

$$\theta_i = 2 * \tan^{-1}(f_i/R_i * \cos(\beta_i/2)) \tag{3}$$

The second step consists in solving the following system of equations, obtained by inverting (2):

$$\begin{cases} T_x\ \cos(\theta_{10}) + T_z\ \sin(\theta_{10}) - R_1\ \cos(\theta_1) = 0 \\ T_y\ \cos(\theta_{20}) + T_z\ \sin(\theta_{20}) - R_1\ \cos(\theta_2) = 0 \\ T_x^2 + T_y^2 + T_z^2 - R_1^2 = 0 \end{cases} \tag{4}$$

This system of equation could possibly be solved analytically, but the results would be difficult to be obtained in our case. A numerical approach is therefore considered here. Finally, the end effector orientations are computed by:

$$\begin{aligned} \eta &= \operatorname{atan2}(T_y, T_z) \\ \mu &= \operatorname{atan2}(T_z, T_x) \end{aligned} \tag{5}$$

To determine the coordinates of the joints B_i, C_i, D_i, E_i and F_i in the reference frame $\mathcal{R}_0$ (Fig. 4), a trilateration approach is considered. This approach is based on the intersection of 3 known spheres to determine the coordinates of the intersection point. In the following, we will show how to determine the coordinates of C_1 and the same approach can be considered to determine the coordinates of the other joints.

The three spheres that are used to determine the coordinates of C_1 are $(S1)$, $(S2)$ and $(S3)$. $(S1)$ has O as the origin and $r_1 = R_1$ as a radius. $(S2)$ has the coordinates of A_1 as a center and $r_2 = \|A_1C_1\|$ as a radius. $(S3)$ has the coordinates of T as a center and $r_3 = \|A_1C_1\|$ as a radius. The distance $\|A_1C_1\|$ is computed as $\|A_1C_1\| = 2R_1\sin(\theta_1/4)$ (Fig. 3). The spheres are defined as:

A trilateration frame $\mathcal{R}_{tri} = (O, \mathbf{x}_{tri}, \mathbf{y}_{tri}, \mathbf{z}_{tri})$ is defined. The $\mathbf{x}_{tri}$ axis is chosen along A_1. The axis $\mathbf{y}_{tri}$ is chosen so that T is in the plane $(\mathbf{x}_{tri}\mathbf{y}_{tri})$:

$$\begin{cases} \mathbf{x}_{tri} = \mathbf{A}_1/\|\mathbf{A}_1\| \\ \mathbf{z}_{tri} = \mathbf{T} \times \mathbf{x}_{tri}/\|\mathbf{T} \times \mathbf{x}_{tri}\| \\ \mathbf{y}_{tri} = \mathbf{z}_{tri} \times \mathbf{x}_{tri} \end{cases} \tag{6}$$

where $\mathbf{A}_1$ represents the vector of coordinates of A_1 in the reference frame. The equations of spheres are:

$$\begin{cases} x^2 + y^2 + z^2 = r_1^2 \\ (x - U)^2 + y^2 + z^2 = r_2^2 \\ (x - V_x)^2 + (y - V_y)^2 + z^2 = r_3^2 \end{cases} \tag{7}$$

where V_x and V_y are the coordinates of T in the trilateration frame. They are computed as $V_x = \mathbf{T}^T\mathbf{x}_{tri}$ and $V_y = \mathbf{T}^T\mathbf{y}_{tri}$. The variable $U = R_1$ represents the coordinate of A_1 in $\mathcal{R}_{tri}$. The coordinates of C_1 in the trilateration frame $\mathcal{R}_{tri}$ can be computed analytically by solving (7). Finally, the coordinates of C_1 in the reference frame are computed as $[^0x\ ^0y\ ^0z]^T = [\mathbf{x}_{tri}\ \mathbf{y}_{tri}\ \mathbf{z}_{tri}][x\ y\ z]^T$.

3.3 Workspace

The geometric workspace of the robot is depicted in Fig. 5. It has been obtained by varying the joint variables $\beta_i \in [0, \pi]$. The geometric parameters of the robot (Fig. 5) have been chosen so that the obtained workspace covers a required workspace defined as a 40° cone with its head pointing to the RCM.

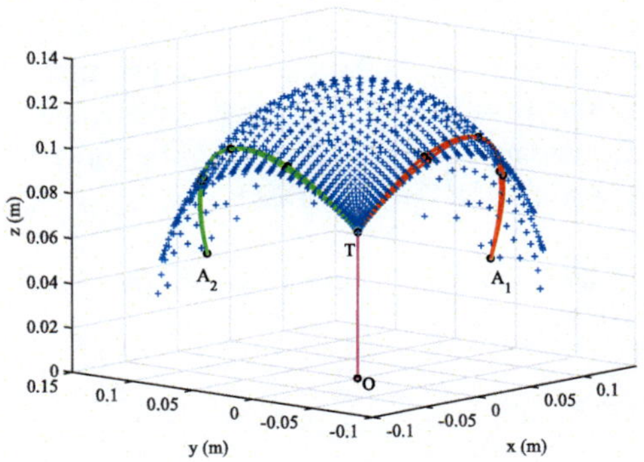

Fig. 5. Geometric workspace obtained using these parameters: R_1 = 133.9 mm, R_2 = 133.5 mm, f_1 = 200 mm, f_2 = 181.6 mm

3.4 Prototype

A prototype of the robot has been developed using 3D printing. Two incremental rotary optical encoders with a resolution of 500 CPR have been integrated in A_1 and A_2. The overall structure has 1100 g weight and a volume similar to a hemisphere of approximately 7.8 dm^3 (Fig. 6).

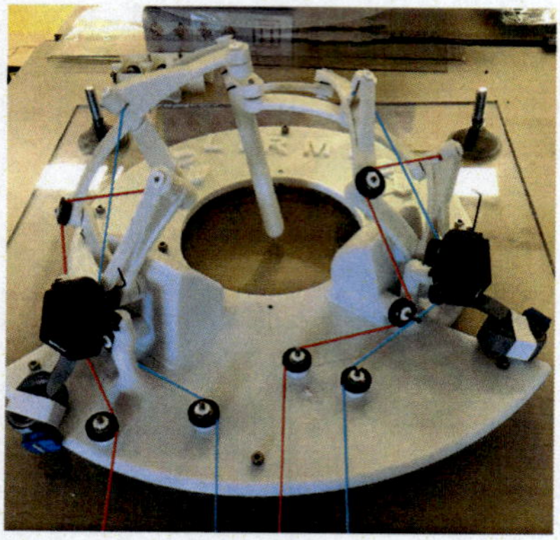

Fig. 6. Prototype of the mechanism.

4 Conclusion and Perspectives

A robotic assistant for percutaneous interventions based on spherical RCM tensegrity mechanism has been proposed in this paper. A geometric approach to determine the inverse and direct kinematic models have been developed. The geometric workspace of the robot is shown as a result of the approach, and a prototype has been mounted for future evaluations. Future work will deal with the differential kinematic modeling, singularity analysis as well as static model determination that are necessary to control the mechanism. Besides and as observed in the actual prototype, structural improvements have to be performed to enhance the vertical stiffness of the robot for a robust manipulation and a better guarantee of the RCM constraint.

Acknowledgement. This work was supported by the Investissements d'Avenir (Labex CAMI ANR-11-LABX-0004).

References

1. Hadavand, M., Naish, M.D., Patel, R.V.: A parallel remote center of motion mechanism for needle-based medical interventions. In: 5th IEEE RAS/EMBS International Conference on Biomedical Robotics and Biomechatronics, pp. 1–6 (2014)
2. Kulkarni, P., Sikander, S., Biswas, P., Frawley, S., Song, S.-E.: Review of robotic needle guide systems for percutaneous intervention. Ann. Biomed. Eng., 1–25 (2019)
3. Wang, Y., et al.: Experimental study of the optimum puncture pattern of robot-assisted needle insertion into hyperelastic materials. Proc. Inst. Mech. Eng. J. Eng. Med. **235**(1), 28–43 (2021)
4. Guo, J., Liu, Y., Wang, J., Zeng, C., Huang, J., Liu, C.: The design of compact robotic-assisted needle position system with MPC-based remote control. Complexity, 1–13 (2020)
5. Realpe, J.R.J., Aiche, G., Abdelaziz, S., Poignet, P.: Asynchronous and decoupled control of the position and the stiffness of a spatial RCM tensegrity mechanism for needle manipulation. In: IEEE International Conference on Robotics and Automation, pp. 3882–3888 (2020)
6. Fuller, B. (ed.): Synergetics, Explorations in the Geometry of Thinking. Collier Macmillan (1975)
7. Jing, Y.Z., Ohsaki, M. (eds.): Tensegrity Structures Form, Stability and Symmetry. Springer, Tokyo (2015). https://doi.org/10.1007/978-4-431-54813-3
8. Juan, S.H., Mirats Tur, J.M.: Tensegrity frameworks: static analysis review. Mech. Mach. Theory **43**(7), 859–881 (2008)
9. Guest, S.D.: The stiffness of tensegrity structures. IMA J. Appl. Math. **76**(1), 57–66 (2011)
10. Aksungur, S.: Remote center of motion (RCM) mechanisms for surgical operations. Int. J. Appl. Math. Electron. Comput. **3**(2), 119–126 (2015)
11. Bai, S.: Optimum design of spherical parallel manipulators for a prescribed workspace. Mech. Mach. Theory **45** 200–211 (2010)
12. Molaei, A., Abedloo, E., Taghirad, H.D., Marvi, Z.: Kinematic and workspace analysis of diamond: an innovative eye surgery robot. In: 23rd Iranian Conference on Electrical Engineering, pp. 882–887 (2015)
13. Skelton, R.E., De Oliveira, M.C.: Tensegrity Systems. Springer, New York (2009). https://doi.org/10.1007/978-0-387-74242-7
14. Yixiang, L., Qing, B., Xiaoming, Y., Jiang, W., Bin, Y., Yibin, L.: A review on tensegrity structures-based robots. Mech. Mach. Theory **168**, 104571 (2022)
15. Arsenault, M., Gosselin, C.M.: Kinematic, static, and dynamic analysis of a spatial three-degree-of-freedom tensegrity mechanism. J. Mech. Des. **128**(5), 951–966 (2006)
16. Ji, Z., Li, T., Lin, M.: Kinematics, singularity, and workspaces of a planar 4-bar tensegrity mechanism. J. Robot., 1–10 (2014)
17. Kaufman, J., Lee, M.: Vascular and Interventional Radiology. The Requesites, Mosby (2004)

Wriflex: Design and Kinematic Analysis of a Self-aligning Parallel Wrist

Henrique Simas[1(✉)], Roberto Simoni[1], and Raffaele Di Gregorio[2]

[1] Federal University of Santa Catarina, Florianópolis, Brazil
{henrique.simas,roberto.simoni}@ufsc.br
[2] Department of Engineering, UNIFE, Ferrara, Italy
raffaele.digregorio@unife.it

Abstract. This paper presents a new self-aligning parallel wrist called Wriflex. Wriflex is composed of four legs: one of *RRR* type and the remaining three of 3*PRRS* type that connect a moving platform to a fixed platform (base). Leg *RRR* is a passive-motion-constraint leg designed to generate the spherical movement of the end-effector. The three prismatic (P) pairs, one for each *PRRS* leg, are the only actuated joints and are located on the base. Here, the kinematic analysis of Wriflex is studied. In particular, the forward and the inverse position analyses are solved in closed form. Then, its instantaneous kinematics is addressed till to identify all its singularities. Eventually, its self-aligning capability is proved by employing the Conditioning Index as kinetostatic-performance measure.

1 Introduction

Simoni *et al.* [1] presented a novel class of 3T parallel mechanisms called *Triflex* which are variable-configuration parallel mechanisms with self-aligning that can change their form (base, legs, or moving platform) without changing the moving platform motion characteristics. For self-alignment, additional passive/null degrees of freedom define its portability and manage the robot posture, adapting its installation in unstructured environments [1]. Gosselin and Kong [2,3] and Kim and Tsai [4] introduced the fully decoupled 3-DOF parallel translational manipulators that served as an inspiration to the kinematic structure of the Triflex-I and Triflex-II [5]. Also, Simas et al. [6] presented the family of self-aligning TetraFLEX, 3T1R parallel manipulators, for field-robotics applications such as construction.

Kuo and Dai [7] presented a systematic approach for the type synthesis of non-overconstrained parallel manipulators with a motion constraint leg called the motion constraint generator. Several 2-, 3-, 4-, and 5-DOF parallel manipulators were synthesised. Others authors also have used motion constrained generators to design novel parallel manipulators, *e.g.*: Dai and Kerr [8] used a passive central leg to constrain the motion of a 6-DOF Stewart platform for measure the contact force on the motion platform; Joshi and Tsai [9] synthesised a family of 3-DOF, 4-legged parallel manipulators and studied the Tricept for machining applications; Serrancín et al. [10] proposed a parallel robot with 3 legs, two

O. Altuzarra and A. Kecskémethy (Eds.): ARK 2022, SPAR 24, pp. 340–347, 2022.
https://doi.org/10.1007/978-3-031-08140-8_37

active and one motion constraint leg to impose the motion for helping with bone milling surgeries; Kumar et al. [11] presented the kinematic analysis of a 2-DOF parallel mechanism with two active legs and one motion constraint leg for ankle rehabilitation.

Di Gregorio [12] presented the kinematic analysis of a single-loop translational manipulator and Di Gregorio and Simas [13] presented the dimensional synthesis of the single-loop translational parallel manipulator $\underline{P}RRR$-$\underline{P}R\underline{P}U$[1] inspired on Triflex-II. Also Simas and Di Gregorio [14,15] presented a general technique to evaluate the effects of manufacturing errors on positioning precision during design and as a case study they analysed a special case of the Triflex with perpendicular axes. Also, Simas et al. [6] presented a general method for calibration and error mounting for Tetraflex self-alignment parallel manipulators family.

Based on the systematic approach for the type synthesis of non-overconstrained parallel manipulators with a motion constraint leg combined with the self-alignment concept presented in Triflex and Tetraflex families, and also based on the methods for calibration, analysis of manufacturing and mounting errors, this paper introduces a new PW (*Parallel Wrist*) called *Wriflex*. Wriflex is a 1*RRR*-3$\underline{P}$*RRS* parallel mechanism with three self-aligning legs and one motion constraint leg. The Wriflex moving platform (end-effector) has three rotational degrees of freedom, composing a spherical wrist, with respect to the base frame. Its kinematic chain is asymmetrical and exhibits interesting features compared with others PW presented in the literature, *e.g.*, agile eye [16].

The paper explores the Wriflex PW in terms of the concept of self-alignment for a asymmetrical kinematic structure with prismatic actuation. The geometric developments for Wriflex show contributions on the self-alignment issues, feasible kinematic modeling, explicit expressions for the singularities, interesting range of joint-displacement and control of distance from singularities by means of the CI (*Conditioning Index*).

2 Wriflex Geometry

Figure 1 shows the geometry of the Wriflex, a self-alignment PW composed of four legs: 1*RRR*-3$\underline{P}$*RRS* connecting the platform to the base. The *RRR* leg (L_0) is a passive leg that plays the role of an articulated platform designed to impose the spherical motion to its distal link (end-effector), by three revolute joints with displacement θ_j, with $j = x, y, z$ with rotation axis passing through the origin of the base frame $O_0 - x_0 y_0 z_0$ and the platform frame $O_p - x_p y_p z_p$, where the points O_0 and O_p are coincident. The θ_x rotation axis, parallel to x_0, is perpendicular to the θ_y rotation axis whereas the θ_y rotation axis, belonging to the $y_0 z_0$-plane and rotated by θ_x from y_0, is perpendicular to the θ_z rotation axis, in which is rotated by θ_x and θ_y respectively.

[1] *P*, *R*, and *S* stand for prismatic joint, revolute joint and spherical pair, respectively. The underline indicates an actuated joint.

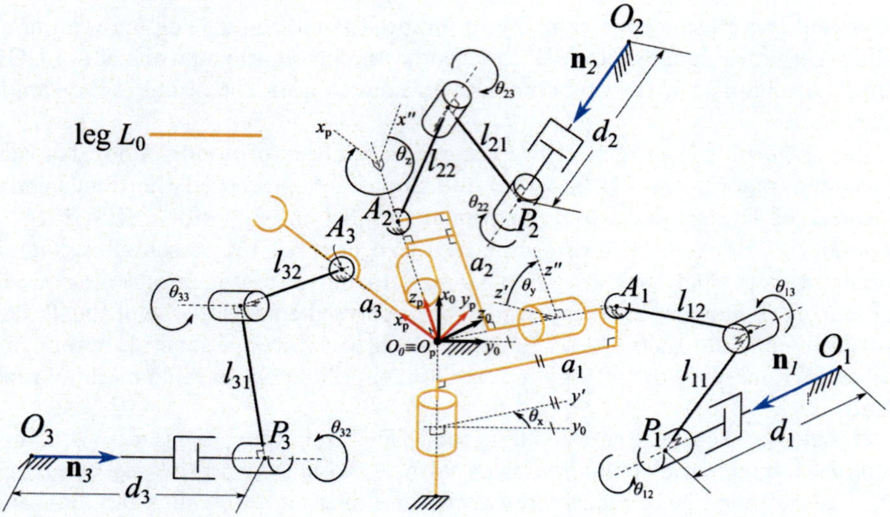

Fig. 1. Wriflex self-aligning parallel wrist architecture.

Leg L_0 has three links with length a_i $i = 1, 2, 3$ that correspond to the segments $\overline{O_0 A_i}$. The remaining legs L_i are $\underline{P}RRS$ serial chains, fixed to the base at point O_i, which has known constant coordinates in $O_0 - x_0 y_0 z_0$, and connected to L_0 by the S-joint, centered in the point A_i. With reference to Fig. 1, in the i-th $\underline{P}RRS$ leg, points P_i and A_i always lie on a plane perpendicular to the unit vector $\mathbf{n}_i = \left[n_{i_x}, n_{i_y}, n_{i_z}\right]^T$ that is parallel both to the P-pair sliding direction and to the axes of the two parallel R-pairs, which are all parallel. The joint variables are d_i, θ_{i2} and θ_{i3}. Eventually, the link lengths are l_{i1} and l_{i2}, respectively.

Regarding the angles from L_0, angle θ_x rotates the frame $O_0 - x_0 y_0 z_0$ around x_0-axis, resulting in frame $O' - x' y' z'$, in the sequence, the angle θ_y rotates frame $O' - x' y' z'$ around y'-axis, resulting in frame $O'' - x'' y'' z''$, and finally consecutively angle θ_y rotates frame $O'' - x'' y'' z''$ around z''-axis, resulting in the platform frame $O_p - x_p y_p z_p$.

In agreement with Kong and Gosselin [3] the instantaneous mobility of a parallel manipulator is given by $M = 6 - c + \sum_{i=1}^{m} R^i$ where, c is the order of wrench system of the moving platform and $R^i = f^i - 6 + c^i$ is the redundant DOF of leg i; moreover, f^i and c^i are the DOF and the wrench system of leg i, respectively. For the Wriflex, c is 3-ζ_∞-system because it is given by the intersection of the wrench systems of all the legs ((RRR): 3-ζ_∞-system; ($PRRRS$): 0-system). Thus, the mobility of the Wriflex is given by

$$M = 6 - c + \sum_{i=1}^{m} R^i = 6 - 3 + 0 = 3 \tag{1}$$

In other words, as the wrench system of the moving platform is a 3-ζ_∞-system, the twist system is 3-ξ_0-system and Wriflex has three rotational degrees of freedom.

3 Position Analysis

The introduced notations in Sect. (2) provide the following closure equations

$$(A_i - O_i) \cdot \mathbf{n}_i = d_i \qquad i = 1, 2, 3 \tag{2}$$

The coordinates of points A_i, for $i = 1, 2, 3$, measured in the base reference (i.e., $O_0 - x_0 y_0 z_0$) can be explicitly written as functions of θ_j for $j = x, y, z$ as follows

$$A_1 = \mathbf{R}_x(\theta_x) A_{1_0} = \overset{\begin{matrix}\mathbf{x}' & \mathbf{y}' & \mathbf{z}'\end{matrix}}{\begin{bmatrix} 1 & 0 & 0 \\ 0 & c_x & -s_x \\ 0 & s_x & c_x \end{bmatrix}} \begin{bmatrix} 0 \\ a_1 \\ 0 \end{bmatrix} = \begin{bmatrix} 0 \\ c_x \\ s_x \end{bmatrix} a_1 \tag{3}$$

$$A_2 = \mathbf{R}_x(\theta_x)\mathbf{R}_y(\theta_y) A_{2_0} = \overset{\begin{matrix}\mathbf{x}'' & \mathbf{y}'' & \mathbf{z}''\end{matrix}}{\begin{bmatrix} c_y & 0 & s_y \\ s_x s_y & c_x & -c_y s_x \\ -c_x s_y & s_x & c_y c_x \end{bmatrix}} \begin{bmatrix} 0 \\ 0 \\ a_2 \end{bmatrix} = \begin{bmatrix} s_y \\ -c_y s_x \\ c_y c_x \end{bmatrix} a_2 \tag{4}$$

$$A_3 = \mathbf{R}_x(\theta_x)\mathbf{R}_y(\theta_y)\mathbf{R}_z(\theta_z) A_{3_0} \tag{5}$$

$$= \overset{\begin{matrix}\mathbf{x}_p & \mathbf{y}_p & \mathbf{z}_p\end{matrix}}{\begin{bmatrix} c_y c_z & -c_y s_z & s_y \\ c_z s_x s_y + c_x s_z & c_x c_z - s_x s_y s_z & -c_y s_x \\ -c_z c_x s_y + s_x s_z & s_x c_z + c_x s_y s_z & c_y c_x \end{bmatrix}} \begin{bmatrix} a_3 \\ 0 \\ 0 \end{bmatrix} = \begin{bmatrix} c_y c_z \\ c_z s_x s_y + c_x s_z \\ -c_z c_x s_y + s_x s_z \end{bmatrix} a_3$$

where $\mathbf{R}_j(\theta_j)$ is the rotation matrix of a frame around j-axis by the angle θ_j, A_{i_0} represents the coordinates of A_i for $\theta_j = 0$ rad. Also c_j and s_j are $\cos(\theta_j)$ and $\sin(\theta_j)$ respectively and the above identification of each column from the rotation matrix shows the projection of the intermediary and final frame axes.

Wriflex has $\mathbf{q} = [d_1, d_2, d_3]^T$ as input vector collecting the active-joint-variables and the orientation of the platform-frame $O_p - x_p y_p z_p$, computed by the consecutive rotation matrix $\mathbf{R}_p(\theta_x, \theta_y, \theta_z) = \mathbf{R}_x(\theta_x)\mathbf{R}_y(\theta_y)\mathbf{R}_z(\theta_z)$ using the vector $\mathbf{p} = [\theta_x, \theta_y, \theta_z]^T$, which is the output vector collecting the adopted orientation parameters.

3.1 Forward Position Analysis

In this case, the forward position analysis is to compute the values of $\mathbf{p}$ compatible with one assigned value of $\mathbf{q}$.

The introduction of Eqs. (3), (4) and (5) into Eq. (2) for leg L_i, and making explicit the c_j and the s_j, for $i = 1 \rightarrow j = x$; $i = 2 \rightarrow j = y$ and $i = 3 \rightarrow j = z$ yields

$$e_1 c_x + f_1 s_x + g_1 = 0 \tag{6}$$
$$e_2 c_y + f_2 s_y + g_2 = 0 \tag{7}$$
$$e_3 c_z + f_3 s_z + g_3 = 0 \tag{8}$$

where $e_1 = a_1 n_{1y}$, $f_1 = a_1 n_{1z}$, $g_1 = -(d_1 + O_1 \cdot \mathbf{n}_1)$, $e_2 = a_2 n_{2z} c_x - a_2 n_{2y} s_x$, $f_2 = a_2 n_{2x}$, $g_2 = -(d_2 + O_2 \cdot \mathbf{n}_2)$, $e_3 = a_3(n_{3x} c_y + n_{3y} s_x s_y - n_{3z} c_x s_y)$, $f_3 = a_3(n_{3y} c_x + n_{3z} s_x + c_x)$ and $g_3 = -(d_3 + O_3 \cdot \mathbf{n}_3)$.

Then, the trigonometric identities $c_j = \frac{1-t_j^2}{1+t_j^2}$ and $s_j = \frac{2t_j}{1+t_j^2}$, where $t_j = \tan\frac{\theta_j}{2}$ with $-\pi < \theta_j < \pi$ when introduced into Eqs. (6), (7) and (8), after some algebraic manipulations, transform them into the quadratic equation

$$(g_i - e_i)t_j^2 + 2 f_i t_j + (g_i + e_i) = 0 \tag{9}$$

which gives the following two solutions $t_j = \frac{f_i + (-1)^n \sqrt{f_i^2 - g_i^2 + e_i^2}}{g_i - e_i}$, $n = 1, 2$. Accordingly, $\theta_j = 2\arctan(t_j)$, $n = 1, 2$.

Employing the approach presented above, Eq. (6) solves for angle θ_x since it only appears as a single variable. With the result obtained for θ_x, Eq. (7) solves the angle θ_y. Eventually, with the values obtained for θ_x and θ_y, Eq. (8) solves the angle θ_z.

3.2 Inverse Position Analysis

The inverse position analysis of Wriflex is the determination of the actuated joint-variables $\mathbf{q}$ compatible with one assigned platform orientation, given by $\mathbf{p}$ or the corresponding $\mathbf{R}_p$ matrix.

$$\mathbf{R}_p(\theta_x, \theta_y, \theta_z) = \begin{bmatrix} h_{11} & h_{12} & h_{13} \\ h_{21} & h_{22} & h_{23} \\ h_{31} & h_{32} & h_{33} \end{bmatrix} = \begin{bmatrix} c_x c_y c_z - s_x s_z & -c_x c_y s_z - s_x s_z & c_x s_y \\ s_x c_y c_z + c_x s_z & -s_x c_y s_z + c_x c_z & s_x s_y \\ -s_y c_z & s_y s_z & c_y \end{bmatrix} \tag{10}$$

Comparing the above matrices, Eq. (10) provides that $\theta_x = \text{atan2}(\pm h_{23}, \pm h_{13})$, $\theta_y = \text{atan2}(\pm\sqrt{(h_{13})^2 + (h_{23})^2}, h_{33})$ and $\theta_z = \text{atan2}(\pm h_{32}, \mp h_{31})$, which, when introduced into Eqs. (3), (4) and (5), give the coordinates of points A_i, $i = 1, 2, 3$, to use in Eq. (2) for the computation of the actuated-joint variables d_i.

4 Instantaneous Kinematics and Singularity Analysis

In this case, the instantaneous kinematics analysis determines the relationship between instantaneous output (operational space) $\dot{\mathbf{p}}$, which uniquely determines the platform angular velocity, and the instantaneous displacements of the actuated-joint variables $\dot{\mathbf{q}}$. Wriflex forward differential kinematics model is obtained by differentiating Eqs. (6), (7) and (8), since the passive leg L_0 excludes the possible presence of constraint singularities. So doing, after some algebraic manipulation the matrix solution come out

$$\begin{bmatrix} a_1(\mathbf{n}_1 \cdot \mathbf{z}') & 0 & 0 \\ a_2(\mathbf{n}_2 \cdot (\mathbf{x}' \times \mathbf{z}'')) & a_2(\mathbf{n}_2 \cdot \mathbf{x}'') & 0 \\ a_3(\mathbf{n}_3 \cdot (\mathbf{x}' \times \mathbf{x}_p)) & a_3(\mathbf{n}_3 \cdot (\mathbf{x}'' \times \mathbf{y}_p)) & a_3(\mathbf{n}_3 \cdot \mathbf{y}_p) \end{bmatrix} \begin{bmatrix} \dot{\theta}_x \\ \dot{\theta}_y \\ \dot{\theta}_z \end{bmatrix} = \begin{bmatrix} \dot{d}_1 \\ \dot{d}_2 \\ \dot{d}_3 \end{bmatrix} \quad (11)$$

where $\mathbf{x}'$, $\mathbf{y}'$, $\mathbf{z}'$, $\mathbf{x}''$, $\mathbf{y}''$, $\mathbf{z}''$, $\mathbf{x}_p$, $\mathbf{y}_p$ and $\mathbf{z}_p$ are unit vectors defined in Eqs. (3), (4) and (5).

Considering the directions assumed by the passive joints of L_0, the platform angular velocity $\omega = [\omega_x, \omega_y, \omega_z]^T$ is obtained as follows [17]

$$\begin{bmatrix} 1 & 0 & s_y \\ 0 & c_x & -c_y s_x \\ 0 & s_x & c_y c_x \end{bmatrix} \begin{bmatrix} \dot{\theta}_x \\ \dot{\theta}_y \\ \dot{\theta}_z \end{bmatrix} = \begin{bmatrix} \omega_x \\ \omega_y \\ \omega_z \end{bmatrix} \quad (12)$$

Considering respectively, $\mathbf{M}\dot{\mathbf{p}} = \dot{\mathbf{q}}$ and $\mathbf{N}\dot{\mathbf{p}} = \omega$ the matrix form of Eq. (11) and Eq. (12), the instantaneous input/output relationship is given by

$$\mathbf{M}N^{-1}\omega = \dot{\mathbf{q}} \quad (13)$$

Equation(13) brings the conclusion that the instantaneous inverse (direct) problem is not determined (i.e., a type-I (type-II) singularity occurs) when $\det(\mathbf{N}) = 0$ (when $\det(\mathbf{M}) = 0$) and that a type-III singularity occurs when $\det(\mathbf{N}) = \det(\mathbf{M}) = 0$.

Accordingly, type-I singularities occur when $\det(\mathbf{N}) = c_y = 0$ (when $\mathbf{N}$ is rank deficient), which provide $\theta_y = \pm\frac{\pi}{2}$ (or the align the 1st and 3rd revolution joints of the leg L_0, see Fig. 1). Type-II singularities are given by $\det(\mathbf{M}) = a_1 a_2 a_3(\mathbf{n}_1 \cdot \mathbf{z}')(\mathbf{n}_2 \cdot \mathbf{x}'')(\mathbf{n}_3 \cdot \mathbf{y}_p) = 0$ which provides the following conditions: $\mathbf{n}_1 \perp \mathbf{z}'$, $\mathbf{n}_2 \perp \mathbf{x}''$ and $\mathbf{n}_3 \perp \mathbf{y}_p$. Eventually, Wriflex does not present Type-III singularities.

5 Simulation Results

This section presents a numerical simulation dealing with inaccuracies from the potential for self-alignment generated by negligence in installing the legs L_i. The inaccuracy of the installation is directly related to the precision in the direction of the vectors $\mathbf{n}_i$. The kinematic performance of the Wriflex will be evaluated by means of the Conditioning Index (CI) [18].

The Wriflex's singularities impose that the angles θ_x, θ_y and θ_z can range in the interval $]-\pi/2, \pi/2[$, whereas, in the limit positions, i.e. $\pm\pi/2$, it characterizes a singular posture with CI=0. Firstly evaluating an appropriate range for the angles θ_x, θ_y and θ_z, it is possible to associate a sphere of radius r that delimits the possible values for such angles within a minimum desired CI.

Selecting $\mathbf{n}_1 = [0, 0, 1]^T$, $\mathbf{n}_2 = [1, 0, 0]^T$, $\mathbf{n}_3 = [0, 1, 0]^T$, $a_1 = a_2 = a_3 = 3$ l.u - *length unit*, $l_{i1} = 4$ l.u. and $l_{i2} = 3.5$ l.u. for $i = 1, 2, 3$, Fig. 2a presents the CI variation inside the sphere volume with $r = 0.7$rad, i.e. for CI > 0.5 requires that $\sqrt{\theta_x^2 + \theta_y^2 + \theta_z^2} \leq 0.7$. Inaccuracies in the orientation of $\mathbf{n}_i$ provide inaccuracies in installing the Wriflex's feet and, consequently, kinematic performance.

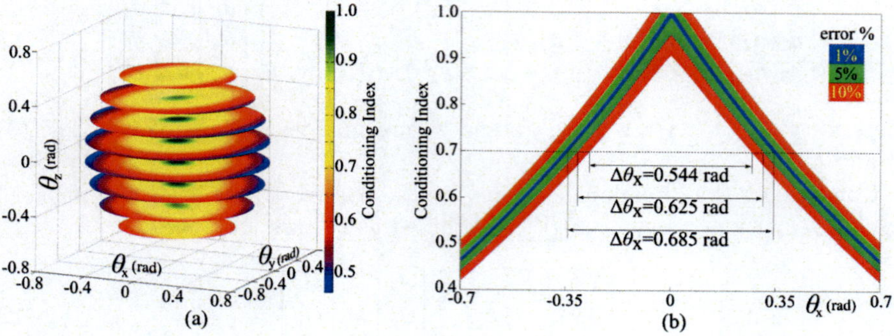

Fig. 2. (a) Sphere with $r = 0.7$ and the respective CI. (b) Sensitivity of the CI for the inaccuracies of 1% (blue) 5% (green) 10% (red).

Figure 2b shows the behavior of CI considering the inaccuracies of 1%, 5% and 10% on the direction of $\mathbf{n}_i$, referred the interval of $\theta_j = 1.4$ rad and defined in Fig. 2a. Taking advantage of symmetry shown in Fig. 2a, choosing CI =0.7 as the minimum allowed value, and $\theta_x = \theta_y = 0$ rad, Fig. 2b also shows that the available range of $\Delta\theta_z$ ($\Delta\theta_j$) decreases as the percentage of imprecision of $\mathbf{n}_i$ increases.

6 Conclusion

This paper has presented a PW called Wriflex. Additional degrees of freedom (self-alignment degrees of freedom) that do not interfere in the platform's motion gives Wriflex dexterity to adapt the active-prismatic joints in any position on the base. Also, the paper has presented the mathematical modeling of Wriflex in terms of positions and instantaneous displacements. The simplicity of Wriflex's kinematics has allowed identifying the singularities and evaluating its kinematic performance as functions of its workspace using the CI. The proposed Wriflex has application as a spherical wrist mechanism mounted in unstructured environments or adapted to 3T parallel robotic systems constituted of platforms with free geometry, e.g. civilian construction, mining, welding, among other field-robotics applications. A further paper will present the dimensional analysis and optimization which result in a CAD model and a prototype of Wriflex.

Funding. This study was funded by CNPq Project 307249/2021-2 and by UNIFE grant no. FAR2020.

Conflicts of Interest. The authors declare that they have no conflict of interest.

References

1. Simoni, R., Simas, H., Martins, D.: Triflex: variable-configuration parallel manipulators with self-aligning. J. Braz. Soc. Mech. Sci. Eng. **37**(4), 1129–1138 (2014). https://doi.org/10.1007/s40430-014-0235-y
2. Gosselin, C., Kong, X.: Cartesian parallel manipulators. US Patent 6,729,202 (2002)
3. Kong, X., Gosselin, C.: Type Synthesis of Parallel Mechanisms. Springer, New York (2007). https://doi.org/10.1007/978-3-540-71990-8
4. Kim, H., Tsai, L.: Design optimization of a cartesian parallel manipulator. J. Mech. Des. **125**(1), 43–52 (2003)
5. Simas, H., Simoni, R., Martins, D.: Triflex II: design and analysis of a self-aligning parallel mechanism with asymmetrical kinematic structure. Meccanica (4), 2991–3002 (2017). https://doi.org/10.1007/s11012-017-0615-3
6. Simas, H., Di Gregorio, R., Simoni, R.: TetraFLEX: design and kinematic analysis of a novel self-aligning family of 3T1R parallel manipulators. J. Field Robot., 1–10 (2022)
7. Kuo, C.-H., Dai, J.S.: Task-oriented structure synthesis of a class of parallel manipulators using motion constraint generator. Mech. Mach. Theory **70**, 394–406 (2013)
8. Dai, J.S., Kerr, D.R.: A six-component contact force measurement device based on the stewart platform. Proc. Inst. Mech. Eng. C J. Mech. Eng. Sci. **214**(5), 687–697 (2000)
9. Joshi, S.A., Tsai, L.-W.: The kinematics of a class of 3-DOF, 4-legged parallel manipulators. J. Mech. Des. **125**(1), 52–60 (2003)
10. Serracín, J.R., Puglisi, L.J., Saltaren, R., Ejarque, G., Sabater-Navarro, J.M., Aracil, R.: Kinematic analysis of a novel 2-DOF orientation device. Rob. Auton. Syst. **60**(6), 852–861 (2012)
11. Kumar, S., Nayak, A., Peters, H., Schulz, C., Müller, A., Kirchner, F.: Kinematic analysis of a novel parallel 2SPRR+1U ankle mechanism in humanoid robot. In: Lenarcic, J., Parenti-Castelli, V. (eds.) ARK 2018. SPAR, vol. 8, pp. 431–439. Springer, Cham (2019). https://doi.org/10.1007/978-3-319-93188-3_49
12. Di Gregorio, R.: Kinematic analysis of a single-loop translational manipulator. In: Parenti-Castelli, V., Schiehlen, W. (eds.) ROMANSY 21 - Robot Design, Dynamics and Control. CICMS, vol. 569, pp. 73–79. Springer, Cham (2016). https://doi.org/10.1007/978-3-319-33714-2_9
13. Di Gregorio, R., Simas, H.: Dimensional synthesis of the single-loop translational parallel manipulator PRRR- PRPU. Meccanica **53**, 481–495 (2018)
14. Simas, H., Di Gregorio, R.: A general technique to evaluate the effects of manufacturing errors on positioning precision during design. In: IDETC/CIE 2016, pp. 1–10 (2016)
15. Simas, H., Di Gregorio, R.: Geometric error effects on manipulators' positioning precision: a general analysis and evaluation method. J. Mech. Robot. (2016). https://doi.org/10.1115/1.4034577
16. Gosselin, C.M., Hamel, J.-F.: The agile eye: a high-performance three-degree-of-freedom camera-orienting device. In: ICRA, vol. 1, pp. 781–786 (1994)
17. Tsai, L.-W.: Robot Analysis: The Mechanics of Serial and Parallel Manipulators. Wiley, New York (1999)
18. Gosselin, C., Angeles, J.: A global performance index for the kinematic optimization of robotic manipulators. J. Mech. Des. **113**(3), 220–226 (1991)

Kinematic Analysis of a Novel Humanoid Wrist Parallel Mechanism

Christoph Stoeffler[1(✉)], Adriano del Rio Fernandez[1], Heiner Peters[2], Moritz Schilling[2], and Shivesh Kumar[2]

[1] University of Bremen, Bremen, Germany
stoeffle@uni-bremen.de
[2] German Research Center for Artificial Intelligence, Robotics Innovation Center, Bremen, Germany
{heiner.peters,moritz.schilling,shivesh.kumar}@dfki.de

Abstract. Wrist mechanisms in humanoid robots play a crucial role by fundamentally influencing the dexterity of robotic hands. Recent designs often exploit parallel mechanisms due to their ability to transmit high loads while offering the possibility to minimize peripheral inertia of the arms. This often comes at the cost of reduced workspace. In this paper, a novel wrist mechanism of type $2S\underline{P}U + 2RSU + 1U$ is presented, with the aim to achieve human-like range of motion with good force transmission capabilities in a compact form. An in-depth kinematic analysis of the new parallel mechanism is presented including solutions to both forward and inverse kinematics and a comparison to the traditional $2S\underline{P}U + 1U$ mechanism is drawn, showing increased dexterity and range of motion.

1 Introduction

A common approach in robotics is to customize the robot's joint modules specifically to their intended tasks. This functional design becomes even more relevant for humanoids, since they possess a complex structure where different parts take different specific tasks, such as walking, grasping, or manipulating. In [8], it is argued that human muscle structures can be considered generally parallel. Recently, parallel kinematic machines (PKMs) found many applications for joint modules in humanoid robots, due to their ability to transmit motor motions non-linearly to the relevant joints (see [9] for a survey). Many different wrist designs have been proposed in the robotics community, where the majority focused on 2-DOF mechanisms - see [7]. Notably, the $2S\underline{P}U + 1U$[1]-topology is often used (e.g. wrist joint in RH5 humanoid [10]) and has for example been extended in [6] for humanoid ankle application. Such mechanisms fall in the class of parallel mechanisms with actuation legs and motion constraint generator legs (see [5] for an extensive study on their synthesis). These designs still suffer from a limited range of motion.

[1] S, P and U stand for spherical, prismatic and universal joint respectively and underscores denote actuated joints.

O. Altuzarra and A. Kecskémthy (Eds.): ARK 2022, SPAR 24, pp. 348–355, 2022.
https://doi.org/10.1007/978-3-031-08140-8_38

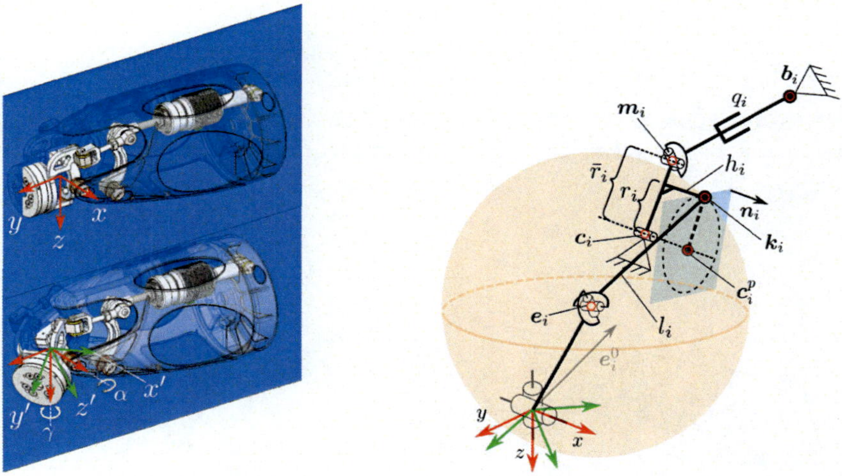

(a) Mechanical design of the new $2S\underline{P}U$ + $2RSU$ + $1U$-mechanism for a humanoid wrist. Due to its symmetry, inverse kinematic descriptions are carried out on one side of the depicted half-plane.

(b) Geometric parameters used in the computation of kinematic relations. Here, one half of the mechanism is depicted by cutting along the $y - z$-axis (see Fig. 1a).

Fig. 1. CAD depiction of new wrist and its schematic

In this paper, we present the kinematic analysis of a novel parallel mechanism of type $2S\underline{P}U + 2RSU + 1U$ for implementing the wrist joint of RH5v2 humanoid (see Fig. 1a) which overcomes the restricted range of motion of a $2S\underline{P}U + 1U$ design used in RH5 humanoid. The mechanism offers increased workspace for *flexion* and *extension* of the hand with a special focus on increasing the flexion capability of the wrist. This mechanism was first employed by the authors in the robotic system *Charlie* [3] as an ankle joint. However, an in-depth analysis of this mechanism is missing in the literature. We present the inverse, forward and differential kinematics solution of this novel mechanism. It is demonstrated that this mechanism can have a maximum of four solutions to its inverse kinematics and eight solutions to the forward kinematics problems. Further, using the differential kinematics, the dexterity analysis of the proposed mechanism is performed and the design is compared with a traditional $2S\underline{P}U + 1U$ design.

2 Architecture and Constraint Equations

Figure 1a shows the CAD design and Fig. 1b shows the mechanism's schematic. It exhibits $n = 4 \times 2 + 1 + 1 = 10$ bodies and $m = 6 \times 2 + 1 = 13$ joints that posses $f_{tot} = 26$ degrees of freedom. Applying the *Grübler-Kutzbach criteria* for spatial mechanisms yields

$$M = 6(n - m - 1) + \sum f_i$$
$$= 6(10 - 13 - 1) + 26 = 2$$

degrees of freedom for the proposed design. The wrist therefore allows movements around the x- and y-axis that we denote subsequently with *inclination* (α) and *tilt* (γ).

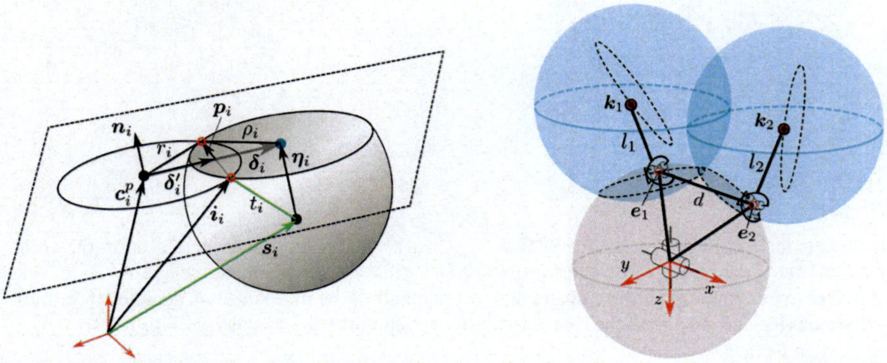

(a) Circle-sphere intersection with two solutions at maximum that can be brought to a circle-circle intersection by projection.

(b) Forward geometric problem: Intersecting sphere of end-effector and cranks l_1 and l_2.

Fig. 2. Geometric details to solve the inverse and forward kinematics

The underlying constraint equations of the mechanism are based on the points $\boldsymbol{k}_i$, that are likewise constrained by actuator lengths and end-effector configuration. From Fig. 1a) we have

$$\left\| \mathbf{c}_i + \frac{\bar{r}_i}{r_i}(\mathbf{k}_i - \mathbf{c}_i^p) - \mathbf{b}_i \right\|^2 - q_i^2 = 0 \quad (1)$$

$$h_i := \left\| \boldsymbol{R}(\alpha, \gamma)\mathbf{e}_i^0 - \mathbf{k}_i \right\|^2 - l_i^2 = 0 \quad (2)$$

with $i \in \{1, 2\}$ and where vectors and scalars are depicted in Fig. 1b. The rotation matrix $\boldsymbol{R}(\alpha, \gamma)$ is parameterized by inclination α and tilt γ. It can generally be of *intrinsic* or *extrinsic* type, employing the matrices $\boldsymbol{R}_z(\gamma)\boldsymbol{R}_x(\alpha)$ or $\boldsymbol{R}_z(\alpha)\boldsymbol{R}_x(\gamma)$, respectively.

3 Kinematics

Computing the forward and inverse kinematics involves to solve for $\boldsymbol{k}_i$, what can be achieved by considering the general problem of a circle-sphere intersection and is depicted in Fig. 2a. Formally, it writes $C(\mathbf{c}_i^p, \mathbf{n}_i, r_i) \cap \mathcal{S}(\mathbf{s}_i, t_i) \rightarrow \mathbf{i}_i^+, \mathbf{i}_i^-$, where superscripts + and − represent the two possible solutions. The parameters r_i,

$\mathbf{n}_i$ and $\mathbf{c}_i^P = \mathbf{c}_i + h_i\mathbf{n}_i$ can be taken from Fig. 1b. By exploiting the dot product together with a scalar multiplication, the vector $\boldsymbol{\eta}_i = \left((\mathbf{c}_i^P - \mathbf{s}_i) \cdot \mathbf{n}_i\right) \mathbf{n}_i$ can be computed which points from $\mathbf{s}_i$ to the plane of intersection. This reduces the problem to a circle-circle intersection and the circle radius the intersected sphere is given by $\rho_i^2 = t_i^2 - \boldsymbol{\eta}_i \cdot \boldsymbol{\eta}_i$. The distance vector between both circle centers is simply obtained by $\boldsymbol{\delta}_i = \mathbf{s}_i + \boldsymbol{\eta}_i - \mathbf{c}_i^P$ as depicted in Fig. 2a. The distance to the line of intersecting points, together with component that accounts for its direction, is then computed by

$$\boldsymbol{\delta}_i' = \frac{\boldsymbol{\delta}_i \cdot \boldsymbol{\delta}_i + r_i^2 - \rho_i^2}{2||\boldsymbol{\delta}_i||} \frac{\boldsymbol{\delta}_i}{||\boldsymbol{\delta}_i||} = \frac{\boldsymbol{\delta}_i}{2} + \boldsymbol{\delta}_i \frac{r_i^2 - \rho_i^2}{2\boldsymbol{\delta}_i \cdot \boldsymbol{\delta}_i} = \frac{\boldsymbol{\delta}_i}{2} + \boldsymbol{\delta}_i \frac{r_i^2 - t_i^2 + \boldsymbol{\eta}_i \cdot \boldsymbol{\eta}_i}{2\boldsymbol{\delta}_i \cdot \boldsymbol{\delta}_i}$$

Additionally, the orthogonal vector to $\boldsymbol{\delta}_i'$ that is pointing to the intersection points and being of length $\sqrt{r_i^2 - \boldsymbol{\delta}_i' \cdot \boldsymbol{\delta}_i'}$ is computed by using the planes normal vector $\mathbf{n}_i$:

$$\mathbf{p}_i = \frac{\boldsymbol{\delta}_i'}{||\boldsymbol{\delta}_i'||} \times \mathbf{n}_i \sqrt{r_i^2 - \boldsymbol{\delta}_i' \cdot \boldsymbol{\delta}_i'} \tag{3}$$

Finally, two intersection points are obtained with

$$\mathbf{i}_i = \mathbf{c}_i^P + \boldsymbol{\delta}_i' \pm \mathbf{p}_i. \tag{4}$$

Inserting $\mathbf{k}_i^+, \mathbf{k}_i^- \leftarrow \mathcal{C}(\mathbf{c}_i^P, \mathbf{n}_i, r_i) \cap \mathcal{S}(\boldsymbol{R}(\alpha, \gamma)\mathbf{e}_i^0, l_i)$ into Eq. 1 and rearranging for q_i gives the solution of inverse kinematics problem. According to the circle-sphere intersection for computing the positions of $\boldsymbol{k}_i$, there exist four inverse solutions to the mechanism that are shown in Fig. 3. Determining the forward kinematics - that is computing the wrist orientation from actuator inputs q_i - presents itself to be more involving. In a first step, the already explained circle-sphere intersection can be exploited again to obtain $\mathbf{m}_i^+, \mathbf{m}_i^- \leftarrow \mathcal{C}(\mathbf{c}_i^P, \mathbf{n}_i, r_i) \cap \mathcal{S}(\mathbf{b}_i, q_i)$ what translates into

$$\mathbf{k}_i = \mathbf{c}_i + \frac{r_i}{\bar{r}_i}(\mathbf{m}_i - \mathbf{c}_i) + h_i\mathbf{n}_i. \tag{5}$$

Once the spatial locations of $\mathbf{k}_i$ are known, the remaining geometric problem to solve, consists of two sphere-sphere intersections, as to be seen in Fig. 2b. By means of *Gröbner bases* of the ideal $\mathcal{I} = \langle h_1, h_2, h_3, h_4 \rangle | \mathcal{I} \subseteq k[t, u, v, w]$ one can solve the system of nonlinear equations, defined by Eq. 2 and augmented by the parametrization of rotations:

$$h_3 := t^2 + u^2 - 1 = 0 \quad \text{where} \quad t = \cos(\alpha),\ u = \sin(\alpha), \tag{6}$$

$$h_4 := v^2 + w^2 - 1 = 0 \quad \text{where} \quad v = \cos(\gamma),\ w = \sin(\gamma). \tag{7}$$

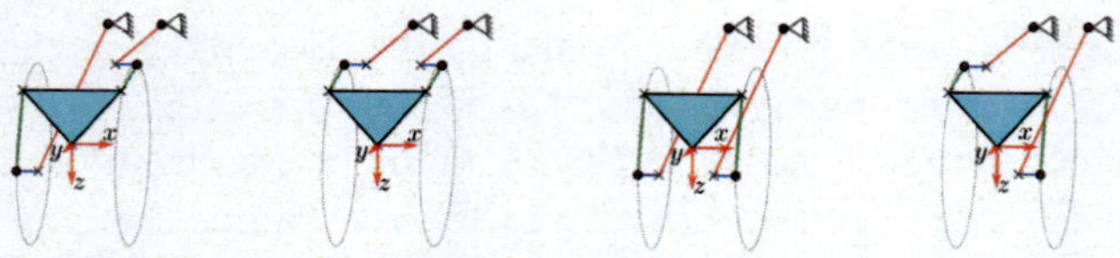

(a) $q_1 = 0.133\,\mathrm{m}$ $q_2 = 0.200\,\mathrm{m}$ (b) $q_1 = 0.133\,\mathrm{m}$ $q_2 = 0.133\,\mathrm{m}$ (c) $q_1 = 0.200\,\mathrm{m}$ $q_2 = 0.200\,\mathrm{m}$ (d) $q_1 = 0.200\,\mathrm{m}$ $q_2 = 0.133\,\mathrm{m}$

Fig. 3. Inverse solutions for the configuration $\alpha = 0$ and $\gamma = 0$. Dashed circles indicate the constraints arising from the revolute joints located in $\boldsymbol{c}_i$.

Sampling through the feasible actuation space (see Table 2) to compute Gröbner bases of the above ideal, revealed an univariate polynomial of degree eight. With four solutions for $\mathbf{k}_i$, a maximum of eight real forward solutions were computed and are depicted in Fig. 4. For the plots of this section and of Sect. 4, the assembly mode of the designed wrist mechanism of Fig. 1a is used for computations[2]. Its geometric parameters are listed in Table 1.

Table 1. Geometric data of the build wrist with variables defined in Fig. 1a - all values in [m], except for $\boldsymbol{n}_i$

$\boldsymbol{b}_1 = [0.015 \quad -0.178 \quad -0.034]^T$	$\boldsymbol{b}_2 = [-0.015 \quad -0.178 \quad -0.034]^T$
$\boldsymbol{c}_1 = [0.015 \quad -0.032 \quad 0.011]^T$	$\boldsymbol{c}_2 = [-0.015 \quad -0.032 \quad 0.011]^T$
$\boldsymbol{e}_1^0 = [0.027 \quad 0 \quad -0.03]^T$	$\boldsymbol{e}_2^0 = [-0.027 \quad 0 \quad -0.03]^T$
$\boldsymbol{n}_1 = [1 \quad 0 \quad 0]^T$	$\boldsymbol{n}_2 = [-1 \quad 0 \quad 0]^T$
$\bar{r}_i = 0.049, \qquad r_i = 0.049,$	$h_i = 0.012, \qquad l_i = 0.045$

With the closed-form solution to inverse kinematics problem, the constraint Jacobian of the mechanism can be computed, which reveals information about workspace conditioning and singularities. Equation 1 can be brought into the form $\boldsymbol{g}(\boldsymbol{x}, \boldsymbol{q}) = \boldsymbol{0}$, differentiation then yields

$$\frac{\partial \boldsymbol{g}(\boldsymbol{x}, \boldsymbol{q})}{\partial \boldsymbol{x}} \dot{\boldsymbol{x}} = -\frac{\partial \boldsymbol{g}(\boldsymbol{x}, \boldsymbol{q})}{\partial \boldsymbol{q}} \dot{\boldsymbol{q}} \Leftrightarrow \dot{\boldsymbol{x}} = -\boldsymbol{J}_x^{-1} \boldsymbol{J}_q \dot{\boldsymbol{q}} = \boldsymbol{J} \dot{\boldsymbol{q}} \tag{8}$$

where the Jacobian $\boldsymbol{J}$ is the product of the inverted *work space Jacobian* $\boldsymbol{J}_x$ and the *joint space Jacobian* $\boldsymbol{J}_q$. The condition index or determinant of the Jacobian (see e.g. [2]) can be used to quantify the mechanism, since its input and output spaces are $\mathbb{R}^2$ and $SO(3)$ respectively.

[2] Software available at https://github.com/dfki-ric/NovelWrist/.

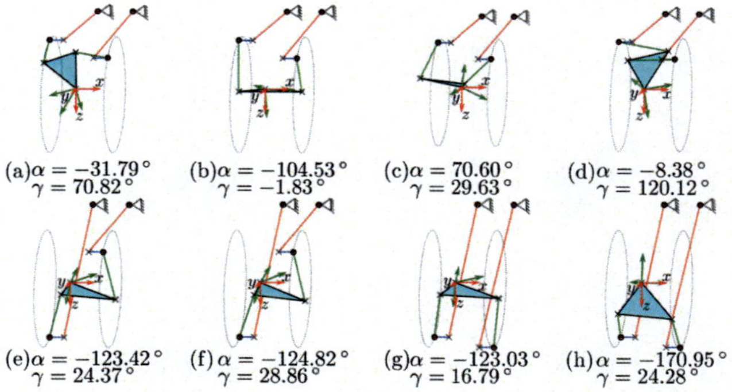

Fig. 4. Forward solutions for the actuator length $q_1 = 0.155\,\text{m}$ and $q_2 = 0.178\,\text{m}$. Dashed circles indicate the constraints arising from the revolute joints located in $\boldsymbol{c}_i$.

4 Performance Analysis

In Fig. 5 the condition index and singularities of the novel mechanism are shown. While the universal joint of the end-effector alone can reach $\alpha, \gamma \in [-\pi, \pi]$, additional constraints of the linkages restrict the workspace and no kinematic solution exists in the white areas. Actuator limits (see Table 2) further restrict motion to the dashed area.

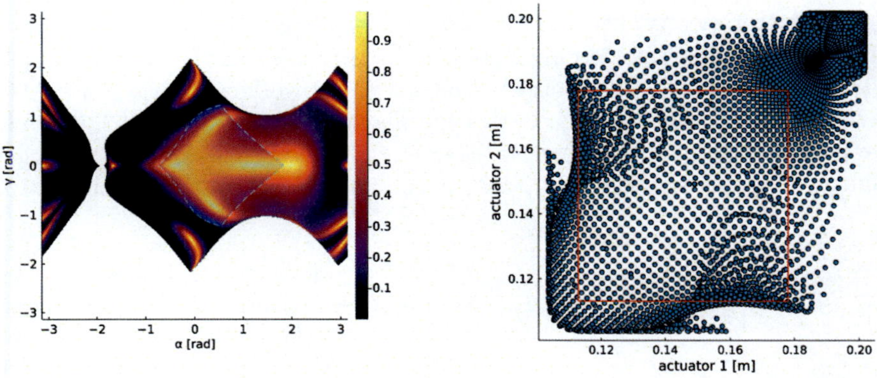

(a) Conditioning of the build mechanism, determined with $1/\text{cond}(\boldsymbol{J}\boldsymbol{J}^T)$.

(b) Configuration space of the mechanism with actuator limits

Fig. 5. Dexterity measure and singularities in task space of the novel wrist

To highlight mechanism improvements, we propose a simple $2S\underline{P}U + 1U$-mechanism as a design to compare against, with the linear actuators acting directly on the end-effector platform. In analogy to the novel mechanism, the inverse kinematics and Jacobian for the comparative design can be computed by rearranging and differentiating $g_i^c = q_i^2 - \left\|\boldsymbol{R}(\alpha, \gamma)\mathbf{e}_i^0 - \mathbf{b}_i\right\|^2$, respectively.

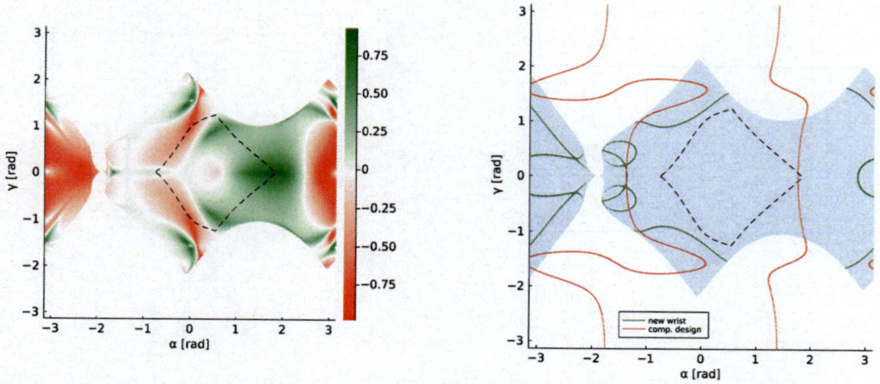

(a) Difference of condition index between new wrist and comparative design. Positive areas indicate that the novel wrist has superior dexterity.

(b) Singularity curves of new wrist and comparative design, showing reduced workspace for the new wrist by grey area.

Fig. 6. Comparison between $2S\underline{P}U + 2RSU + 1U$ and $2S\underline{P}U + 1U$ mechanism

Figure 6a and 6b show differences in the condition indices of the mechanisms and their singularity curves, respectively. It can be noted that the novel design shows similar dexterity for pure tilt movements under zero inclination angle. However, the design aim of increased dexterity for high inclination angles ($\gamma = 0, 1 \leq \alpha \leq 2$) is clearly met. Additionally, the singularity occurring at positive inclination angles is considerably shifted to higher values in the novel design.

Due to limitations in the construction space, the work space was not shifted further in positive tilt direction in this specific application, although this would have been advantageous in terms of dexterity. Finally, Fig. 7 gives an overview of maximal speed and torque in task space under given actuator input and is

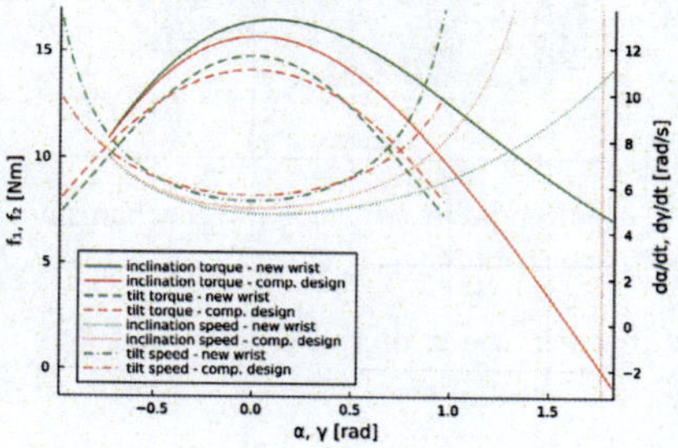

Fig. 7. Torques and speeds for pure inclination ($\gamma = 0$) and pure tilt ($\alpha = 0$).

Table 2. Mechanism specifications and comparative human data [1,4]. Actuator force and speed refer to the rated ones and rotational speeds are those obtained in the configuration of maximal torques.

	Range	Force	Speed
lin. actuators	113–178 mm	262 N	152 mm/s
incl. ($\gamma = 0$)	−42/105°	12/17 Nm	277°/s
tilt ($\alpha = 0$)	−56/56°	12/15 Nm	309°/s
Comparisons		(max.)	
human incl.	−78/95°	25 Nm	n.a.
human tilt	−48/37°	7 Nm	n.a.
incl. $2S\underline{P}U + U$	−42/101°	−2/16 Nm	291°/s
tilt $2S\underline{P}U + U$	−56/56°	8/14 Nm	323°/s

complemented by Table 2. The novel mechanism is closer to the maximal torques in humans and delivers sufficient torque at high tilt angles.

5 Conclusion

In this work, a novel wrist design in the class of parallel 2-DOF mechanisms has been introduced. A closed-form solution of the inverse kinematics and a Gröbner bases approach for the forward kinematics has been presented. This is followed by a performance analysis of the wrist, showing increased dexterity and enlarged workspace compared to a classical 2-DOF mechanism.

Acknowledgements. This work was partially supported from the projects VeryHuman (FKZ 01IW20004) and TransFIT (FKZ 50RA1701) funded by the German Aerospace Center (DLR) with federal funds from the Federal Ministry of Education and Research (BMBF) and Federal Ministry of Economic Affairs and Energy (BMWi) respectively.

References

1. Christensen, J.M., et al.: Man-Systems Integration Standards NASA-STD-3000, Vol. II Rev. B (1995)
2. Merlet, J.P.: Jacobian, manipulability, condition number and accuracy of parallel robots. In: Proceedings of 12th International Symposium of Robotics Research (2005)
3. Kuehn, D., Schilling, M., Stark, T., Zenzes, M., Kirchner, F.: System design and testing of the hominid robot Charlie. J. Field Robot. (2016). https://doi.org/10.1002/rob.21662
4. Plewa, K., Potvin, J.P., Dickey, J.P.: Wrist rotations about one or two axes affect maximum wrist strength. Appl. Ergon. **53**, 152–160 (2016)
5. Kuo, C.H., Dai, J.S.: Task-oriented structure synthesis of a class of parallel manipulators using motion constraint generator. Mech. Mach. Theory **70**, 394–406 (2013)
6. Kumar, S., Nayak, A., Peters, H., Schulz, C., Müller, A., Kirchner, F.: Kinematic analysis of a novel parallel 2SPRR+1U ankle mechanism in humanoid robot. In: Lenarcic, J., Parenti-Castelli, V. (eds.) ARK 2018. SPAR, vol. 8, pp. 431–439. Springer, Cham (2019). https://doi.org/10.1007/978-3-319-93188-3_49
7. Bajaj, N.M., Spiers, A.J., Dollar, A.M.: State of the art in artificial wrists: a review of prosthetic and robotic wrist design. IEEE Trans. Robot. (2019). https://doi.org/10.1109/TRO.2018.2865890
8. Ceccarelli, M., Russo, M., Morales-Cruz, C.: Parallel architectures for humanoid robots. Robotics (2020). https://doi.org/10.3390/robotics9040075
9. Kumar, S., Wöhrle, H., de Gea Fernández, J., Müller, A., Kirchner, F.: A survey on modularity and distributivity in series-parallel hybrid robots. Mechatronics **68**, 102367 (2020)
10. J. Eßer et al.: Design, analysis and control of the series-parallel hybrid RH5 humanoid robot. In: 2020 IEEE-RAS 20th International Conference on Humanoid Robots, pp. 400–407 (2021). https://doi.org/10.1109/HUMANOIDS47582.2021.9555770

Biomechanics

Smooth Wrapping of Stretchable Thick Strands Over a Surface - The 2D Case

Katharina Müller(✉) and Andres Kecskemethy

Chair of Mechanics and Robotics, University of Duisburg-Essen, Lotharstr. 1, 47057 Duisburg, Germany
{katharina.mueller,andres.kecskemethy}@uni-due.de

Abstract. The problem of wrapping of a strand over a surface is ubiquitous, such as in biomechanics for muscle length and muscle length rate estimation, or in robotics when wires are guided over non-circular surfaces. Hereby, continuous solutions are sought, as discrete solutions generate jumps in the position which lead to spikes in velocity and acceleration transmission. Up to now, the continuous problem has been solved only for infinitesimally thin lines using a chain of geodesics. In continuation of previous papers on a completely novel approach for continuous wrapping of thick strands of constant length, this paper regards now the problem of extendible midlines including cross-contraction, which is shown to generate again fundamentally new transmission equations including perturbation theory. Due to the complexity of the ensuing differential equations, only the 2D case is regarded, i.e. where the "surface" is a curve. However, the results can be transferred to the 3D case with general surfaces as the normal curvature along the geodesic line. The effectiveness of the approach is shown for the case of a parabolic convex muscle on an elliptical surface.

1 Introduction

The present topic has evolved from the biomechanical question of how muscles wrap around surfaces in musculoskeletal simulations, as needed for computing muscle forces in terms of muscle lengths and length rates. Here, it is common practice to model muscles as thin massless lines and to define the muscle path using via points or by wrapping them over smooth surfaces. A very efficient method using geodesic differential geometry for multi-surface, infinitesimal thickness wrapping has been presented in [6], showing the great computational advantage of continuous over discretized methods. However, when the muscle thickness is non negligible, the correct placement of the muscle paths creates a new problem. Existing approaches try to solve this problem by lifting the surface such that it fits the centerline of the muscle [2], but this is only applicable to constant surfaces and strands of constant thickness. Another way is to use FEM models which are, however, computationally expensive and also discontinuous. Other discretizations as e.g. the bead chain discretization method of Franci and Parenti-Castelli [1] are intuitive and well-implementable but again lead to jerky

O. Altuzarra and A. Kecskéthy (Eds.): ARK 2022, SPAR 24, pp. 359–366, 2022.
https://doi.org/10.1007/978-3-031-08140-8_39

force and acceleration transmission. In a new approach of the present authors, we targeted to model thick strands as curves of half-thickness b over the strand midline length s and to compute the transmission equations for wrapping and sliding in a continuous manner. In this setting, we started with a strand of constant length and "conicity" (i.e. aperture angle α) over arbitrary surfaces in [3] and then developed the equations for strands of constant length but variable aperture angle $\alpha \neq$ const in [4], showing computational speed improvements of factors 200 to 800 compared to the bead method. Hereby, the continuous method is shown to result as the envelope of the bead chain for inter-bead distance limiting to zero. This contribution now regards the effects of stretching and contraction, yielding again completely new kinematical transmission equations. Due to the complexity of the problem, only the 2D case is again regarded. The idea for the algorithm was already briefly outlined at ECCOMAS 2021 [5] and is developed here in depth. The developed methods can also be used in robotics where strands of non-constant thickness wrap over surfaces.

2 Model Derivation

A thick strand wrapped as illustrated in Fig. 1 consists of a wrapped segment AB in the middle enclosed by two free, undeformed segments at both ends. O and I are the free ends, Q_A and Q_B the points where the strand gets in contact with the surface. A and B are the corresponding points on the centerline, marking the transition between wrapped and unwrapped centerline segments.

Let the strand and the surface geometry be given and the movement of the end I be limited to one degree of freedom. Let the end O be fixed to consider the effects of elongation alone. The sought quantities are, thus, (a) the midline length s_{OA} of the free end between O and A; (b) the location of the contact points Q_A and Q_B which are defined by the surface coordinates σ_A and σ_B; and (c) the location of the end I expressed by the joint variable s_I. These four variables define the current state of the wrapped strand and can be subsumed into one state vector at A and one at B as

$$\underline{z}_A = \begin{bmatrix} s_{OA} \\ \sigma_A \end{bmatrix} \quad \text{and} \quad \underline{z}_B = \begin{bmatrix} s_I \\ \sigma_B \end{bmatrix}. \tag{1}$$

As already introduced in [5], the algorithm computing the state variables during the wrapping process consists of two nested integrations. In the external integration, the constraint matching takes place, while in the internal integration the wrapped segment length s_{OB} and its rate of change are computed for each external step.

2.1 Description of the Stretchable Thick Strand

The idea of the stretchable thick strand is illustrated here for the example of a parabolic strand (Fig. 2), but can be generalized later to any other case. The half-width b over the midline path length s is defined by

$$b(s) = \frac{b_E}{\ell\,(2\lambda - 1)} \left(2\lambda\, s - \frac{s^2}{\ell}\right) \tag{2}$$

starting at zero for $s = 0$ and with half-width b_E at the end of the strand for $s = \ell$ with ℓ being the strand's total length. The position of the vertex of the parabola is defined by the parameter $\lambda > 0.5$.

As a first rough consideration of the deformation of the strand due to stretching, the lateral contraction is exemplarily introduced following Poisson's ratio idea. For simplicity and without loss of generality, the change in the area of the cross-section related to the length change is used, leading to

$$\frac{A\,(\ell) - A\,(\ell_0)}{A\,(\ell_0)} = (1 - 2\nu)\,\frac{\ell - \ell_0}{\ell_0} \tag{3}$$

with the Poisson's ratio ν, the actual strand length ℓ and the unstretched length ℓ_0 and

$$A\,(\ell) = \int_0^{\ell} b\,(s)\;\mathrm{d}s = \frac{b_E\,\ell}{2\lambda - 1}\,\frac{3\lambda - 1}{3}. \tag{4}$$

Let the strand geometry for the unstretched case be defined by the parameters b_E^0, λ and ℓ_0. After substituting $\ell = \ell_0(1 + \varepsilon)$ and Eq. (4) into Eq. (3), the actual parameter b_E for the current longitudinal strain ε can be calculated by

$$b_E(\varepsilon) = b_E^0 \left(1 - 2\nu \frac{\varepsilon}{1 + \varepsilon}\right).0 \tag{5}$$

Note that this expression works also for large values of ε as long as $\nu < 0.5$, while the original Poisson's ratio derivation is defined only for infinitesimal ε.

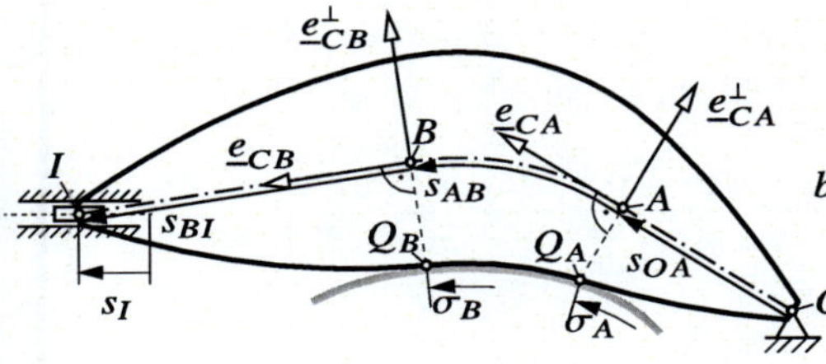

Fig. 1. Thick, convex strand with free endpoints O and I wrapped on a convex surface between A and B

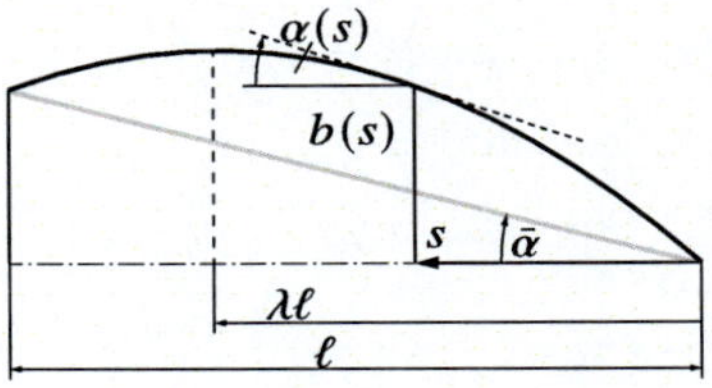

Fig. 2. Parabolic example for a convex, curved strand geometry The bigraph B_2

2.2 Implicit Velocity Constraints at A and B

To obtain the state variables $\underline{z}_A$ and $\underline{z}_B$ for the current configuration, four constraint equations are formulated. The length constraints stipulate that the distance between points O and A must correspond to the midline length s_{OA} and that the sum of the individual segment lengths s_{OA}, s_{AB} and s_{BI} must correspond to the total length ℓ of the strand. In addition, the midlines of the free segments OA and BI must be perpendicular to the cross-sections at A and B, while the surface contact at the points Q_A and Q_B is tangential. Thus, there result one length constraint and one orthogonality constraint at A and B each as

$$\underline{g}_A = \begin{bmatrix} g_{1A} \\ g_{2A} \end{bmatrix} = \begin{bmatrix} s_{OA} - \underline{r}_{OA}(\sigma_A, s_{OA}, \varepsilon) \cdot \underline{e}_C(\sigma_A, s_{OA}, \varepsilon) \\ \underline{r}_{OA}(\sigma_A, s_{OA}, \varepsilon) \cdot \underline{e}_C^{\perp}(\sigma_A, s_{OA}, \varepsilon) \end{bmatrix} = 0 \tag{6}$$

$$\underline{g}_B = \begin{bmatrix} g_{1B} \\ g_{2B} \end{bmatrix} = \begin{bmatrix} \underline{r}_{BI}(\sigma_B, s_{OB}, \varepsilon) \cdot \underline{e}_C(\sigma_B, s_{OB}, \varepsilon) + s_{OB}(\varepsilon) - \ell(\varepsilon) \\ \underline{r}_{BI}(\sigma_B, s_{OB}, \varepsilon) \cdot \underline{e}_C^{\perp}(\sigma_B, s_{OB}, \varepsilon) \end{bmatrix} = 0 \tag{7}$$

where $\underline{r}_{OA}$ and $\underline{r}_{BI}$ are the vectors from the points O to A and from B to I, respectively. The implicit constraints at velocity level apply as

$$\dot{\underline{g}}_A = \frac{\partial \underline{g}_A}{\partial \underline{z}_A}\, \dot{\underline{z}}_A + \frac{\partial \underline{g}_A}{\partial \varepsilon}\, \dot{\varepsilon} \stackrel{!}{=} 0 \quad \text{and} \quad \dot{\underline{g}}_B = \frac{\partial \underline{g}_B}{\partial \underline{z}_B}\, \dot{\underline{z}}_B + \frac{\partial \underline{g}_B}{\partial \varepsilon}\, \dot{\varepsilon} \stackrel{!}{=} 0, \tag{8}$$

where $\stackrel{!}{=}$ means a required condition. At end A, the derivatives of $\underline{e}_{CA}$ and $\underline{e}_{CA}^{\perp}$ with respect to the rotation φ around the center of curvature are

$$\frac{\partial \underline{e}_{CA}}{\partial \varphi} = -\underline{e}_{CA}^{\perp} \quad \text{and} \quad \frac{\partial \underline{e}_{CA}^{\perp}}{\partial \varphi} = \underline{e}_{CA}. \tag{9}$$

At B, the change in the strand length in the wrapped segment s_{AB} leads to a change in the total strand length s_{OB} at B and thus to a change in φ_B as

$$\frac{\partial \underline{e}_{CB}}{\partial \varphi} = \underline{e}_{CB}^{\perp} \left[-1 + \frac{\partial \alpha}{\partial s} \frac{\partial s_{OB}}{\partial \varphi} \right] \quad \text{and} \quad \frac{\partial \underline{e}_{CB}^{\perp}}{\partial \varphi} = \underline{e}_{CB} \left[1 - \frac{\partial \alpha}{\partial s} \frac{\partial s_{OB}}{\partial \varphi} \right]. \tag{10}$$

As derived in [4], the arc length ODE for $\alpha \neq$ const and non-constant curvature radius ($\rho \neq$ const) to compute the wrapping path between O and B is

$$\frac{\partial s}{\partial \sigma} = \frac{\partial s}{\partial \varphi} \frac{\mathrm{d}\varphi}{\mathrm{d}\sigma} = \frac{\rho(\sigma)\, \cos(\alpha(s)) + b(s)}{1 + b(s)\, \frac{\partial \alpha}{\partial s}} \frac{\mathrm{d}\varphi}{\mathrm{d}\sigma} \tag{11}$$

leading to

$$\frac{\partial \underline{e}_{CB}}{\partial \varphi} = -\underline{e}_{CB}^{\perp} \left[1 - X_B \left.\frac{\partial \alpha}{\partial s}\right|_B \right] \quad \text{and} \quad \frac{\partial \underline{e}_{CB}^{\perp}}{\partial \varphi} = \underline{e}_{CB} \left[1 - X_B \left.\frac{\partial \alpha}{\partial s}\right|_B \right] \tag{12}$$

with

$$\mathcal{X}_B = \frac{\partial s_{OB}}{\partial \varphi} = \left. \frac{\rho \cos\alpha + b}{1 + b\frac{\partial \alpha}{\partial s}} \right|_B . \tag{13}$$

Additionally applying

$$\frac{\partial \underline{r}_{OQA}}{\partial \varphi} = \rho_A \, \underline{e}_t^A \,, \quad \frac{\partial \underline{r}_{QBI}}{\partial \varphi} = -\rho_B \, \underline{e}_t^B, \tag{14}$$

$$\frac{\partial \ell}{\partial \varphi} = 0 \quad \text{and} \quad \frac{\partial b}{\partial \varphi} = \frac{\partial b}{\partial s} \frac{\partial s}{\partial \varphi} = \tan\alpha \, \mathcal{X}_B \ , \tag{15}$$

and $\underline{r}_{OQA} = \underline{r}_{Q,A} - \underline{r}_O$ and $\underline{r}_{QBI} = \underline{r}_I - \underline{r}_{Q,B}$, it holds for the Jacobians

$$\mathrm{J}_A = \frac{\partial \underline{g}_A}{\partial \underline{z}_A} = \begin{bmatrix} 1 - \underline{r}_{OQA} \cdot \underline{e}_{CA}^{\perp} \left.\frac{\partial \alpha}{\partial s}\right|_A & -\rho_A \cos\alpha_A + \underline{r}_{OQA} \cdot \underline{e}_{CA}^{\perp} \\ \tan\alpha_A - \underline{r}_{OQA} \cdot \underline{e}_{CA} \left.\frac{\partial \alpha}{\partial s}\right|_A & -\rho_A \sin\alpha_A + \underline{r}_{OQA} \cdot \underline{e}_{CA} \end{bmatrix} \begin{bmatrix} 1 & 0 \\ 0 & \frac{\mathrm{d}\varphi}{\mathrm{d}\sigma} \end{bmatrix} \tag{16}$$

$$\begin{aligned} \mathrm{J}_B &= \frac{\partial \underline{g}_B}{\partial \underline{z}_B} \\ &= \begin{bmatrix} \underline{u}_I \cdot \underline{e}_{CB} & -\rho_B \cos\alpha_B - \underline{r}_{QBI} \, \underline{e}_{CB}^{\perp} \left(1 - \mathcal{X}_B \left.\frac{\partial \alpha}{\partial s}\right|_B\right) + \mathcal{X}_B \\ \underline{u}_I \cdot \underline{e}_{CB}^{\perp} & -\rho_B \sin\alpha_B + \underline{r}_{QBI} \, \underline{e}_{CB} \left(1 - \mathcal{X}_B \left.\frac{\partial \alpha}{\partial s}\right|_B\right) - \tan\alpha_B \, \mathcal{X}_B \end{bmatrix} \begin{bmatrix} 1 & 0 \\ 0 & \frac{\mathrm{d}\varphi}{\mathrm{d}\sigma} \end{bmatrix} . \end{aligned} \tag{17}$$

Differentiating the constraints Eq. (6) and Eq. (7) with respect to ε leads to

$$\frac{\partial g_{1A}}{\partial \varepsilon} = -\left[\underline{r}_{Q,A} - \underline{r}_O\right] \cdot \underline{e}_C^{\perp} \left.\frac{\partial \alpha}{\partial \varepsilon}\right|_A \tag{18}$$

$$\frac{\partial g_{2A}}{\partial \varepsilon} = -\left[\underline{r}_{Q,A} - \underline{r}_O\right] \cdot \underline{e}_C \left.\frac{\partial \alpha}{\partial \varepsilon}\right|_A + \frac{\partial b_A}{\partial \varepsilon} \tag{19}$$

$$\frac{\partial g_{1B}}{\partial \varepsilon} = \left[\underline{r}_I - \underline{r}_{Q,B}\right] \cdot \underline{e}_C^{\perp} \left.\frac{\partial \alpha}{\partial \varepsilon}\right|_B + \frac{\partial s_{OB}}{\partial \varepsilon} - \ell_0 \tag{20}$$

$$\frac{\partial g_{2B}}{\partial \varepsilon} = -\left[\underline{r}_I - \underline{r}_{Q,B}\right] \cdot \underline{e}_C \left.\frac{\partial \alpha}{\partial \varepsilon}\right|_B - \frac{\partial b_B}{\partial \varepsilon} \tag{21}$$

at end A and at end B, respectively.

Note that while at A the derivatives of cross-section radius b and angle α with respect to ε can be computed directly from the strand geometry at this point, at B also the change of the strand length at B due to the length change in the wrapped segment s_{OB} caused by the strand deformation has to be taken into account as

$$\left.\frac{\partial \alpha}{\partial \varepsilon}\right|_B = \left.\frac{\partial \alpha}{\partial \varepsilon}\right|_{s_{OB}=\text{const}} + \frac{\partial \alpha}{\partial s} \frac{\partial s_{OB}}{\partial \varepsilon} \quad \text{and} \quad \frac{\partial b_B}{\partial \varepsilon} = \left.\frac{\partial b_B}{\partial \varepsilon}\right|_{s_{OB}=\text{const}} + \frac{\partial b}{\partial s} \frac{\partial s_{OB}}{\partial \varepsilon} . \tag{22}$$

This finally leads to

$$\frac{\partial g_{1B}}{\partial \varepsilon} = \underline{r}_{QBI} \cdot \underline{e}_C^{\perp} \left\{ \left.\frac{\partial \alpha}{\partial \varepsilon}\right|_{s_{OB}=\text{const}} + \frac{\partial \alpha}{\partial s}\frac{\partial s_{OB}}{\partial \varepsilon} \right\} + \frac{\partial s_{OB}}{\partial \varepsilon} - \ell_0 \tag{23}$$

$$\frac{\partial g_{2B}}{\partial \varepsilon} = - \underline{r}_{QBI} \cdot \underline{e}_C \left\{ \left.\frac{\partial \alpha}{\partial \varepsilon}\right|_{s_{OB}=\text{const}} + \frac{\partial \alpha}{\partial s}\frac{\partial s_{OB}}{\partial \varepsilon} \right\} - \left.\frac{\partial b_B}{\partial \varepsilon}\right|_{s_{OB}=\text{const}} - \tan\alpha \frac{\partial s_{OB}}{\partial \varepsilon}. \tag{24}$$

2.3 Change of Wrap Length s_{OB} for Infinitesimal Stretch Change $d\varepsilon$

For the computation of the change of wrap length s_{OB} for infinitesimal length increase $d\varepsilon$, one needs a first-order perturbation-theory approach. Let the mid-line path length be described as

$$\begin{aligned} s(\sigma, d\varepsilon) &= s_1(\sigma) + s_2(\sigma)\, d\varepsilon \\ s_1(\sigma) &\widehat{=} s_{OB} \\ s_2(\sigma) &\widehat{=} \frac{\partial s_{OB}}{\partial \varepsilon} = \kappa_{OB}^{\varepsilon}. \end{aligned} \tag{25}$$

Introducing the function $f(s, \sigma, \varepsilon)$ representing the generic wrap increment Eq. (11)

$$f(s, \sigma, \varepsilon) = \frac{\partial s}{\partial \sigma} = \frac{\rho(\sigma)\, \cos(\alpha(s(\varepsilon))) + b(s(\varepsilon))}{1 + b(s(\varepsilon))\frac{\partial \alpha}{\partial s}} \frac{d\varphi}{d\sigma}, \tag{26}$$

one obtains the two coupled ODEs which can be solved, for $\kappa_{OB}^{\varepsilon} = s_2(\sigma)$, simultaneously with the initial conditions $s_{1,0} = s_{OA}$ and $s_{2,0} = \partial s_{OA}/\partial \varepsilon - \partial s/\partial \sigma|_A \cdot \partial \sigma_A/\partial \varepsilon$

$$\frac{d}{d\sigma} s_1 = f(s_1, \sigma; \varepsilon) \qquad \text{and} \qquad \frac{d}{d\sigma} s_2 = \left.\frac{\partial f}{\partial s}\right|_{\varepsilon=\text{const}} s_2 + \left.\frac{\partial f}{\partial \varepsilon}\right|_{s=\text{const}} \tag{27}$$

with

$$\frac{\partial f}{\partial s} = \frac{\tan\alpha - 2\rho \sin\alpha \frac{\partial \alpha}{\partial s} - b\left[\rho \sin\alpha \left(\frac{\partial \alpha}{\partial s}\right)^2 + (\rho \cos\alpha + b)\frac{\partial^2 \alpha}{\partial s^2}\right]}{\left(1 + b\frac{\partial \alpha}{\partial s}\right)^2} \tag{28}$$

$$\frac{\partial f}{\partial \varepsilon} = \frac{\left(-\rho \sin\alpha \frac{\partial \alpha}{\partial \varepsilon} + \frac{\partial b}{\partial \varepsilon}\right)\left(1 + b\frac{\partial \alpha}{\partial s}\right) - (\rho \cos\alpha + b)\left(\frac{\partial b}{\partial \varepsilon}\frac{\partial \alpha}{\partial s} + b\frac{\partial^2 \alpha}{\partial s\, \partial \varepsilon}\right)}{\left(1 + b\frac{\partial \alpha}{\partial s}\right)^2}. \tag{29}$$

For illustration purposes, the coupled ODEs in Eq. (27) are compared to the explicit solution when wrapping a cone on a circe. In this case, the differential equation Eq. (11) simplifies to

$$ds = (\rho \cos\alpha + \tan\alpha\, s)\, d\varphi \tag{30}$$

which can be integrated directly. Separation of variables leads to

$$s_{OB}(\varphi) = \frac{\rho \cos\alpha}{\tan\alpha}\left(e^{\tan\alpha\,\varphi} - 1\right) + s_{OA}\,e^{\tan\alpha\,\varphi} \tag{31}$$

for the initial values $s_0 = s_{OA}$, $\varphi_0 = 0$. Direct derivation with respect to ε results in

$$\begin{aligned}\frac{\partial s_{OB}}{\partial \varepsilon}(\varphi) = {} & \frac{\rho \frac{\partial \cos\alpha}{\partial \varepsilon}}{\tan\alpha}\left(e^{\tan\alpha\,\varphi} - 1\right) - \frac{\rho \cos\alpha}{\tan^2\alpha}\frac{\partial \tan\alpha}{\partial \varepsilon}\left(e^{\tan\alpha\,\varphi} - 1\right) \\ & + \frac{\rho \cos\alpha}{\tan\alpha}\frac{\partial \tan\alpha}{\partial \varepsilon}\varphi\,e^{\tan\alpha\,\varphi} + s_{OA}\frac{\partial \tan\alpha}{\partial \varepsilon}\varphi\,e^{\tan\alpha\,\varphi} + \frac{\partial s_{OA}}{\partial \varepsilon}e^{\tan\alpha\,\varphi}.\end{aligned} \tag{32}$$

Using perturbation theory, one obtains for $\rho = \text{const}$, $\alpha = \text{const}$ and $\sigma \equiv \varphi$

$$\frac{\mathrm{d}}{\mathrm{d}\varphi}s_1 = \rho \cos\alpha + \tan\alpha\, s_1 \tag{33}$$

$$\frac{\mathrm{d}}{\mathrm{d}\varphi}s_2 = \tan\alpha\, s_2 + \rho\frac{\partial \cos\alpha}{\partial \varepsilon} + \frac{\partial \tan\alpha}{\partial \varepsilon}s_1 \tag{34}$$

which by direct integration (first of Eq. (33), then of Eq. (34)) leads again to

$$\begin{aligned}\frac{\partial s}{\partial \varepsilon}(\varphi) = {} & \frac{\rho \frac{\partial \cos\alpha}{\partial \varepsilon}}{\tan\alpha}\left(e^{\tan\alpha\,\varphi} - 1\right) - \frac{\rho \cos\alpha}{\tan^2\alpha}\frac{\partial \tan\alpha}{\partial \varepsilon}\left(e^{\tan\alpha\,\varphi} - 1\right) \\ & + \frac{\rho \cos\alpha}{\tan\alpha}\frac{\partial \tan\alpha}{\partial \varepsilon}\varphi\,e^{\tan\alpha\,\varphi} + s_{OA}\frac{\partial \tan\alpha}{\partial \varepsilon}\varphi\,e^{\tan\alpha\,\varphi} + \frac{\partial s_{OA}}{\partial \varepsilon}e^{\tan\alpha\,\varphi}.\end{aligned} \tag{35}$$

3 Computational Results

For a comparison regarding computation time and accuracy, the presented approach was applied to the example of a curved strand wrapping on an ellipsoidal surface curve as shown in Fig. 3 for the start and the end configuration. Point O is fixed while point I can slide vertically so that the boundary constraints are fulfilled for any ε.

The presented method was implemented in Matlab using the Runge-Kutta routine ode45 for integration of the velocity constraints in the range $[\varepsilon_0, \varepsilon_E]$ and compared with iterations at position level using equidistant Newton steps.

One can recognize that the integration method is more than twice as fast with respect to the iteration method, with still excellent accuracy at the end of the integration. Due lack of space and the increasing complexity in the implementation of the bead chain method no comparisons were carried out with respect to this method. However, in [4], it was already shown that a computation time benchmark using an implementation of the bead method resulted in computational savings by a factor of 200–800 for the case of constant strand length.

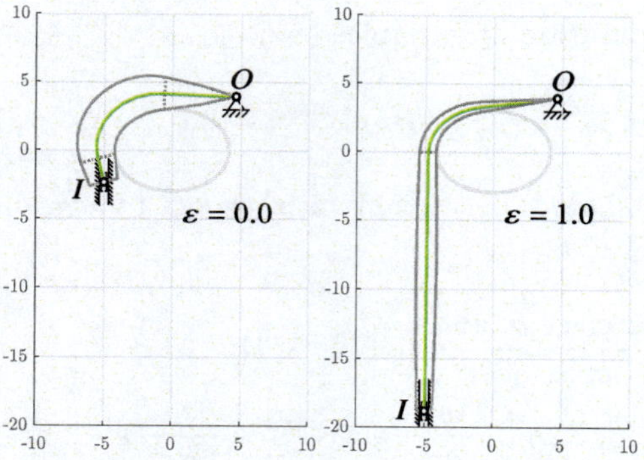

Model and numeric parameters	
ε_0	0.0
ε_E	1.0
$\Delta\varepsilon$	0.01
tolNewton	$1.0e-08$
tolODErel	$1.0e-08$
tolODEabs	$1.0e-08$

Method	CPU time [s]	Accuracy
Integration	0.5800	1.4e-08
Iteration	1.1378	-4.9e-11

Computations performed on Intel(R) Core(TM) i7-10850H CPU @ 2.70GHz

Fig. 3. Start and end configuration, parameters and results of the numerical computations

4 Conclusions

Based on the continuous and fast method for wrapping a thick, convex strand over a frictionless convex surface introduced in preliminary publications, this contribution extends the existing method and presents the detailed derivation of the effects of stretching and lateral contraction. Future work will include initial free end O freely movable, wrapping over multiple surfaces and the spatial contact surface case.

References

1. Franci, R., Parenti-Castelli, V.: A new tool to investigate the interactions between elastic fibers and rigid bodies. In: IFToMM 2007: Proceedings of Twelfth World Congress in Mechanism and Machine Science, Besançon, France (2007)
2. Garner, B.A., Pandy, M.G.: The obstacle-set method for representing muscle paths in musculoskeletal models. Comput. Methods Biomech. Biomed. Eng. **3**(1), 1–30 (2000)
3. Müller, K., Kecskemèthy, A.: A continuous and computationally efficient method for wrapping a “thick” strand over a surface — the planar single-surface case. In: IFToMM WC 2019. MMS, vol. 73, pp. 709–718. Springer, Cham (2019). https://doi.org/10.1007/978-3-030-20131-9_71
4. Müller, K., Kecskemethy, A.: A New approach for continuous wrapping of a thick strand on a surface — the planar case with constant length and free ends. In: Lenarčič, J., Siciliano, B. (eds.) ARK 2020. SPAR, vol. 15, pp. 293–302. Springer, Cham (2021). https://doi.org/10.1007/978-3-030-50975-0_36
5. Müller, K., Kecskemethy, A.: Planar wrapping and stretching of a thick strand on a surface by continuous integration. In: ECCOMAS Thematic Conference on Multibody Dynamics, Budapest, Hungary, 12–15 December 2021 (2021)
6. Scholz, A., Sherman, M., Stavness, I., Delp, S., Kecskeméthy, A.: A fast multi-obstacle muscle wrapping method using natural geodesic variations. Multibody Syst. Dyn. **36**(2), 195–219 (2015). https://doi.org/10.1007/s11044-015-9451-1

Exploiting Reciprocity Between Constraints and Instantaneous Motion to Reconstruct Individual Knee Kinematics

Michele Conconi(✉), Nicola Sancisi, and Vincenzo Parenti-Castelli

University of Bologna, Bologna, Italy
michele.conconi@unibo.it

Abstract. Musculoskeletal models are limited in their clinical application by the complexity of their personalization. In this work we present a novel and general approach for the reconstruction of the individual joint kinematics that may provide a reference for the personalization of more complex models. The approach relies on the principle of virtual work and obtains the joint motion by imposing reciprocity between instantaneous helical axis and joint constraints. As a case of study, we investigate a cadaveric knee, whose motion was experimentally measured for validation. Comparison between computed and experimental motion resulted in maximum mean rotational and translational errors of $2.6 \pm 0.9°$ and 1.2 ± 0.7 mm, respectively, supporting the efficacy of the presented approach.

Keywords: Musculoskeletal model personalization · Knee · Reciprocity

1 Introduction

Despite the flourishing of musculoskeletal models in the last thirty years, their impact in clinical applications is normally limited to the general assessment of physiological or pathological behavior, with rare direct use in treatments [1]. One cause is the difficulty in defining effective personalization strategies to cope with the high variability among individuals. Indeed, the more sophisticated a musculoskeletal model is, being thus capable to capture the fine interaction between human tissues, the more complex the experimental measurements needed for the tuning of its parameters become [2–5]. The transition of these models from the laboratory to the clinic is therefore limited.

In many cases [6–9], parameter tuning is performed by minimizing the differences between predicted and reference kinematics. However, measuring joint motion is either invasive (bone-pin markers; dual fluoroscopy), or inaccurate (skin markers), and in any case it requires long experimental procedures not suitable for the clinical practice. As an alternative, we have shown how it is possible to reconstruct the natural kinematics of a joint from a representation of its articular surfaces, which can be obtained from a simple MRI [10, 11]. The natural kinematics, defined as the natural motion of the joint in physiologically loaded conditions, is indeed obtained as the envelope of poses that maximize joint congruence, optimizing the distribution of the contact loads. This

O. Altuzarra and A. Kecskéméthy (Eds.): ARK 2022, SPAR 24, pp. 367–374, 2022.
https://doi.org/10.1007/978-3-031-08140-8_40

approach has been used to obtain a reference motion for the personalization of musculoskeletal models [12–14], but assumes the modeled articulations are healthy; thus its application to pathological conditions rises some concerns. Moreover, its application to highly non-conforming joints such as the knee, though successful, revealed to be more complicated. due to the higher importance of the role of ligaments in stabilizing the natural motion in this case.

In this study we present an alternative and more general approach to reconstruct the individual joint kinematics from a tridimensional representation of patient anatomy. Recently, we have shown theoretically and experimentally how the reciprocity between constraints and natural motion implies that ligament and contact forces must all intersect the instantaneous helical axis of joint motion [15, 16]. This characteristic is more general, since it is based on the only assumption a natural unresisted motion of the joint exists, which is generally true for both healthy and pathological diarthrodial joints, excluding severe complications that prevent the joint to move without applying high loads, or severe laxity in which joint motion not guided by articular constraints may exist. With the new approach, once articular surfaces and ligament insertions have been reconstructed from MRI, it is its thus theoretically possible to reconstruct the motion from the personalized articular constraints by imposing reciprocity. The method is presented and tested on a specimen, for which the natural motion was known from experimental measurements.

2 Materials and Metods

2.1 Experimental Acquisition

We analyzed a cadaveric specimen (age: 63 years, height: 162.5 cm; weight: 72 kg). First, the whole leg was acquired through a CT scanner (GE Medical Systems, Helical Mode, pixel spacing 0.50/0.50, slice thickness 0.62 mm) and bone models of the femur and tibia were reconstructed by segmentation through the freeware software MITK. Then, an MRI of the knee was acquired (3T GE Medical Systems, COR 3D merge, pixel spacing 0.31/0.31, slice thickness 1.5 mm, TR = 33.5 ms, TE = 10.86 ms, flip angle 8°) and both bone and cartilage models of the femur and tibia were segmented. In addition, origin and insertion of anterior cruciate (ACL), posterior cruciate (PCL), superficial, deep-anterior and deep-posterior medial collateral (MCL, MCLa, and MCLp respectively), lateral collateral (LCL), and anterolateral (ALL) ligaments were identified. MR and CT data were aligned registering bone to bone by an iterative-closest-point (ICP) algorithm.

The specimen was declared free from anatomical defect by an orthopedic surgeon, who skeletonized the leg leaving the joint capsulae intact. A stereophotogrammetric system (Vicon Motion Systems, Ltd., Oxford, UK) was used to measure the tibia and femur relative motion by means of two trackers directly fixed to the bones, thus introducing no soft-tissue artifacts. Each tracker included three markers. The same system was used to digitize bone surfaces, and data from medical images was aligned to VICON's one, again registering bone model to bone point clouds through ICP. The specimen was mounted on a test rig for in vitro analysis of the knee joint motion [17]: the femur was connected to the rig and was passively flexed, while the tibia was free to move according to its unresisted motion. Anatomical reference systems were defined based on [18] and femoro-tibial relative motion was parametrized using the center of the femoral reference system to track the translation and the Cardanic-angle sequence z-x-y proposed in [19] for the orientation.

2.2 Numerical Model

For the sake of clarity, we briefly report the analysis presented in [15]. $\boldsymbol{T}$ and $\boldsymbol{W}_i$ are respectively the twist representing the knee instantaneous helical axis (IHA) and the wrench corresponding to the ith articular constraint. In particular, if at each instant the origin of the reference system is chosen on the axis of $\boldsymbol{T}$ (Fig. 1), it is possible to write

$$\mathbf{T} = t\begin{bmatrix} \mathbf{u}_t \\ h_t\mathbf{u}_t \end{bmatrix},\ \mathbf{W}_i = w_i\begin{bmatrix} \mathbf{u}_i \\ (P_i - O) \times \mathbf{u}_i + h_i\mathbf{u}_i \end{bmatrix} \tag{1}$$

where $\boldsymbol{u}_t$ and $\boldsymbol{u}_i$ are the screw unit vectors of the IHA and the ith wrench, respectively; h_t and h_i are the screw pitches of the IHA and the i-th wrench, respectively; P_i is a generic point on the line of the ith wrench; t and w_i are the screw amplitudes of the IHA and the i-th wrench, respectively.

By introducing the interchange operator $\boldsymbol{\Delta} = \begin{bmatrix} 0 & \mathbf{I}_3 \\ \mathbf{I}_3 & 0 \end{bmatrix}$, where $\boldsymbol{I}_3$ is the 3×3 unit matrix, and 0 is the 3×3 zero matrix, the reciprocity can be expressed as

$$(\mathbf{W}_i)^T\boldsymbol{\Delta}\ \mathbf{T} = \underbrace{(h_t + h_i)\mathbf{u}_t \cdot \mathbf{u}_i}_{I} + \underbrace{\mathbf{u}_t \cdot ((P_i - O) \times \mathbf{u}_i)}_{II} = 0, \quad i = 1, \dots, n \tag{2}$$

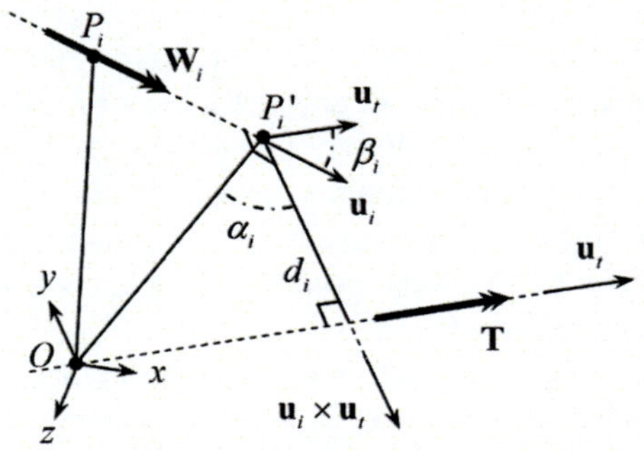

Fig. 1. The screws representing the knee IHA (T) and the ith wrench (Wi) at the generic instant of time

As discussed in [15], Eq. (2) simplifies to

$$(\mathbf{W}_i)^T\boldsymbol{\Delta}\ \mathbf{T} = \mathbf{u}_t \cdot ((P_i - O) \times \mathbf{u}_i)\ = 0,\ i = 1, \dots, n \tag{3}$$

Equation (3) is satisfied if and only if $(P_i - O)$, $\boldsymbol{u}_i$ and $\boldsymbol{u}_t$ belong to the same plane, that is $\boldsymbol{T}$ and $\mathbf{W}_i$ are incident (Fig. 2.a) or parallel (Fig. 2.b), this latter condition conflicting with knee physiology. This means that the lines of action of each articular constraint that guides the natural motion must all cross the IHA.

To reconstruct the knee motion, we started from an experimental pose (for instance the pose measured from CT or MRI, or in this case the first pose of the reference motion) and impose an increment of 1° of flexion. The remaining five motion components are determined through an optimization algorithm, using the simplex method and implemented in C++, as described below.

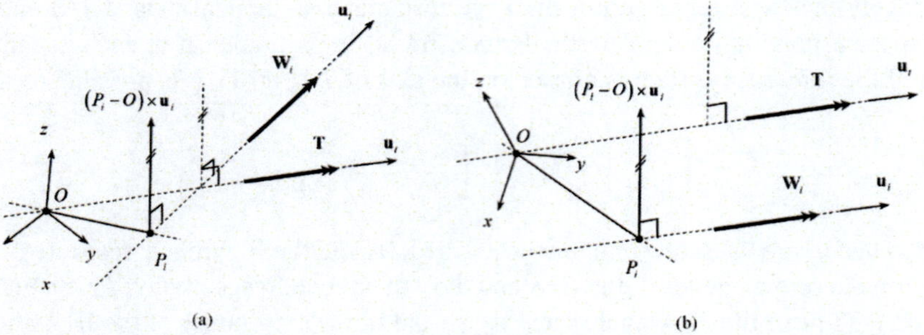

Fig. 2. Representation of the two geometrical conditions that satisfy Eq. (3): incident (a) and parallel (b) screws

The lines of action of ligament forces are determined as those passing from the centroid of corresponding origin and insertion, thus changing with the relative position of the femur with respect to the tibia. Contact areas are determined based on a proximity criterion, assuming as influence threshold 5 mm, corresponding to the average thickness of the menisci [20]. The directions of the medial (MAC) and lateral (LAC) contact forces are found as the average of the normals of the condylar contact surfaces, weighted proportionally to their distance from the tibial plateau, and applied to the geometrical centroid of the condylar contact areas. At any tested femoro-tibial pose, the optimizer computes the IHA from the previous to the present pose and the direction of the constraint lines of action. The objective function returns the sum of the distance between wrench and twist line. To force the contact on both condyles while preventing unphysiological indentation, a penalty term is added the objective measure when minimum condylar distance is above 0 mm or below −0.25 mm.

To validate the approach, numerical predictions are compared with experimental data, by computing the mean absolute error (MAE) for each motion component over the flexion angle. To evaluate the work done by the different constraints, wrench-to-twist distance is determined, as well as ligament isometry, defined as the percent difference between maximum (L_{Max}) and minimum (L_{Min}) ligament length during flexion:

$$\%\Delta L = \frac{L_{Max} - L_{Min}}{L_{Max}} \cdot 100 \quad (4)$$

3 Results

The comparison between computed and experimental data is shown in Fig. 3. A graphical representation of the twist-wrench reciprocity is provided in Fig. 4 for several flexion angles. The MAE for each motion components is reported in Table 1. Table 2 shows the average wrench-to-twist distances, while Table 3 reports the isometry of each fiber in the computed motion. Finally, Table 4 reports the mean indentation on the two condylar contacts.

Table 1. Mean absolute error MAE for each optimized motion component

	AA [°]	IE [°]	AP [mm]	PD [mm]	ML [mm]
MAE	0.6 ± 0.5	2.6 ± 0.9	1.1 ± 1.2	1.2 ± 0.7	0.6 ± 0.4

Table 2. Mean wrench-to-twist distance for all ligament and contact constraints implemented in the model

	ACL	PCL	MCL	MCLa	MCLp	LCL	ALL	MAC	LAC
d_{mean} [mm]	2.4 ± 1.8	4.10 ± 3.0	1.22 ± 1.08	2.9 ± 1.9	1.9 ± 1.6	1.7 ± 1.8	2.6 ± 2.2	1.7 ± 1.5	1.4 ± 1.1

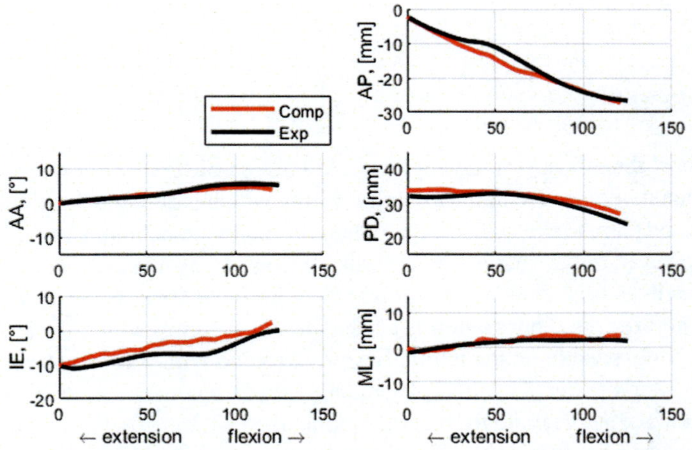

Fig. 3. Comparison between knee kinematics computed by imposing reciprocity (red) and experimental motion (black). The abduction/adduction (AA), internal/external rotation (IE), anterior/posterior (AP), proximal/distal (PD), and medial/lateral (ML) translations are plotted vs the flexion/extension.

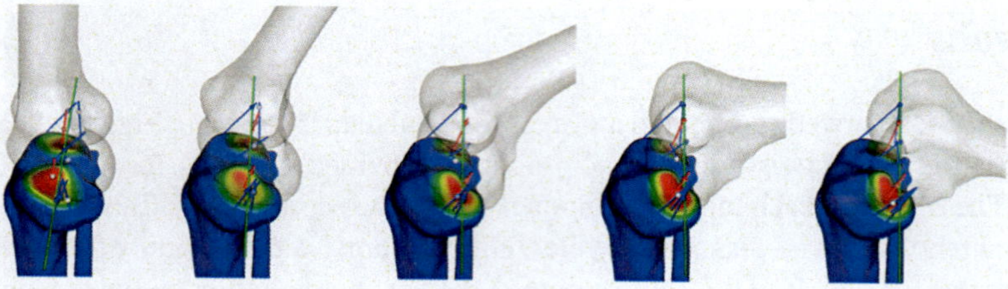

Fig. 4. Representation of the IHA (green line) and of the lines of action of the ligament (blue) and contact (red) wrenches every 30° of the computed knee motion. Spheres represent the closest point to the IHA on each wrench and their colour goes from white (d > 5 mm) to the one of the corresponding line (d = 0 mm). Colours on the tibia are proportional to the tibiofemoral relative distance (red: distance ≤ 0 mm; blue: distance > 7 mm).

Table 3. Ligament isometry for each considered fiber

	ACL	PCL	MCL	MCLa	MCLp	LCL	ALL
$\%\Delta L$	9.9	22.4	2.2	16.6	7.1	5.6	13.2

Table 4. Mean indentation on each condylar contact

	MAC	LAC
Mean indentation [mm]	-0.39 ± 0.69	-0.06 ± 0.17

4 Discussion

Results support the efficacy of the approach. Maximum differences between computed and experimental motion are below 3° and 1.5 mm, which is suitable for biomechanical applications and comparable with measurements of joint kinematics and with other models which, however, require parameter optimization on a measured kinematics [6–9]. The average wrench-to-twist distance is comparable with what obtained in previous experiments [15] and residual indentation is acceptable. Coherently with knee anatomy, the higher error is observable on the internal-external rotation, the less constrained motion components after flexion. In terms of isometry, some ligaments show an elongation above the normal 5% threshold for natural motion, the PCL resulting close to physiological limits for ligament tensioning. However, fibers are simply taken as the central fiber of each ligament bundle, which is in general not the case for the most isometric fiber, as shown in the literature. Future improvements will test the feasibility of a recursive optimization, in which location of the fibers will be optimized as well.

This work has limitations. Only one specimen was analyzed. Next activities will test the approach on other five in-vitro and five in-vivo knees. A careful sensitivity analysis needs to be performed, to test the influence of model parameters on the outcomes, particularly of the starting pose. Ligaments are modeled with a single or reduced number of fibers. More complex representation of ligament structure may improve the results. All the constraints have the same weight on the determination of the final motion, while in a real knee some structures may resist more than other.

5 Conclusions

We presented and validated a novel method for the reconstruction of individual knee kinematics based on joint morphology. The method exploits the reciprocity between joint constraints and IHA to reconstruct the unresisted femoro-tibial motion. The presented approach is general and thus extendable to other articulations and may provide a valid reference for personalization of more sophisticated models. In particular, the tibio-talar and subtalar joint will be tested in the near future.

References

1. Smith, S.H., Coppack, R.J., van den Bogert, A.J., Bennett, A.N., Bull, A.M.: Review of musculoskeletal modelling in a clinical setting: current use in rehabilitation design, surgical decision making and healthcare interventions. Clin. Biomech. **83**, 105292 (2021)
2. Beillas, P., Papaioannou, G., Tashman, S., Yang, K.H.: A new method to investigate in vivo knee behavior using a finite element model of the lower limb. J. Biomech. **37**(7), 1019–1030 (2004)
3. Harris, M.D., et al.: A combined experimental and computational approach to subject-specific analysis of knee joint laxity. J. Biomech. Eng. **138**(8), 081004 (2016)
4. Jung, Y., Phan, C.B., Koo, S.: Intra-articular knee contact force estimation during walking using force-reaction elements and subject-specific joint model. J. Biomech. Eng. **138**(2), 021016 (2016)
5. Sancisi, N., Parenti-Castelli, V.: A sequentially-defined stiffness model of the knee. Mech. Mach. Theory **46**(12), 1920–1928 (2011)
6. Sancisi, N., Parenti-Castelli, V.: A 1-dof parallel spherical wrist for the modelling of the knee passive motion. Mech. Mach. Theory **45**(4), 658–665 (2010)
7. Blankevoort, L., Huiskes, R.: Validation of a three-dimensionalmodel of the knee. J. Biomech. **29**(7), 955–961 (1996)
8. Guess, T.M., Thiagarajan, G., Kia, M., Mishra, M.: A subject specific multibody model of the knee with menisci. Med. Eng. Phys. **32**(5), 505–515 (2010)
9. Mootanah, R., et al.: Development and validation of a computational model of the knee joint for the evaluation of surgical treatments for osteoarthritis. Comput. Methods Biomech. Biomed. Eng. **17**(13), 1502–1517 (2014)
10. Conconi, M., Leardini, A., Parenti-Castelli, V.: Joint kinematics from functional adaptation: a validation on the tibio-talar articulation. J. Biomech. **48**(12), 2960–2967 (2015)
11. Conconi, M., Sancisi, N., Parenti-Castelli, V.: Prediction of individual knee kinematics from an MRI representation of the articular surfaces. IEEE Trans. Biomed. Eng. **68**(3), 1084–1092 (2020)
12. Smale, K.B., et al.: Effect of implementing magnetic resonance imaging for patient-specific OpenSim models on lower-body kinematics and knee ligament lengths. J. Biomech. **83**, 9–15 (2019)
13. Martelli, S., et al.: The relationship between tibiofemoral geometry and musculoskeletal function during normal activity. Gait Posture **80**, 374–382 (2020)

14. Conconi, M., Montefiori, E., Sancisi, N., Mazzà, C.: Modeling musculoskeletal dynamics during gait: evaluating the best personalization strategy through model anatomical consistency. Appl. Sci. **11**(18), 8348 (2021)
15. Conconi, M., Sancisi, N., Parenti-Castelli, V.: The geometrical arrangement of knee constraints that makes natural motion possible: theoretical and experimental analysis. J. Biomech. Eng. **141**(5), 051001 (2019)
16. Conconi, M., Sancisi, N., Parenti-Castelli, V.: The geometrical arrangement of joint constraints that makes natural motion possible: experimental verification on the ankle. In: Lenarčič, J., Siciliano, B. (eds.) ARK 2020. SPAR, vol. 15, pp. 109–116. Springer, Cham (2021). https://doi.org/10.1007/978-3-030-50975-0_14
17. Forlani, M., Sancisi, N., Conconi, M., Parenti-Castelli, V.: A new test rig for static and dynamic evaluation of knee motion based on a cable-driven parallel manipulator loading system. Meccanica **51**(7), 1571–1581 (2015). https://doi.org/10.1007/s11012-015-0124-1
18. Tashman, S., Anderst, W.: In-vivo measurement of dynamic joint motion using high speed biplane radiography and CT: application to canine ACL deficiency. J. Biomech. Eng. **125**(2), 238–245 (2003)
19. Grood, E.S., Suntay, W.J.: A joint coordinate system for the clinical description of three-dimensional motions: application to the knee. J. Biomech. Eng. **105**(2), 136–144 (1983)
20. Braz, P., Silva, W.: Meniscus morphometric study in humans. J. Morphol. Sci. **27**(2), 62–66 (2010)

Generation of Parametric Gait Patterns

Jakob Ziegler(✉), Hubert Gattringer, and Andreas Müller

Institute of Robotics, JKU Linz, Altenberger Street 69, 4040 Linz, Austria
{jakob.ziegler,hubert.gattringer,a.mueller}@jku.at

Abstract. In this paper, a methodology to generate realistic gait patterns is presented. Human gait motion capture data is used along with a kinematic model of the human lower extremity to derive a parametric and time-continuous analytical description of the walking motion. This allows for reproduction of individual recorded gait cycles and for generating new artificial gait cycles. A data pool of about 5700 reproduced gait cycles from 120 participants walking at different velocities is used to generate trajectories of human lower limb joints. Walking motions of simulated male or female persons can thus be synthesized with a prescribed gait speed. The method shall serve as scientific basis for research focused on rehabilitation, motion assistance and simulations.

1 Introduction

As a part of the demographic change, life expectancy is currently increasing. Rehabilitation of movement disorders and motion assistance has already become a major societal issue to be addressed. Maintaining the momentum of the last few years, robotic systems and methods are gaining increased attention as a complement and support to traditional motion analysis and therapy approaches. Thereby, it is crucial to intensify research in this field with focus on the solid foundation for a comprehensive dataset of normative reference motion patterns. Such reference patterns might not only allow for detecting pathologies or for monitoring the therapy progress of a patient, they are also important for the design of wearable robotic systems or novel motion assistance approaches. A certain walking velocity can be achieved by either separately adjusting step length or step frequency, or by a combination of both. Thus, it is evident that gait speed has significant impact on the gait pattern [10] and a multitude of recorded walking trials at various speeds would therefore present a reasonable basis for a reference dataset. However, human gait is very person-specific and innately shows significant variations from one individual to another. We present an approach to overcome this challenge by generating artificial gait patterns based on the data of recorded walking trials.

2 Motion Data

Predominantly, optical motion capture systems with reflective markers are used to record human motion. In order to increase measurement reproducibility and

O. Altuzarra and A. Kecskéméthy (Eds.): ARK 2022, SPAR 24, pp. 375–382, 2022.
https://doi.org/10.1007/978-3-031-08140-8_41

comparability, markers are typically placed on easily palpable anatomical landmarks. Various approaches, especially different kinematic models, are used in the literature to extract the time evolution of joint angles from motion capture data. We followed the idea to utilize existing data of human motion, which then are mapped to the same kinematic human body model to increase the comparability of the extracted joint trajectories of different datasets. In this paper, focus is given to the biomechanical modeling of the human lower body and the analysis and mathematical description of human gait patterns. To this end, motion datasets of healthy participants performing normal gait on even ground, where also the raw marker data is available, are used. This is assumed to best reflect natural walking, while there are known differences in the gait pattern when walking on an instrumented treadmill [15]. Our approach aims at modeling the influence of gait speed and sex onto the gait pattern. In spite of the vast amount of research on human motion analysis, motion datasets meeting the described requirements are relatively rare. However, the datasets of Fukuchi et al. [4], Lencioni et al. [11] and Schreiber et al. [13] were deemed suitable.

A minimum gait speed of $0.69\,\mathrm{ms}^{-1}$ is recommended for comparison of EMG activities and joint trajectories [6]. Around a gait speed of $2.0\,\mathrm{ms}^{-1}$ humans usually change the locomotion mode from walking to running [12]. The valid speed range is thus set to 0.69–$2.0\,\mathrm{ms}^{-1}$. Based on these limits the data of 120 participants (55 female, 65 male), performing about 5700 complete gait cycles, are investigated. One complete gait cycle refers to the movement between the first ground contact of a leg to the next ground contact of the same leg and thus consists of a stance phase followed by a swing phase. The corresponding distance covered is called *stride*, i.e. one *step* of each leg. All complete gait cycles are analyzed individually for male and female participants. To improve the comparability of gait parameters all data are normalized with respect to leg length to get non-dimensional (n.d.) quantities [7].

3 Kinematic Lower Body Model

The human lower body is modeled as a kinematic chain with the body segments pelvis, thigh, shank and foot. Note that the human body is usually assumed to be symmetrical with respect to the sagittal plane. Thus, in the following only the right leg is investigated as the methods can easily be transferred to the left leg.

Following the recommendation in [16], the inertial frame and all body-fixed frames are aligned so that in the standard anatomical position every x-axis is pointing anteriorly (forwards), every y-axis superiorly (upwards) and every z-axis to the right side of the body. Subscripts P, T, S and F will be used subsequently to refer to pelvis, thigh, shank and foot, respectively. Hip, knee and ankle joints are modeled as spherical joints with three degrees of freedom (DOF). Together with the three translational and three rotational DOF of the pelvis, which is the basis of the kinematic chain, the model of the lower extremity has thus 15 DOF. An estimation of the hip joint position can be done with a predictive approach via regression formulas based on anthropometric measures. A recent review [8] states that the method of [5], which was used in this paper, and which is based on pelvis width, pelvis depth and leg length, gives the highest accuracy. A more detailed description of the position of the other body-fixed coordinate frames with respect to marker data can be found in a previous publication of the authors [17].

The pose of the first of k bodies in a serial kinematic chain, described in the inertial frame, is represented by

$$\mathbf{T}_{\mathrm{I1}} = \begin{bmatrix} \mathbf{R}_{\mathrm{I1}} & {}_{\mathrm{I}}\mathbf{r}_1 \\ \mathbf{0}_{1\times 3} & 1 \end{bmatrix} \in \mathrm{SE}(3). \tag{1}$$

The transformation $\mathbf{T}_{\mathrm{Ik}}$ between the inertial frame and segment k can be calculated recursively as $\mathbf{T}_{\mathrm{Ik}} = \mathbf{T}_{\mathrm{I1}}\mathbf{T}_{12}\ldots\mathbf{T}_{(\mathrm{k}-1)\mathrm{k}}$, that is the combination of relative rotations $\mathbf{R}_{(\mathrm{k}-1)\mathrm{k}}$ and translations ${}_{\mathrm{k}-1}\mathbf{r}_{(\mathrm{k}-1)\mathrm{k}}$ between consecutive segments of the kinematic chain with

$$\mathbf{T}_{(\mathrm{k}-1)\mathrm{k}} = \begin{bmatrix} \mathbf{R}_{(\mathrm{k}-1)\mathrm{k}} & {}_{\mathrm{k}-1}\mathbf{r}_{(\mathrm{k}-1)\mathrm{k}} \\ \mathbf{0}_{1\times 3} & 1 \end{bmatrix}. \tag{2}$$

The configuration of the model can be described by $d = 15$ generalized coordinates, summarized in the vector $\mathbf{q} = [q_1, q_2, ..., q_d]^\mathsf{T}$ containing the individual DOF, which means $\mathbf{T}_{\mathrm{Ik}} = \mathbf{T}_{\mathrm{Ik}}(\mathbf{q})$.

Given a set of markers attached to body k in the chain, the (constant) position of the jth marker measured in the body-fixed frame is denoted ${}_{\mathrm{k}}\mathbf{p}_{\mathrm{k},j}$. Its position measured in the inertial frame, denoted ${}_{\mathrm{I}}\mathbf{m}_{\mathrm{k},j}$, is then determined by

$$\begin{bmatrix} {}_{\mathrm{I}}\mathbf{m}_{\mathrm{k},j} \\ 1 \end{bmatrix} = \mathbf{T}_{\mathrm{Ik}}(\mathbf{q}) \begin{bmatrix} {}_{\mathrm{k}}\mathbf{p}_{\mathrm{k},j} \\ 1 \end{bmatrix}, \tag{3}$$

which is equivalent to ${}_{\mathrm{I}}\mathbf{m}_{\mathrm{k},j} = {}_{\mathrm{I}}\mathbf{r}_{\mathrm{k}}(\mathbf{q}) + \mathbf{R}_{\mathrm{Ik}}(\mathbf{q})\,{}_{\mathrm{k}}\mathbf{p}_{\mathrm{k},j}$.

4 Motion Mapping

Within the framework of human motion analysis the term *motion mapping* denotes the use of a kinematic or dynamic model of the human body along with motion capture data in order to reproduce the motion. Goal of this procedure is to obtain a description of the underlying movement, e.g. the time evolution of the joint angles, joint torques or muscle forces. This paper focuses on kinematics, while the forces and torques causing an observed movement are not investigated at this point.

A serial kinematic model as described above with known geometry can be set to any valid configuration by defining $\mathbf{q}(t)$ at time t. In this case $\mathbf{q}$ contains the positional DOF of the pelvis and the rotational DOF of pelvis and body joints. Cardan angles α, β and γ according to the rotation sequence x-y'-z'' are thereby used to describe rotations. Given a set of $\{\mathbf{m}_{1,n}, \ldots, \mathbf{m}_{M,n}\}$ measured marker positions for measurement sample n the optimal model configuration $\mathbf{q}_n^*$ can be found in a least squares sense by solving the optimization problem

$$\mathbf{q}_n^* = \arg\min_{\mathbf{q}_n} \sum_{m=1}^{M} w_m \|\mathbf{m}_{m,n} - \hat{\mathbf{m}}_m(\mathbf{q}_n)\|^2, \tag{4}$$

where w_m is the individual marker weight and $\{\hat{\mathbf{m}}_1(\mathbf{q}_n), \ldots, \hat{\mathbf{m}}_M(\mathbf{q}_n)\}$ are the estimated marker positions, calculated by means of the forward kinematics of the model according to Eq. (3), depending on the current configuration $\mathbf{q}_n$.

In general, movement tasks related to locomotion consist of repetitive patterns. Thus, it is reasonable to describe these movement tasks as periodic signals, which can be expressed quite easily by Fourier series. For joint coordinate $q_i(t)$ the Fourier series of order l, with period T, is

$$q_i(t) = a_{i,0} + \sum_{k=1}^{l} \left[a_{i,k} \cos(k\omega t) + b_{i,k} \sin(k\omega t)\right], \tag{5}$$

where $\omega = 2\pi/T$ and $\mathbf{a}_i = [a_{i,0}, \ldots, a_{i,l}]^\intercal$ and $\mathbf{b}_i = [b_{i,1}, \ldots, b_{i,l}]^\intercal$ are the $2l + 1$ Fourier coefficients. With this parameterization it follows that for a gait cycle c consisting of $n = 1, \ldots, N_c$ samples the joint coordinates $\mathbf{q} = \mathbf{q}(t, \boldsymbol{\pi}_c)$ are now dependent on trajectory parameters denoted with $\boldsymbol{\pi}_c = [\mathbf{a}_{c,1}{}^\intercal, \ldots, \mathbf{a}_{c,d}{}^\intercal, \mathbf{b}_{c,1}{}^\intercal, \ldots, \mathbf{b}_{c,d}{}^\intercal]^\intercal$ corresponding to the Fourier coefficients. Thus, the nonlinear optimization problem in Eq. (4) changes to

$$\boldsymbol{\pi}_c^* = \arg\min_{\boldsymbol{\pi}_c} \sum_{n=1}^{N_c} \sum_{m=1}^{M} w_m \|\mathbf{m}_{m,n} - \hat{\mathbf{m}}_{m,n}(\boldsymbol{\pi}_c)\|^2. \tag{6}$$

This time-continuous parameterization shows robustness against missing data or noise, with errors within the typical range of soft tissue artifacts [17]. By solving Eq. (6) for all $c = 1, \ldots, C$ gait cycles, the values of the kth coefficients of the jth DOF are gathered in $\mathbf{a}_{j,k} = [a_{1,j,k}, \ldots, a_{C,j,k}]^\intercal$ and $\mathbf{b}_{j,k} = [b_{1,j,k}, \ldots, b_{C,j,k}]^\intercal$.

5 Parametric Gait Generation

Any desired gait speed v can theoretically be achieved by an infinite number of combinations of stride length l and stride frequency f as $v = lf$. Nevertheless, it appears that humans do not choose this combination in a completely arbitrary way. The walk ratio, i.e. stride length divided by stride frequency, provides an indication of that. Although the magnitude of this value varies from person to person, it seems to be relatively stable over a wide range of walking speeds [14]. We describe the relationship between gait cycle duration and gait speed with $T = (c_1 v^{c_2})^{-1}$ as it was already observed empirically that the stride frequency complies with the power law $f \propto v^\beta$ or $l \propto v^{1-\beta}$, respectively [2,9]. A constant walk ratio would mean that $l/f = vT^2 = \text{const}$, which is the case when $c_2 = 0.5$. Fitting procedures of the v-T-curve are carried out for the data of the male and female participants (Fig. 1a).

Denote $\hat{\mathbf{a}}_{j,k} = \boldsymbol{\Theta}\boldsymbol{\kappa}_{a,j,k}$ and $\hat{\mathbf{b}}_{j,k} = \boldsymbol{\Theta}\boldsymbol{\kappa}_{b,j,k}$ with $\boldsymbol{\Theta} = [\boldsymbol{\Theta}_1{}^\intercal, \ldots, \boldsymbol{\Theta}_C{}^\intercal]^\intercal$, where $\boldsymbol{\Theta}_c = \left[1 \; v_c \; v_c^2\right]$ contains the gait speed v_c of gait cycle c. By defining errors $\mathbf{e}_a = \mathbf{a}_{j,k} - \hat{\mathbf{a}}_{j,k}$ and $\mathbf{e}_b = \mathbf{b}_{j,k} - \hat{\mathbf{b}}_{j,k}$ the optimal values of the regression coefficients can be calculated in a least squares sense as $\boldsymbol{\kappa}_{a,j,k}^* = (\boldsymbol{\Theta}^\intercal\boldsymbol{\Theta})^{-1}\boldsymbol{\Theta}^\intercal\mathbf{a}_{j,k}$

and $\boldsymbol{\kappa}^*_{b,j,k} = (\boldsymbol{\Theta}^\mathsf{T}\boldsymbol{\Theta})^{-1}\boldsymbol{\Theta}^\mathsf{T}\mathbf{b}_{j,k}$, respectively. For a desired (n.d.) gait speed v_d the Fourier coefficients defining the artificial motion can thus be calculated by

$$\hat{a}_{j,k} = \begin{bmatrix}1 \; v_d \; v_d^2\end{bmatrix} \boldsymbol{\kappa}^*_{a,j,k} \tag{7}$$

$$\hat{b}_{j,k} = \begin{bmatrix}1 \; v_d \; v_d^2\end{bmatrix} \boldsymbol{\kappa}^*_{b,j,k}. \tag{8}$$

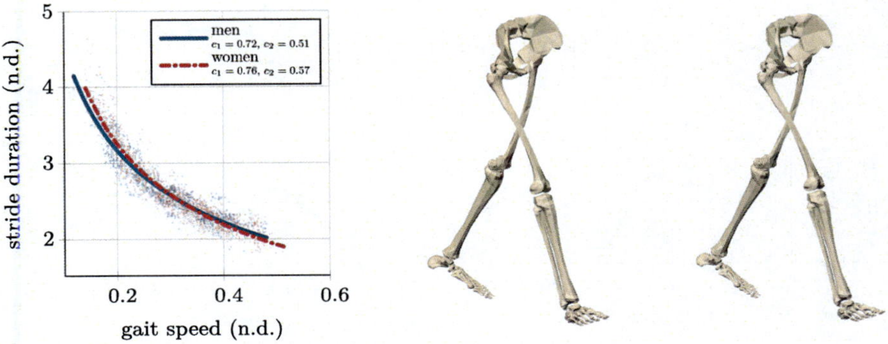

(a) (b) (c)

Fig. 1. Gait generation: v-T curves fitted to male (*squares*) and female (*dots*) data (**a**) and comparison of artificially created gait patterns of a male (**b**) and a female (**c**) person walking at a gait speed of 1.5 ms^{-1}. Note that (**b**) and (**c**) show an animation and are therefore best viewed in the PDF version of this document or online [1]. (see Supplementary Material)

6 Results and Discussion

As it is shown in Fig. 1a the female population seems to walk at slightly lower cadence and therefore longer strides than the male population, primarily at slower gait speeds. Although the two curves are obviously very similar, generally one would agree with the statement that men and women walk differently. Therefore, the great majority of information about differences of male and female gait must be contained in the joint angle trajectories. A direct comparison of artificially created gait patterns for a male and a female person, generated with the methodology introduced in this paper, is visualized in Fig. 1b and c. Note that this figure shows an animation which is best viewed in the PDF version of this document or online [1]. It is apparent that the female gait pattern can best be distinguished from the male counterpart by looking at the pelvis. More pronounced pelvic motion of women is consistently reported in the literature and might be partially explained by the greater pelvis width to femoral length ratio of women [3]. It can equally be observed by analyzing the present data (Fig. 2b, d and f), especially with regard to the rotation around the x-axis (pelvic obliquity). In Fig. 2a, c and e the hip, knee and ankle rotation around the z-axis of

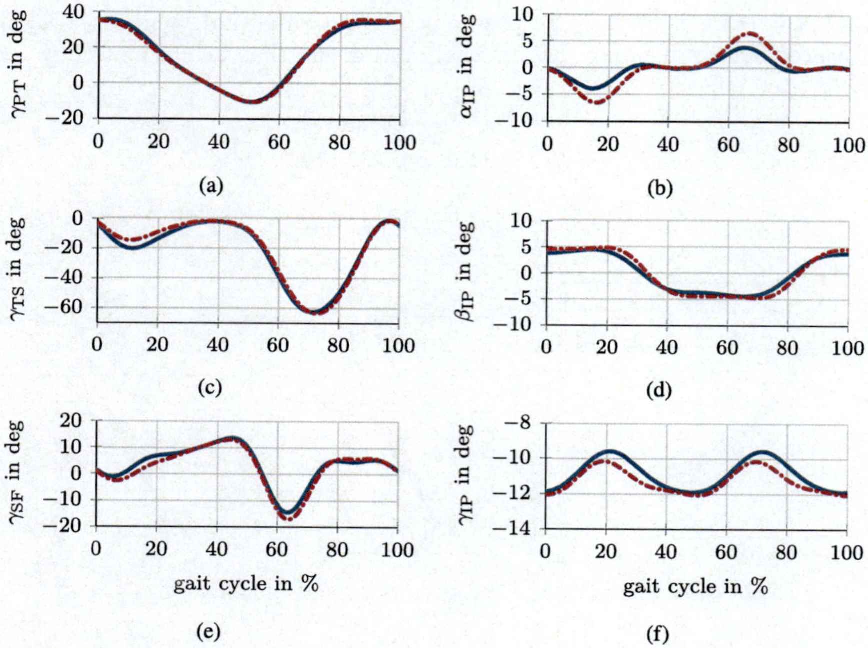

Fig. 2. γ_{PT}, γ_{TS} and γ_{SF} corresponding to the z-axis orientation of the hip (**a**), knee (**c**), and ankle (**e**) joint, as well as the three rotational DOF of the pelvis (**b,d,f**) for a simulated man (*solid line*) and woman (*dash-dotted line*) walking at $1.5\,\mathrm{ms}^{-1}$

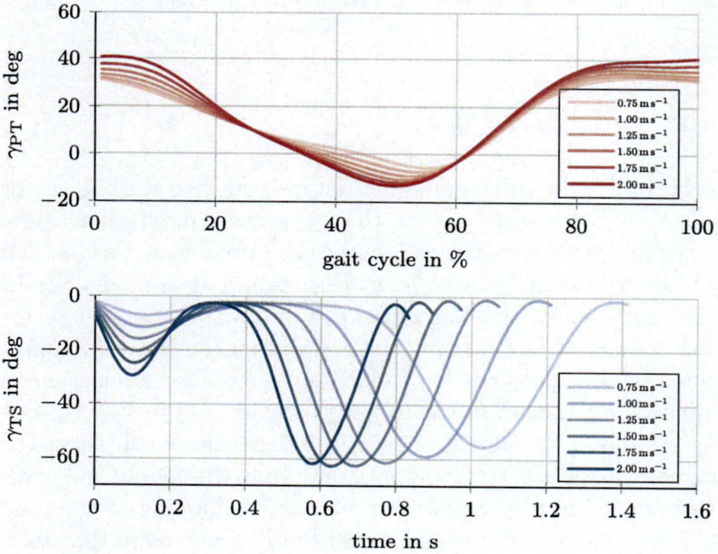

Fig. 3. Synthesized trajectories of the hip angle around the z-axis (γ_{PT}) of a male (top) and of the knee angle around the z-axis (γ_{TS}) of a female (bottom) person. Darker color indicates faster gait speed.

the right leg are compared, corresponding to the flexion and extension of these joints and being mainly responsible for the forward propulsion of the body. As can be seen, overall the rotations around the z-axis of the lower limb joints of the two persons are very similar, especially regarding the hip joint. Nevertheless, a difference of the knee angle evolution during early stance phase is visible. In this phase the man shows greater joint range of motion compared to the woman, suggesting higher maximum knee flexion. A possible explanation is that, due to the fact that man are generally heavier than women, greater shock absorption after initial contact is required during male gait. Figure 3 shows that on the one hand the gait cycle duration is decreasing and on the other hand the angular range of motion is increasing with faster gait, especially for joint angles corresponding to flexion and extension. This is consistent with the general consensus in the literature that stride length and stride frequency both increase monotonically with increasing gait speed.

The introduced methodology enables an easy generation of physiologically consistent gait patterns in terms of joint angle trajectories for simulated male or female individuals. Thereby, the prescribed gait speed can be arbitrarily chosen, although for reasonable results it should be within 0.69–2.0 ms^{-1}. Our contribution aims at providing an easily applicable solution for the parametric generation of time-continuous motions, which is deemed particularly useful for the development and simulation of novel (robotic) approaches regarding rehabilitation and gait assistance.

Acknowledgements. This work has been supported by the Linz Institute of Technology (LIT) and the COMET-K2 Center for Symbiotic Mechatronics of the Linz Center of Mechatronics (LCM) funded by the Austrian federal government and the federal state of Upper Austria.

References

1. Generation of Parametric Gait Patterns (video). https://youtu.be/qRqPaddAPTI
2. Bertram, J.E., Ruina, A.: Multiple walking speed-frequency relations are predicted by constrained optimization. J. Theor. Biol. **209**(4), 445–453 (2001). https://doi.org/10.1006/jtbi.2001.2279
3. Bruening, D.A., Frimenko, R.E., Goodyear, C.D., Bowden, D.R., Fullenkamp, A.M.: Sex differences in whole body gait kinematics at preferred speeds. Gait Posture **41**(2), 540–545 (2015). https://doi.org/10.1016/j.gaitpost.2014.12.011
4. Fukuchi, C.A., Fukuchi, R.K., Duarte, M.: A public dataset of overground and treadmill walking kinematics and kinetics in healthy individuals. PeerJ **2018**(4), e4640 (2018). https://doi.org/10.7717/peerj.4640
5. Harrington, M.E., Zavatsky, A.B., Lawson, S.E., Yuan, Z., Theologis, T.N.: Prediction of the hip joint centre in adults, children, and patients with cerebral palsy based on magnetic resonance imaging. J. Biomech. **40**(3), 595–602 (2007). https://doi.org/10.1016/j.jbiomech.2006.02.003
6. van Hedel, H.J., Tomatis, L., Müller, R.: Modulation of leg muscle activity and gait kinematics by walking speed and bodyweight unloading. Gait Posture **24**(1), 35–45 (2006). https://doi.org/10.1016/j.gaitpost.2005.06.015
7. Hof, A.L.: Scaling gait data to body size. Gait Posture **4**(3), 222–223 (1996). https://doi.org/10.1016/0966-6362(95)01057-2

8. Kainz, H., Carty, C.P., Modenese, L., Boyd, R.N., Lloyd, D.G.: Estimation of the hip joint centre in human motion analysis: a systematic review. Clin. Biomech. **30**(4), 319–329 (2015). https://doi.org/10.1016/j.clinbiomech.2015.02.005
9. Kuo, A.D.: A simple model of bipedal walking predicts the preferred speed-step length relationship. J. Biomech. Eng. **123**(3), 264–269 (2001). https://doi.org/10.1115/1.1372322
10. Lelas, J.L., Merriman, G.J., Riley, P.O., Kerrigan, D.C.: Predicting peak kinematic and kinetic parameters from gait speed. Gait Posture **17**(2), 106–112 (2003). https://doi.org/10.1016/S0966-6362(02)00060-7
11. Lencioni, T., Carpinella, I., Rabuffetti, M., Marzegan, A., Ferrarin, M.: Human kinematic, kinetic and EMG data during different walking and stair ascending and descending tasks. Sci. Data **6**(1), 1–10 (2019). https://doi.org/10.1038/s41597-019-0323-z
12. Neptune, R.R., Sasaki, K.: Ankle plantar flexor force production is an important determinant of the preferred walk-to-run transition speed. J. Exp. Biol. **208**(5), 799–808 (2005). https://doi.org/10.1242/jeb.01435
13. Schreiber, C., Moissenet, F.: A multimodal dataset of human gait at different walking speeds established on injury-free adult participants. Sci. Data **6**(1), 1–7 (2019). https://doi.org/10.1038/s41597-019-0124-4
14. Sekiya, N., Nagasaki, H.: Reproducibility of the walking patterns of normal young adults: test-retest reliability of the walk ratio (step-length/step-rate). Gait Posture **7**(3), 225–227 (1998). https://doi.org/10.1016/S0966-6362(98)00009-5
15. Watt, J.R., Franz, J.R., Jackson, K., Dicharry, J., Riley, P.O., Kerrigan, D.C.: A three-dimensional kinematic and kinetic comparison of overground and treadmill walking in healthy elderly subjects. Clin. Biomech. **25**(5), 444–449 (2010). https://doi.org/10.1016/j.clinbiomech.2009.09.002
16. Wu, G., et al.: ISB recommendation on definitions of joint coordinate system of various joints for the reporting of human joint motion - part I: ankle, hip, and spine. J. Biomech. **35**(4), 543–548 (2002). https://doi.org/10.1016/S0021-9290(01)00222-6
17. Ziegler, J., Reiter, A., Gattringer, H., Müller, A.: Simultaneous identification of human body model parameters and gait trajectory from 3D motion capture data. Med. Eng. Phys. **84**, 193–202 (2020). https://doi.org/10.1016/j.medengphy.2020.08.009

Parametrization of Compliant, Object-Level Controllers from Human Demonstrations

Elena Galbally Herrero(✉), Adrian Piedra, Cynthia Brosque, and Oussama Khatib

Stanford University, Stanford, USA
elenagal@stanford.edu

Abstract. For decades, robots have been able to reliably follow precise trajectories making them ideal tools for assembly lines and other structured environments. However, pre-programmed motions fail under uncertainty and are unsafe around humans, making them inadequate for unstructured environments. This paper presents a framework to generate safe, robust, and generalizable robot behaviors for contact tasks where compliance plays a key role. First, we collect task data from haptic demonstrations. Then, we segment the data into a sequence of compliant primitives. Finally, we extract the key parameters required for a robot to perform each of the primitive actions using interpretable, model-based controllers.

This method was experimentally validated on a steel bolting task using a 7-DOF Franka Panda robot. By recombining the primitives, we were also able to screw a cap onto containers of different sizes, placed in arbitrary configurations, using two different 7-DOF manipulators. The results show that our method generates position and orientation invariant, robot-agnostic controllers.

1 Introduction

As humans, we are constantly learning new skills. In particular, when it comes to manipulating new objects, we are extremely effective at generalizing to new scenarios and using compliance to our advantage. Compliance is key to generating robust behaviors by reducing our need to rely on precise motions or the speed and accuracy of our sensory information [10]. Inspired by humans, we have developed a data-efficient approach to learn robust, modular, and generalizable compliant primitives. Each of these primitives is defined by a compliant frame, its parametrization, and a set of conditions. We define a compliant frame as an origin and three directions in space about which we control motion and compliance. The origin of the frame specifies where to resolve our control forces and moments. This representation of the autonomous robot skills allows for a reduced number of parameters, robot-agnostic behaviors, and physical interpretability [3].

O. Altuzarra and A. Kecskémthy (Eds.): ARK 2022, SPAR 24, pp. 383–395, 2022.
https://doi.org/10.1007/978-3-031-08140-8_42

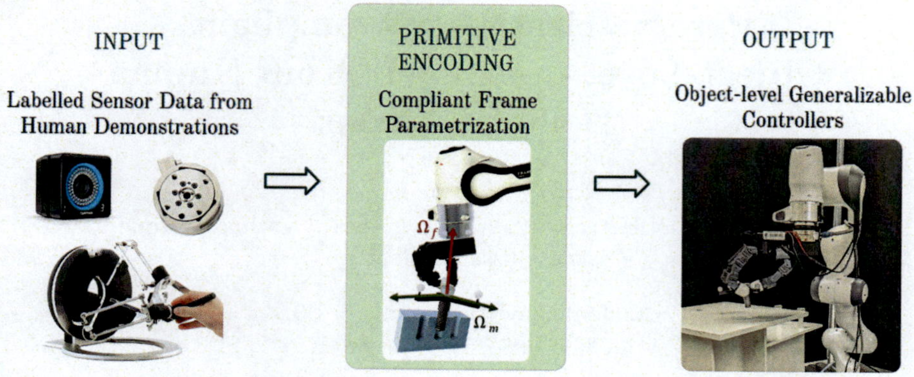

Fig. 1. The proposed method takes as input sensor data from human demonstrations of a complex manipulation task and encodes the human strategy into a reduced set of parameter. The output is a set of operational space controllers that can run on manipulators with different kinematics.

The low dimensionality of the latent space makes the proposed approach sample efficient, allowing us to learn each skill from as few as 10 human demonstrations. When learning a new task, we record human demonstrations and use a weakly supervised algorithm to segment the data into a sequence of skills that the robot can execute [5]. By recombining existing skills, we avoid the need to program new tasks completely from scratch.

The main contributions that differentiate this work from previous literature are: the parameterization of primitives using compliant frames, a method to go from human data to operational space control parameters, a framework and hardware setup for haptic demonstration data collection, and a public data set for the explored tasks. Compared to other end-to-end black box approaches [8,13,19], we show improved sample efficiency and interpretability. Additionally, by applying the method to two different robots and objects, we were able to show that our primitive encoding is able to generalize (Fig. 1).

2 Encoding Human Strategies

In this section, we want to understand what is happening to an object during a human demonstration and then encode the manipulation strategy in such a way that the robot can achieve the same object behavior. In order to understand what is happening to the object, we need to collect data and express it in a frame that makes it as easy as possible to interpret. Once we have understood the object behavior, we must express it as a set of parameters that can be fed into a robot controller. To generate robot-agnostic behaviors, we want to collect object-level data and run object-level controllers.

This paper defines a compliant primitive as a parametrization of a six degrees of freedom (DOF) controller defined with respect to a frame attached to the manipulated object. Directions of motion or compliance at the frame are specified such that they result in the desired object behavior. Prior literature [21] uses the notion of object-centric, task-axes controllers. In contrast to the work presented here, the controller parameters are extracted exclusively from visual data. We believe taking into account force data is advantageous when dealing with contact or occlusion.

2.1 What We Have: Task Data

Details on how to collect object-level data can be seen in Sect. 3. Here we detail the specific inputs considered during the experiments: contact information, pose, and velocity data. We record only object-level data because it allows the encoded behaviors to be agnostic to both the platform they were recorded on, as well as the one used to execute them.

We use a 6-DOF force/torque sensor to collect the forces and moments that the manipulated objects are subject to. The setup must be calibrated to remove the effects of inertial forces. Additionally, note that using a non-rigid grasp introduces noise and unmodeled nonlinear effects to the contact data. However, the use of a non-rigid grasp (e.g. gripper) can make the control approach more scalable than a rigid attachment by enabling us to switch easily between objects. Furthermore, using a gripper allows for regrasping. This is critical in tasks involving screwing, where the joint limits of the robot do not allow for full tightening. In this work, we tested both options (see Table 1): a rigid 3D-printed attachment for the bottle cap task and a 4-finger Allegro gripper for the steel bolting task.

The collected data for pose, velocity and force is defined as the following vector $\boldsymbol{y} \in \mathbb{R}^{19}$:

$$\boldsymbol{y} = [\boldsymbol{y}_{obj}, \dot{\boldsymbol{y}}_{obj}, \boldsymbol{F}_c] \tag{1}$$

where: $\boldsymbol{y}_{obj} = [x, y, z, \lambda_1, \lambda_2, \lambda_3, \lambda_4]$ is the object pose (note that λ_i are the components of a quaternion representing the orientation), $\dot{\boldsymbol{y}}_{obj} = [v_x, v_y, v_z, \omega_x, \omega_y, \omega_z]$ is a velocity vector obtained from post processing and using a transformation matrix to map the velocities from $\mathbb{R}^4$ to $\mathbb{R}^3$, and $\boldsymbol{F}_c = [f_x, f_y, f_z, \tau_x, \tau_y, \tau_z]$ are the contact forces and moments (Fig. 2).

The object pose can be obtained either from a motion capture system or from the robot's end-effector pose. If the data was collected through direct demonstration by instrumenting the world, we would have to use motion capture. Note that, when using a non-rigid grasp, motion capture is also useful to compensate for any relative motion between the gripper and object (e.g. due to slipping).

2.2 What We Want: A Parametrized Task-Space Robot Controller

We use task space controllers to ensure robot independent behaviors. The Operational Space control framework [11] is an object-level control framework for torque-controlled robots that relies on precise models of the robot and allows for a dynamically consistent operation of the robot. Let the robot have a task to fulfill, described by the task Jacobian J_t, the task coordinates x_t, and the

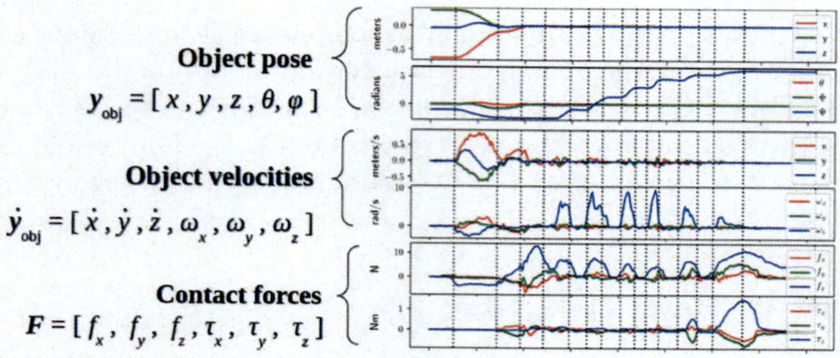

Fig. 2. Sample sensor data from human demonstrations.

associated task velocity $\dot{x}_t$, such that $\dot{x}_t = J_t\dot{q}$, where q represents the robot generalized coordinates and $\dot{q}$ represents the robot generalized velocities.

The equation of motion of the robot in free space is:

$$M(q)\ddot{q} + b(q, \dot{q}) + g(q) = \Gamma \tag{2}$$

where $M(q)$ is the robot mass matrix, $b(q, \dot{q})$ represents the Coriolis and centrifugal forces, $g(q)$ is the robot gravity vector, and Γ is the motor torques. In the following, we will drop the dependencies in q and $\dot{q}$ for better readability.

We can obtain the operational space equation of motion for the task by multiplying the previous equation by the transpose of the dynamically consistent inverse of the task Jacobian defined by:

$$\bar{J}_t = M^{-1}J_t^T\Lambda_t, \tag{3}$$

where $\Lambda_t = (J_t M^{-1} J_t^T)^{-1}$ is the task inertia matrix. To be precise, the dynamically consistent inverse of the Jacobian is its inverse inertia weighted pseudo-inverse. After multiplying Eq. (2) by $\bar{J}_t$ we get the operational space equation of motion:

$$\Lambda_t\ddot{x}_t + \mu_t + p_t = F_t \tag{4}$$

where $\mu_t = \bar{J}_t^T b - \dot{J}_t\dot{q}$ is the task space Coriolis and centrifugal, and $p_t = \bar{J}_t^T g$ is the gravity projected onto the task space. The task control torques will then be $\Gamma_t = J_t^T F_t$.

When there is a contact, the operational space equation of motion (5) becomes:

$$\Lambda_t\ddot{x}_t + \mu_t + p_t + F_c = F_t \tag{5}$$

where F_c is the contact force. There are many approaches to solving for F_c, some of which include using a particle filter or an analytical observer and have been explored in prior work [4].

In order to design a task controller for this system, we separate the task space into the force space and motion space, $F_t = F_f + F_m$. The force space corresponds to the constrained directions of motion, and the motion space corresponds to the directions in which the motion is free. The force and motion spaces projections are defined by block diagonal matrices Ω_t and Ω_m. Where:

$$\tau = J^T[\Lambda\Omega_m F_m{}^* + \Lambda\Omega_f F_f{}^*] \tag{6}$$

The control force $F_m{}^* = [f_m, m_m] \in \mathbb{R}^6$ is composed of a linear and angular component. Take $f_m{}^* = K_p(x - x_d) - K_v\dot{x}$, then x_d is the parameters we are trying to extract from the data. The same applies to the angular components.

2.3 What to Do: Define a Frame and Extract Parameters

Below is a summary of how to select the object frame and extract the key task parameters that allow us to generate robot-agnostic behaviors. Section 4 presents experimental results using the outlined method:

1. Express all of the sensor data in a single frame. More info on how to select the frame is in Sect. 4.1
2. Segment the data into primitives using an HMM and physical priors. See previous work [5] for more details.
3. Select the directions of motion and compliance (Fig. 5) and extract the controller inputs. See example for the align and screw primitives in Sect. 4.2
4. Improve or tune the controller parametrization by adding new demonstrations of the primitive to the data set.

3 Haptic Data Collection

The data collection setup should easily adapt to different tasks, and the data should provide insights into human strategy in a way that is relevant for robot implementation while remaining platform-independent by avoiding joint space representations.

We believe that a haptic interface such as the one illustrated in Fig. 3 is ideal for collecting meaningful demonstration data by forcing the operator to use and plan with the capabilities and limitations of the robot instead of their own. Another benefit over direct human demonstrations is that the setup is not task-specific, enabling data collection for a wide variety of single arm or two-arm manipulation tasks without modifications to the hardware or software setup. Furthermore, haptically collected data is more realistic since it includes gripper slippage instead of using rigid attachments between the force sensor and the manipulated object.

Haptic Controller. When designing a haptic controller we want to achieve accurate force and velocity tracking on the robot side while providing high-fidelity feedback to the human. Some of the most common type of haptic controllers include impedance-based [6], admittance-based [15], hybrid approaches combining the two [12,17], and adaptive control laws [18,20]. However, in this work we choose to use an approach that relies on independent, autonomous behaviors on both the robot and haptic side [9]. This control strategy has shown to be effective at dealing with communication delays and uncertainties.

Haptic Experiments. Our flexible haptic demonstration framework allows us to effortlessly switch tasks or collect data using new manipulators. In contrast, instrumenting the world is a much more rigid approach that does not scale well to new objects. Using a Force Dimension haptic device we collected data for both bolting and screwing caps. We developed a basic user interface (UI) that provides the operator with a real-time plot of the contact forces and moments as well as the ability to easily select which sensor variables they want to log for future analysis (Fig. 3). One further advantage of this setup is that it allow us to readily extend our data collection framework to bimanual tasks. In future work, we hope to explore a bimanual approach for the bottle cap screwing task.

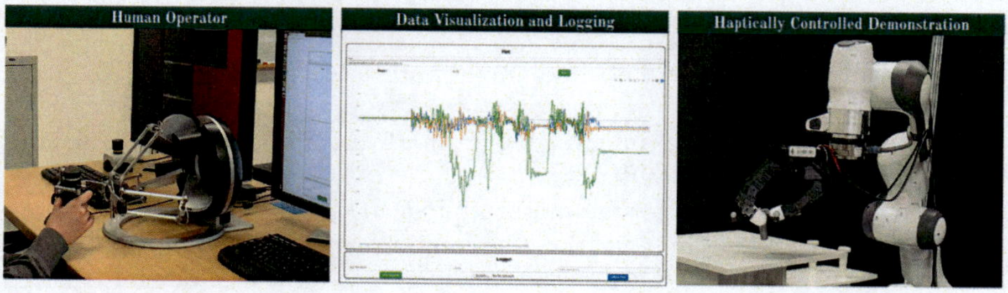

Fig. 3. Left to right: operator haptically demonstrating the task, UI with real-time plot of the contact forces and option to log the desired demonstration sensor data to a file, and haptically controlled robot executing the task.

Haptics gives us a way to keep the human in the loop and learn from recovery strategies. Keeping the human in the loop is particularly valuable for safety critical tasks as well as improving autonomy through continued learning. In the specific case of construction, haptics enables operators to perform the task remotely while increasing safety and maintaining their job [1].

4 Experimental Validation

In this paper we explore steel beam bolting [2, 16] and placing caps on containers of different sizes [3]. Both of these tasks can be performed using the same set of primitives: free space motion, make contact, align, and screw. In this section we will focus on the last two, since they are the most complex in terms of contact constraints (Fig. 4).

Table 1. Experiments

Task	Setup	Moving objects	Fixed objects
Bolting	1 × (Panda + Allegro)	Screw	Threaded hole
Bottle	1 × (Panda + Allegro)	Cap	Bottle
Bottle	1 × (LBR iiwa + Rigid grasp)	Cap	Bottle

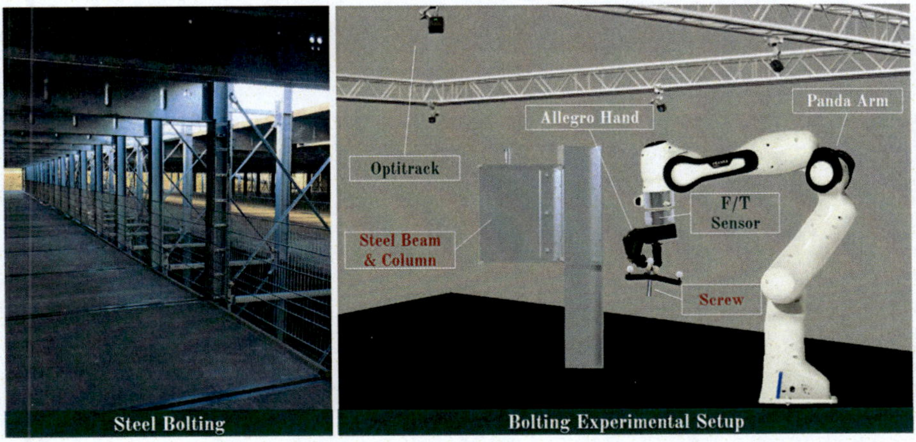

Fig. 4. Bolting image from prefabricated parking structure, courtesy of our industry partner Goldbeck. The bottom right render shows the experimental setup: sensors in green, objects in red, and robotic elements in white.

4.1 Frame Selection

We aim to minimize the negative effects of contact forces on the task and maximize isotropy. In order to meet this criteria, we will locate the frame origin (operational point) such that, if the null space task has more than 1-DOF, the null space motions resulting for the contact forces improve the performance of the main task. Otherwise, if the null space only has 1-DOF, locate it at the COM for free space motions and at the center of the contact surface/line or the contact point for tasks involving contact.

An example in which the null space has more than 1-DOF is an alignment primitive where we only control forces and leave the moments to the null space task. Assume the object is a container cap. In this case, the null space motion would tend to rotate the cap about the control point. Therefore, it would be preferable to place it at the center of the initial contact line (which is our desired axis of rotation) rather than the center of the cap.

Regarding the axes of the frame, we will select the directions about which we control motion and compliance by using the formulation described in Sect. 2.2 and illustrated in Fig. 5 for the bolting experiment.

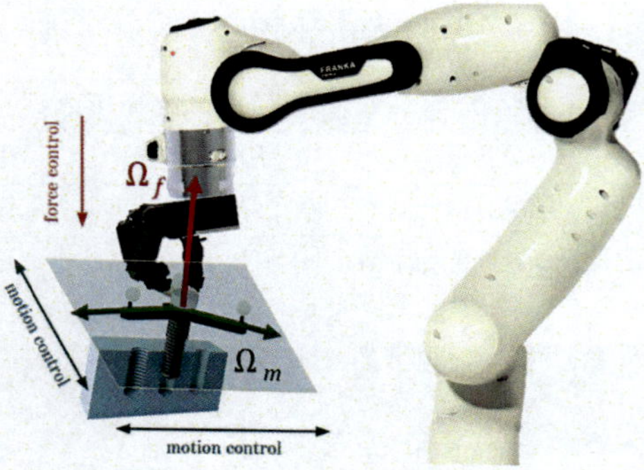

Fig. 5. The force space for translations is the line represented by the red arrow and the motion space for translations is the orthogonal plane represented by the green arrows. In force space we control compliance and in motion space we design a force to control the robot motion.

4.2 Primitive Parameters

Below we explore the controllers used for the main contact primitives involved in the two tasks explored in this paper: an alignment primitive where we control for zero moments and a screw primitive where we control linear force in the direction perpendicular to the cap and accelerations in the orthogonal plain. The specific values used during the experiments were extracted from the segmented and labelled data [5] and can be found on Table 2.

4.2.1 Align

During the align primitive, the robot attempts to rotate the cap/screw until it is aligned with the bottle/threaded. The vector of desired forces is expressed as

Table 2. Gain and threshold values used on the real kuka experiments

Parameter	Value	Units	Parameter	Value	Units
$k_{p,pos}$	40	$1/s^2$	V_{max}	0.1	m/s
$k_{v,pos}$	10	1/s	$\epsilon_{x_{err}}$	0.02	m
$k_{p,ori}$	4	1	$\epsilon_{\dot{x}_{err}}$	0.0001	m/s
$k_{v,ori}$	0.5	1/s	$\epsilon_{\|M_{sen}\|}$	0.1	N * m
$k_{p,slide}$	1.5	1	$\epsilon_{\|F_{err}\|}$	0.2	N
$k_{p,f}$	0.25	1	$\epsilon_{\|\omega_{xy}\|}$	0.01	rad/s
$k_{v,f}$	3	kg/s^2	$\epsilon_{F_{sen}(z)}$	−1	N
$k_{i,f}$	1	1/s	$\epsilon_{q_{err,screw}}$	1.5	rad
$k_{p,m}$	2	1	$\epsilon_{M_{sen}(z}$	−0.5	N * m
$k_{v,m}$	1	$kg * m^2/s$	$\epsilon_{q^*_{err,screw}}$	0.5	rad
$k_{i,m}$	1	1/s	$\epsilon_{M_{sen,CT}(z)}$	1	N * m
$k_{v,joint}$	10	1/s	$\epsilon_{q_{EE,CT}}$	0.4	rad
$k_{v,screw}$	4	1/s	$t_{screw,max}$	1	s

$$F_{des} = R_{EE}^{0}(F_z + k_{p,slide}\mathbf{u}_{slide}), \tag{7}$$

whereas the vector of error in the forces is expressed as

$$F_{err} = -R_{EE}^{0}(F_{sen} - F_{des}). \tag{8}$$

The vector of joint torques that are applied to the robot is computed as follows:

$$\tau = J_{v,cap}^{T} F_x + J_{\omega,cap}^{T} F_r, \tag{9}$$

where

$$F_x = F_{des} - k_{p,f} F_{err} - k_{v,f} \dot{x}_{err} - k_{i,f} \int F_{err}, \tag{10}$$

$$F_r = \cancelto{0}{M_{des}} - k_{p,m} M_{err} - k_{v,m}\omega - k_{i,m} \int M_{err}, \tag{11}$$

$$\omega_{xy} = (R_{EE}^{0})^{T}\omega. \tag{12}$$

In order to keep the controller stable, we reset $\int F_{err}$ and $\int M_{err}$ to zero once the terms are too large. Furthermore, we clip the control force value F_x to $3 \times F_{des}$.

The align state is finished if the following condition is true:

$$\begin{aligned} \mathbf{1}\big[&(||M_{sen}|| \leq \epsilon_{||M_{sen}||}) \wedge (||F_{err}|| \leq \epsilon_{F_{err}}) \\ &\wedge (||\omega_{xy}|| \leq \epsilon_{||\omega_{xy}||}) \wedge (F_{sen}(z) \leq \epsilon_{F_{sen}(z)})\big]. \end{aligned} \tag{13}$$

4.2.2 Screw

During screw, the two objects are aligned and the robot attempts to screw one onto the other. The vector of joint torques is computed as follows

$$\begin{aligned} \tau &= \underbrace{J_{v,cap}^{T} R_{EE}^{0} \begin{bmatrix} 0 \\ 0 \\ f_z \end{bmatrix}}_{F_x} + N_{v,cap}^{T} F_{o}ri, \\ &= J_{v,cap}^{T} F_x + N_{v,cap}^{T} F_{o}ri, \end{aligned} \tag{14}$$

where

$$\ddot{q} = -k_{v,joint}\dot{q}, \tag{15}$$

$$\ddot{q}_{EE} = -k_{v,screw}\dot{q}_{err,screw}, \tag{16}$$

$$q_{err,screw} = q_{EE} - q_{EE,des}. \tag{17}$$

The screw state has four possible transitions: either it is still running and self-transitions to the screw state, the screwing finished successfully and the controller advances to the release state, the threads of the cap and bottle were crossed and the controller must return to the rewind state, or the screwing failed and the controller must return to the align state.

The screw state is finished if the following condition is true:

$$\mathbf{1}\left[(q_{err,screw} \leq \epsilon_{q_{err,screw}}) \wedge (M_{sen}(z) \leq \epsilon_{M_{sen}(z)})\right]. \tag{18}$$

The screw state failed if the following condition is true:

$$\mathbf{1}\left[(q_{err,screw} \leq \epsilon_{q^{*}_{err,screw}}) \wedge (t_{screw} \geq t_{screw,max})\right], \tag{19}$$

There was cross-threading if the following condition is true:

$$\mathbf{1}\left[(M_{sen}(z) \geq M_{sen,CT}(z)) \wedge (q_{EE} \geq q_{EE,CT})\right]. \tag{20}$$

Note that the transition conditions were defined using heuristics. Nevertheless, the specific threshold values were learned automatically from the labelled haptic data.

4.3 Robot and Object Agnostic Behavior

When implementing the primitive, we need a robot model. All of the control equations require robot-specific parameters such as jacobians and mass matrices. However, by using frames attached to the object, the strategy itself does not depend on the robot. For example, resolving for zero moments to align the cap does not depend on the robot used. This is illustrated by the simulation results shown in Fig. 6. The sensed moments and forces for the alignment primitive are analogous for both the Kuka and Panda. The observed behavior was also

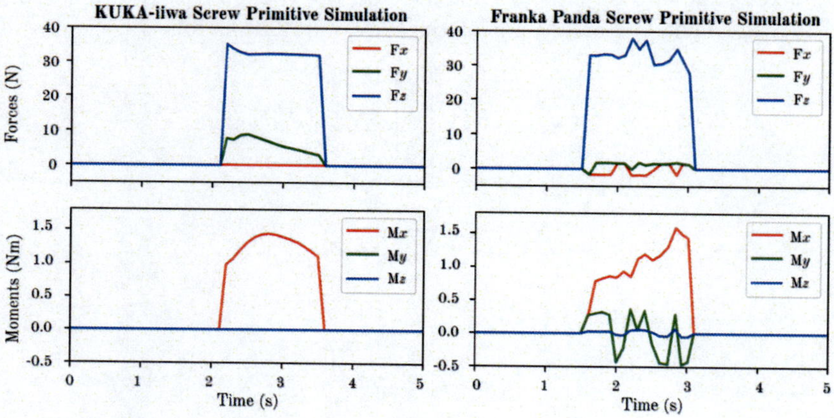

Fig. 6. Simulation experiments: filtered contact forces and moments in simulation for the screw task using two different robot models.

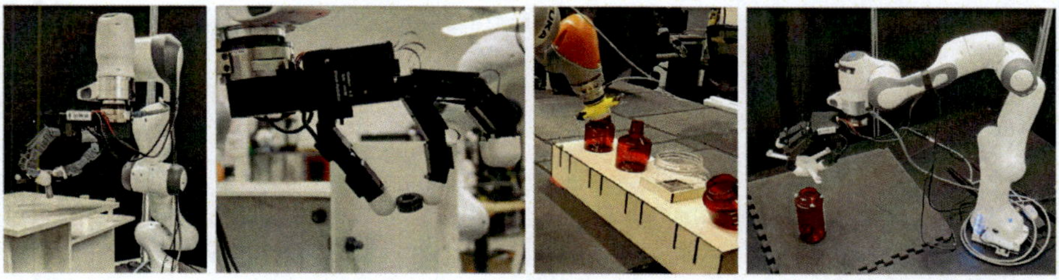

Fig. 7. The primitive parametrization presented in this paper is agnostic to both the specific kinematics of the robotic platform and the object being manipulated. Video of data collection and task execution experiments: video

equivalent, and no gain tuning or controller adaptation was required. We simply fed our controller a different robot model. The hardware experiments in Fig. 7 illustrate the same robot-agnostic behavior.

5 Conclusions

This work provides a method to extract robot control parameters from human demonstration data. The experiments show the success of our method for two different tasks: steel bolting and placing caps on containers. In particular, the results demonstrate that parametrizing compliant primitives as frames is sample efficient, low dimensional, and leads to robot-agnostic generalizable behaviors. By recombining compliant primitives we are able to move away from preprogrammed motions that fail in unstructured environments and plan at a higher level of abstraction [14]. Future work will explore the use of haptics not only to collect data during the initial task learning phase, but also to keep the human in the loop [7] and collect failure recovery strategies that will allow us to refine the primitive parameters and reduce failure rates over time.

Acknowledgements. Thank you to everyone at the Stanford Robotics Lab! Special thanks as well to Toki Migimatsu for his invaluable help with the initial exploration of this work. Thank you also to Mikael for his guidance and to Marco Speziali for his help with rendering and grasping advice.

References

1. Brosque, C., Galbally, E., Khatib, O., Fischer, M.: Human-robot collaboration in construction: opportunities and challenges. In: Proceedings of the 2nd International Congress on Human-Computer Interaction, Optimization and Robotic Applications, HORA 2020 (2020). https://doi.org/10.1109/HORA49412.2020.9152888
2. Choi, T., Do, H., Park, D., Park, C., Kyung, J.: Bolting with the industrial dual-arm robot. In: 2014 11th International Conference on Ubiquitous Robots and Ambient Intelligence, URAI 2014, pp. 484–485 (2014). https://doi.org/10.1109/URAI.2014.7057464

3. Edmonds, M., et al.: A tale of two explanations: enhancing human trust by explaining robot behavior. Technical report (2019). http://robotics.sciencemag.org/
4. Galbally, E., Jorda, M.: Real time collision detection and identification for robotic manipulators (2018). https://arxiv.org/abs/1802.00546v1
5. Herrero, E.G., Ho, J., Khatib, O.: Understanding and segmenting human demonstrations into reusable compliant primitives. In: 2021 IEEE/RSJ International Conference on Intelligent Robots and Systems (IROS), pp. 9437–9444 (2021). https://doi.org/10.1109/IROS51168.2021.9636523. https://ieeexplore.ieee.org/document/9636523/
6. Hogan, N.: Impedance control: an approach to manipulation. In: Proceedings of the American Control Conference, vol. 1, pp. 304–313 (1984). https://doi.org/10.23919/ACC.1984.4788393
7. Hoque, R., et al.: LazyDAgger: reducing context switching in interactive imitation learning (2021)
8. Jang, E., Vijayanarasimhan, S., Pastor, P., Ibarz, J., Levine, S.: End-to-end learning of semantic grasping (2017). https://arxiv.org/abs/1707.01932v3
9. Jorda, M.: Robust robotic manipulation for effective multi-contact and safe physical interactions. Ph.D. thesis, Stanford University (2020)
10. Khansari, M., Klingbeil, E., Khatib, O.: Adaptive human-inspired compliant contact primitives to perform surface–surface contact under uncertainty. Int. J. Robot. Res. **3513**, 1651–1675 (2016). https://doi.org/10.1177/0278364916648389. https://doi-org.stanford.idm.oclc.org/10.1177/0278364916648389
11. Khatib, O.: A unified approach for motion and force control of robot manipulators: the operational space formulation. IEEE J. Robot. Autom. **3**(1), 43–53 (1987). https://doi.org/10.1109/JRA.1987.1087068
12. Kim, H., Kwon, J., Oh, Y., You, B.J., Yang, W.: Weighted hybrid admittance-impedance control with human intention based stiffness estimation for human-robot interaction. In: IEEE International Conference on Intelligent Robots and Systems, pp. 6926–6931 (2018). https://doi.org/10.1109/IROS.2018.8594435
13. Kordia, A.H., Melo, F.S.: An end-to-end approach for learning and generating complex robot motions from demonstration. In: 16th IEEE International Conference on Control, Automation, Robotics and Vision, ICARCV 2020, pp. 1008–1014 (2020). https://doi.org/10.1109/ICARCV50220.2020.9305399
14. Lopes, M., Melo, F., Montesano, L., Santos-Victor, J.: Abstraction levels for robotic imitation: overview and computational approaches. Stud. Comput. Intell. **264**, 313–355 (2010). https://doi.org/10.1007/978-3-642-05181-4_14. https://link.springer.com/chapter/10.1007/978-3-642-05181-4_14
15. Maples, J.A., Becker, J.J.: Experiments in force control of robotic manipulators, pp. 695–702 (1986). https://doi.org/10.1109/ROBOT.1986.1087590
16. Nam, H., Choi, W., Ryu, D., Lee, Y., Lee, S.H., Ryu, B.: Design of a bolting robot for constructing steel structure. In: International Conference on Control, Automation and Systems, ICCAS 2007, pp. 1946–1949 (2007). https://doi.org/10.1109/ICCAS.2007.4406667
17. Ott, C., Mukherjee, R., Nakamura, Y., Mukherjee, R., Nakamura, Y.: A hybrid system framework for unified impedance and admittance control. J. Intell. Robot. Syst. **78**, 359–375 (2015). https://doi.org/10.1007/s10846-014-0082-1. www.sarcos.com
18. Park, J., Khatib, O.: A haptic teleoperation approach based on contact force control. Int. J. Robot. Res. **25**(5-6), 575–591 (2016). https://doi.org/10.1177/0278364906065385. https://journals.sagepub.com/doi/abs/10.1177/0278364906065385

19. Rahmatizadeh, R., Abolghasemi, P., Boloni, L., Levine, S.: Vision-based multi-task manipulation for inexpensive robots using end-to-end learning from demonstration. In: Proceedings of the IEEE International Conference on Robotics and Automation, pp. 3758–3765 (2018). https://doi.org/10.1109/ICRA.2018.8461076
20. Shahdi, A., Sirouspour, S.: Adaptive/robust control for time-delay teleoperation. IEEE Trans. Rob. **25**(1), 196–205 (2009). https://doi.org/10.1109/TRO.2008.2010963
21. Sharma, M., Kroemer, O.: Generalizing object-centric task-axes controllers using keypoints. Technical report (2013). https://deepai.org/publication/generalizing-object-centric-task-axes-controllers-using-keypoints

Linkages

Geometric Model for Serial-Chain Robot Inverse Kinematics in Case of Two Translational DoF, Spatial Rotation and Functional Redundancy

Moritz Schappler(✉), Tobias Blum, and Tim-David Job

Institute of Mechatronic Systems, Leibniz University Hannover, Hannover, Germany
moritz.schappler@imes.uni-hannover.de

Abstract. Geometric formulations for the inverse kinematics problem (IKP) in robotics are often set up in the full Cartesian space of three translational and three rotational coordinates (3T3R). When transferring this to tasks with spatial rotation like 3T2R, 2T3R or 2T2R, the result is usually not defined in a minimal set of independent coordinates. Removing the excluded operational space coordinates completely from the expressions is interesting from a theoretical point of view and simplifies further calculations. This can be achieved by formulating a 2R residual using the *Z*-*Y*′-*X*″ Tait-Bryan angles and a 2T residual derived by the projection of the pointing direction on a plane. In this paper, the minimal-coordinate IKP is derived for 2T2R and 2T3R tasks on position level with application to a gradient-projection scheme. Limitations of the redundant coordinate are considered within the nullspace.

Keywords: Serial-link robot · Inverse kinematics · Functional redundancy · Geometric model · Nullspace projection · 2T2R · 2T3R · Coordinate inequality constraint

1 Introduction and State of the Art

The inverse kinematics of robot manipulators in tasks with reduced degrees of freedom (DoF) has been investigated in the context of *functional redundancy*, where operational space and joint space have more DoF than the task space [17]. Some tasks with axis-symmetric tools like welding [2,9] or drilling [19,21] require three translational and two rotational DoF (3T2R), where six-axis robots are redundant.

In some applications, the process is also *independent of the feed* in the tool axis' direction. This results in a task with five DoF with full orientation (2T3R) or four DoF with rotational symmetry (2T2R), which can lead to one or two redundant DoF. In waterjet cutting the jet can have an effective cutting range of several centimeters [1]. The laser target acquisition task e.g. allows arbitrary motion in beam direction [3]. Other laser appliances like laser cutting [4] or

O. Altuzarra and A. Kecskémthy (Eds.): ARK 2022, SPAR 24, pp. 399–408, 2022.
https://doi.org/10.1007/978-3-031-08140-8_43

remote laser welding [5] can have a variable distance with a focus area or an adjustable focus point. Further examples are drilling with a dedicated feed axis of the drilling tool [10], spraying with a range of distance [6] or the usage of a camera with a specific depth of field [7]. It should be noted that in most of these 2T applications, the feed can not be disregarded as for the 2R case, but merely has to be *limited to an acceptable range* [7]. A related case are 2T constraints in the remote center of motion (RCM) problem [13,14]. Exemplary *2T applications specifically of parallel mechanisms* are 2T3R medical needle holders [11], laser satellite tracking (implemented as 3T2R in [3]), 2T2R oscillating screens [18] or simulation of a 2T3R spinal cord in bionics [20].

The solution of the inverse kinematic problem (IKP) with functional redundancy can be obtained with the well-established *nullspace-projection method* [17]. A requirement to use the method is the proper construction of the IK residual and the IK Jacobian. One way is to add a virtual joint into the kinematic structure and to augment the manipulator Jacobian by one column [2]. By adding a virtual prismatic joint the 3T2R task from [2] can be transferred to a 2T2R task like in [5] or for the RCM problem in combination with an RCM-constrained Jacobian in [13]. The *augmentation approach* has the drawback that it has a higher computational cost and can lead to an ill-conditioned Jacobian [9]. The need of a *minimal-coordinate approach* is emphasized for the RCM problem in [14]. Another option to adapt the Jacobian is to reduce it instead of augmenting it by removing a row of the "task frame Jacobian" [22]. *Reducing the Jacobian* requires calculating the full residual and Jacobian first and removing the redundant row subsequently [16].

Further solutions for the functional redundancy are based on the orthogonal decomposition of the twist [8] or constructing a cone or pyramid with a range of tolerances for tilt angles and positions on the tool axis [6]. Another approach uses a *functional relationship between task DoF and redundant DoF* for an optimized performance index via Monte-Carlo simulation [19]. Identifying the functional relationship requires pre-defining a functional space. For the considered 2T2R case with two redundant DoF a *closed-form solution* is harder to obtain, if possible at all. *Cascaded optimization* utilizes an inner loop that solves the standard IK problem and an outer loop optimizing the performance index [21]. Cascaded optimization and global optimization in general have higher computational cost and do not exploit the *nullspace-projection method*, which is favorable due to *local optimality* and efficiency, provided a feasible formulation is chosen. In [12] e.g. only a scalar potential is used as a constraint for a cone-like task similar to 2T2R, termed "field of view". This leads to issues with the differentiability [12] and therefore with IK convergence.

Despite their occasional occurrences in literature, the kinematics of tasks with 2T3R and 2T2R DoF are not *systematically investigated for functional redundancy* yet and methods for a general modeling of these tasks are sparse. Further, the problem is mostly formulated on *velocity level* which requires attention when handling the nonlinear orientation residual, since the residual's Jacobian is not the differential kinematics Jacobian. In this paper a general minimal-coordinate geometric approach for the *position-level* IKP for 2T3R and 2T2R tasks is presented. The focus lies on local optimization, as motivated above. The contributions of this paper are:

- a general kinematic modeling approach for the IKP of 2T3R and 2T2R tasks,
- an application to the inverse kinematics of serial robots with functional redundancy and redundant coordinate limitations, shown by a simulation example.

The outline of this paper is as follows. In Sect. 2 the proposed kinematic modeling approach for a serial chain is shown. The application to inverse kinematics and the resolution of functional redundancy is given in Sect. 3. Section 4 demonstrates the results of the simulation examples and Sect. 5 concludes the paper.

2 Inverse Kinematic Model for Serial Chains

For robots in 2T2R and 2T3R tasks one translational DoF needs to be removed from the kinematic equations to obtain an expression of minimal dimension. This DoF is defined to be the displacement along the end effector's z axis. The established formulation for the translational part of the inverse kinematics (IK) residual that relates the robot's joint coordinates $\boldsymbol{q}$ and operational space coordinates $\boldsymbol{x}$ is

$$\tilde{\boldsymbol{\delta}}_{\mathrm{t}}(\boldsymbol{q},\boldsymbol{x}) = {}_{(0)}\boldsymbol{r}_{DE} = -{}_{(0)}\boldsymbol{r}_{D} + {}_{(0)}\boldsymbol{r}_{E}(\boldsymbol{q}) = -\boldsymbol{x}_{\mathrm{t}} + {}_{(0)}\boldsymbol{r}_{E}(\boldsymbol{q}) \in \mathbb{R}^3 \tag{1}$$

with $\boldsymbol{x}^{\mathrm{T}} = \left[\boldsymbol{x}_{\mathrm{t}}^{\mathrm{T}}\, \boldsymbol{x}_{\mathrm{r}}^{\mathrm{T}}\right] = \left[r_{0D,x}\; r_{0D,y}\; r_{0D,z}\; \varphi_x\; \varphi_y\; \varphi_z\right] \in \mathbb{R}^6$ and $\boldsymbol{x}_{\mathrm{r}}$ as Tait-Bryan angles. Coordinate systems are defined for the robot base $(\mathrm{CS})_0$, the actual end effector pose $(\mathrm{CS})_E$ and the desired pose $(\mathrm{CS})_D$, expressed with $\boldsymbol{x}$. The residual (1) is not feasible for the 2T case since it would correspond to an exact position adjustment. To obtain translational redundancy, new coordinates must be chosen that allow an elimination of the redundant DoF in the residual. The goal is to choose the coordinates $\boldsymbol{x}_{\mathrm{t}}$ and $\boldsymbol{x}_{\mathrm{r}}$ so that they can be reduced via selection matrices to $\boldsymbol{y}_{\mathrm{t}} = \boldsymbol{P}_{y,\mathrm{t}}\,\boldsymbol{x}_{\mathrm{t}}$ for the 2T case and $\boldsymbol{y}_{\mathrm{r}} = \boldsymbol{P}_{y,\mathrm{r}}\,\boldsymbol{x}_{\mathrm{r}}$ for the 2R case. By stacking the translational and rotational coordinates the combinations 2T2R and 2T3R (and also 3T2R or 3T3R) can be created.

In this approach the translational task *minimal* coordinates are chosen as

$$\boldsymbol{y}_{\mathrm{t}} = \left[r_{0D',x}\; r_{0D',y}\right]^{\mathrm{T}} = \boldsymbol{P}_{y,\mathrm{t}}\left[r_{0D',x}\; r_{0D',y}\; \lambda_{D'}\right]^{\mathrm{T}} \in \mathbb{R}^2, \tag{2}$$

where the index D' in $r_{0D',x}$ and $r_{0D',y}$ refers to the intersection of the z axis of $(\mathrm{CS})_D$ with the x-y plane of $(\mathrm{CS})_0$ shown in Fig. 1a and $\lambda_{D'} = {}_{(D)}r_{DD',z}$ is the distance to this intersection point which is derived in the following.

The point E' is obtained similarly from $(\mathrm{CS})_E$ by setting up a line equation

$$\boldsymbol{g}(\boldsymbol{q}) = {}_{(0)}\boldsymbol{r}_{E}(\boldsymbol{q}) + \lambda\,{}_{(0)}\boldsymbol{e}_{z}^{E}(\boldsymbol{q}) \tag{3}$$

from the vector of location ${}_{(0)}\boldsymbol{r}_E(\boldsymbol{q})$ in the direction of the z axis $\boldsymbol{e}_z^E(\boldsymbol{q})$ of $(\mathrm{CS})_E$. The intersection ${}_{(0)}\boldsymbol{r}_{E'}$ of the line in (3) with the x-y plane of $(\mathrm{CS})_0$ is obtained by

$$\lambda(\boldsymbol{q}) = \lambda_{E'}(\boldsymbol{q}) = \left(0 - r_{0E,z}(\boldsymbol{q})\right) / e^E_{z,z}(\boldsymbol{q}) = -r_{0E,z}(\boldsymbol{q})/e^E_{z,z}(\boldsymbol{q}). \tag{4}$$

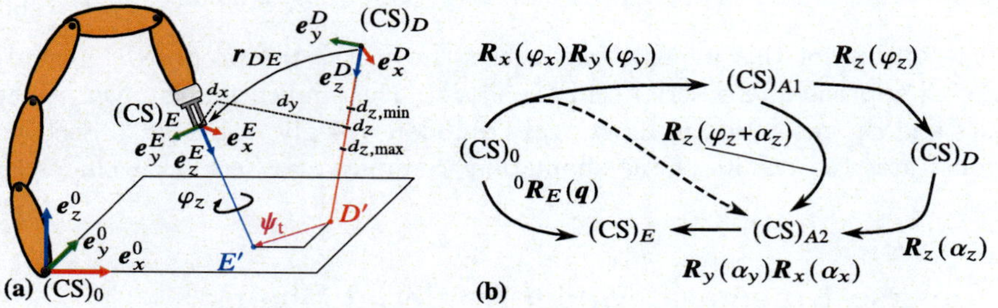

Fig. 1. (**a**) Geometric approach for the IK's position residual and (**b**) orientation residual

Inserting $\lambda_{E'}$ in $\boldsymbol{g}(\boldsymbol{q})$ gives ${}_{(0)}\boldsymbol{r}_{E'}$, which is also sketched in Fig. 1a. The variable $\lambda_{E'}$ can be understood as the distance from $(\mathrm{CS})_E$ to E'.

With the new coordinates corresponding to (2), the translational residual

$$\boldsymbol{\delta}_{\mathrm{t}}(\boldsymbol{q},\boldsymbol{x}) = \left[-r_{0D',x}(\boldsymbol{x})+r_{0E',x}(\boldsymbol{q}),\ \ -r_{0D',y}(\boldsymbol{x})+r_{0E',y}(\boldsymbol{q}),\ \ -\lambda_{D'}(\boldsymbol{x})+\lambda_{E'}(\boldsymbol{q})\right]^{\mathrm{T}} \tag{5}$$

is now defined. The reduced residual for 2T tasks can be obtained through

$$\boldsymbol{\psi}_{\mathrm{t}}(\boldsymbol{q},\boldsymbol{y}) = \begin{bmatrix}1&0&0\\0&1&0\end{bmatrix}\boldsymbol{\delta}_{\mathrm{t}}(\boldsymbol{q},\boldsymbol{x}) = \boldsymbol{P}_{\psi,\mathrm{t}}\boldsymbol{\delta}_{\mathrm{t}}(\boldsymbol{q},\boldsymbol{x}) \ \in \mathbb{R}^2. \tag{6}$$

The peculiarity of the kinematic model lies in the *sole dependency on the reduced coordinates* $\boldsymbol{y}$. For the solution of the IKP shown in Sect. 3 the gradient of the residual w.r.t. $\boldsymbol{q}$ is needed. It follows by partial derivation (and by ignoring the third row) to

$$\begin{aligned}\frac{\partial}{\partial \boldsymbol{q}}\boldsymbol{\psi}_{\mathrm{t}}(\boldsymbol{q},\boldsymbol{x}) &= \boldsymbol{P}_{\psi,\mathrm{t}}\left(\frac{\partial}{\partial \boldsymbol{q}}{}_{(0)}\boldsymbol{r}_{0E'}(\boldsymbol{q})\right) = \boldsymbol{P}_{\psi,\mathrm{t}}\left(\frac{\partial}{\partial \boldsymbol{q}}{}_{(0)}\boldsymbol{r}_{0E}(\boldsymbol{q}) + \frac{\partial}{\partial \boldsymbol{q}}\left(\lambda_{E'}(\boldsymbol{q}){}_{(0)}\boldsymbol{e}_z^E(\boldsymbol{q})\right)\right)\\ &= \boldsymbol{P}_{\psi,\mathrm{t}}\left(\frac{\partial}{\partial \boldsymbol{q}}{}_{(0)}\boldsymbol{r}_{0E}(\boldsymbol{q}) + \frac{\partial \lambda_{E'}(\boldsymbol{q})}{\partial \boldsymbol{q}}{}_{(0)}\boldsymbol{e}_z^E(\boldsymbol{q}) + \lambda_{E'}(\boldsymbol{q})\frac{\partial {}_{(0)}\boldsymbol{e}_z^E(\boldsymbol{q})}{\partial \boldsymbol{q}}\right).\end{aligned} \tag{7}$$

Since $\lambda_{E'}(\boldsymbol{q})$ from (4) is the quotient of $f_1(\boldsymbol{q}) = -r_{0E,z}(\boldsymbol{q})$ and $f_2(\boldsymbol{q}) = e^E_{z,z}(\boldsymbol{q})$, dependent on $\boldsymbol{q}$, it's partial derivative is calculated by the quotient rule for differential calculus. The term $\partial \boldsymbol{e}_z^E(\boldsymbol{q})/\partial \boldsymbol{q}$ in f_2 can be obtained either by symbolic derivation using computer algebra systems or by a relation with the rotational part of the geometric Jacobian, which can be derived with the methods from

[16]. The rotational residual is calculated as in [16] using *Z-Y′-X″* Tait-Bryan angles

$$\boldsymbol{\delta}_{\mathrm{r}}(\boldsymbol{q},\boldsymbol{x}) = \boldsymbol{\alpha}\left({}^{D}\boldsymbol{R}_E(\boldsymbol{q},\boldsymbol{x}_{\mathrm{r}})\right) = \boldsymbol{\alpha}\left({}^{0}\boldsymbol{R}_D^{\mathrm{T}}(\boldsymbol{x}_{\mathrm{r}}){}^{0}\boldsymbol{R}_E(\boldsymbol{q})\right) = [\alpha_z,\alpha_y,\alpha_x]^{\mathrm{T}} \in \mathbb{R}^3. \quad (8)$$

The series of elementary rotations is depicted in Fig. 1b. By explicitly adding two intermediate frames $(\mathrm{CS})_{A1}$ and $(\mathrm{CS})_{A2}$ that share the same z axis with the desired frame $(\mathrm{CS})_D$, the purpose of this angle convention becomes apparent. Since the rotation around this axis is the redundant DoF, these frames with elementary rotations φ_z and α_z can be omitted for the reduced residual, which is sketched by the dashed line to $(\mathrm{CS})_{A2}$. Analogously to (6), multiplying a permutation matrix leads to the reduced residual $\boldsymbol{\psi}_{\mathrm{r}}(\boldsymbol{q},\boldsymbol{y}) = \boldsymbol{P}_{\psi,\mathrm{r}}\boldsymbol{\delta}_{\mathrm{r}}(\boldsymbol{q},\boldsymbol{x}) \in \mathbb{R}^2$ for 2R tasks. For the derivation of the gradient $\boldsymbol{\psi}_{\mathrm{r},\partial\boldsymbol{q}} = \partial\boldsymbol{\psi}_{\mathrm{r}}/\partial\boldsymbol{q}$ and details of this approach refer to [15,16].

Similar to the coordinate definitions $\boldsymbol{y}$ and $\boldsymbol{x}$, the full residual $\boldsymbol{\psi}(\boldsymbol{q},\boldsymbol{y})^{\mathrm{T}} = \left[(\boldsymbol{P}_{\psi,\mathrm{t}}\boldsymbol{\delta}_{\mathrm{t}}(\boldsymbol{q},\boldsymbol{x}))^{\mathrm{T}}\ (\boldsymbol{P}_{\psi,\mathrm{r}}\boldsymbol{\delta}_{\mathrm{r}}(\boldsymbol{q},\boldsymbol{x}))^{\mathrm{T}}\right]$ is obtained by stacking the translational and rotational residual in any combination depending on the task DoF (2T2R, 2T3R, ...). This is done by choosing the according permutation matrix, e.g. by $\boldsymbol{P}_{\psi,\mathrm{t}}$=$\boldsymbol{P}_{\psi,\mathrm{2T}}$, from

$$\boldsymbol{P}_{\psi,\mathrm{3T}} = \boldsymbol{I}_3, \quad \boldsymbol{P}_{\psi,\mathrm{2T}} = \begin{bmatrix}1&0&0\\0&1&0\end{bmatrix}, \quad \boldsymbol{P}_{\psi,\mathrm{3R}} = \boldsymbol{I}_3, \quad \boldsymbol{P}_{\psi,\mathrm{2R}} = \begin{bmatrix}0&1&0\\0&0&1\end{bmatrix}. \quad (9)$$

The constraint/residual definition above can be interpreted as a generalized formal description of an *equality task* for the task coordinates $\boldsymbol{y}$, as e.g. discussed in [12]. A possible *range limitation* for the redundant T and/or R coordinate (introduced e.g. in [7]) builds upon this definition and corresponds to an *inequality constraint* (or *set-based task* in [12]) and will be discussed in the next section.

3 Inverse Kinematics and Functional Redundancy

An analytic closed-form solution to the implicit inverse kinematics problem $\boldsymbol{\psi}(\boldsymbol{q},\boldsymbol{y})$ can only exist in a parameterized form (since $\dim(\boldsymbol{q})>\dim(\boldsymbol{y})$) and for certain kinematic conditions [17]. Therefore the solution of the IKP is obtained through the Newton-Raphson method derived by the Taylor series of the IK residual $\boldsymbol{\psi}(\boldsymbol{q},\boldsymbol{y})$ to

$$\boldsymbol{\psi}(\boldsymbol{q}^{k+1},\boldsymbol{y}) = \boldsymbol{\psi}(\boldsymbol{q}^{k},\boldsymbol{y}) + \boldsymbol{\psi}_{\partial\boldsymbol{q}}(\boldsymbol{q}^{k},\boldsymbol{y})(\boldsymbol{q}^{k+1}-\boldsymbol{q}^{k}) \stackrel{!}{=} \boldsymbol{0} \quad (10)$$

with the IK Jacobian matrix $\boldsymbol{\psi}_{\partial\boldsymbol{q}}(\boldsymbol{q}^{k},\boldsymbol{y})$. With † for pseudo inverse, the increment

$$\Delta\boldsymbol{q}^{k} = (\boldsymbol{q}^{k+1}-\boldsymbol{q}^{k}) = -\left(\boldsymbol{\psi}_{\partial\boldsymbol{q}}(\boldsymbol{q},\boldsymbol{y})\right)^{\dagger}\boldsymbol{\psi}(\boldsymbol{q}^{k},\boldsymbol{y}) \quad (11)$$

of the joint angles can be used in an iterative algorithm to move towards the solution. Secondary tasks $\boldsymbol{v}$=$h_{\partial\boldsymbol{q}}$ can be optimized in the nullspace of the operational task via

$$\Delta\boldsymbol{q} = \Delta\boldsymbol{q}_{\mathrm{T}} + \Delta\boldsymbol{q}_{\mathrm{N}} = -\boldsymbol{\psi}_{\partial\boldsymbol{q}}^{\dagger}\boldsymbol{\psi} + \boldsymbol{N}\boldsymbol{v} \quad \text{with} \quad \boldsymbol{N} = \boldsymbol{I} - \boldsymbol{\psi}_{\partial\boldsymbol{q}}^{\dagger}\boldsymbol{\psi}_{\partial\boldsymbol{q}} \quad (12)$$

using the gradient-projection method. To focus on the geometric derivation of the residual, using multiple optimization criteria as shown in [12] is not considered.

The distance d_z of robot end effector to workpiece in pointing direction is obtained from the vector ${}_{(D)}\boldsymbol{r}_{DE}^{\mathrm{T}} = [d_x, d_y, d_z]$ shown in Fig. 1a. The reference distance of the robot can be limited e.g. by the task specification to be within the bounds $d_{z,\min}$ to $d_{z,\max}$. To respect these bounds the modified hyperbolic potential function

$$h_{d_z,\mathrm{hyp}}(d_z) = \frac{(d_{z,\max} - d_{z,\min})^2}{8}\left(\frac{1}{(d_z - d_{z,\min})^2} + \frac{1}{(d_z - d_{z,\max})^2}\right) \geq 1 \tag{13}$$

from [21] can be used. Cubic splines serve for activation of the potential beyond zero at the thresholds $d_{z,\mathrm{thr,min}}$ and $d_{z,\mathrm{thr,max}}$, which provides a continuously differentiable

$$h_{\mathrm{dist}}(h_{d_z}) = \begin{cases} h_{d_z,\mathrm{hyp}}(h_{d_z}) & \text{for} \quad d_z < d_{z,\mathrm{sw,min}} \quad \text{or} \quad d_z > d_{z,\mathrm{sw,max}} \\ 0 & \text{for} \quad d_{z,\mathrm{sw,min}} < d_{z,\mathrm{thr,min}} \leq d_z \leq d_{z,\mathrm{thr,max}} \\ h_{d_z,\mathrm{spline}}(h_{d_z}) & \text{otherwise (cubic spline interpolation).} \end{cases} \tag{14}$$

A symbolic derivation of (14) w.r.t. $\boldsymbol{q}$ is not feasible. Instead, the gradient is computed with the projection of the known expression $\partial h(d_z)/\partial d_z$ as

$$h^*_{\partial \boldsymbol{q}} = (\partial h/\partial d_z)\ (\partial d_z/\partial \boldsymbol{q}). \tag{15}$$

The asterisk denotes the origin of the term from a projection approach instead of a symbolic derivation. Using symbolic derivation leads to a different expression $h_{\partial \boldsymbol{q}}$, but after projection into the nullspace of the IK Jacobian in (12), both gradients are equal, i.e. $\boldsymbol{N} h_{\partial \boldsymbol{q}} = \boldsymbol{N} h^*_{\partial \boldsymbol{q}}$. The second term in (15) comes from the last row of the rotated translational part ${}_{(D)}\boldsymbol{J}_{\mathrm{t}}$ of the robot's Jacobian. A rotation from default base frame $\boldsymbol{J}_{\mathrm{t}}$:=${}_{(0)}\boldsymbol{J}_{\mathrm{t}}$ into desired frame $(\mathrm{CS})_D$ is performed with

$${}_{(D)}\boldsymbol{J}_{\mathrm{t}} = {}^{D}\boldsymbol{R}_{0\,(0)}\boldsymbol{J}_{\mathrm{t}} = {}^{D}\boldsymbol{R}_{0}\frac{\partial}{\partial \boldsymbol{q}}{}_{(0)}\boldsymbol{r}_{E} = \frac{\partial}{\partial \boldsymbol{q}}{}^{D}\boldsymbol{R}_{0}({}_{(0)}\boldsymbol{r}_{D} + {}_{(0)}\boldsymbol{r}_{DE}) = \frac{\partial}{\partial \boldsymbol{q}}{}_{(D)}\boldsymbol{r}_{DE}. \tag{16}$$

The relations hold since ${}^{D}\boldsymbol{R}_0$ and ${}_{(0)}\boldsymbol{r}_D$ are only dependent on the desired pose $\boldsymbol{x}$ and not on the joint configuration $\boldsymbol{q}$ and are invariant to the operator $\partial/\partial \boldsymbol{q}$. Generally spoken, the Jacobian relation is invariant to the reference frame. Therefore $\partial d_z/\partial \boldsymbol{q}$ in (15) can be obtained as the last row of the Jacobian ${}_{(D)}\boldsymbol{J}_{\mathrm{t}}$ from (16).

Other performance criteria can be included like a potential from [9] based on the squared deviation of the joint coordinates $\boldsymbol{q}$ from their mid-range $\bar{\boldsymbol{q}}$ with

$$h_1(\boldsymbol{q}) = \frac{1}{2}(\boldsymbol{q} - \bar{\boldsymbol{q}})^{\mathrm{T}}\boldsymbol{W}_1(\boldsymbol{q} - \bar{\boldsymbol{q}}), \quad h_{1,\partial \boldsymbol{q}} = \frac{\partial h_1}{\partial \boldsymbol{q}} = \boldsymbol{W}_1(\boldsymbol{q} - \bar{\boldsymbol{q}}) \quad \text{and} \quad \boldsymbol{W}_1 = \boldsymbol{I}. \tag{17}$$

Both criteria are combined to one potential $h_{\mathrm{total}} = h_{\mathrm{dist}} + h_1$. Other criteria like singularity avoidance [9] or hyperbolic weight of joint limit distance [21] can be used as well. If e.g. for 2T2R tasks a two-DoF functional redundancy arises, a task-priority scheme [12] can be used to incorporate several objectives simultaneously.

4 Evaluation

A simulation study was performed with the proposed algorithm for an industrial robot kinematics of type KUKA KR 30-3. The initial robot pose is shown in Fig. 2a together with the desired pose $(\mathrm{CS})_D$ and *possible* limitations for the redundant coordinate d_z in the 2T tasks. Different settings (combining fixed/free/limited translation 3T/2T/2T* and fixed/free rotation 3R/2R) of the IK algorithm are compared with summed optimization of the criterion h_1 from (17) and h_{dist} from (14). The resulting final poses for each setting are shown in Fig. 2b–e. The evolution of the redundant translational coordinate d_z (in the 2T case) and the redundant rotational coordinate φ_z regarding the z axis (in the 2R case) are plotted in Fig. 3a–b and the optimization criterion h_1 in Fig. 3c. The IK progress is normalized from 0 to 100%. However, the number of iterations k in (11) varies from 16 (3T2R), over 150 (2T2R*, with d_z limitation) and 166 (2T3R*) to 181 (2T2R, without d_z limitation). It depends on the maximum step size (for holding the linear approximation), damping (of least-squares), convergence criteria (set for $\boldsymbol{\psi}$ and $\Delta \boldsymbol{q}_{\mathrm{N}}$ to 10^{-6}) and other meta parameters in the implementation of (11). The computation time remains low with about 24 µs per iteration for the 2T methods and 63 µs for the 3T2R method (due to higher impact of overhead) using a MATLAB-MEX Linux implementation on an Intel

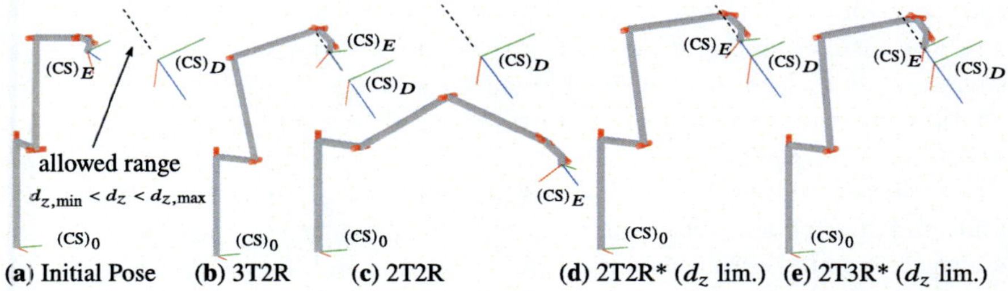

Fig. 2. (**a**) Robot in initial pose and (**b–e**) in final poses for different IK settings

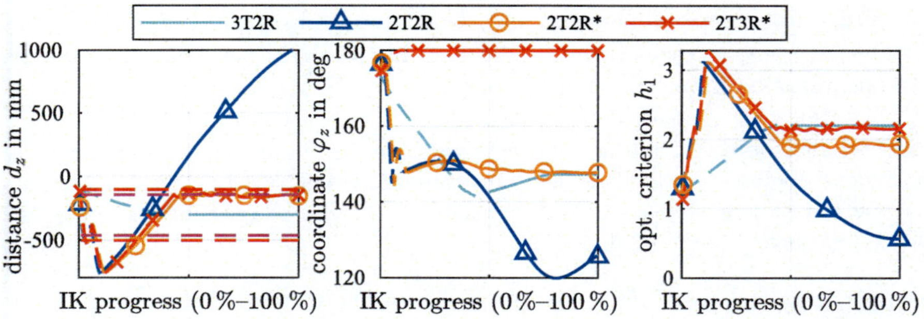

Fig. 3. (**a**) Target distance d_z with threshold $d_{z,\mathrm{thr,min/max}}$ and limit $d_{z,\mathrm{min/max}}$, (**b**) end effector rotation φ_z corresponding to $(\mathrm{CS})_E$ and (**c**) joint limit optimization criterion h_1 over the convergence of the IK algorithm. Dashed lines mark $\boldsymbol{\delta} \not\approx \mathbf{0}$

i5-7500 CPU. Using the method online in robot control is therefore possible, as shown by others.

In the 2T2R case, without limiting the coordinate d_z, the functional redundancy of degree two is used only for a far reaching elongation of the arm in Fig. 2c. This strongly improves h_1, but is infeasible regarding collisions, singularities and possible process restrictions. This is improved by the method 2T2R* which uses $h_1 + h_{\text{dist}}$ at the cost of a degraded h_1, see Fig. 2d. Using a fixed orientation φ_z with 2T3R* does not allow to improve h_1 further since the one degree of functional redundancy is already used for h_{dist} which is at it's limit as can also be seen in Fig. 2e. Finally, for comparison, the case of 3T2R in Fig. 2b is included with a shifted $(\text{CS})_D$ in the middle of $d_{z,\text{min}}$ and $d_{z,\text{max}}$, which allows optimization of h_1 to a poorer result than the other methods due to the reduced range of self-motion. While the overall convergence of all methods is clear from Fig. 3, a remaining problem is visible at the 2T3R* case. Leaving the linear approximation leads to oscillations in the nullspace motion between the criteria h_{dist} and h_1, i.e. a degraded convergence in the second half of the motion, which may be solved by further tuning of the damping parameters.

5 Conclusion and Remark on Parallel Robots

The proposed formulation for the definition of the inverse kinematics problem can be used for tasks with two translational degrees of freedom and spatial motion, especially 2T2R and 2T3R. The method is targeted at a numeric implementation with an analytic derivation and can be embedded in larger frameworks, such as task-priority inverse kinematics of [12]. Extensions such as the limitation of the redundant coordinate with a projection method make it feasible in practical applications.

The minimal dimension of the residual and the elimination of dependent operational space coordinates allow an efficient transfer to gradient projection schemes for *parallel robots with functional redundancy* to extend works like [3] and [10] on 2T2R tasks. The first leg chain's kinematic constraints then correspond to the residual $\boldsymbol{\psi}$ in this paper and the constraints of following leg chains have to be defined relative to the first, leading leg [15]. The functionally redundant direct and inverse kinematics matrices ("Jacobians") can then be obtained by differentiating the constraints. The position-level IK can then be used for pose optimization in synthesis or control.

Acknowledgements. The authors acknowledge the support by the Deutsche Forschungsgemeinschaft (DFG) under grant number 341489206. MATLAB code to reproduce the results is available at GitHub under free license at `github.com/SchapplM/robotics-paper_ark2022_2T2R`.

References

1. Bahls, T., Fröhlich, F.A., Hellings, A., Deutschmann, B., Albu-Schäffer, A.O.: Extending the capability of using a waterjet in surgical interventions by the use of robotics. IEEE Trans. Biomed. Eng. **64**(2), 284–294 (2017)

2. Baron, L.: A joint-limits avoidance strategy for arc-welding robots. Paper Presented at International Conference on Integrated Design and Manufacturing in Mechanical Engineering, Montreal, Canada (2000)
3. Chen, Y., McInroy, J.E., Yi, Y.: Optimal, fault-tolerant mappings to achieve secondary goals without compromising primary performance. IEEE Trans Rob Autom **19**(4), 680–691 (2003)
4. Dolgui, A., Pashkevich, A.: Manipulator motion planning for high-speed robotic laser cutting. Int. J. Prod. Res. **47**(20), 5691–5715 (2009)
5. Erdős, G., Kardos, C., Keméňy, Z., Kovàcs, A., Vàncza, J.: Process planning and offline programming for robotic remote laser welding systems. Int. J. Comput. Integr. Manuf. **29**(12), 1287–1306 (2015)
6. From, P.J., Gravdahl, J.T.: A real-time algorithm for determining the optimal paint gun orientation in spray paint applications. IEEE Trans. Autom. Sci. Eng. **7**(4), 803–816 (2010)
7. Gueta, L.B., Cheng, J., Chiba, R., Arai, T., Ueyama, T., Ota, J.: Multiple-goal task realization utilizing redundant degrees of freedom of task and tool attachment optimization. In: International Conference on Robotics and Automation (ICRA), pp. 1714–1719. IEEE (2011)
8. Huo, L., Baron, L.: Kinematic inversion of functionally-redundant serial manipulators: application to arc-welding. Trans. Can. Soc. Mech. Eng. **29**(4), 679–690 (2005)
9. Huo, L., Baron, L.: The joint-limits and singularity avoidance in robotic welding. Ind. Rob. Int. J. **35**(5), 456–464 (2008)
10. Kobler, J.P., Kotlarski, J., Lexow, G.J., Majdani, O., Ortmaier, T.: Design optimization of a bone-attached, redundant and reconfigurable parallel kinematic device for skull surgery. In: International Conference on Robotics and Automation (ICRA), pp. 2364–2371. IEEE (2014)
11. Kumar, N., Piccin, O., Bayle, B.: A task-based type synthesis of novel 2T2R parallel mechanisms. Mech. Mach. Theor. **77**, 59–72 (2014)
12. Moe, S., Antonelli, G., Teel, A.R., Pettersen, K.Y., Schrimpf, J.: Set-based tasks within the singularity-robust multiple task-priority inverse kinematics framework: general formulation, stability analysis, and experimental results. Front. Robot. AI **3**, 16 (2016)
13. Sadeghian, H., Zokaei, F., Jazi, S.H.: Constrained kinematic control in minimally invasive robotic surgery subject to remote center of motion constraint. J. Intell. Rob. Syst. **95**(3–4), 901–913 (2019)
14. Sandoval, J., Poisson, G., Vieyres, P.: A new kinematic formulation of the RCM constraint for redundant torque-controlled robots. In: International Conference on Robotics and Automation (IROS). IEEE (2017)
15. Schappler, M., Tappe, S., Ortmaier, T.: Modeling parallel robot kinematics for 3T2R and 3T3R tasks using reciprocal sets of Euler angles. MDPI Robot. **8**(3), 68 (2019)
16. Schappler, M., Tappe, S., Ortmaier, T.: Resolution of functional redundancy for 3T2R robot tasks using two sets of reciprocal Euler angles. In: Proceedings of the 15th IFToMM World Congress (2019)
17. Sciavicco, L., Siciliano, B.: Modelling and Control of Robot Manipulators. Springer, London (2000). https://doi.org/10.1007/978-1-4471-0449-0
18. Ye, W., Fang, Y., Guo, S., Qu, H.: Type synthesis of 2R2T parallel mechanisms based on motion equivalent chain method. Proc. Inst. Mech. Eng., Part C J. Mech. Eng. Sci. **228**(17), 3209–3217 (2014)

19. Zanchettin, A.M., Rocco, P., Robertsson, A., Johansson, R.: Exploiting task redundancy in industrial manipulators during drilling operations. In: ICRA, pp. 128–133. IEEE (2011)
20. Zhu, S.J., Huang, Z., Zhao, M.Y.: Singularity analysis for six practicable 5-DoF fully-symmetrical parallel manipulators. Mech. Mach. Theor. **44**(4), 710–725 (2009)
21. Zhu, W., Qu, W., Cao, L., Yang, D., Ke, Y.: An off-line programming system for robotic drilling in aerospace manufacturing. Int. J. Adv. Manuf. Technol. **68**(9–12), 2535–2545 (2013)
22. Žlajpah, L.: On orientation control of functional redundant robots. In: International Conference on Robotics and Automation (ICRA), pp. 2475–2482. IEEE (2017)

The Distance Geometry of the Generalized Lobster's Arm

Federico Thomas(✉) and Josep M. Porta

Institut de Robòtica i Informàtica Industrial (CSIC-UPC), ETSEIB, Diagonal 647, Pavelló E, 08028 Barcelona, Spain
f.thomas@csic.es, porta@iri.upc.edu

Abstract. This paper proposes a distance-based formulation to solve the inverse kinematics of what is known as the *generalized Lobster's arm*: a 6R serial chain in which all consecutive revolute axes intersect. Since the solution of the inverse kinematics of a general 6R serial chain comes down to finding the roots of a 16th-degree polynomial, one might think that this polynomial also contains the solutions to the inverse kinematics of 6R serial chains with special geometric parameters as a mere particular case. Nevertheless, under certain geometric circumstances various problems can appear. Some are of numerical nature, but others are fundamental problems of the used method. For that reason, it is still useful to study 6R chains with special geometric parameters, especially when the new formulation leads to a simpler solution, gives new insights, and provides new connections with other problems, as is the case in this paper.

1 Introduction

In 1841, in a communication addressed to the Philosophical Society of Cambridge, Robert Willis (1800–1875) showed that the joints of a common crab's claw work in the same way as those of what we would today classify as a 5R kinematic chain [1]. Willis' description appeared later summarized at the end of his influential book "*Principles of Mechanism*" [2, pp. 461–463]. This description was accompanied by the drawing in Fig. 1. He observed that the crab's claw is composed of six rigid bodies (denoted by A, B, C, D, E, and F in the drawing) connected in series through five revolute joints (denoted by 1, 2, 3, 4 and 5 in the drawing). What makes the arrangement of these five joint axes remarkable is that any two consecutive rotation axes in the chain intersect.

O. Altuzarra and A. Kecskémthy (Eds.): ARK 2022, SPAR 24, pp. 409–417, 2022.
https://doi.org/10.1007/978-3-031-08140-8_44

In 1979, J. Duffy and S. Derby, as a result of a suggestion by K. H. Hunt—who was aware of Willis' observations—studied the inverse kinematics (i.e., the determination of joint angles required to move the end-effector to a desired position and orientation) of what they called the *generalized lobster arm* [3]. This arm is a 6R kinematic chain where every two consecutive axes intersect (Fig. 2). The resolution of this problem was seen as an intermediate step worth to be solved before addressing the same problem for the general 6R arm which, few years earlier, was ranked as the "*Mount Everest of kinematic problems*" by F. Freudenstein [4]. Curiously enough, as we will see later, the arrangement of joints in the generalized lobster arm does not provide much simplifications with respect to the general 6R arm, at least in the number of solutions. J. Duffy and S. Derby showed, using a long and complicated process, how to reduced this inverse kinematics problem to the computation of the roots of a 24th-degree polynomial, which is now seen as an outdated result.

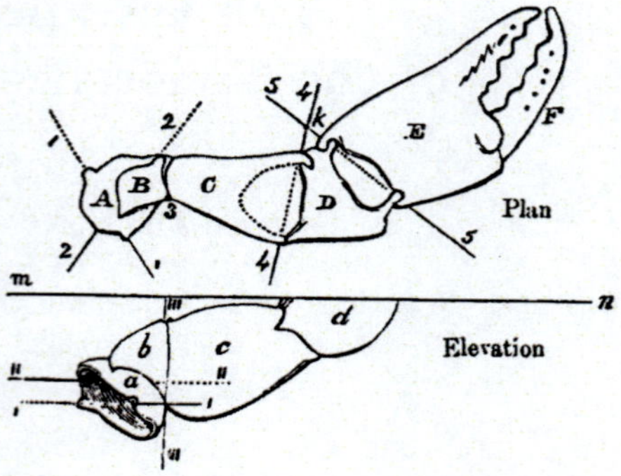

Fig. 1. Willis' drawing of the common crab's claw used to explain how its joint axes are arranged (adapted from [2, p. 462]).

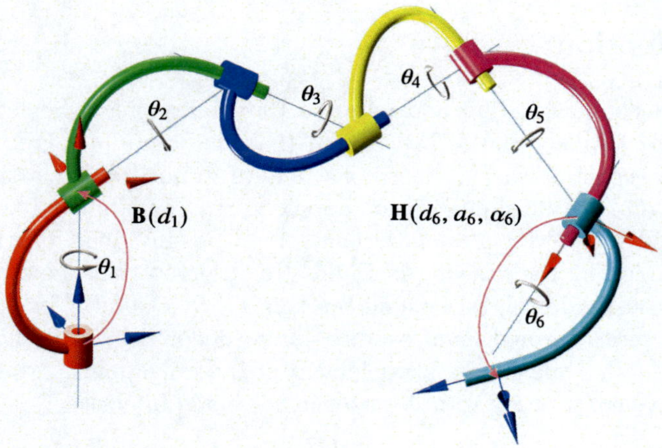

Fig. 2. A generalized lobster's arm, as defined by Duffy and Derby, is a 6R serial kinematic chain where all consecutive rotation axes intersect. In terms of standard DH-parameters, this means that $a_i = 0$ for $i = 1, \ldots, 5$.

In 1992, V. Murthy and K. J. Waldron revisited the problem in [5]. They solved it including a further generalization: the intersection between the second and the third axis and between the fourth and the fifth axis were no longer required. The standard DH-parameters of this kind of 6R serial chain appear in Table 1, where the parameters marked with an asterisk are actually irrelevant because they can be incorporated in the base and the hand transformations. In Fig. 2, these two transformations are represented by $\mathbf{B}(d_1)$ and $\mathbf{H}(d_6, a_6, \alpha_6)$, respectively. Murthy and Waldron reduced the resulting system of equations to a single univariate polynomial equation of degree 16. They also demonstrated that the used elimination process introduced no extraneous roots. This implied that the end-effector could reach a given position and orientation in at most sixteen different ways. This was an important improvement with respect to the 24th-degree polynomial solution of Duffy and Derby.

Table 1. Standard DH-parameters of the 6R serial chain analyzed by Murthy and Waldron [5]. The generalized lobster's arm can be seen as a particular case of this chain in which $a_2 = a_4 = 0$.

Link	θ_i	d_i	a_i	α_i
1	θ_1	*	0	α_1
2	θ_2	d_2	a_2	α_2
3	θ_3	d_3	0	α_3
4	θ_4	d_4	a_4	α_4
5	θ_5	d_5	0	α_5
6	θ_6	*	*	*

Six years earlier, in 1986, E. J. F. Primrose had already proved that the general 6R robot could have up to 16 inverse kinematic solutions. However, due to the complexity of the formulas, he could not come up with a way to remove the 16 redundant solutions of a polynomial equation of degree 32 [6]. This was considered as an important landmark in the history of kinematics. Nevertheless, in 1984, H.-Y. Lee already devised a method to explicitly obtain the 16th-degree polynomial in his Master Thesis written in Chinese [7]. This method became known to the western world when it appeared four years later in [8,9]. Subsequently, Raghavan and Roth [10,11] reformulated it in a cleared way using DH parameters. Since then, many other methods and variations have appeared leading to an extensive literature on the topic that we will not review here. However, it is worth observing that most methods reduce the problem to a single univariate polynomial in the one of the joints half angle tangent, and all the remaining joint variables follow from linear equations once the roots of the univariate polynomials are found.

Certain values of the kinematic parameters may reduce the degree of the resultant polynomial. Nevertheless, the intersections between the revolute axes in the 6R serial chain studied by Murthy and Waldron also lead to a 16th-degree polynomial. Unfortunately, Murthy and Waldron did not discuss the case in which all consecutive axes intersect. Thus, to the best of our knowledge, the question concerning the number of possible solutions for the inverse kinematics of the generalized lobster's arm remained open. We clarify this point in Sect. 2.

Since the roots of a 16th-degree polynomial equation gives the solution of the general case, one might think that it also contains the solutions to 6R serial chains with special geometric parameters as a mere particular case. Nevertheless, as it was pointed out in [13], under certain geometric circumstances, various problems appear. Some are of numerical nature, but others are fundamental problems of the used method. For that reason it is still useful to study 6R chains with special geometric parameters, such as the one discussed in this paper.

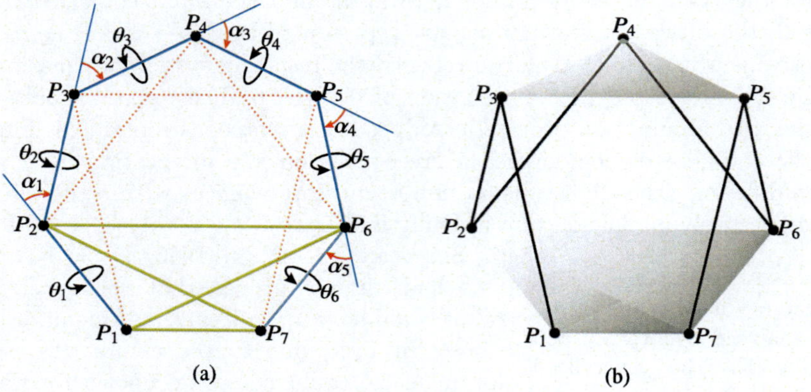

Fig. 3. (a) Bar-and-joint framework associated with the generalized lobster's arm. The lengths of the bars in blue and dotted red are determined by the geometric parameters of the arm; and those in green, by the closure condition. (b) 4-3 Gough-Stewart platform whose associated bar-and-joint framework has the same topology as that of the generalized lobster's arm.

The rest of this paper is organized as follows. Section 2 describes how the inverse kinematics problem of the generalized lobster's arm can be reformulated as the position analysis of seven points in $\mathbb{R}^3$ where some of their pairwise distances are known. This comes out to be equivalent to solving the forward kinematics of a 4-3 Stewart-Gough platform. Then, Sect. 3 shows how to solve this problem by computing a distance inversion in a strip of tetrahedra and Sect. 4 presents an example. Finally, the main contributions are summarized in Sect. 5.

2 A Distance-Based Formulation

From a purely geometric point of view, a generalized lobster's arm can be described as the bar-and-joint framework depicted in Fig. 3(a). The points $P_2, \ldots, P_6$ correspond to the intersections between the six rotation axes and P_1 and P_7 can arbitrarily be taken on the first and the last rotation axes provided that they do not coincide with P_2 and P_6, respectively. The origin of the reference frames at the base and at the end-effector can be placed at P_1 and P_7, respectively. Then, we have that

$$|P_i P_{i+1}|^2 = d_i^2, \quad i = 1, \ldots, 6. \tag{1}$$

These distances are associated with the bars in solid blue in Fig. 3(a). Moreover, since the angle between consecutive joint axes is known and constant, we also have, using the cosine rule for supplementary angles, that

$$|P_i P_{i+2}|^2 = d_i^2 + d_{i+1}^2 + 2d_i d_{i+1} \cos\alpha_i, \quad i = 1, \ldots, 5. \tag{2}$$

These distances are associated with the bars in dotted red in Fig. 3(a).

When the end-effector of the lobster's arm is fixed at a given location with respect to its base reference frame, the set of points $\{P_1, P_2, P_6, P_7\}$ defines a tetrahedron whose edge lengths are all known. The corresponding closing bars are depicted in light green in Fig. 3(a).

As a result of the above representation, the inverse kinematics problem of the generalized lobster's arm comes down to obtaining all the possible 3D embeddings of the bar-and-joint framework in Fig. 3(a). In general, one must be careful on how these embeddings are performed because they must respect the orientations of the involved tetrahedra [15] but, since in this case the framework contains just one tetrahedron, this can be ignored as it is done, for example, in [16].

Another interesting outcome of formulating the problem in terms of distances is that we can straightforwardly conclude that the inverse kinematics of the generalized lobster's arm and the forward kinematics of the 4-3 Gough-Stewart platform in Fig. 3(b) are equivalent position analysis problems. It is worth noting that this equivalency is not related to the series-parallel duality studied in [5]. Therefore, since the forward kinematics of this particular Gough-Stewart has up to 16 real solutions [15], so has the inverse kinematics of the generalized lobster's arm. In other words, the fact that any set of consecutive revolute axes of a 6R arm intersect at arbitrary angles does not reduce the number of solutions of its inverse kinematics. Some reductions are however obtained in the case that the axes intersect at right angles. Mavroidis and Roth gave a detailed investigation of these latter cases in [14].

3 The Resolution Process

Although it is not straightforward to see it at first glance, it is not difficult to express $d_{57}^2 = |P_5P_7|^2$ (a known distance) as a function of $d_{36}^2 = |P_3P_6|^2$ (an unknown distance). This is a one-to-many mapping whose inversion leads to a 16th-degree polynomial in the unknown distance. Each real root of this polynomial leads, using a sequence of trilaterations, to a valid configuration of the bar-and-joint framework. A detailed explanation of this method can be found in [17]. Next, we just give a brief summary.

First of all observe that, if we remove the bar P_5P_7 and we add the bar P_3P_6 in the bar-and-joint framework in Fig. 3, the resulting framework can be seen as a strip of four tetrahedra: $P_7P_1P_2P_6$, $P_1P_2P_6P_3$, $P_2P_6P_3P_4$, and $P_6P_3P_4P_5$. Each tetrahedron shares a face with the following one in the strip. For example, the first and the second one share the face $P_1P_2P_6$. Now, let us suppose that the two neighboring tetrahedra in Fig. 4(left) belong to this strip. The squared distance between P_l and P_m can be expressed as (see [12] for details):

$$d_{l,m}^2 = \frac{2}{D(i,j,k)} \left(D(i,j,k,l;i,j,k,m)\Big|_{s_{l,m}=0} \pm \sqrt{D(i,j,k,l)\, D(i,j,k,m)} \right). \quad (3)$$

where the terms of the form $D(\cdot)$ stand for Cayley-Menger determinants or bideterminants, and the $\pm$ sign accounts for the two possible solutions depending on the relative orientation between the two tetrahedra.

If one of the points in the set $\{P_i, P_j, P_k\}$ does not belong to any other tetrahedron in the strip, it can be removed from the strip provided that a bar connecting P_l and P_m, with length given by Eq. (3), is added [see Fig. 4(right)]. This reduces the number of tetrahedra in the strip by one. Then, by repeating this operation three times, we end up with an expression for the known squared distance d_{57}^2 as a function of the unknown squared distance d_{36}^2. This expression contains radicals, and singularity factors associated with the shared faces between consecutive tetrathedra in the strip, that can be easily cleared to finally obtain a closure polynomial of 16th-degree (see [17] for details).

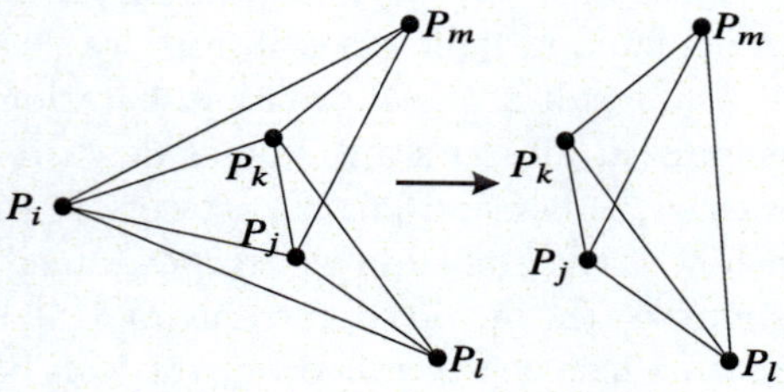

Fig. 4. Substitution rule in a strip of tetrahedra that permits reducing by one the number of tetrahedra in the strip.

4 Example

Let us consider the 6R kinematic chain with the standard DH-parameters given in Fig. 5(a). We can use the method described in the previous section to determine the joint angles required to move the end-effector to the location, with respect to the base reference frame, defined, for instance, by

$$\mathbf{E} = \mathbf{R}_x(-2.4019)\, \mathbf{R}_z(0.7047)\, \mathbf{T}(0, 0, -2.2744). \quad (4)$$

For this particular example, we obtain 14 real solutions. The corresponding join angles appear numbered in Fig. 5(b) and the corresponding graphical representation of the arm configuration in Fig. 5(c).

Alternatively to the described method, we could straightforwardly use the implementation of the celebrated Manocha-Canny's method [18], available at [19]. Unfortunately, this implementation delivers 18 solutions, six of them being erroneous. These spurious solutions appear in Fig. 5(c) without number, and the missed solutions are marked with an asterisk.

(a)

Link	θ_i	d_i	a_i	α_i
1	θ_1	1.0000	0	1.2661
2	θ_2	1.0000	0	1.9058
3	θ_3	1.5969	0	2.3505
4	θ_4	1.2247	0	1.6116
5	θ_5	1.0000	0	1.3181
6	θ_6	1.0000	0	2.7411

(b)

sol.	θ_1	θ_2	θ_3	θ_4	θ_5	θ_6
1	0.3529	-2.1181	-0.7380	0.1745	-0.7304	1.9820
2	0.3321	-1.5990	0.9899	-0.1750	0.8327	0.1452
3	0.3293	2.4415	0.7239	-0.1752	1.7406	1.9539
4	0.2668	-0.6467	-0.6026	0.1863	-1.4280	0.9062
5	0.2633	2.8899	0.5955	-0.1873	1.4031	1.6535
6	0.4520	0.5264	-1.2307	0.1923	-2.3244	-0.2700
7	0.1679	3.0171	-1.4765	0.2258	-1.3087	-2.9700
8*	0.0836	-0.1470	1.6851	-0.2728	1.7280	-1.2528
9*	1.8664	-0.4997	-1.7695	1.1613	2.3125	-2.0415
10	2.4236	-3.1147	1.4208	-1.5051	-1.2302	-3.0822
11	3.0179	-3.0882	-0.5596	1.7591	1.2549	1.5303
12	3.0816	-1.9218	-0.9766	1.7763	2.1124	2.4124
13	-2.4803	2.1982	-0.6337	1.7982	0.7401	0.8245
14	-2.5240	2.3406	0.8787	-1.8062	-0.7739	2.2421
	1.8664	2.7047	-2.0793	-0.1201	-3.1357	2.5302
	0.0836	2.2749	-0.4562	-1.3347	1.5828	-2.8315
	2.9407	-2.6665	0.8593	-1.9783	2.2324	2.6025
	2.6512	-3.1377	-0.2420	1.5604	-3.0005	0.7550
	154.9700	-119.1354	-6.8803	-179.6604	144.9683	1.5549
	130.3419	-26.1367	-76.4750	90.6894	-162.2357	1.5229

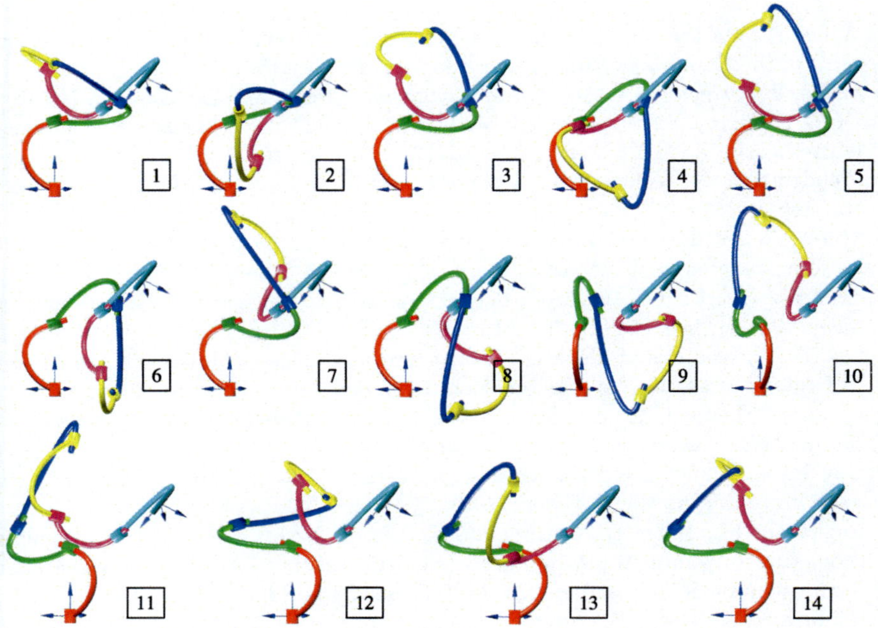

(c)

Fig. 5. Using the method described in this paper, 14 real solutions are obtained for the inverse kinematics of the arm with the DH-parameters given in (a) and the end-effector located at Eq. (4). These solutions are numbered from 1 to 14 in (b), and the corresponding configurations are depicted to visually verify their correctness in (c).

5 Conclusion

We have derived a 16th-degree polynomial whose roots determine the inverse kinematics solutions of the generalized lobster's arm. Contrarily to other methods that also derive closing polynomials, the one obtained here needs no variable eliminations. Moreover, the polynomial variable is a distance, and all its coefficients also come from operations with distances. No angles are involved at any point. We have also shown how Manocha and Canny's method fails to deliver the correct solutions for the analyzed 6R serial arm. As Dietmaiyer already observed, this is a common problem for general methods when some parameters are set to zero.

Acknowledgements. This work was partially supported by the Spanish Government through project PID2020-117509GB-I00/AEI/10.13039/50110001103.

References

1. Willis, R.: On the arrangement of the joints of crustaceous animals. In: Transactions of the Cambridge Philosophical Society, 10 May 1841
2. Willis, R.: Principles of Mechanism. Longmans, Green, and Co., London (1870)
3. Duffy, J., Derby, S.: Displacement analysis of spatial 7R mechanism: a generalized lobster's arm. ASME J. Mech. Des. **101**(2), 224–231 (1979)
4. Freudenstein, F.: Kinematics: past, present and future. Mech. Mach. Theor. **8**(2), 151–160 (1973)
5. Murthy, V., Waldron, K.J.: Position kinematics of the generalized lobster arm and its series-parallel dual. ASME J. Mech. Des. **114**(3), 406–413 (1992)
6. Primrose, E.J.F.: On the input-output equation of the general 7R mechanism. Mech. Mach. Theor. **21**(2), 509–510 (1986)
7. Lee, H.-Y.: Analysis of a spatial mechanism (in Chinese), Master Thesis, Beijing Institute of Posts and Telecommunications (1984)
8. Lee, H.-Y., Liang, C.G.: A new vector theory for the analysis of spatial mechanisms. Mech. Mach. Theor. **23**(3), 209–217 (1988)
9. Lee, H.-Y., Liang, C.G.: Displacement analysis of the general spatial 7-link 7R mechanism. Mech. Mach. Theor. **23**(3), 219–226 (1988)
10. Raghavan, M., Roth, B.: Kinematic analysis of the 6R manipulator of general geometry. In: Miura, H., Arimoto, S. (eds.) Proceedings of the 5th International Symposium on Robotics Research. MIT Press, Cambridge, MA (1990)
11. Raghavan, M., Roth, B.: Inverse kinematics of the general 6R manipulator and related linkages. ASME J. Mech. Des. **115**(3), 502–508 (1993)
12. Rojas, N., Thomas, F.: The closure condition of the double banana and its application to robot position analysis. In: Proceedings of the IEEE International Conference on Robotics and Automation, pp. 4641–4646 (2013)
13. Dietmaier, P.: Displacement analysis of the homogeneous orthogonal 7-link 7R space mechanism. In: Lenarčič, J., Stifter, S. (eds.) Advances in Robot Kinematics, pp. 302–309. Springer, Wien (1991). https://doi.org/10.1007/978-3-7091-4433-6_34
14. Mavroidis, C., Roth, B.: Structural parameters which reduce the number of manipulator configurations. ASME J. Mech. Des. **116**(1), 3–10 (1994)

15. Porta, J.M., Thomas, F.: Yet another approach to the Gough-Stewart platform forward kinematics. In: Proceedings of the IEEE International Conference in Robotics and Automation, Brisbane, Australia (2018)
16. Thomas, F.: A distance geometry approach to the singularity analysis of 3R robots. ASME J. Mech. Robot. **8**(1), 011001 (2015)
17. Thomas, F., Porta, J.M.: Closure polynomials for strips of tetrahedra. In: Lenarčič, J., Merlet, J.-P. (eds.) Advances in Robot Kinematics, pp. 303–312 (2016)
18. Manocha, D., Canny, J.: Efficient inverse kinematics for general 6R manipulators. IEEE Trans. Robot. Autom. **10**(1), 648–657 (1994)
19. Manocha, D.: Inverse kinematics code for serial manipulators. GAMMA Research Group. http://gamma.cs.unc.edu/software/downloads/IK/ik-1.0.tar.gz. Accessed Jan 2022

Degenerate Cases in the Inverse Kinematics Problem of a General 6R Serial Robot

Aritra Das[1(✉)], Anirban Nag[2], and Subir Kumar Saha[2]

[1] Jadavpur University, Jadavpur, Kolkata, India
aritradas.habra@gmail.com

[2] Indian Institute of Technology Delhi, Delhi, India
anirbannag@iitd.ac.in, saha@mech.iitd.ac.in

Abstract. This paper highlights some of the degeneracies encountered in the inverse kinematics problem (IKP) of a general 6R (revolute jointed) serial robot. It is known that algebraic eliminations introduce certain mathematical degeneracies, encountering which an algorithm may fail. The degeneracy conditions are often overlooked or remains unreported when an algorithm is proposed. Awareness of these conditions becomes important when these algorithms form important cogs of bigger objectives like design, path-planning or control. Kinematic problems often lead to matrices whose entries are polynomials in a single variable. These can, in turn, be treated as polynomials whose coefficients are matrices. Degeneracy study of such systems can be attempted akin to their scalar counterparts, by investigating the conditions that result in the determinant of coefficient matrix associated with the leading term evaluating to zero. The same conditions have been analysed for the IKP of a general 6R, and some of the degenerate cases presented in this paper.

1 Introduction

The IKP of a general 6R (revolute jointed) robot has interested researchers for the past several decades. It is well established that the upper-bound on the number of solutions to this problem is 16. The principal approaches adopted in literature to solve this problem have been algebraic elimination based techniques [5,6,11] and continuation [12]. Algebraic elimination techniques involves reducing a coupled multi-variate system of polynomial equations to a univariate polynomial through a series of manipulations involving basic mathematical operations. The sequential application of the said operations makes these techniques more amenable to implementation in a generic programming environment rather than the use of advanced mathematical techniques which form the basis of the other approaches. As a result the execution of these algorithms happens at near real-time speeds. On the other hand algebraic elimination techniques are often an assemblage of a special set of operations aimed at obtaining the solution to the problem in general. In the process certain mathematical conditions

O. Altuzarra and A. Kecskémthy (Eds.): ARK 2022, SPAR 24, pp. 418–425, 2022.
https://doi.org/10.1007/978-3-031-08140-8_45

are introduced, which may lead to the failure of these algorithms. The study of these failure cases is important, to not only understand their causes, but to also explore wayouts if there exists any. The study is critical because, apart from the specific failure points, the system may become ill-conditioned at the vicinity of these points leading to erroneous results. This paper explores the failure conditions of the algorithm proposed by Manocha and Canny [5] to solve the IKP of the general 6R robot. Interestingly the need to propose a new algorithm as done in [5] arose from the numerical instability of a very similar algorithm proposed by Rahavan and Roth [11]. Incidentally the formulation presented in [5] lead to a quadratic lambda-matrix problem [4]. Although it is enticing to do an analogous study similar to the scalar quadratic equation, the two cases do not have a one-to-one correspondence due to matrix computations being non-commutative. However, there are conditions where some analogy can be established, e.g., the degeneracy of the leading term. A case at hand is the IKP of the "generalised lobster arm" as presented by Murthy and Waldron [9]. Interesting the authors in [9] chose not to use the algorithm presented in [11]. A possible reason for this may be the failure of the algorithms presented in [5,11] to solve this particular architecture. The motivation of reporting the present results stems from the fact that an algorithm can never be identified as a general, unless it can address all the "special" cases that it may encounter, or at least identify them so as to avoid the pitfalls. Although the authors do report the existence of "singular pencils" in [5], they do not point out the exact failure conditions. Further, the proposition to solve a generalised eigenvalue problem, when the leading coefficient matrix of the lambda-matrix problem becomes rank deficient is misleading because the numerical issues continue to persist in this case also.

The present paper is organised in the following manner: Sect. 2 presents a brief background of the IKP of the general 6R robot. Section 3 explores the possible degeneracies that may occur in the existing algorithms. The paper is concluded in Sect. 4.

2 Mathematical Background

The present paper explores the possible degenerate cases in the IKP of the general 6R robot solved using algebraic eliminations (Fig. 1). The algorithm proposed in [5] has been a benchmark for this problem in the literature. The same is revisited here for investigation. The algorithm is briefly presented here for the sake of continuity. The IKP involves determining the joint angles $\theta_i, (i = 1, \ldots, 6)$ for a given end-effector (EE) pose[1]. The homogeneous transformation matrices leading upto the EE can be written as:

[1] The DH parameters [2] are used as per convention with a_i, d_i representing the link length and link offset, respectively, α_i is the twist angle and θ_i is the joint angle. It may be noted that in this paper $\sin\theta_i \equiv s_i$ and $\cos\theta_i \equiv c_i$.

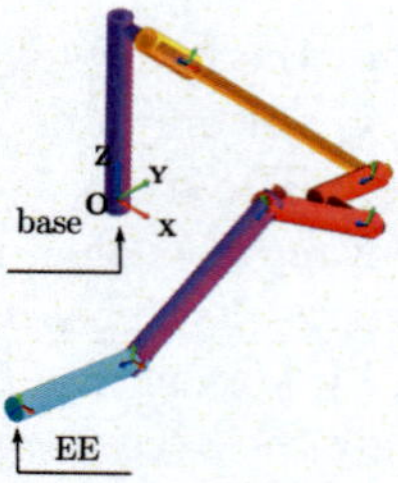

Fig. 1. Schematic of a 6R robot. The generation of the figures and cross-validation of the work has been done using the `Roboanalyzer` software [1].

$$\prod_{i=1}^{6} \mathbf{T}_i = \mathbf{T}_{EE}, \tag{1}$$

where $\mathbf{T}_i$ is the homogeneous transformation matrix relating frame $\{i+1\}$ to $\{i\}$ and $\mathbf{T}_{EE}$ is the homogeneous transformation matrix representing the end-effector pose. The expressions of the transformation matrices are identical to those presented in [5] and are skipped in this paper for brevity. Raghavan and Roth [11] arranged Eq. (1) in the following manner:

$$\mathbf{T}_3\mathbf{T}_4\mathbf{T}_5 = \mathbf{T}_2^{-1}\mathbf{T}_1^{-1}\mathbf{T}_{EE}\mathbf{T}_6^{-1}. \tag{2}$$

The re-arrangement helped in reducing the complexity of the equations. Further it was observed that columns 3 and 4 of the right-hand side of Eq. (2) was devoid of the variable θ_6. Equating the entries of the said columns in Eq. (2), six equations in five unknowns $\theta_1, \ldots, \theta_5$ were arrived at. Following [5, 11] the three equations obtained by equating the third columns of the resulting matrices in Eq. (2) are treated as a vector $\mathbf{p}$. Similarly, the equations obtained by equating the fourth columns are arranged in a vector $\mathbf{l}$. In total, 14 scalar equations are generated. It has been proven that the equations generated by $\mathbf{p} \cdot \mathbf{p}, \mathbf{p} \cdot \mathbf{l}, \mathbf{p} \times \mathbf{l}, (\mathbf{p} \cdot \mathbf{p})\mathbf{l} - 2(\mathbf{p} \cdot \mathbf{l})\mathbf{p}$, retain the same monomial structure as $\mathbf{p}, \mathbf{l}$. As is the practice in most algebraic elimination technique, all the unknowns except one are eliminated so as to obtain a minimal univariate polynomial. In this case too, the additional equations are first used to eliminate the variables θ_1 and θ_2 to obtain the following relation:

$$\mathbf{\Sigma} \quad [s_4 s_5 \quad s_4 c_5 \quad c_4 s_5 \quad c_4 c_5 \quad s_4 \quad c_4 \quad s_5 \quad c_5 \quad 1]^\top = \mathbf{0}, \tag{3}$$

where $\mathbf{\Sigma} \in \mathbb{R}^{6\times 9}$, and the entries are linear combinations in s_3, c_3. Using the substitutions, $s_i = \frac{2x_i}{1+x_i^2}$, and $c_i = \frac{1-x_i^2}{1+x_i^2}$, where, $x_i = \tan\frac{\theta_i}{2}$, Eq. (3) is modified to:

$$\mathbf{\Sigma}' \quad [x_4^2 x_5^2 \quad x_4^2 x_5 \quad x_4^2 \quad x_4 x_5^2 \quad x_4 x_5 \quad x_4 \quad x_5^2 \quad x_5 \quad 1]^\top = \mathbf{0}, \tag{4}$$

where $\mathbf{\Sigma}'_{ij} = \alpha_{ij} x_3^2 + \beta_{ij} x_3 + \gamma_{ij}$, $\mathbf{\Sigma}' \in \mathbb{R}^{6\times 9}$. The constants α_{ij}, β_{ij}, γ_{ij}, are functions of the DH parameters and the pose variables. Finally, x_4, x_5 is dialytically eliminated from Eq. (4) by constructing a square-system of the form:

$$\mathbf{\Sigma}''[x_4^3x_5^2 \quad x_4^3x_5 \quad x_4^3 \quad x_4^2x_5^2 \quad x_4^2x_5 \quad x_4^2 \quad x_4x_5^2 \quad x_4x_5 \quad x_4 \quad x_5^2 \quad x_5 \quad 1]^\top = \mathbf{0}, \tag{5}$$

where $\mathbf{\Sigma}'' \in \mathbb{R}^{12\times12}$ is of the form:

$$\mathbf{\Sigma}'' = \begin{pmatrix} \mathbf{\Sigma}' & \mathbf{0} \\ \mathbf{0} & \mathbf{\Sigma}' \end{pmatrix}. \tag{6}$$

Since Eq. (5) is a system of over-constrained equations, a non-trivial solution is obtained by imposing the condition $\det(\mathbf{\Sigma}'') = 0$. While Raghavan and Roth proposed a direct expansion of the determinant of $\mathbf{\Sigma}''$ to obtain a 16-degree univariate polynomial in x_3, it has often been observed that numerical errors propagate through the intermediate steps resulting in a non-minimal polynomial leading to inaccurate results. To circumvent the problem Manocha and Canny proposed an eigenproblem approach to find the roots of x_3. Following Eqs. (4, 6), $\mathbf{\Sigma}''$ can be written as:

$$\Sigma'' = \mathbf{A}x_3^2 + \mathbf{B}x_3 + \mathbf{C}, \tag{7}$$

where $\mathbf{A}$, $\mathbf{B}$, $\mathbf{C} \in \mathbb{R}^{12\times12}$, its entries being functions of the DH parameters and variables. Equation (5) allows the formulation of a generalised eigenproblem [3] of the form $\mathbf{P}x_3 = \lambda\mathbf{Q}x_3$, where $\mathbf{P}$, $\mathbf{Q} \in \mathbb{R}^{24\times24}$ are purely numeric matrices[2]. The eigenvalues λ_i, $i = 1, \ldots, 24$ are identical to the values of x_3 obtained from solving the equation $\det(\mathbf{\Sigma}'') = 0$. It has been observed that out of the 24 eigenvalues 8 are purely imaginary being the roots of the polynomial $(1+x_3^2)^4$ [5]. The remaining 16 eigenvalues correspond to the solution of the IKP of the general 6R robot.

3 Degenerate Conditions

The degeneracy conditions of the algorithms can be determined analytically by observing Eq. (7), which is a quadratic equation in x_3 with coefficients as matrices. Drawing analogy from a scalar quadratic equation, the algorithm is bound to fail or suffer from serious ill-condition issues when $\mathbf{A}$ is singular or nearly singular. Although a generalised eigenproblem approach is suggested in [5] as a possible circumvention of the ill-conditioning problem when $\mathbf{A}$ is close to being singular, it has been observed that the same does not help in many cases.

3.1 Analytical Determination of the Degenerate Conditions

It is evident that the algorithm would encounter problems when $\det(\mathbf{A}) = 0$. Symbolic computation of the determinant proved to be computationally expensive, hence an alternative approach was adopted by decomposing $\mathbf{A}$ into a lower

[2] Following [5], $\mathbf{P} = \begin{pmatrix} \mathbf{1} & \mathbf{0} \\ \mathbf{0} & \mathbf{A} \end{pmatrix}$, $\mathbf{Q} = \begin{pmatrix} \mathbf{0} & \mathbf{1} \\ -\mathbf{B} & -\mathbf{C} \end{pmatrix}$, where $\mathbf{1} \in \mathbb{R}^{12\times12}$ is the identity matrix.

triangular matrix **L** and an upper triangular matrix **U** so that the determinant computation can now be represented as:

$$\det(\mathbf{A}) = \det(\mathbf{L})\det(\mathbf{U}). \tag{8}$$

The symbolic decomposition was made anemable by treating the actual entries of **A** which are functions of the DH parameter as simpler atomic expressions such that Eq. (8) can be written as:

$$\det\begin{pmatrix}\mathbf{a} & \mathbf{0}\\ \mathbf{0} & \mathbf{a}\end{pmatrix} = \det\begin{pmatrix}1 & & & \\ & 1 & & \mathbf{0}\\ & & \ddots & \\ l_{ij} & & & 1\end{pmatrix}\det\begin{pmatrix}u_{11} & & & \\ & u_{22} & & u_{ij}\\ & & \ddots & \\ & \mathbf{0} & & u_{ii}\end{pmatrix}, \quad i = j = 12, \tag{9}$$

where a_{ij} of **a** are atomic representations of $\mathbf{A}_{ij}$ (see Eq. 7). The LU decomposition shown in Eq. (9) has been accomplished in the symbolic environment of `MATLAB 2021a` [7]. Since the diagonal elements of the matrix **L** are unity, det(**A**) becomes:

$$\det(\mathbf{A}) = \prod_{i=1}^{12} u_{ii}. \tag{10}$$

It is interesting to note that det(**A**) would evaluate to zero, when any one of the element of u_{ii} evaluates to zero. Further, the nature of the decomposition is such that u_{ii} basically forms the pivot elements of a Gaussian elimination routine. It is therefore natural to point out that the conditions leading to each u_{ii} becoming zero are possible causes of failure for the said algorithm. The conditions can be checked in a sequential manner because each u_{ii+1} exists *iff* $u_{ii} \neq 0$. The first three diagonal elements of **U** are obtained as follows:

$$u_{11} = a_{11}, \quad u_{22} = (a_{11}a_{22} - a_{12}a_{21})/a_{11} = (a_{11}a_{22} - a_{12}a_{21})/u_{11}, \tag{11a}$$

$$u_{33} = \frac{a_{13}a_{22}a_{31} - a_{12}a_{23}a_{31} - a_{13}a_{21}a_{32} + a_{11}a_{23}a_{32} + a_{12}a_{21}a_{33} - a_{11}a_{22}a_{33}}{u_{22}}. \tag{11b}$$

It is observed from Eq. (11a) that $a_{11} \neq 0$ is the first necessary condition for the existence of det(**A**). However it is not a sufficient condition because each of the remaining terms u_{ii} may evaluate to zero for certain combinations of the DH parameters and the end-effector pose. Due to the importance of the first pivot element a_{11} in the whole analysis some of the observations leading to it being zero are discussed hereafter. Denoting the set of all DH parameters[3] by $\boldsymbol{\delta}$ and the pose parameters by $\boldsymbol{\rho}$ the quantity a_{11} can be represented as:

$$a_{11} = f(\boldsymbol{\delta}, \boldsymbol{\rho}). \tag{12}$$

Substituting the twist angle $\alpha_1 = 0$, in Eq. (12) a_{11} is obtained as $a_{11} = a_1 g_1(\boldsymbol{\delta}, \boldsymbol{\rho})$. It is therefore evident that for the concurrent condition of $\alpha_1 = 0$ and $a_1 = 0$, a_{11}

[3] $\boldsymbol{\delta} \equiv [a_i, d_i, \alpha_i]^\top,\ i = 1, \ldots, 6, \quad \boldsymbol{\rho} \equiv [\mathbf{l}, \mathbf{m}, \mathbf{n}, p_x, p_y, p_z]^\top$.

would vanish. This is a condition of total failure of the algorithm as the matrices **A**, **B**, **C** in Eq. (7) degenerate to null-matrices with all entries being zero. The resulting architecture has been shown in Fig. 3a. The second factor $g_1(\boldsymbol{\delta}, \boldsymbol{\rho})$ can also degenerate in multiple ways. It is difficult to analyse the exact degeneracy conditions due to g_1 owing to the large number of architecture and pose variables contained in the expression. However, an attempt has been made to analyse the same at a subset level. In order to do so a specific architecture is chosen retaining the position variables and equating the resulting expression to zero, such that:

$$g_1(\boldsymbol{\delta}, \boldsymbol{\rho}) \equiv h(p_x, p_y, p_z)|_{\delta*} = 0. \tag{13}$$

Equation (13) represents a surface[4] in the Cartesian space over which g_1 would vanish thereby leading to a degeneracy of the formulation. Similarly, investigation of u_{22} in Eq. (11a) also leads to certain conditions that result in $\det(\mathbf{A}) = 0$. A possible scenario arises by substituting $\alpha_6 = 0$ in the expression of u_{22} so as to obtain:

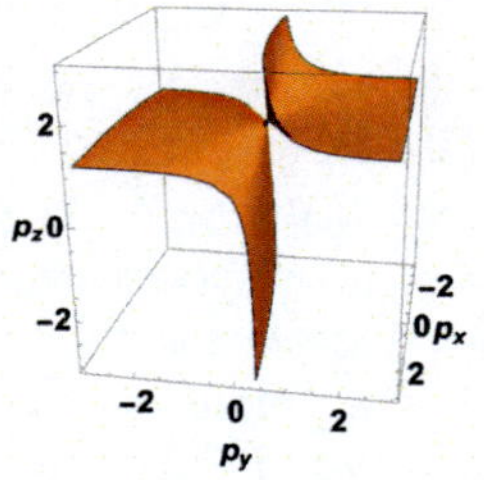

Fig. 2. Surface in the Cartesian space corresponding to Eq. (13)

$$u_{22}|_{\alpha_6=0} = (a_6(l_y n_x - l_x n_y) + n_y p_x - n_x p_y) g_2(\boldsymbol{\delta}, \boldsymbol{\rho}). \tag{14}$$

Equation (14), shows that apart from the architecture parameters, for certain combinations of the end-effector pose, there maybe a degeneracy in the algorithm. An evident example is the case when $n_x = 0$, $n_y = 0$ in Eq. (14). The physical condition for this degeneracy corresponds to the condition when there is a rotation of 0 or π radians about the **X** or **Y** axes. Some of the other cases that has been found out[5] are:

- $\alpha_5 = 0, a_5 = 0$ lead to vanishing of u_{22}
- $\alpha_5 = 0, d_3 = d_4 = d_5 = 0,\ a_3 = a_4 + a_5, \alpha_6 = \pi/2$ lead to vanishing of u_{33}.

Schematic depiction of the manipulators corresponding to some of these failure cases have been presented in Fig. 3.

[4] The idea of studying degenerate surfaces was used in [10] to study the degeneracy conditions in the algorithm of the forward kinematic problem of the Stewart platform manipulator.

[5] The discrete cases presented in this paper are meant as pointers that need to be care of while adopting elimination based solution techniques and are not exhaustive.

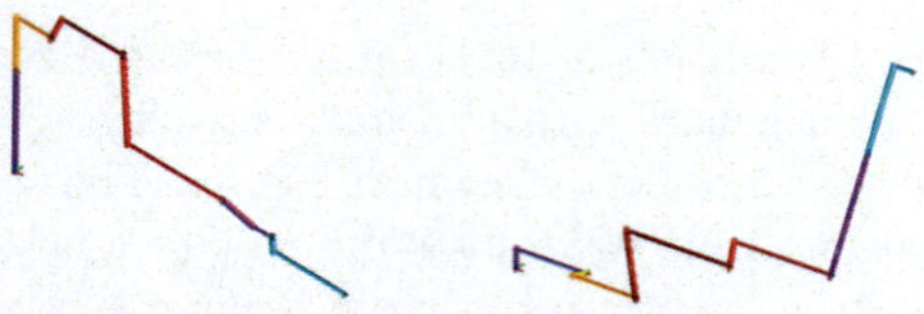

(a) $\alpha_1 = 0, a_1 = 0$ (b) $\alpha_5 = 0, a_5 = 0$

Fig. 3. Illustrative examples of robots corresponding to the failure cases

3.2 The Generalised Lobster Arm

A well known example from the literature where the IKP algorithm proposed in [5,11] would fail is the generalized lobster arm [9]. The reason for failure can be investigated on similar lines as explained above. Using the DH parameters of the generalised lobster arm as reported in [9], it is seen that a_{11} (see Eq. 11a) is obtained as:

$$u_{11} = \sin\alpha_1(n_y p_x - n_x p_y)h'(\delta) \tag{15}$$

It is evident that $\alpha_1 = 0, \pi$ would lead to u_{11} evaluating to zero leading to the determinants of the coefficients matrices **A**, **B**, **C** vanishing. Since δ for this robot is fixed, the only other way u_{11} may evaluate to zero is through certain combinations of the pose variables. It can be ascertained therefore that the failure of the algorithm is principally extrinsic due to the formulation it has adhered to, rather than intrinsic which may arise due to some special architecture[6]. As a result alternate techniques should exist that would provide the solution to this IKP, an example being the algorithm presented in [9]. For example, choosing a pose for which $u_{11} = 0$, such as, a point belonging to the degenracy surface like the one shown in Fig. 2, would result in the determinants of the coefficient matrices corresponding to Eq. (7) invariably vanishing. For the same architecture on the other hand, the determinants of the coefficient matrices are of the order of $\det(\mathbf{A}) \approx 10^{-30}$, $\det(\mathbf{B}) \approx 10^{-5}, \det(\mathbf{C}) \approx 10^{-30}$, when the points do not belong to the degenerate surface. It is therefore imperative that even though the algorithm presented [5] can be used to solve the IKP of this particular robot, the results may suffer from numerical inaccuracies.

4 Conclusion

This paper presents some of the degenerate cases that may arise while solving the IKP of the general 6R robot, where the DH parameters can assume arbitrary values. It is well known that algebraic elimination techniques introduce conditions that may lead to a breakdown in the algorithm. On the other hand, the computations involved in these algorithms make them more amenable towards

[6] The failure could have been termed intrinsic if u_{11} identically evaluated to zero for $\alpha_1 \neq 0, \pi$. It may also be argued that intrinsic failures would invariably change the cardinality of the solution space as explained in [8].

real-time implementation which makes them attractive candidates while being used as modules of path-planning or control problems. Identifying the architecture and the pose parameters that may lead to issues in the solution of the IKP therefore becomes important. It has been shown that for a given architecture, there may exist surfaces comprising of the pose parameters over which the algorithm may degenerate. Using the example of the generalised lobster arm it has been shown that for points which do not belong to the surface or are far away from it, the algorithm may still work, albeit with numerical inaccuracies. It remains to be investigated that whether the so called failures can actually be avoided by identifying the surfaces or regions where the algorithm is more susceptible to numerical issues. In such a case the same algorithm may be used to solve the IKP of robots which it has seemingly failed up until now.

References

1. Chittawadigi, R., Saha, S.K.: Roboanalyzer v7.5 (2009). http://www.roboanalyzer.com
2. Denavit, J., Hartenberg, R.S.: A kinematic notation for lower-pair mechanisms based on matrices. J. Appl. Mech. **22**, 215–221 (1955)
3. Golub, G.H., Van Loan, C.F.: Matrix Computations, 3rd edn. Johns Hopkins University Press, Baltimore, USA (2012)
4. Higham, N.J., Kim, H.M.: Numerical analysis of a quadratic matrix equation. IMA J. Numer. Anal. **20**(4), 499–519 (2000). https://doi.org/10.1093/imanum/20.4.499
5. Manocha, D., Canny, J.F.: Real time inverse kinematics for general 6R manipulators. In: ICRA, pp. 383–389 (1992)
6. Manocha, D., Zhu, Y.: A fast algorithm and system for the inverse kinematics of general serial manipulators. In: Proceedings of the 1994 IEEE International Conference on Robotics and Automation, pp. 3348–3353. IEEE (1994)
7. MATLAB: version 9.10.0 (R2021a). The MathWorks Inc., Natick, Massachusetts (2021)
8. Mavroidis, C., Roth, B.: Structural parameters which reduce the number of manipulator configurations. J. Mech. Des. **116**(1), 3–10 (1994). https://doi.org/10.1115/1.2919373
9. Murthy, V., Waldron, K.J.: Position kinematics of the generalized lobster arm and its series-parallel dual. J. Mech. Des. **114**(3), 406–413 (1992). https://doi.org/10.1115/1.2926566
10. Nag, A., Safar, V., Bandyopadhyay, S.: A uniform geometric-algebraic framework for the forward kinematic analysis of 6-6 Stewart platform manipulators of various architectures and other related 6-6 spatial manipulators. Mech. Mach. Theor. **155**, 104090 (2021). https://doi.org/10.1016/j.mechmachtheory.2020.104090
11. Raghavan, M., Roth, B.: A general solution for the inverse kinematics of all series chains. In: Proceedings of the 8th CISM-IFTOMM Symposium on Robots and Manipulators (1990)
12. Tsai, L.W., Morgan, A.P.: Solving the kinematics of the most general six- and five-degree-of-freedom manipulators by continuation methods. J. Mech. Trans. Autom. Des. **107**(2), 189–200 (1985). https://doi.org/10.1115/1.3258708

The Spatial Pantograph and Its Mass Balance

Volkert van der Wijk(✉)

Department of Precision and Microsystems Engineering,
Delft University of Technology, Delft, Netherlands
v.vanderwijk@tudelft.nl

Abstract. This paper presents the spatial version of the well-known pantograph linkage. While the traditional pantograph is a planar parallelogram linkage with 1 internal degree of freedom (DoF) and four moving links connected with solely revolute pairs, the general spatial pantograph is a linkage with 3 internal DoFs in which the four moving links are connected with spherical joints and two additional out-of-plane links are connected with universal joints. The out-of-plane links constrain the linkage to maintain the essential parallelogram and also constrain all links to move similarly. This is necessary for full controllability and also for inherent force balance, which is obtained for specific conditions on the mass distribution of the links. The force balance conditions are given and also a reduced version of the spatial pantograph linkage with 2 internal DoFs is presented which has five moving links connected with solely universal joints.

Keywords: Pantograph linkage · Spatial kinematics · Inherent force balance · Static balance

1 Introduction

The planar pantograph linkage remains a fascinating linkage, originally invented by Christoph Scheiner in 1603 for a painter to copy the view of a landscape scaled on canvas (for unskilled painters supposingly...). Among numerous other applications, the pantograph showed especially interesting for the design of inherently balanced linkages and it can be regarded a key element in (inherent) mass balancing as it is at the basis of all solutions [7]. This is due to the properties of similarity of the pantograph and since pantographs are embodiments of the principal vectors, which are fixed length vectors describing the motion of masses with respect to their common center of mass (CoM) [5].

Since principal vectors are not limited to planar motions [3,6], it is the question how they can be embodied to obtain a spatial pantograph. Of spatial pantographs not much is known yet. Moving planar pantographs spatially is regularly seen for instance as a manipulator arm [1] or to balance a spatial linkage by connecting the common CoM of the linkage directly to the base [2,4,9]. However, in these situations the complete pantograph system is moved spatially as

O. Altuzarra and A. Kecskémthy (Eds.): ARK 2022, SPAR 24, pp. 426–433, 2022.
https://doi.org/10.1007/978-3-031-08140-8_46

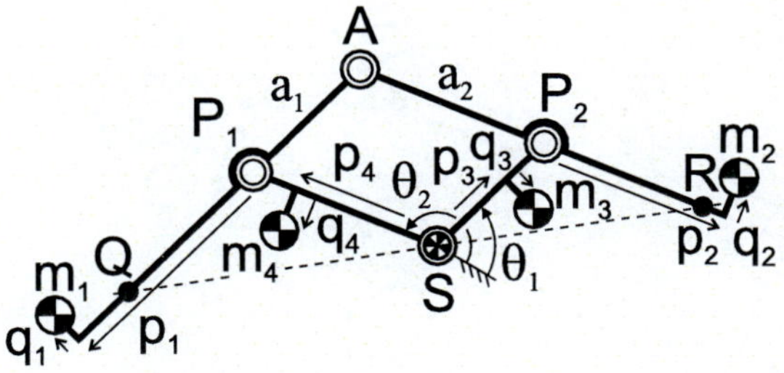

Fig. 1. General planar 2-DoF principal vector linkage or planar pantograph with 1 internal DoF and with the common CoM of the four moving masses in joint and base pivot S for all motion. P_1 and P_2 are the principal points of the linkage and Q and R are a set of similarity points [5].

a single rigid body instead of spatial relative motions within the pantograph itself. Also the synthesis of spatial linkages based on planar pantographs such as umbrella canopies and movable roofs [8,10] do not incorporate any spatial kinematics of the pantograph itself. In [5,9] a true kinematically spatial pantograph linkage was found and integrated with a Delta robot arm for force balance. The design however is limited, complex, and gained few fundamental insight. The parallelepiped mechanism in [11] shows significant correspondence to a spatial pantograph although this was not observed and investigated.

The goal of this paper is to present the general kinematic design of the spatial pantograph with its inherent balance conditions. This is done step by step starting with the balanced planar pantograph and at the end a reduced kinematic solution is presented as well.

2 General Kinematic Solution

Figure 1 shows the general planar version of the inherently balanced pantograph, a principal vector linkage consisting of the four links SP_1, P_1A, AP_2, and P_2S with lengths a_1 and a_2 connected with revolute pairs in the joints S, P_1, A, and P_2 [5]. The planar pantograph has 1 internal degree of freedom (DoF) and with an additional revolute pair in S as the base pivot the pantograph has 2-DoF motion capability as illustrated with θ_1 and θ_2. Each link i has a general mass distribution with the link CoM defined by p_i and q_i as illustrated. For specific conditions of the mass distributions, the force balance conditions $m_1p_1 = m_2a_1 + m_3p_3$, $m_1q_1 = m_3q_3$, $m_2p_2 = m_1a_2 + m_4p_4$, and $m_2q_2 = m_4q_4$, the common CoM of the four masses m_i is stationary in joint S for all motions.

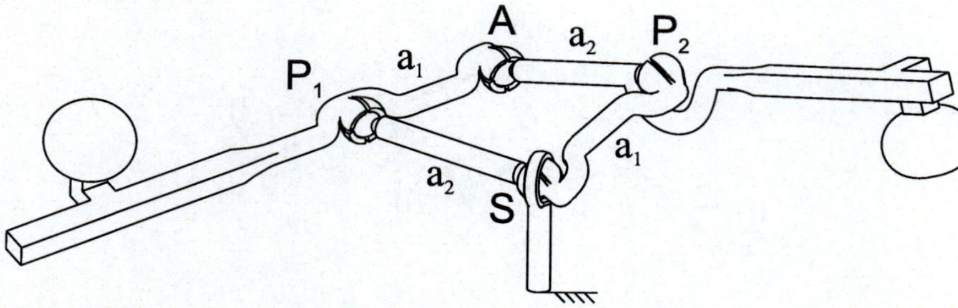

Fig. 2. Solely converting the revolute pairs in Fig. 1 into spherical joints is not sufficient to obtain a spatial pantograph as the joints must be constrained to lay in a single plane during motion.

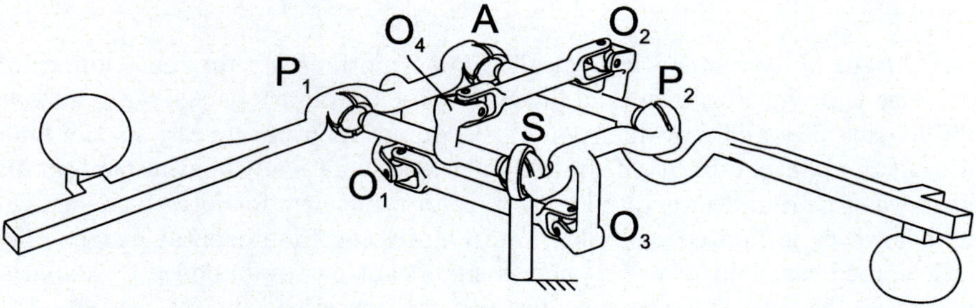

Fig. 3. The general 6-DoF spatial pantograph linkage with 3 internal DoFs and a spherical base pivot in S. The four parallelogram links are connected with spherical joints and the two additional parallel out-of-plane links are connected with universal joints.

Turning the planar pantograph into a spatial pantograph is not as simple as converting the revolute pairs into spherical joints. As illustrated in Fig. 2, with solely spherical joints in S, P_1, A, and P_2 the linkage becomes overmobile (with 9 DoFs as calculated with the mobility formula) and the essential pantograph condition that the joints need to lay within a single plane to have a parallelogram linkage is lost. Also the condition for similar motions of opposite links is lost since the links have individual internal rotational mobility along the axis through their joints. Therefore also no inherent balance solution is possible.

Figure 3 presents the general spatial pantograph linkage which, with respect to Fig. 2, has the two additional links O_1O_3 and O_2O_4 at an offset from the parallelogram plane with universal joints in O_1, O_2, O_3, and O_4. Link O_1O_3 is parallel to links P_1S and AP_2 and the specific illustrated orientation of the universal joints in O_1 and O_3 is necessary to have link O_1O_3 move fully similarly with the two parallel links regarding the rotation about the link's axis through O_1 and O_3. Link O_2O_4 is parallel to links AP_1 and P_2S and also here the specific illustrated orientation of the universal joints in O_2 and O_4 is necessary to have

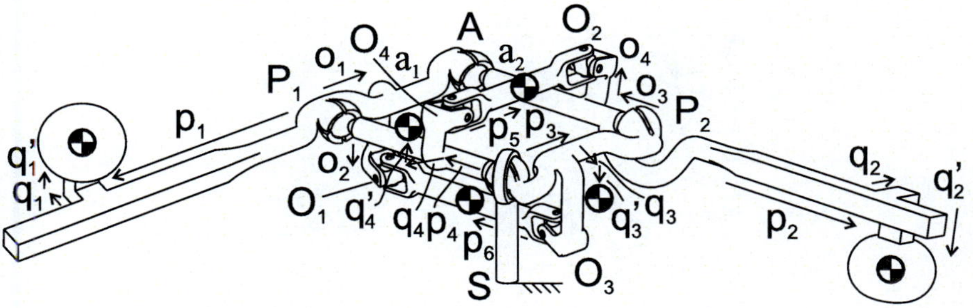

Fig. 4. The general spatial pantograph linkage with its mass parameters for inherent balance.

link O_2O_4 move fully similarly with its two parallel links regarding the rotation about the link's axis through O_2 and O_4.

The spatial pantograph has 3 internal DoFs, which can be observed from link P_2S moving with 3-DoF relative to link P_1S due to their spherical joint in S. With another spherical joint in S as the base pivot, the spatial pantograph obtains 6-DoF motion capability. Both out-of-plane links constrain the parallelogram plane during motion and they also constrain the pairs of opposite links to rotate synchronously and similarly for the rotations about the links' longitudinal axes, the lines through their joints, such that there are no links with individual internal rotational mobility. For instance by link O_1O_3 the rotation of link AP_1 about the link's axis through A and P_1 is similar to the rotation of link P_2S about its axis through P_2 and S. Since only one of the two out-of-plane links, either O_1O_3 or O_2O_4, is already sufficient to constrain the parallelogram plane during motion, the linkage has one overconstraint for which the mobility formula calculates only 5 DoFs. This is because of the four universal joints. For a normally constrained design one of the universal joints in O_1, O_2, O_3, and O_4 must be changed into a spherical joint, no matter which one.

The spatial pantograph in Fig. 3 can be regarded a general and minimal kinematic solution since it has the minimum number of elements needed for full spatial motion capability of each link and such that all elements, including the cross shafts of the universal joints, move similarly with respect to one another. If the out-of-plane links would be connected with spherical joints instead of universal joints then they would have internal rotational mobility along their axes through their joints and they would not constrain the parallelogram plane properly, for which then more additional out-of-plane links would be needed.

In Fig. 4 the general spatial pantograph is shown with the mass parameters of the links. With respect to the planar pantograph in Fig. 1 the masses of the four links SP_1, P_1A, AP_2 and P_2S also have an out-of-plane position defined with q_i' in the directions as illustrated, q_1' and q_4' to one side and q_2' and q_3' to the other side. The link O_1O_3 is located at a distance o_1 from P_1S in the direction along P_1A and at a distance o_2 normal to P_1A as illustrated, and has a mass m_6 of which the CoM is located at a distance p_6 from O_3 along O_1O_3 as illustrated.

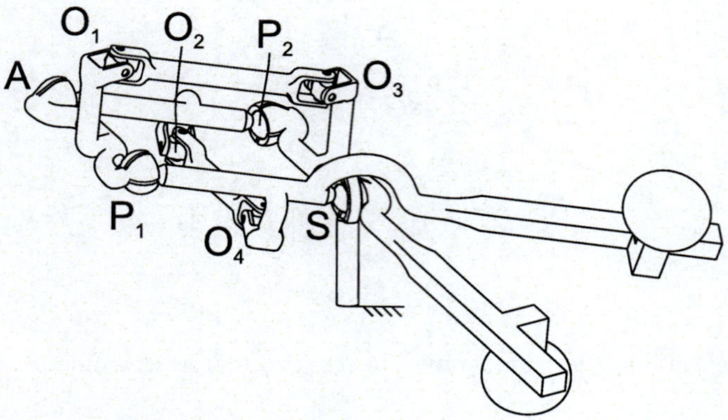

Fig. 5. Alternative configuration of the general spatial pantograph obtained by just reassembling the parts of the spatial pantograph in Fig. 3 differently. This configuration can also be regarded as a 6-DoF parallel manipulator with joint A as the 6-DoF end-effector which is force balanced with just two countermasses.

Similarly the link O_2O_4 is located at a distance o_3 from P_2S in the direction along P_2A and at a distance o_4 normal to P_2A as illustrated, and has a mass m_5 of which the CoM is located at a distance p_5 from O_4 along O_2O_4 as illustrated. The mass of the cross shafts in O_1 and O_3 is included in m_6 and the mass of the cross shafts in O_2 and O_4 is included in m_5.

The balance conditions for which the common CoM of all elements is in joint S for any pose and motion of the spatial pantograph linkage can be derived similarly as for the planar pantograph [5] and result in:

$$m_1p_1 = m_2a_1 + m_3p_3 + m_5p_5 + m_6o_1 \quad (1)$$
$$m_1q_1 = m_3q_3 \quad (2)$$
$$m_2p_2 = m_1a_2 + m_4p_4 + m_5o_3 + m_6p_6 \quad (3)$$
$$m_2q_2 = m_4q_4 \quad (4)$$
$$m_1q_1' = m_3q_3' + m_6o_2 \quad (5)$$
$$m_2q_2' = m_4q_4' + m_5o_4 \quad (6)$$

For these six balance conditions the spatial pantograph is inherently balanced. With respect to the planar pantograph there are two additional balance conditions, the latter two, defining the balance in the out-of-plane direction. It was numerically verified that realistic solutions close to the illustration as shown exist.

It was found not possible to have links O_1O_3 and O_2O_4 have a general link CoM since an offset of their link CoMs from the lines through their joints cannot be compensated for by other links.

Figure 5 shows another configuration of the general spatial pantograph in Fig. 3 which was obtained by reassembling the same parts differently. The balance conditions for this configuration are equal to those for the configuration

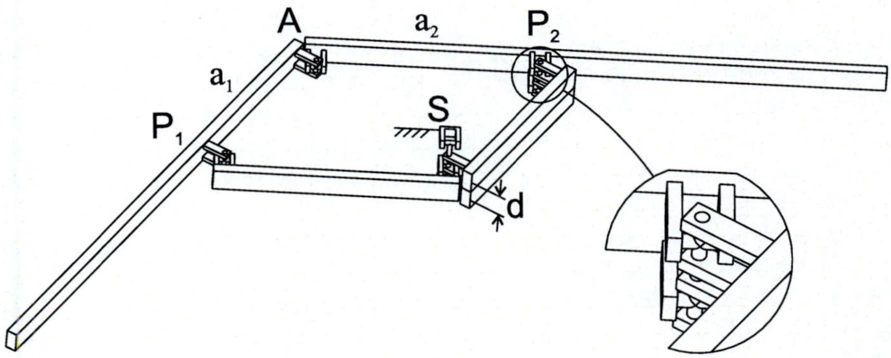

Fig. 6. Spatial pantograph linkage with 3-DoF, a reduced version of the general spatial pantograph with 2 internal DoFs based on solely universal joints and a revolute joint as base pivot.

in Fig. 3. It is also possible to obtain this configuration by just changing the parameter values of the configuration in Fig. 4 for which the link masses m_3 and m_4 then have their CoM beyond S with negative values for p_3 and p_4. This configuration has strong correspondence with the balanced parallelepiped mechanism presented in [11] when considering joint A as a 3-DoF end-effector, with as a significant difference that this configuration is balanced with just two countermasses whereas the parallelepiped mechanism requires three countermasses. Moreover, since there is 6-DoF motion capability about joint A, the general spatial pantograph in Fig. 5 can also be considered a 6-DoF parallel manipulator balanced with just two countermasses with joint A as the 6-DoF end-effector and with all 6 actuators on the base.

In general it is interesting to observe from the two configurations in Figs. 3 and 5 that where the planar pantograph is based on a single parallelogram, for a spatial pantograph - and for principal vector linkages to move spatially in general - there seems to be the need for a certain out-of-plane construction of again parallelograms. Links O_1O_3 and O_2O_4 can be seen as forming two additional parallelograms with which the linkage obtained three orthogonal parallelogram constructions spanning the 6-DoF space. That's an interesting aspect for continuing the research with.

3 Kinematic Solution with Solely Universal Joints

Figure 6 presents another spatial pantograph linkage, which is a reduced version of the general spatial pantograph in Fig. 3 with only 2 internal DoFs and 1 DoF due to the revolute joint as base pivot in S. It consists of five moving links connected solely with universal joints. Between S and P_2 there are two parallel links of which one can be considered the out-of-plane link as in the general spatial pantograph, however here they are both constructed out of plane as they are nicely aligned and symmetrically placed with respect to the parallelogram plane, which however is not necessary. Also here the out-of-plane link constrains the other links to a parallelogram linkage. Due to the universal joints the links have no internal rotational mobility along their axis through their joints, for which no second out-of-plane link is needed. However, these two rotational mobilities are also not free to be moved, which explains why this architecture has a reduction of DoFs. To move correctly, all the universal joints need to be oriented similarly, as illustrated.

The spatial pantograph in Fig. 6 has equal mobility to the spatial pantograph that was integrated in the Delta robot arm for balancing the Delta robot in [5, 9]. Since this design is significantly simpler with fewer elements, it might be an interesting solution for balancing the Delta robot and comparable manipulators in which one or more of the arms need to be turned into a spatial pantograph.

4 Conclusion

The spatial version of the planar pantograph linkage was presented as a linkage with 3 internal degrees of freedom (DoFs) based on four moving links connected with four spherical joints and two out-of-plane links connected with four universal joints. The out-of-plane links constrain the parallelogram for spatial motions and they also constrain all parallel links to move similarly. When connected with a spherical joint as a base pivot, a 6-DoF spatial pantograph mechanism is obtained. It was shown that the spatial pantograph is inherently force balanced for a specific mass distribution of the links, for which the six force balance conditions were given. Also a reduced spatial pantograph linkage with 2 internal DoFs was presented that is simpler with only five moving links in total connected with six universal joints.

Acknowledgements. This research was conducted in 2018 during a 3.5-month research visit by the author to the Robotics Laboratory at Laval University, Quebec, Canada, for which the author would like to thank Clément Gosselin for his kindness and the use of the lab's 3D printing style.

References

1. Briot, S., Arakelian, V.H., Guégan, S.: PAMINSA: a new family of partially decoupled parallel manipulators. Mech. Mach. Theor. **44**, 425–444 (2009)
2. Dobrovol'skii, V.V.: On the motion of the center of mass of a four-bar linkage. Mechanisms **3**, 233–234 (1968)
3. Fischer, O.: Theoretische grundlagen für eine Mechanik der lebenden Körper. Teubner, Leipzig (1906)
4. Russo, A., Sinatra, R., Xi, F.: Static balancing of parallel robots. Mech. Mach. Theor. **40**, 191–202 (2005)
5. Van der Wijk, V.: Methodology for analysis and synthesis of inherently force and moment-balanced mechanisms - theory and applications (dissertation). University of Twente (2014) https://doi.org/10.3990/1.9789036536301

6. Wijk, V.: Spatial orientations of principal vector planes for inherent dynamic balancing. In: Lenarcic, J., Parenti-Castelli, V. (eds.) ARK 2018. SPAR, vol. 8, pp. 302–309. Springer, Cham (2019). https://doi.org/10.1007/978-3-319-93188-3_35
7. Van der Wijk, V.: The Grand 4R Four-Bar Based Inherently Balanced Linkage Architecture for synthesis of shaking force balanced and gravity force balanced mechanisms. Mech. Mach. Theor. **150**, 103815 (2020)
8. Van der Wijk, V.: Inherent gravity force balance in moving architecture with new designs. In: Proceedings of the First International Symposium on Architecture, Technology and Innovation 2020 (2020)
9. van der Wijk, V., Herder, J.L.: Dynamic balancing of Clavelś delta robot. In: Kecskeméthy, A., Müller, A. (eds) Computational Kinematics (2009). https://doi.org/10.1007/978-3-642-01947-0_39
10. Van der Wijk, V., Kiper, G., Yaşir, A.: Synthesis and experiments of inherently balanced umbrella canopy designs. In: Proceedings of the TrC-IFToMM Symposium on Theory of Machines and Mechanisms, no.23 (2015)
11. Wu, Y., Gosselin, C.M.: Design of reactionless 3-DOF and 6-DOF parallel manipulators using parallelepiped mechanisms. IEEE Trans. Robot. **21**(5), 821–833 (2005)

A New Family of Overconstrained P5R-Mechanisms

Zijia Li(✉)

Academy of Mathematics and Systems Science, Chinese Academy of Sciences, Beijing, China
lizijia@amss.ac.cn

Abstract. This paper constructs overconstrained P5R closed linkages by merging a P4R linkage and a Goldberg 5R linkage. The construction uses bond theory and motion polynomial factorization. To our knowledge, this linkage is new, and it is the only known overconstrained P5R linkage that has no parallel joint-axes or concurrent joint-axes.

Keywords: Goldberg 5R linkages · P4R linkages · Motion polynomials · Bond theory · Dual quaternions

1 Introduction

In kinematics, closed overconstrained linkages with only rotation joints, e.g., Bennett 4R linkages, Goldberg 5R linkages, Bricard 6R linkages, of mobility one have been considered by many authors (for instance, [2,4,8,27,28], and reviews as [1,5,6,9,15]).

This paper mainly focuses on closed linkages with a prismatic joint, and others are rotation joints, e.g., the P4R linkage, the P3R linkage. It is known that many examples of these linkages have parallel axes, e.g., parallel rotation joint-axes (adjacent [3] or not). More precisely, we consider a particular type of P5R linkages. The spherical projection of the rotation axes has five directions. Namely, there are no parallel rotation joint-axes. One can use the construction method of hybrid overconstrained linkages [26,27] to produce P5R linkages, i.e., merging two 4-bar linkages when they share a joint that has the same motion. However, the known 4-bar linkage with the prismatic joint must have parallel axes, even after removing one rotation joint. If we take 2P4R as the target, it is possible to construct a 2P4R linkage without parallel joint-axes. For instance, one construction is to merge a PPRR linkage and a spherical linkage when they share a joint. However, it seems that the P5R in this paper is the only known example without parallel joint-axes or concurrent joint-axes. For instance, many linkages with prismatic joints are studied in [3,13,19,21,24,25], but there are no examples of P5R linkages without parallel joint-axes.

O. Altuzarra and A. Kecskémthy (Eds.): ARK 2022, SPAR 24, pp. 434–441, 2022.
https://doi.org/10.1007/978-3-031-08140-8_47

Our main tools are the bond theory and the factorization of a motion polynomial based on dual quaternions. Using particular linear motion polynomials, we make a Goldberg 5R linkage and a P4R linkage sharing two joints. After merging these two linkages, we can construct the P5R linkage. The main novelty is to find that these two linkages can share two joints where the bond theory and the factorization of motion polynomial is used.

The rest of this paper is organized as follows. Section 2 gives the preliminaries we need, i.e., dual quaternions, the factorization of a motion polynomial, and the bond theory. Section 3 introduces our motivation. Finally, Sect. 4 contains the main result and examples.

2 Preliminaries

This paper uses the two tools (Bond theory and motion polynomial factorization) based on dual quaternions to obtain the new P5R linkage.

2.1 Dual Quaternions

Dual quaternions $\mathbb{DH}$ form an algebra which is also an 8-dimensional real vector space and is generated by $1, \epsilon, \mathbf{i}, \mathbf{j}, \mathbf{k}, \epsilon\mathbf{i}, \epsilon\mathbf{j}, \epsilon\mathbf{k}$ (see [11]). Following the isomorphism described in [14, Sect. 2.4], we can represent a rotation by a dual quaternion in the Study quadric (dual quaternions h with $h\overline{h} \in \mathbb{R}^*$, where $\overline{h}$ is the quaternion conjugate of h) of the form $\left(\cot\left(\frac{\phi}{2}\right) - h\right)$, where ϕ is the rotation angle and h is a dual quaternion such that $h^2 = -1$ depending only on the rotation axis. We can represent a prismatic motion of a P-joint by a dual quaternion in the Study quadric of the form $(t - \epsilon h)$, where t is the scale of the translation distance and h is a vector quaternion representing the direction such that $h^2 = -1$.

Let L denote a P5R linkage given by one quaternion h_1 and five dual quaternions $h_2, \ldots, h_6$ such that $h_i^2 = -1$ for $i = 1, \ldots, 6$. A configuration (see [17]) is a 6-tuple $(t_1, \ldots, t_6)$, such that the closure condition holds Eq. (1). Let K_L denote the set of all configurations of L. We say L is movable when the configuration set K_L is one-dimensional.

$$(t_1 - \epsilon h_1)(t_2 - h_2)(t_3 - h_3)(t_4 - h_4)(t_5 - h_5)(t_6 - h_6) \in \mathbb{R}\backslash\{0\} \tag{1}$$

2.2 Factorization of a Motion Polynomial

We start by recalling the definition of the motion polynomial from the paper [11]. A motion polynomial P is a monic polynomial of degree n in the Study quadric with $P\bar{P} \in \mathbb{R}[t]$. In general, there exist a factorization of P as

$$P = (t - h_1)(t - h_2) \cdots (t - h_n) \tag{2}$$

where h_i are dual quaternions in the Study quadric. One of the central applications is to construct closed mechanisms by merging two different factorizations of a motion polynomial. Unfortunately, the number of factorizations can be more than two or infinitely many. Therefore, the primary strategy is to find a motion polynomial of proper degrees (quartic or higher degree) and its two proper factorizations. In addition, the two factorizations should give two open chains for merging to a closed-loop. Our main contribution is to use special linear motion polynomials for merging two linkages (a P4R and a special Goldberg 5R linkage with all zero offsets) based on the bond theory [12,20]. Namely, the motion polynomials for prismatic joints are considered comparing to [7,16].1 The derived P5R linkage is obtained by removing the two shared joints of a P4R linkage and a special Goldberg 5R linkage.

2.3 Bond Theory

We start by recalling the bond theory from the paper [10,11,20]. We denote a closed nR (or P) linkage of mobility 1 by $L = (h_1, \ldots, h_n)$. The configuration curve is $K_L \subset (\mathbb{P}^1_{\mathbb{R}})^n$ which is defined by Eq. (1). We denote the bond set

$$B := \{(t_1, \ldots, t_n) \in K_{\mathbb{C}} \mid (t_1 - h_1)(t_2 - h_2) \cdots (t_n - h_n) = 0\}. \tag{3}$$

where $K_{\mathbb{C}} \subset (\mathbb{P}^1_{\mathbb{C}})^n$ be the Zariski closure of K_L. The set B is a finite set containing conjugate complex points because the Eq. (1) is real. In addition, by Theorem 2 in [12], for a bond point $\beta = (t_1, \ldots, t_n)$, there are two indices $i, j \in [n]$, $i < j$, such that $t_i^2 + 1 = t_j^2 + 1 = 0$. The main property of the bond point is that it can shorter the Eq. (1) [12, Corollary 12], i.e., the equality

$$(t_i - h_i)(t_{i+1} - h_{i+1}) \cdots (t_j - h_j) = (t_j - h_j)(t_{j+1} - h_{j+1}) \cdots (t_i - h_i) = 0 \tag{4}$$

holds. We visualize these bond points by *bond diagrams*, where the connections represent the number of bond points (conjugate pairs).

In Figure 1, we show the linkages used in our construction and our new linkages with bond diagrams.

3 Motivation

Using the strategy of hybrid linkages from [26], one can construct new 6R, P5R, 2P4R linkages by merging two candidate linkages of 4R linkages (including Bennett 4R linkages, spherical 4R linkages, planar 4R linkages), P3R linkages, or 2P2R linkages where all rotation joints are parallel. A generalization of hybrid linkages with Goldberg 5R linkages is well studied in [28,28,29]. However, these hybrid linkages with P joints always have parallel joint-axes (two axes have the same direction), and they are well-reviewed [3]. We aim to construct a P5R linkage without parallel joint-axes, i.e., any two rotation joints axes' directions are different. The candidates for our construction are the P4R linkages [22,23] and Goldberg 5R linkages. Notice that it is easy to construct a P5R linkage with

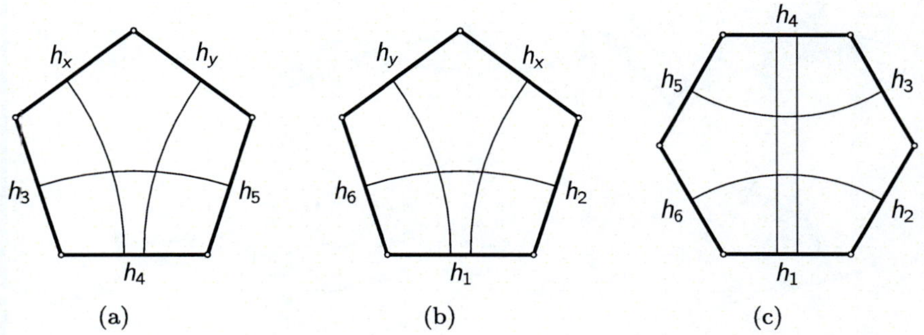

Fig. 1. Bond diagrams for the Goldberg 5R linkages (a), the P4R linkages (b), the P5R linkages (c)

five orientations of rotation axes (without parallel joint-axes) by merging a P4R linkage [22,23] and a spherical 5R linkage. As the spherical 5R has mobility 2, the P5R has mobility one by the mobility counting [9]. Notice that there are three concurrent axes in this type of P5R linkages. This paper is more interested in P5R linkages without three concurrent axes.

4 The Main Results

In the beginning, let us make our aim clear. Namely, we need to merge a P4R linkage and a special Goldberg 5R linkage (all zero offsets). For this purpose, we need to set the conditions that the merged linkage is still mobile. For a P4R linkage, without loss of generality, we assume that the motion polynomial of each revolute joint has norm t^2+1. If we want to merge two joints of a P4R with two of a Goldberg 5R, we also need that the motion polynomials of two neighbor joints of the Goldberg 5R have norm t^2+1, for some free parameter t. Then we can construct a P5R linkage by combining these linkages with removing the common two joints. With this construction, the rotation axes of the P5R linkage have five orientations, and no two neighboring rotation joints are necessarily concurrent (a particular case can make the concurrent).

Our construction of such P5R linkages without concurrent axes can be detailed as follows:

I. We choose a Goldberg 5R linkage $L_0 = [h_3, h_4, h_5, h_y, h_x]$ such that its two neighboring joints h_x, h_y have the same norm of motion polynomials, e.g., norm polynomial as t^2+1.
II. Use these two joints h_x, h_y to construct a P4R linkage $L_1 = [h_1, h_2, h_x, h_y, h_6]$, where h_1 is the prismatic joint. By the necessary conditions of P4R linkages, the direction of h_1 should divide the twist angle of h_x, h_y by half. The h_2 and h_6 is obtained by translating h_x, h_y along a translation. This procedure contributes the two rotation axes h_2 and h_6 and the translation h_1.
III. Return the 6R linkage $[h_1, h_2, h_3, h_4, h_5, h_6]$.

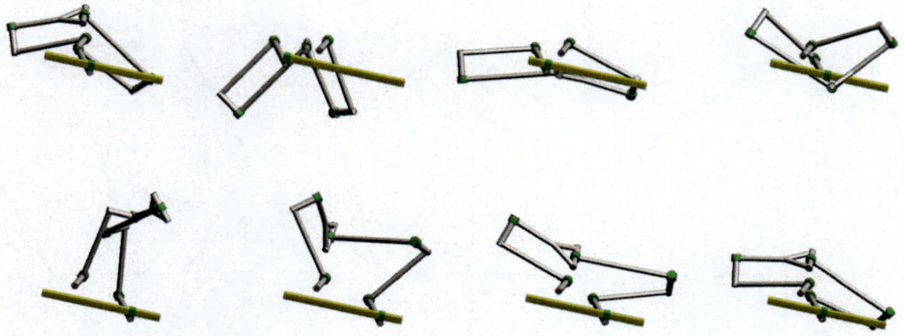

Fig. 2. Nine configurations of the P5R-linkage in the example.

There are two options in the procedure I (merging two Bennett linkages or using factorization of motion polynomials), which give the same result. For instance, a possible motion polynomial can be taken as $(t-a-bh_3)(t-h_4)(t-h_4)$, where $h_3^2 = h_4^2 = -1$ and a, b are arbitrary real numbers but not $0, \pm 1$. Similarly, for procedure II, we can use factorization of a motion polynomial of degree three to construct a P4R linkage. Namely, take a motion polynomial defined by $m = (t-h_x)(t-h_y)(t-h_6)$ where h_6 and h_y are paralell and the product $(t-h_y)(t-h_6)$ generates a circular translation. In total, the m defined a general Darboux motion (of degree three) as in [18] with an open chain of 3R, and another factorization defines an open chain of PR. The closed loop by join the two open chains is a P4R linkage in the procedure II. In addition, by the motion polynomial factorization, there will be factorizations of motion polynomials

$$(t^2+1) - c(t+d)h_1)(t-h_2)(t-a-bh_3) = (t+h_6)(t-a+bh_5)(t+h_4)^2 \quad (5)$$

with arbitrary real numbers a, b, c, d. This factorization also certify the mobility of our P5R linkage.

We add one numerical example of the P5R to show the mobility and the demenstration. In the following, we will only show the Plücker coordinates for each step in our construction procedure, i.e., the screw systems of the Goldberg 5R linkage, the P4R linkage and the P5R linkage. In Fig. 2, we show eight configurations of the P5R linkage.

Goldberg 5R

$$h_3 = \mathbf{i}\left(\frac{1}{3}+\frac{4}{9}\epsilon\right) + \mathbf{j}\left(\frac{2}{3}-\frac{4}{9}\epsilon\right) + \mathbf{k}\left(\frac{2}{3}+\frac{2}{9}\epsilon\right),$$

$$h_4 = \mathbf{j} - \epsilon\mathbf{i} - 2\epsilon\mathbf{k},$$

$$h_5 = \mathbf{i}\left(\frac{45}{103}-\frac{2212}{10609}\epsilon\right) + \mathbf{j}\left(\frac{22}{103}-\frac{52380}{10609}\epsilon\right) - \mathbf{k}\left(\frac{90}{103}+\frac{13910}{10609}\epsilon\right),$$

$$h_y = \mathbf{i}\left(\frac{390}{1751}+\frac{5083075}{3066001}\epsilon\right) - \mathbf{j}\left(\frac{1526}{1751}-\frac{4386210}{3066001}\epsilon\right) + \mathbf{k}\left(\frac{45}{103}+\frac{362242}{180353}\epsilon\right),$$

$$h_x = \mathbf{i}\left(\frac{14}{51} - \frac{1739}{2601}\epsilon\right) + \mathbf{j}\left(\frac{46}{51} - \frac{2122}{2601}\epsilon\right) - \mathbf{k}\left(\frac{1}{3} + \frac{422}{153}\epsilon\right).$$

P4R:

$$h_1 = \left(\frac{2612}{5253}\mathbf{i} + \frac{160}{5253}\mathbf{j} + \frac{32}{309}\mathbf{k}\right)\epsilon,$$
$$h_2 = \mathbf{i}\left(\frac{14}{51} + \frac{12779}{267903}\epsilon\right) + \mathbf{j}\left(\frac{46}{51} - \frac{160766}{267903}\epsilon\right) - \mathbf{k}\left(\frac{1}{3} + \frac{24970}{15759}\epsilon\right),$$
$$h_6 = \mathbf{i}\left(\frac{390}{1751} + \frac{1575771}{3066001}\epsilon\right) - \mathbf{j}\left(\frac{1526}{1751} - \frac{4698330}{3066001}\epsilon\right) + \mathbf{k}\left(\frac{45}{103} + \frac{1512134}{541059}\epsilon\right).$$

New P5R: Take $L := [h_1, h_2, h_3, h_4, h_5, h_6]$. The Denavit-Hartenberg parameters listed are the twist distances a_i, offsets d_i, $c_i := \cos\alpha_i$, the Bennett ratios $b_i = a_i/\sin\alpha_i$, where α_i the twist angles, for $i = 1, \ldots, 6$. Notice that we have a P-joint that reduces the number of invariant geometric parameters. Therefore, the numerical invariant Denavit-Hartenberg parameters of L are:

$$c_u = \left\{\frac{2\sqrt{1545}}{309}, \frac{8}{17}, \frac{2}{3}, \frac{22}{103}, -\frac{8}{17}, \frac{2\sqrt{1545}}{309}\right\},$$
$$b_v = \left\{-\frac{3559}{1545}, -\frac{19}{5}, -\frac{19}{5}, \frac{3559}{1545}\right\},$$
$$d_w = \left\{-\frac{2312}{4635}, 0, -\frac{2312}{4635}\right\},$$

where $u = 1, \ldots, 6$, $v = 2, \ldots, 5$, $w = 3, 4, 5$. Here it is not strange that we have less than 18 geometric parameters. Because we only considered the invariant geometric parameters during the motion of the P5R linkage. It is helpful to use fewer parameters to count the exact dimension of the family for this particular P5R linkages. Notice that we have a P-joint that reduces the number of invariant geometric parameters. The parameters listed are the twist distances a_i, offsets d_i, $c_i := \cos\alpha_i$, the Bennett ratios $b_i = a_i/\sin\alpha_i$, where α_i the twist angles. Based on the symbolic equations of the Goldberg 5R linkage [6] and the P4R linkage [22,23], the symbolic equations of the geometric parameters for the P5R linkage are:

$$c_1 = c_6,\ c_2 = -c_5,\ b_2 = -b_5,\ b_3 = b_4,\ d_4 = 0,\ d_3 = d_5,$$
$$(1 - c_1^2)(a_4 - a_3)^2 = b_3^2 c_1^2 (c_3 + c_4)^2, \cos(\alpha_3 - \alpha_4) = 1 - 2c_1^2. \tag{6}$$

The linear relations might change when the orientations are changed.

Remark 1. Notice that we can use an isomer trick developed by Wohlhart [29] to construct new linkages when there is a Bennett condition between three adjacent rotation axes in a loop. Then the trick is to replace the middle joint among the three axes with another joint, making a Bennett loop with these three joints. The mobility of the isomerized mechanism does not change, and the geometric parameters need a switch of two twist angles. By the geometric parameters shown above, we can use the isomer trick to get another mechanism which is still in the family based on the symbolic equations of the geometric parameters. For instance, one can switch the c_3 and c_4, and the h_4 can be replaced by

$$h_4' = \mathbf{i}\left(\frac{56}{1545} - \frac{586507}{477405}\right) + \mathbf{j}\left(\frac{260}{309} - \frac{750722}{477405}\right) - \mathbf{k}\left(\frac{833}{1545} + \frac{1211024}{477405}\epsilon\right).$$

Acknowledgements. We would like to thank Tiago Guerreiro, Hans-Peter Schröcker and Josef Schicho for discussion and helpful remarks.

References

1. Baker, J.: A comparative survey of the Bennett-based, 6-revolute kinematic loops. Mech. Mach. Theor. **28**, 83–96 (1993)
2. Baker, J.E.: An analysis of the Bricard linkages. Mech. Mach. Theor. **15**(4), 267–286 (1980)
3. Baker, J.E.: Overconstrained six-bars with parallel adjacent joint-axes. Mech. Mach. Theor. **38**(2), 103–117 (2003)
4. Bennett, G.T.: A new mechanism. Engineering **76**, 777–778 (1903)
5. Chen, Y., You, Z.: Spatial Overconstrained Linkages - The Lost Jade. In: Koetsier, T., Ceccarelli, M. (eds.) Explorations in the History of Machines and Mechanisms, History of Mechanism and Machine Science, vol. 15, pp. 535–550. Springer, Netherlands (2012)
6. Dietmaier, P.: Einfach übergeschlossene Mechanismen mit Drehgelenken. Habilitation thesis, Graz University of Technology (1995)
7. Gallet, M., Koutschan, C., Li, Z., Regensburger, G., Schicho, J., Villamizar, N.: Planar linkages following a prescribed motion. arXiv preprint arXiv:1502.05623 (2015)
8. Goldberg, M.: New five-bar and six-bar linkages in three dimensions. Trans. ASME **65**, 649–656 (1943)
9. Guerreiro, T., Li, Z., Schicho, J.: Classification of higher mobility linkages. arXiv e-prints (2021)
10. Hegedüs, G., Li, Z., Schicho, J., Schröcker, H.P.: The theory of bonds II: closed 6R linkages with maximal genus. J. Symbol. Comput. **68**, 167–180 (2015)
11. Hegedüs, G., Schicho, J., Schröcker, H.P.: Factorization of rational curves in the Study quadric. Mech. Mach. Theor. **69**, 142–152 (2013)
12. Hegedüs, G., Schicho, J., Schröcker, H.P.: The theory of bonds: a new method for the analysis of linkages. Mech. Mach. Theor. **70**, 407–424 (2013)
13. Hsu, K.L., lon Ting, K.: Overconstrained mechanisms derived from RPRP loops. J. Mech. Des. **140**, 062301 (2018)
14. Husty, M.L., Schröcker, H.P.: Algebraic geometry and kinematics. In: Emiris, I.Z., Sottile, F., Theobald, T. (eds.) Nonlinear Computational Geometry, The IMA Volumes in Mathematics and its Applications, vol. 151, chap. Algebraic Geometry and Kinematics, pp. 85–107. Springer, New York (2010). https://doi.org/10.1007/978-1-4419-0999-2_4
15. Li, Z.: Closed linkages with six revolute joints. Phd thesis, Johannes Kepler University Linz (2015)

16. Li, Z., Scharler, D.F., Schröcker, H.P.: Factorization results for left polynomials in some associative real algebras: state of the art, applications, and open questions. J. Comput. Appl. Math. **349**, 508–522 (2019)
17. Li, Z., Schicho, J.: A technique for deriving equational conditions on the Denavit-Hartenberg parameters of 6R linkages that are necessary for movability. Mech. Mach. Theor. **94**, 1–8 (2015)
18. Li, Z., Schicho, J., Schröcker, H.P.: 7R darboux linkages by factorization of motion polynomials. In: Proceedings of the 14th World Congress on the Theory of Machines and Mechanisms, pp. 841–847 (2015)
19. Li, Z., Schicho, J., Schröcker, H.P.: Spatial straight line linkages by factorization of motion polynomials. J. Mech. Robot. **8**(2), 021002 (2016)
20. Li, Z., Schicho, J., Schröcker, H.P.: A survey on the theory of bonds. IMA J. Math. Control. Inf. **35**, 279–295 (2018)
21. Liu, K., Yu, J., Kong, X.: Structure synthesis and reconfiguration analysis of variable-degree-of-freedom single-loop mechanisms with prismatic joints using dual quaternions. J. Mech. Robot. **14**(2), 1 (2021)
22. Mavroidis, C., Beddows, M.: A spatial overconstrained mechanism that can be used in practical applications. In: Proceedings of the Fifth Applied Mechanisms & Robotics Conference, Cincinnati, ASME, New York. Citeseer (1997)
23. Mavroidis, C., Roth, B.: New and revised overconstrained mechanisms. J. Mech. Des. **117**(1), 75–82 (1995)
24. Pfurner, M., Kong, X., Huang, C.: Complete kinematic analysis of single-loop multiple-mode 7-link mechanisms based on Bennett and overconstrained RPRP mechanisms. Mech. Mach. Theor. **73**, 117–129 (2014)
25. Pfurner, M., Stigger, T., Husty, M.L.: Algebraic analysis of overconstrained single loop four link mechanisms with revolute and prismatic joints. Mech. Mach. Theor. **114**, 11–19 (2017)
26. Waldron, K.J.: Hybrid over constrained linkages. J. Mech. **3**(2), 73–78 (1968)
27. Waldron, K.J.: Overconstrained linkages. Environ. Planning B-planning Des. **6**, 393–402 (1979)
28. Wohlhart, K.: Merging two general Goldberg 5R linkages to obtain a new 6R space mechanism. Mech. Mach. Theor. **26**, 659–668 (1991)
29. Wohlhart, K.: On isomeric overconstrained space mechanisms. In: Proceeding of IFToMM 8, pp. 153–158 (1991)

Motion Planning

Trajectory Planner for Type II Singularities Avoidance Based on Output Twist Screws

José L. Pulloquinga[1(✉)], Rafael J. Escarabajal[1], Marina Vallés[1], Ángel Valera[1], and Vicente Mata[2]

[1] Department of Systems Engineering and Automation, Institute of Automation and Industrial Informatics, Universitat Politècnica de València, 46022 Valencia, Spain
{jopulza,raessan2}@doctor.upv.es, {mvalles,giuprog}@isa.upv.es

[2] Department of Mechanical Engineering and Materials, University Institute of Concerted Research of Mechanical and Biomechanical Engineering, Universitat Politècnica de València, 46022 Valencia, Spain
vmata@mcm.upv.es

Abstract. Motion planning of manipulators is a major discipline in robotics that involves the generation of feasible trajectories within the workspace so that neither collisions nor singularities are encountered during its execution. The singularity issue is of paramount importance in parallel robots (PRs) which, despite their capability of performing very accurate movements, suffer from the so-called Type II singularities. These singular configurations occur within the workspace, and a PR loses control when approaching such configurations. In this paper, an algorithm for Type II singularities avoidance in PRs is proposed that allows to i) create a new trajectory from an initial to a final configuration that is non-singular, and ii) adapt an existing trajectory that goes through singular configurations so that the new generated trajectory evades them with minimum modifications of the original path. The new trajectory may serve as a reference position for safe control of a PR, and it is used in an experiment on a real 4-DOF PR meant for knee rehabilitation, with a position controller to follow the trajectory. The results show how the initially unstable behavior when following an original singular trajectory can be fixed after modifying that trajectory with the proposed algorithm.

1 Introduction

Trajectory generation for robotics [1] entails a careful design in which many elements must be considered including the design of smooth motions, where velocity and acceleration profiles must be taken into account, the choice of joint versus operational space satisfying robot kinematic constraints [2] and the obstacles avoidance [3].

O. Altuzarra and A. Kecskémthy (Eds.): ARK 2022, SPAR 24, pp. 445–452, 2022.
https://doi.org/10.1007/978-3-031-08140-8_48

Within the category of kinematic constraints, it is worth talking about singularities [4,5]. They are configurations of the robot in which the number of degrees of freedom (DOFs) change instantaneously. For a serial robot, singularities typically occur when one or more joints reach their limit [6], and are usually easy to detect and tackle. However, for a parallel robot (PR), this issue is more involved. Singularities in PRs were first studied by Gosselin and Angeles [7] who defined two categories for singularities: Type I (or Inverse Kinematic) singularities occur when the Inverse Jacobian matrix is rank deficient and the mobile platform of the PR loses at least one DOF. Type II (or Forward Kinematic) singularity imply the gain of at least one DOF as the determinant of the Forward Jacobian matrix reaches 0. The Type II singularity presents an unexpected dependency between the input parameters despite all the actuators being locked. In this situation, the mobile platform of the PR cannot bear external forces.

The main problem of using the numerical determinant of the Forward Jacobian matrix for estimating the proximity to a Type II singularity is its lack of physical interpretation [8]. Recent approaches have been analyzed to characterize them using Screw Theory as a basis [9]. The studies in [10–12] are based on the idea of defining a transmission index (TI) that gives information about the contribution of the actuators to the force of the end effector. Pulloquinga et al. [13] proposed an angle between two instantaneous screw axes from the Output Twist Screws (OTSs) as a proximity detector of Type II singularities ($\Omega_{i,j}$), and in [14] the $\Omega_{i,j}$ angle was successfully used to release a PR from Type II singular configurations using a vision system to measure the actual pose of the PR. However, that study did not analyze the alternative of avoiding a Type II singularity during the trajectory planning phase.

The contribution of this paper is the design of a planner algorithm which, based on the computation of the $\Omega_{i,j}$, is able to offline generate a reference free of Type II singularities ready to be tracked by a PR, either from an original trajectory or by linearly connecting two different configurations, with minimum modification of the initial trajectory. This planner is very convenient since it exempts the designer of the trajectory from any concerns about singularity issues while defining the trajectory. Another advantage of the algorithm is that no extra instrumentation is required since the computation is performed offline, affecting the reference signal which will remain free of singularities, and thus the kinematic model will stay valid during the process.

Section 2 presents the mathematical foundations in terms of the Type II singularities and its characterization by the angles $\Omega_{i,j}$. In Sect. 3, a detailed description of the developed planning algorithm is exposed. Section 4 describes the architecture employed for the robot and the experiments and the results obtained. The comparison of the performance of the PR before and after including the planning algorithm is also depicted. Finally, the conclusions are exposed in Sect. 5.

2 Mathematical Foundation

2.1 Singularities

In a PR, a set of constraint equations ($\boldsymbol{\phi}$) defines the relationship between the active generalized coordinates or inputs ($\boldsymbol{q_{ind}}$) and the DOFs of the mobile platform or outputs ($\boldsymbol{x}$) [7]. Taking time derivatives of $\boldsymbol{\phi}$, the kinematics behavior of a PR is defined as follows:

$$\boldsymbol{J_I}\dot{\boldsymbol{q}}_{\boldsymbol{ind}} + \boldsymbol{J_D}\dot{\boldsymbol{x}} = \boldsymbol{0} \tag{1}$$

$\boldsymbol{J_I}$ is the Inverse Jacobian matrix and $\boldsymbol{J_D}$ is the Forward Jacobian matrix. For a non-redundant PR both matrices are square matrices ($F \times F$), where F stands for the number of DOFs of the mobile platform.

A Type I singularity occurs when the $\boldsymbol{J_I}$ matrix becomes rank deficient. In this case, the outputs lose at least one DOF. In a Type II singularity, the $\boldsymbol{J_D}$ matrix is rank deficient ($\|\boldsymbol{J_D}\| = 0$) and the mobile platform gains at least one DOF due to an unexpected dependency between the inputs. In this configuration, all actuators are locked ($\dot{\boldsymbol{q}}_{\boldsymbol{ind}} = \boldsymbol{0}$). However, the mobile platform can move ($\dot{\boldsymbol{x}} \neq 0$) if an external action is applied, i.e., control over the PR is lost, which is potentially dangerous for the user or the PR itself. In [15], the two types of singularities are deeply analyzed.

The calculation of $\|\boldsymbol{J_D}\|$ is a simple numerical method to detect Type II singularities. However, the elements of the $\boldsymbol{J_D}$ matrix are not homogeneous in dimensions and $\|\boldsymbol{J_D}\|$ is not able to identify the active elements responsible for the Type II singularity.

2.2 Angle Between Two Output Twist Screws

The motion of the end effector of a PR is produced by the combined action of several actuators, which renders it difficult to identify the individual contribution of each actuator. Based on Screw Theory, Takeda and Funabashi [11] proposed separating the motion of a point of the end effector(\$) in F OTSs $\$_O$.

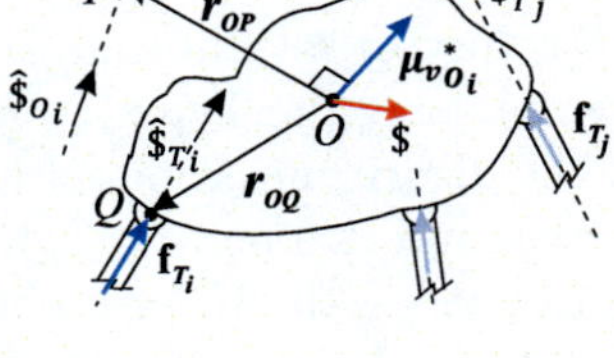

Fig. 1. Motion decomposition for a PR.

The $\$_{O_i}$ is found by solving the reciprocal product ($\circ$) with the Transmission Wrench Screw (TWS) provided to the end effector considering that the all actuators are locked except one (see Fig. 1), i.e., solving the system:

$$\hat{\$}_{O_i} \circ \hat{\$}_{T_j} = 0 \; (i, j = 1, 2, \ldots, F, i \neq j) \tag{2}$$

with

$$\hat{\$}_O = (\boldsymbol{\mu_{wO}}; \boldsymbol{\mu^*_{vO}}) = (\boldsymbol{\mu_{wO}}; h\boldsymbol{\mu_{wO}} + \boldsymbol{r} \times \boldsymbol{\mu_{wO}}) \tag{3}$$

where $\hat{\$}_{T_j}$ is the unitary screw of each TWS of the actuators, $\boldsymbol{\mu_{wO}}$ is the instantaneous screw axis, h is the screw's pitch and $\boldsymbol{r}$ is the minimal distance between the selected point of the end effector and $\boldsymbol{\mu_{wO}}$.

Wang et al. [12], verified that for a singular configuration at least two $\hat{\$}_O$ are linearly dependent, i.e., both $\boldsymbol{\mu_{wO}}$ and $\boldsymbol{\mu^*_{vO}}$ are parallel. Thus, the angle between two $\boldsymbol{\mu_{wO}}$s ($\Omega_{i,j}$) was proposed as a Type II singularity proximity detector in [13], where a singular configuration is verified by the equality of the components $\boldsymbol{\mu^*_{vO}}$. The angle $\Omega_{i,j}$ between $\boldsymbol{\mu_{wO_i}}$ and $\boldsymbol{\mu_{wO_j}}$ is defined as:

$$\Omega_{i,j} = \arccos\left(\boldsymbol{\mu_{wO_i}} \cdot \boldsymbol{\mu_{wO_j}}\right) (i, j = 1, 2, \ldots, F, i \neq j) \tag{4}$$

The Eqt. 4 considers that the constraint wrench screws are always linear independent, i.e., there are no constraint singularities. The index $\Omega_{i,j}$ presents two advantages with respect to the $\|\boldsymbol{J_D}\|$:

- Physical meaning since it is an angular magnitude.
- When the $\Omega_{i,j} = 0$ the indices i and j identify actuators responsible for the singular configuration (the unexpected dependency between the inputs parameters).

The index $\Omega_{i,j}$ has been used to release a 4-DOF PR from a Type II singularity in an online application in [14] but, to the best of the authors' knowledge, no application using $\Omega_{i,j}$ for trajectory generation has been published, so this paper presents a novel algorithm for obtaining non-singular trajectories in the planning stage.

3 Methodology for Trajectory Planner

Considering a discretized trajectory in several samples (N_{ptos}), this study proposes to plan a trajectory by analyzing the Type II singularities in each sample. Fig. 2 shows an overview of the algorithm proposed for one iteration. The trajectory planner first takes a reference trajectory ($\boldsymbol{x_r}$) and calculates the correspondent joint trajectory ($\boldsymbol{q_{ind_r}}$) using the Inverse Kinematic model. Next, the $\|\boldsymbol{J_D}\|$ and the minimum element of $\boldsymbol{v\Omega_r}$ are calculated for each sample from $\boldsymbol{x_r}$, which are named $\|\boldsymbol{J_D}\|_r$ and $min\Omega_r$. The $\boldsymbol{v\Omega_r}$ includes all $\Omega_{i,j}$ produced by the F actuators. If $\|\boldsymbol{J_D}\|_r$ or $min\Omega_r$ are below a predefined threshold, that point is considered to be a singular configuration and the $\boldsymbol{q_{ind_r}}$ is deviated to generate the non-singular trajectory ($\boldsymbol{q}_{\boldsymbol{ind}_{\boldsymbol{d}}}$). Solving the Forward Kinematics problem for $\boldsymbol{q}_{\boldsymbol{ind}_{\boldsymbol{d}}}$, a non-singular reference in configuration space ($\boldsymbol{x}_{\boldsymbol{d}}$) is achieved. In order to ensure that the $\boldsymbol{x}_{\boldsymbol{r}}$ is only modified in the proximity of a Type II singularity, the $\boldsymbol{x}_{\boldsymbol{d}}$ is used as a feedback signal for the next iteration, with $\boldsymbol{x}_{\boldsymbol{d}} = \boldsymbol{x}_{\boldsymbol{r}}$ as initial condition.

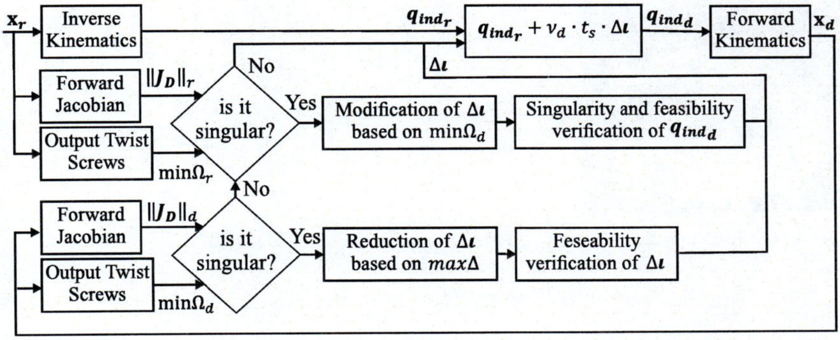

Fig. 2. Block diagram for the trajectory planner algorithm.

In Fig. 2, v_d is the avoidance velocity constant, t_s is the sample time used for the trajectory discretization, and $\boldsymbol{\Delta\iota}$ stands for an integer accumulator for the deviation required in the F actuators. The modification of $\boldsymbol{\Delta\iota}$ is controlled by the minimum element ($min\Omega_d$) of the $\boldsymbol{v\Omega_d}$, that stores the $\Omega_{i,j}$ calculated for $\boldsymbol{x_d}$. Based on the sub-indices of $min\Omega_d$, the corresponding two elements of $\boldsymbol{\Delta\iota}$ are maintained, increase or decrease by one (the actuators stop, go forward or go backwards), and then a verification process is performed to check the non-singular condition of the trajectory and its physical feasibility. While $\boldsymbol{x_d}$ is singular $\boldsymbol{\Delta\iota}$ is different from zero, and after $\boldsymbol{x_d}$ avoids a Type II singularity $\boldsymbol{\Delta\iota}$ returns to zero if and only if $\boldsymbol{x_r}$ is non-singular. In this case, $\boldsymbol{\Delta\iota}$ is reduced by pairs starting from the elements with the maximum deviation ($max\Delta$).

4 Results and Discussion

4.1 Experimentation Setup

The sequence of steps conducted for the experiments comprise i) the generation of an initial trajectory for rehabilitation purposes (more specifically, for hip flexion) where a singular configuration arises halfway, ii) the execution of the planner, implemented in MATLAB, to obtain the non-singular version of the trajectory, and iii) the performance on the actual 4-DOF PR of both initial and adapted trajectory, with a perturbation in terms of external forces to check its behavior.

The 4-DOF PR has two translational movements (x_m, z_m) in the tibiofemoral plane, one rotation (ψ) around the coronal plane and one rotation (θ) around the tibiofemoral plane [16]. The external limbs have a UPS configuration, while the central one has an RPU configuration (see Fig. 3).

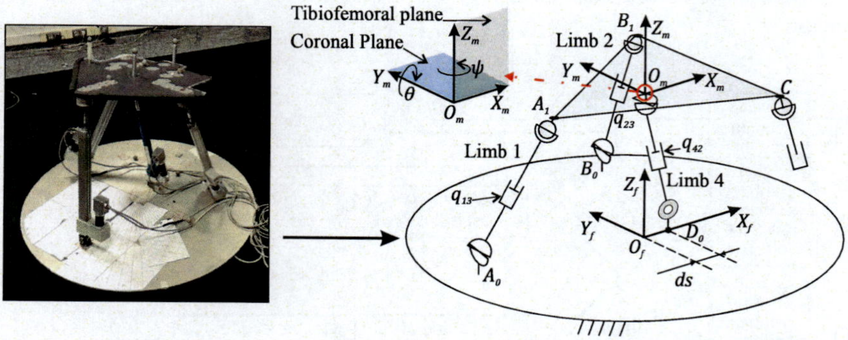

Fig. 3. Mechanical representation of the 3UPS + RPU PR.

The letters R, U, S and P stand for revolute, universal, spherical and prismatic joints, respectively, and "_" indicates the actuated joint. The pose of the mobile platform is defined as $\boldsymbol{x} = [x_m \ z_m \ \theta \ \psi]^T$ and the active joints that control the movement of the robot are prismatic actuators with names $\boldsymbol{q_{ind}} = [q_{13} \ q_{23} \ q_{33} \ q_{42}]^T$. The control system runs a PID controller in Robot Operating System 2 (ROS2) using the C++ programming language, at a rate of 100 Hz.

4.2 Trajectory Planner Performance

The trajectory planner algorithm generates a non-singular trajectory by deviating the trajectory of the linear actuators q_{33} and q_{42} since the limbs 3 and 4 are the ones involved in $min\Omega_r$. Fig. 4a presents the original reference for the actuator q_{33_r} and the non-singular trajectory q_{33_d} generated by the proposed algorithm, where the maximum $\boldsymbol{\Delta_\iota}$ was $[0 \ 0 \ 56 \ -57]^T$. It verifies that the q_{33_d} introduces a smooth deviation with respect to q_{33_r} to avoid a Type II singularity with a maximum value of 5 mm. In Fig. 4b, the actual cartesian position z_m measured with an external vision system (which is described in [14]) is plotted for both trajectories and the their responses differ when an external perturbation at instant $t = 18s$ occurs. The singular trajectory yields to the force and moves unexpectedly, while the adapted trajectory remains stiff.

In Fig. 5a, the values of $\|\boldsymbol{J_D}\|_d$ are compared to $\|\boldsymbol{J_D}\|_r$ to verify the effectiveness of the planner in the task. Fig. 5b depicts the values of $min\Omega_d$ with respect to the original $min\Omega_r$. The results show that both indicators remain above the thresholds $\|\boldsymbol{J_D}\|_{lim}$ and Ω_{lim} for the whole modified trajectory, meaning that the singularity gets evaded.

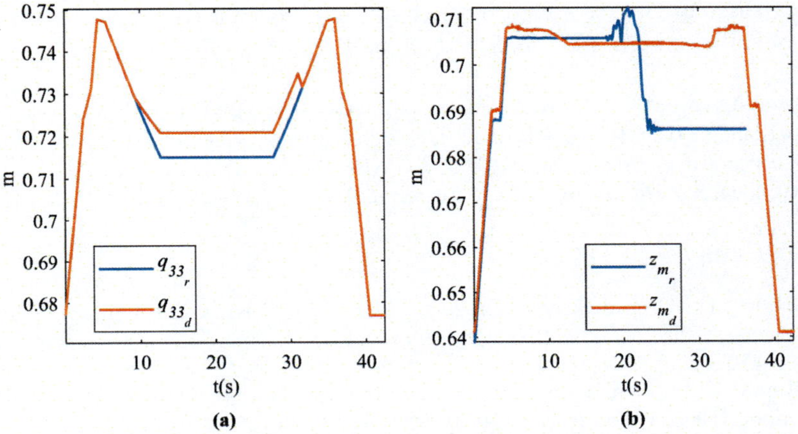

Fig. 4. Tracking of the location in (a) q_{33} and (b) z_m

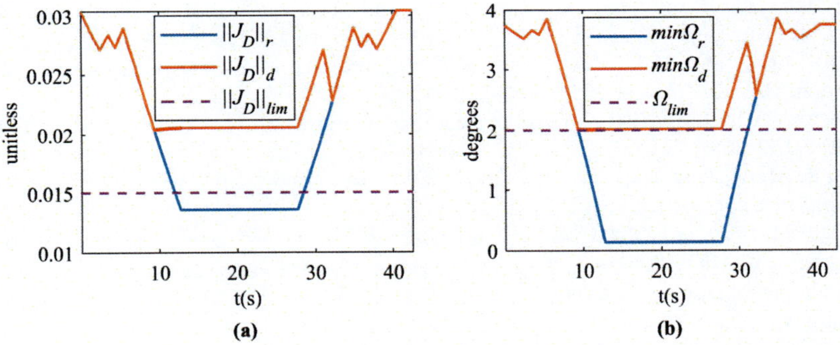

Fig. 5. (a) $\|\boldsymbol{J_D}\|$ and (b) $\boldsymbol{min\Omega}$ for the designed trajectory

5 Conclusion

Based on the idea of OTS and the index $\Omega_{i,j}$, an offline algorithm for non-singular trajectory planning has been designed. An experiment has been developed in which an original trajectory has Type II singularity halfway, and the proposed algorithm detects and generates a new trajectory to avoid this configuration with a maximum deviation of 5 mm. This algorithm represents a safe method to generate trajectories for applications where human-robot interaction happens. The proposed trajectory planner has been defined for a general PR and has been succesfully applied on a knee rehabilitation PR in this study. However, the algorithm requires to be implemented in different PRs to verify the advantages of using the $\Omega_{i,j}$ in trajectory planning. In future work, the algorithm is going to be adapted for online Type II singularities avoidance. The feedback will be replaced by an external sensor that measures the real location of the PR. In addition, since a force controller modifies the position trajectory without considering the

Type II singularities, this algorithm could be used as a complement to reduce the probability to reach a singular configuration.

Acknowledgements. This research was partially funded by Fondo Europeo de Desarrollo Regional (PID2021-125694OB-I00), cofounded by Vicerrectorado de Investigación de la Universitat Politècnica de València (PAID-11-21) and by Programa de Ayudas de Investigación y Desarrollo de la Universitat Politècnica de València (PAID-01-19).

References

1. Carbone, G., Gomez-Bravo, F. (eds.): Motion and Operation Planning of Robotic Systems. MMS, vol. 29. Springer, Cham (2015). https://doi.org/10.1007/978-3-319-14705-5
2. Stilman, M.: Global manipulation planning in robot joint space with task constraints. IEEE Trans. Robot. **26**(3), 576–584 (2010)
3. Khatib, O.: Real-time obstacle avoidance for manipulators and mobile robots. In: Proceedings 1985 IEEE International Conference on Robotics and Automation, vol. 2, pp. 500–505 (1985)
4. Bohigas, O., Manubens, M., Ros, L.: Singularities of Robot Mechanisms. MMS, vol. 41. Springer, Cham (2017). https://doi.org/10.1007/978-3-319-32922-2
5. Pai, D.K., Leu, M.C.: Genericity and singularities of robot manipulators. IEEE Trans. Robot. Autom. **8**(5), 545–559 (1992)
6. Karger, A.: Singularity analysis of serial robot-manipulators. J. Mech. Des. **118**(4), 520–525 (1996)
7. Gosselin, C., Angeles, J.: Singularity analysis of closed-loop kinematic chains. IEEE Trans. Robot. Autom. **6**(3), 281–290 (1990)
8. Voglewede, P.A., Ebert-Uphoff, I.: Measuring "closeness" to singularities for parallel manipulators. In: Proceedings of the IEEE International Conference on Robotics and Automation, vol. 2004, pp. 4539–4544. Institute of Electrical and Electronics Engineers Inc. (2004)
9. Davidson, J.K., Hunt, K.H., Pennock, G.R.: Robots and screw theory: applications of kinematics and statics to robotics. J. Mech. Des. **126**(4), 763 (2004)
10. Yuan, M.S.C., Freudenstein, F., Woo, L.S.: Kinematic analysis of spatial mechanisms by means of screw coordinates Part 2-analysis of spatial mechanisms. J. Manuf. Sci. Eng., Trans. ASME **93**(1), 67–73 (1971)
11. Takeda, Y., Funabashi, H.: Motion transmissibility of in-parallel actuated manipulators. JSME Int. J. Series C **38**(4), 749–755 (1995)
12. Wang, J., Wu, C., Liu, X.J.: Performance evaluation of parallel manipulators: motion/force transmissibility and its index. Mech. Mach. Theor. **45**(10), 1462–1476 (2010)
13. Pulloquinga, J.L., et al.: Experimental analysis of Type II singularities and assembly change points in a 3UPS+RPU parallel robot. Mech. Mach. Theor. **158**, 104242 (2021)
14. Pulloquinga, J.L., et al.: Vision-based hybrid controller to release a 4-DOF parallel robot from a type II singularity. Sensors **21**(12), 4080 (2021)
15. Briot, S., Khalil, W.: Dynamics of Parallel Robots. Springer, From rigid bodies to flexible elements (2015)
16. Araujo-Gómez, P., Mata, V., Díaz-Rodríguez, M., Valera, A., Page, A.: Design and kinematic analysis of a novel 3UPS/RPU parallel kinematic mechanism with 2T2R motion for knee diagnosis and rehabilitation tasks. J. Mech. Robot. **9**(6), 061004 (2017)

Kinematically Adapted Sampling-Based Motion Planning Algorithm for Robotic Manipulators

Amir Shahidi(✉), Thomas Kinzig, Mathias Hüsing, and Burkhard Corves

Institute of Mechanism Theory, Machine Dynamics and Robotics IGMR, RWTH Aachen University, Eilfschornsteinstraße 18, 52062 Aachen, Germany
shahidi@igmr.rwth-aachen.de
https://www.igmr.rwth-aachen.de

Abstract. We introduce modifications to the sampling-based motion planning approach in robotics to adapt the method to the positioning problem of robotic manipulators. The proposed method combines the information from configuration and task spaces of the mechanisms to cluster and subsequently reduce the number of the samples. The clustering process results in construction of a graph, dubbed *kinematic graph* $\mathfrak{G}_k(\mathfrak{V}_k, \mathfrak{E}_k)$. We present a step by step instruction of the construction of the kinematic graph. The kinematic graph introduces interesting advantages to the planning algorithm. For instance, planning using the kinematic graph will sort the possibilities of sudden configuration changes, due to the surjection of forward kinematics function for robotic manipulators, in planning phase out. Moreover, combination of information from configuration and task spaces can be utilized to form the cost and heuristic functions for the heuristic search algorithms, like A*. Furthermore, the clustering and reduction of the number of the samples has direct effect on the *solution depth*, that is, the shortest path found by the search algorithm. This in turn reduces the expense of computation and worst-time complexity of the search algorithm. Finally, the information from the vertices of the kinematic graph can be used for a BOOLEAN collision check of the sampled configuration, without the need of extra calls on the forward kinematics function.

1 Background

Planning feasible motions for robotic systems have been the focus of numerous studies in the robotic community. The majority of planning problems rely on the configuration space ($\mathcal{C}$ for short) of the robot [15,16]. Per definition, the entire configuration of the robot can be represented as a point $\boldsymbol{q} \in \mathcal{C}$. Conventional motion planning algorithms, like optimisation-based planning (e.g., potential fields [7]) and combinatorial planning [11] (e.g., cell decomposition [2]), are conducted on an exact and explicit description of $\mathcal{C}$, which result in computationally expensive and sophisticated mathematical description of the planning problem.

O. Altuzarra and A. Kecskéméthy (Eds.): ARK 2022, SPAR 24, pp. 453–461, 2022.
https://doi.org/10.1007/978-3-031-08140-8_49

This in turn makes the scalability of such algorithms to higher dimension challenging. Moreover, the transformation of obstacles (naturally defined in work space ($\mathcal{W}$ for short)) into $\mathcal{C}$ can be very complicated [20] (the subspace of $\mathcal{C}$ covered with obstacle is known as $\mathcal{C}_{\text{obs}}$) and such algorithms demand a complex collision detection [10].

The expenses of planning in explicit description of $\mathcal{C}$ can be circumvented by sampling-based planning algorithms. In such procedure, the collision detection is reduced to simple BOOLEAN tests on (probabilistic or deterministic) samples from $\mathcal{C}$. The admissible sets of the configurations (i.e., $\boldsymbol{q} \in \mathcal{C}_{\text{free}} \triangleq \mathcal{C} \setminus \mathcal{C}_{\text{obs}}$), that results in connection of initial state of the robot to a final state that satisfies the goal posture conditions, will result in the path τ. (For details on sampling-based motion planning algorithms, the interested readers are referred to [11,14].) As these algorithms rely on samples from $\mathcal{C}$, to evaluate their performance a common evaluation criteria is used, referred to the *completeness* of the algorithm. Complete algorithms report, in finite amount of time, if a solution exist, and return one if there exist one [11]. Only combinatorial algorithms can fulfill the requirements of this definition, as they rely on exact definition of $\mathcal{C}$. Therefore, for algorithms that rely on discrete samples of $\mathcal{C}$ a weaker definition applies, called *resolution completeness*, that is, the algorithm reports the existence of the solution, respectively the solution, depending on the sampling resolution.

The sampling of $\mathcal{C}$ can be performed through *probabilistic* or *deterministic* methods [11,20]. Algorithms relying on the former methods increase the resolution and thus give rise to another concept of completeness, namely *probabilistic completeness*, that is, the probability of finding the solution, if there exists one, approaches to one, as the runtime increases. The algorithms relying on the latter methods can further be divided to *regular* and *irregular* sampling techniques. Due to the discrete character of the finite samples, to evaluate the quality for representation of $\mathcal{C}$, that is, the coverage of $\mathcal{C}$, there is yet another criteria to be considered, namely *denseness*; the closure of the dense samples yield $\mathcal{C}$. Sampling-based algorithms with dense samples are resolution complete.

The samples drawn from the above-mentioned methods result in the construction of a graph $\mathfrak{G}(\mathfrak{V}, \mathfrak{E})$, where $\mathfrak{V}$ represent the vertices of $\mathfrak{G}$ (i.e., the drawn samples), and $\mathfrak{E}$ represents the edges on $\mathfrak{G}$ (i.e., the connectivity information of the samples). Of course special care should be taken in the neighbouring methods and the computation of costs and heuristics. A discrete search algorithm, like Dijkstra's algorithm [4], A* [5] or their modern versions ARA* [13], LPA* [9] or D* [8,21] for deterministic respectively RRT or RDT [12] and PRM [6] for probabilistic methods, on this graph realises the local planning of the algorithm.

As can be derived from the discussion above, the state of research in the field of sampling-based motion planning has focused primarily on $\mathcal{C}$ of the robot. However, the tasks of the robotic systems are defined in task space ($\mathcal{T}$ for short), being the configuration space of the task at hand ($\mathcal{W}$ is generally a subspace of $\mathcal{T}$). Hence, in many robotic applications, it is desireable to find a suitable motion in $\mathcal{T}$. Although sampling-based planning algorithms are of discrete nature, and an explicit transformation of obstacles from $\mathcal{W}$ into $\mathcal{C}$ is not necessary, the discrete transformations from $\mathcal{W}$ into $\mathcal{C}$ shall be followed for performing BOOLEAN collision checks. This in turn proves to be problematic for robotic systems with an open-chain topology, since the forward kinematics function $\mathcal{K} : \mathcal{C} \rightarrow \mathcal{T}$ is surjective. Thus, a pure planning in $\mathcal{T}$ can guarantee neither a collision free motion of the robot (configurational collision in $\mathcal{W}$) nor a feasible motion in consideration of actuator limits (strictly speaking, configuration changes cannot be sorted out).

In this paper, we present the idea of combining the information from both $\mathcal{C}$ and $\mathcal{T}$ of the robot for the construction of a novel graph structure dubbed *kinematic graph* $\mathfrak{G}_k(\mathfrak{V}_k, \mathfrak{E}_k)$, where the vertices $\mathfrak{V}_k$ inherit the information from both $\mathcal{C}$ and $\mathcal{T}$, and edges $\mathfrak{E}_k$ are originated from connectivity information from $\mathcal{C}$. The kinematic graph $\mathfrak{G}_k$ shall be constructed a priori. The case study of this paper focuses on path planning problems for the cases, where $\mathcal{T}$ is a two or three dimensional EUCLIDEAN space, i.e., $\mathbb{R}^2$ or $\mathbb{R}^3$. Strictly speaking, the positioning problem of an articulated robotic arm will be put in focus. This covers a majority of robotic structures and applications, such as planar robotic systems or Scara robot. In a wide range, it is also a common practice in robotics to consider the positioning (the task of the reginal structure, i.e., articulated arm) and orientation (the task of the local structure, i.e., wrist) problems separately, but in conjunction with each other, e.g., for robots that demonstrate decoupled structures [1].

It is worthwhile to mention that there have been recent attempts to utilize $\mathcal{T}$ information in sampling-based planning problem, for instance, to compute the heuristics in EUCLIDEAN space of $\mathcal{T}$ using breadth-first search [3], or to balance exploitation (enhancing existing solutions) and exploration (search for new solutions) [18]. Also a hierarchial path planner based on exact representation of $\mathcal{W}$ for collision avoidance in conjunction with local planner on $\mathcal{C}$ was developed in [17].

2 Kinematic Graph

The idea of the kinematic graph $\mathfrak{G}_k$ is to meet the challenges of sampling-based motion planning for open-chain mechanisms due to the surjection of $\mathcal{K}$ by introducing information from $\mathcal{T}$ directly into $\mathfrak{V}_k$. Therefore both $\mathcal{C}$ and $\mathcal{T}$ of the robot should be sampled. To illustrate the steps to be taken to generate $\mathfrak{G}_k$, let us consider the structure of a two degrees of freedom (dof) planar mechanism (see Fig. 1(a)). The $\mathcal{C}$ sampling is conducted as grid-based sampling (see Fig. 1(b)). These samples are called *nodes* in the following. The $\mathcal{T}$-sampling is realised by convex discretisation of the EUCLIDEAN space of the environment of the mechanism via *voxels.* The voxels are characterised by their *centroid.* Observe that here $\mathcal{T} \equiv \mathcal{W} \subset \mathbb{R}^2$. Also note that for the case of this example, a relatively poor sampling resolution ($\mathcal{C}_{\text{res}} = \frac{\pi}{8}$, $\mathcal{T}_{\text{res}} = 0.3m$) is considered to elaborate the key ideas of $\mathfrak{G}_k$ visually. Then we apply $\mathcal{K} : \mathcal{C} \rightarrow \mathcal{T}$, see Fig. 1(c).

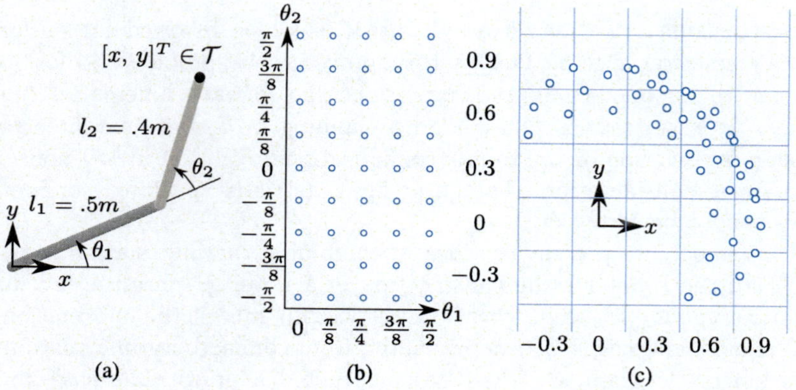

Fig. 1. (a) Structure of the two dof planar mechanism; (b) $\mathcal{C}$ samples (nodes); (c) Task space samples (voxels) and the transformed nodes from $\mathcal{C}$ (via forward kinematics $\mathcal{K}$).

This transformation distributes the nodes in $\mathcal{T}$ in a cluttered fashion. Nevertheless it is notable that the number of samples in $\mathcal{T}$, representing the "approximated" $\mathcal{T}$ of the mechanism (i.e., the occupied voxels, see also Fig. 2), is less than (or in rare cases equal to) the ones representing $\mathcal{C}$. This relation always holds for mechanisms with an open-chain topology. A close investigation of Fig. 2 reveals multiple facts. The basic fact (and probably the least interesting one for experts in robotics) is the surjection of $\mathcal{K}$. Another fact to be drawn from Fig. 2 is that, as stated in Sect. 1, direct planning in $\mathcal{T}$ delivers no guarantee on configurational feasibility of the motion, as the mechanism in "elbow-down configuration" may collide with the $\mathcal{W}$-obstacle, should the planned path have traversed to voxels "a" or "b". Same holds for "elbow-up configuration" when traversing to voxels "c", "d", or "e".

Beside these, this figure reveals the interesting fact that, based on $\mathcal{K}$-mapping, one does not need as many samples for the planning algorithm for mechanisms as there are in $\mathcal{C}$. On the contrary, a group of samples from $\mathcal{C}$ proves to be equivalent considering their transformation to discretised $\mathcal{T}$. This gives rise to the idea of *clustering* the samples in $\mathcal{C}$ based on their "immediate neighbourhood" in $\mathcal{C}$ and their *posture* (here; position) information in $\mathcal{T}$. This algorithm shall be followed throughout the samples from $\mathcal{C}$, and the samples are to be clustered accordingly. These clusters serve as the vertices $\mathfrak{V}_k$ of the kinematic graph. Hence, the vertices $\mathfrak{V}_k$ contain information from both $\mathcal{C}$ and $\mathcal{T}$. For the case of the above two dof planar mechanism, $\mathcal{C}$ clusters are shown in Fig. 3(b). Although these clusters seem to be cluttered in $\mathcal{C}$, their $\mathcal{T}$ representation shall determine their concept well, see Fig. 4(a); *each cluster belongs to merely one voxel in* $\mathcal{T}$.

After clustering is done, the information of clusters, i.e., the vertices $\mathfrak{V}_k$ of the kinematic graph, can be shrunk to the necessary level. A natural choice for abstracting the information from $\mathcal{C}$ is using the average of the samples in the cluster as a representative "moved" sample of $\mathcal{C}$, see Fig. 3(c). The same can

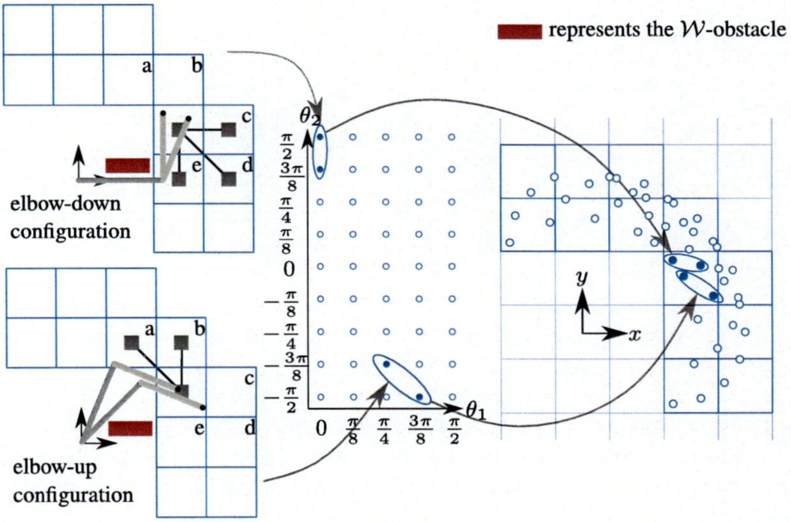

Fig. 2. The voxels in operation space can be reached with different configurations of the mechanism (for instance, in this example: elbow-up and elbow-down).

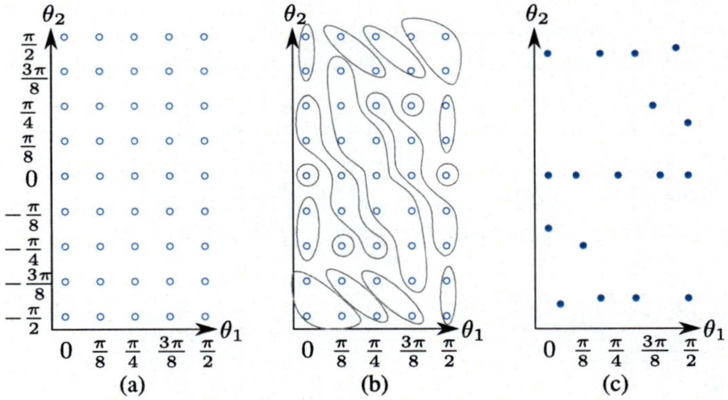

Fig. 3. Representation of vertices $\mathfrak{V}_k$ of the kinematic graph in $\mathcal{C}$; (a) Original deterministic regular samples of $\mathcal{C}$; (b) Clustered samples in $\mathcal{C}$; (c) The $\mathcal{C}$ clusters shall be represented by their average value. These "averaged" samples form the vertices $\mathfrak{V}_k$ of the kinematic graph.

be applied to posture information of the transformed samples into $\mathcal{T}$. However, there are two possibilities; we can either use the average of the postures of the clustered nodes, or identify all the clusters in a voxel by its centroid, see Fig. 4(b). Observe that the kinematic graph abstracts the samples from $\mathcal{C}$ to the clusters of the samples from $\mathcal{C}$ that reach the voxels in $\mathcal{T}$ with different configurations. Thus, each voxel may contain several clusters. The edges $\mathfrak{E}_k$ of the kinematic graph should be formed, based on neighbourhood of the nodes in clusters, i.e.

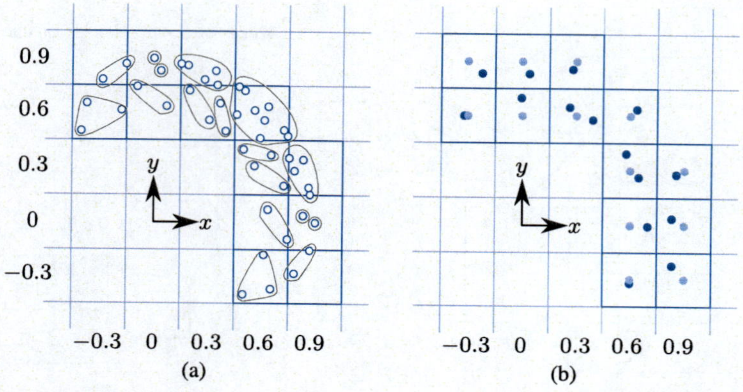

Fig. 4. Representation of vertices $\mathfrak{V}_k$ of the kinematic graph in $\mathcal{T}$; (a) Clustered samples in $\mathcal{T}$; (b) The $\mathcal{T}$ clusters may be represented by their average value (shown in dark blue). Alternatively, the centroids of the voxels can be used as the identifiers for the clusters (show in light blue). However, since a sampled approximation of $\mathcal{T}$ is used, the boundary nodes may not be reachable in the later case. These "averaged" samples form the vertices $\mathfrak{V}_k$ of the kinematic graph. (The nodes might overlap.)

$\mathfrak{V}_k$, in $\mathcal{C}$. In the illustrative example above, the two clusters in the current voxel of the mechanism and their neighbours are shown exemplarily in Fig. 2. Heuristic search algorithm can be applied to explore the kinematic graph $\mathfrak{G}_k$. The worst-time complexity of search algorithms is $O(b^d)$, where d represents the shortest path between start and goal vertex, called *solution depth.* The direct effect of enormous reduction in the number of the samples through clustering (think about it in the limit, i.e., high resolution) should be obvious to the reader.

3 Conclusive Notes

Let us consider two simple examples shown in Figs. 5 and 6. For these examples we consider sampling resolution as follows: $\mathcal{C}_{\text{res}} = 0.00\bar{5}\pi$, $\mathcal{T}_{\text{res}} = 0.03m$. The initial configurations are shown in solid black line. A plain A* search algorithm is used to explore the kinematic graph $\mathfrak{G}_k$. The progress of the configuration during the path is from light blue to dark blue. Since $\mathfrak{V}_k$ contains $\mathcal{C}$ and $\mathcal{T}$ information, it is possible to utilize different combinations of costs and heuristics from both spaces for the local planning. Note that, in the case of the example shown in Fig. 6, it is *not* possible to plan a feasible path using just $\mathcal{T}$ information. The figures demonstrate the differences in computed path, both in the quality of the generated path and the progress of the configuration of the mechanism through the path. Nevertheless optimasl combinations of these costs and heuristics, and also definition of best suited ones based on $\mathcal{T}$ information, is still an open issue left for future research.

When the goal posture of the manipulator is given in $\mathcal{T}$, the probability that the goal posture coincides with a posture information involved in a vertex

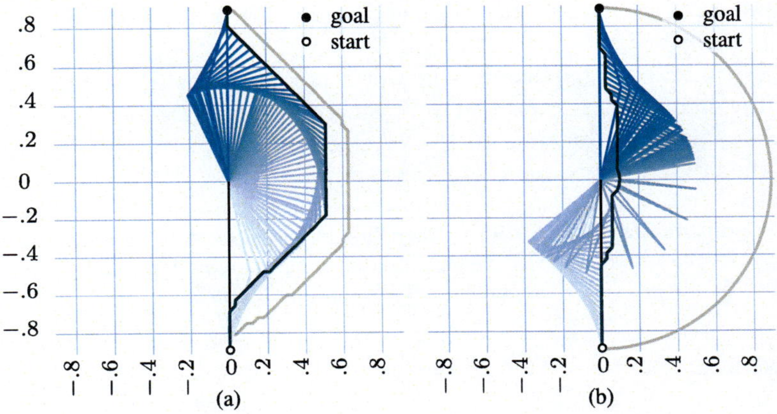

Fig. 5. An exemplary path of the two dof planar mechanism. Costs and heuristics; (a) Petrol: combination of EUCLIDEAN distance in $\mathcal{T}$ and linear manipulability of the mechanism. Gray: combination of distance in $\mathcal{C}$ and linear manipulability of the mechanism; (b) Petrol: EUCLIDEAN distance in $\mathcal{T}$. Gray: distance in $\mathcal{C}$.

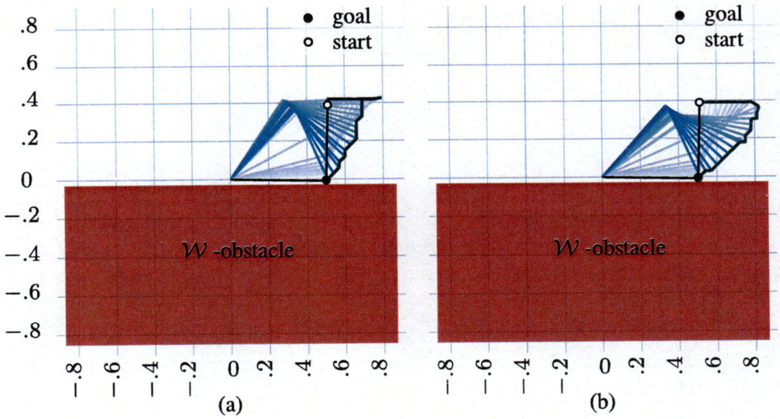

Fig. 6. An exemplary path of the two dof planar mechanism, in presence of operation space obstacle. Costs and heuristics; (a) Combination of EUCLIDEAN distance in $\mathcal{T}$ and linear manipulability of the mechanism; (b) EUCLIDEAN distance in $\mathcal{T}$.

of $\mathfrak{G}_k$ is relatively low. However, the final result can be utilized as a "perfect" initial guess for numerical solution of the "real" goal posture. A better practice is, however, to feed the post process path from the planning algorithm to a kinematic control loop, like the one presented in our previous work [19].

Yet another issue, that belongs to our ongoing investigation, is the optimal mapping of the $\mathcal{W}$-obstacles into $\mathfrak{G}_k$. Nonetheless, the primary results prove, that for single query algorithms, the obstacle avoidance using $\mathfrak{G}_k$ can outperform the conventional obstacle avoidance routines, as the $\mathcal{W}$ information of the configuration of the manipulator is already contained in $\mathfrak{V}_k$, that means, no calls on $\mathcal{K}$ are necessary.

As a final note, an important issue worth mentioning is that, in practice, the goal of the manipulators is defined in $\mathcal{T}$. Surjection of $\mathcal{K}$ leads to the well-known challenge of multiple answers to the inverse problem $\mathcal{K}^{-1}$. This in turn may render the use of $\mathcal{C}$-based heuristics impractical.

Acknowledgements. The Authors would like to thank for the kind support of German Research Foundation DFG (Deutsche Forschungsgemeinschaft) under Germany's Excellence Strategy EXC-2023 Internet of Production 390621612.

References

1. Angeles, J.: Fundamentals of Robotic Mechanical Systems. MES, vol. 124. Springer, Cham (2014). https://doi.org/10.1007/978-3-319-01851-5
2. Brooks, R.A., Lozano-Perez, T.: A subdivision algorithm in configuration space for find path with rotation. IEEE Trans. Syst. Man Cybern. **SMC-15**(2), 224–233 (1985)
3. Cohen, B., Chitta, S., Likhachev, M.: Single-and dual-arm motion planning with heuristic search. Int. J. Robot. Res. **33**(2), 305–320 (2014)
4. Dijkstra, E.W., et al.: A note on two problems in connexion with graphs. Numer. Math. **1**(1), 269–271 (1959)
5. Hart, P.E., Nilsson, N.J., Raphael, B.: A formal basis for the heuristic determination of minimum cost paths. IEEE Trans. Syst. Sci. Cybern. **4**(2), 100–107 (1968)
6. Kavraki, L.E., Svestka, P., Latombe, J.C., Overmars, M.H.: Probabilistic roadmaps for path planning in high-dimensional configuration spaces. IEEE Trans. Robot. Autom. **12**(4), 566–580 (1996)
7. Khatib, O.: Real-time obstacle avoidance for manipulators and mobile robots. In: Autonomous Robot Vehicles, pp. 396–404. Springer, New York (1986). https://doi.org/10.1007/978-1-4613-8997-2_29
8. Koenig, S., Likhachev, M.: D* Lite. Aaai/iaai **15** (2002)
9. Koenig, S., Likhachev, M., Furcy, D.: Lifelong planning A*. Artif. Intell. **155**(1–2), 93–146 (2004)
10. Koren, Y., Borenstein, J., et al.: Potential field methods and their inherent limitations for mobile robot navigation. In: ICRA, vol. 2, pp. 1398–1404 (1991)
11. LaValle, S.M.: Planning Algorithms. Cambridge University Press (2006)
12. LaValle, S.M., et al.: Rapidly-exploring random trees: a new tool for path planning (1998)
13. Likhachev, M., Gordon, G.J., Thrun, S.: ARA*: anytime A* with provable bounds on sub-optimality. In: Advances in Neural Information Processing Systems, pp. 767–774 (2004)
14. Lindemann, S.R., LaValle, S.M.: Current issues in sampling-based motion planning. In: Dario, P., Chatila, R. (eds.) Robotics Research. The Eleventh International Symposium. STAR, vol. 15, pp. 36–54. Springer, Heidelberg (2005). https://doi.org/10.1007/11008941_5
15. Lozano-Perez, T.: A simple motion-planning algorithm for general robot manipulators. IEEE J. Robot. Autom. **3**(3), 224–238 (1987)
16. Lozano-Pérez, T., Wesley, M.A.: An algorithm for planning collision-free paths among polyhedral obstacles. Commun. ACM **22**(10), 560–570 (1979)

17. Mesesan, G., Roa, M.A., Icer, E., Althoff, M.: Hierarchical path planner using workspace decomposition and parallel task-space RRTs. In: 2018 IEEE/RSJ International Conference on Intelligent Robots and Systems (IROS), pp. 1–9. IEEE (2018)
18. Rickert, M., Sieverling, A., Brock, O.: Balancing exploration and exploitation in sampling-based motion planning. IEEE Trans. Robot. **30**(6), 1305–1317 (2014)
19. Shahidi, A., Hüsing, M., Corves, B.: Kinematic control of serial manipulators using Clifford algebra. IFAC-PapersOnLine **53**(2), 9992–9999 (2020). https://doi.org/10.1016/j.ifacol.2020.12.2717. 21st IFAC World Congress
20. Siciliano, B., Khatib, O. (eds.): Springer Handbook of Robotics. Springer, Cham (2016). https://doi.org/10.1007/978-3-319-32552-1
21. Stentz, A., et al.: The focussed D* algorithm for real-time replanning. In: IJCAI, vol. 95, pp. 1652–1659 (1995)

Robotic Analysation Methodology for Multidirectional Additive Manufacturing Process Configuration

Markus Schmitz(✉), Florian Menz, Mathias Hüsing, and Burkhard Corves

Institute of Mechanism Theory, Machine Dynamics and Robotics,
Eilfschornsteinstraße 18, 52062 Aachen, Germany
schmitzm@igmr.rwth-aachen.de

Abstract. Multidirectional additive manufacturing is increasingly used in WAAM processes. Especially in pure object manipulation, the weight and size of the manufactured part during the process is relevant. Depending on the robot and the associated configuration of the welding head pose and end effector, different volumes can be produced. This work presents an efficient calculation methodology, which on the one hand clarifies the influence of the configuration on the printable volume, and on the other hand allows an optimization of e.g. the welding head in the working space of the robot. The results show a pronounced relevance of a reasonable configuration especially for multidirectional additive manufacturing under the constraints of the WAAM process. Furthermore, an optimal welding head pose for the kuka kr6 could be found.

1 Introduction to Multidirectional Additive Manufacturing

Multidirectional Additive Manufacturing (MDAM) is increasingly used in Wire-Arc Additive Manufacturing (WAAM) processes. The core of MDAM is pure object manipulation. The robot moves an object below a welding/printing head. If MDAM is to be extended to the application in WAAM processes by means of object manipulation, this is accompanied by the loss of one degree of freedom around the vertical axis of the welding head (compare with 5D printing [1]). This is due to the following reasons: A wire must be fed mainly from the front, or at least from a defined direction with respect to the direction of movement. Sensor technology used for monitoring and controlling the process should continuously analyse the weld seam and must be constantly aligned with the welding process.

The resulting challenges and algorithms for path and trajectory planning have already been summarized by Schmitz et al. [2] and Parmer et al. [3].

In addition to the MDAM pre-processing, the arrangement of the robot, the welding head and the substrate platform has a significant influence on the welding result. Depending on the configuration of these elements, this can have an influence on the size, shape and quality of the future object. In the following, a

O. Altuzarra and A. Kecskémethy (Eds.): ARK 2022, SPAR 24, pp. 462–469, 2022.
https://doi.org/10.1007/978-3-031-08140-8_50

method is presented to analyse the so-called printable volume, which can theoretically be printed with a fixed configuration. The results can be used to optimize the poses of components depending on process-specific requirements.

2 Configuration Analysation of MDAM Processes

The execution of the MDAM WAAM process requires a 6 DoF robot, a welding head and, as an interface, a substrate plate (end effector of the robot). In this respect, the selection of a robot, the positioning of the welding head in the workspace of the robot and the positioning, size and orientation of the end effector on the flange of the robot must be answered. Together with the desired component part and the manufacturing process complex influencing relationships result.

The focus of the investigations from a robotic point of view is the printable volume, which is derived for a defined additive manufacturing process and configuration of the robot, the substrate plate and the welding head. On the one hand, for a specific hardware configuration and a defined welding process, it can be determined, whether the desired components can be produced or not. On the other hand, the hardware can also be optimally selected and configured based on the component types to be produced in a specific manufacturing process.

The variance of the expected results for analysis or optimization of the printable volume is large due to the number of possible influencing variables. This work is primarily intended to enable and optimize the WAAM process for the MDAM. Therefore, in the following, the optimum welding head pose and the resulting printable volume will be calculated and analysed for a parameter set relevant in the WAAM process (five possible building directions, material: Ti6A14V, Kuka kr6 (payload 6 kg)). Subsequently, the influencing variables are to be varied, starting from the WAAM parameter set, in order to analyse the respective influences on the printable volume and the associated welding head pose.

Starting from a fixed parameter set, the pose of the welding head can be varied. Different printable volumes (in size and quality) can be expected for various poses. The analysis methodology is generally based on standard workspace analyses commonly used in robotics. Unlike this, the method used in this work discretizes a space in the object coordinate system above a substrate plate attached to the robot end effector for a given welding head. A printable volume can be represented as a composite of voxel elements for specific manufacturing scenarios. Each volume element has a fixed position ($\mathbf{ve}_{x,y,z}$) in the object coordinate system, but can be approached via different orientations (a, o). A volume element with a defined build orientation and welding direction is described by a coordinate system ($VE_{ve,o,a}$).

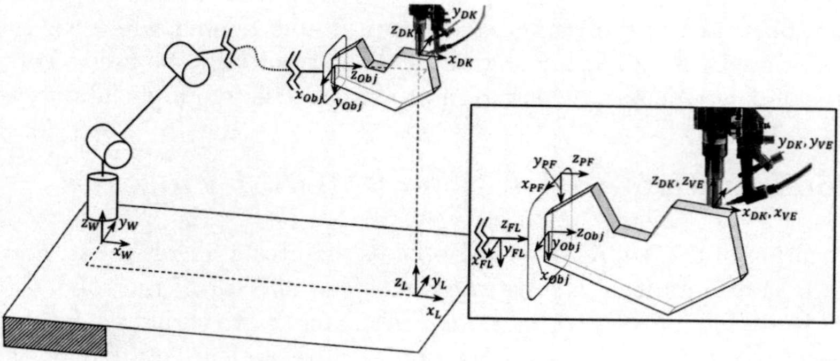

Fig. 1. Introduced coordinate systems and their relationships to the analysis of the printable volume.

The entire setup is described by the following coordinate systems: welding head DK, world W, object Obj, flange of the robot FL and substrate plate PF. The projection of DK into the X-Y plane of W is denoted by L (cf. Fig. 1). The assumption of the downhand position for the welding head is defined: $\mathbf{e}_{z,DK} = \mathbf{e}_{z,W} \Rightarrow \mathbf{e}_{z,L} = \mathbf{e}_{z,W}$. To print a volume element, approached in a defined orientation, the coordinate systems DK and $VE_{ve,o,a}$ must match (necessary condition). ${}^{Obj}\mathbf{T}_{VE_{ve,a,o}}$ describes the transformation from the coordinate system of the volume element to the object coordinate system:

$$ {}^{Obj}\mathbf{T}_{VE_{ve,o,a}} = \begin{bmatrix} \mathbf{I} & {}^{Obj}\mathbf{r}_{VE_{ve}O,ObjO} \\ \mathbf{0}^T & 1 \end{bmatrix} \begin{bmatrix} {}^{VE_{ve}}\mathbf{R}_{VE_{ve,o}} & \mathbf{0} \\ \mathbf{0}^T & 1 \end{bmatrix} \begin{bmatrix} {}^{VE_{ve,o}}\mathbf{R}_{VE_{ve,o,a}} & \mathbf{0} \\ \mathbf{0}^T & 1 \end{bmatrix} \tag{1} $$

with

$$ {}^{Obj}\mathbf{r}_{VE_{ve}O,ObjO} = \begin{bmatrix} ve_x \; ve_y \; ve_z \end{bmatrix} \tag{2} $$

${}^{VE_{ve,o}}\mathbf{R}_{VE_{ve,o,a}}$ describes the rotation defined by the orientation of the volume element around the axis of the welding head (influenced and varied in the process by maintaining the correct wire feed direction). ${}^{VE_{ve}}\mathbf{R}_{VE_{ve,o}}$ describes the rotation of the object coordinate system to a defined build orientation. Of interest for the analysis is the calculation of the inverse kinematics of the robot. For this purpose, the transformation between world coordinate system and object coordinate system ${}^{W}\mathbf{T}_{Obj}$ must be determined:

$$ {}^{W}\mathbf{T}_{Obj}(ve, o, a) = {}^{W}\mathbf{T}_{DK}\, {}^{DK}\mathbf{T}_{VE_{ve,o,a}}\, {}^{VE_{ve,o,a}}\mathbf{T}_{Obj} = {}^{W}\mathbf{T}_{DK}\, {}^{Obj}\mathbf{T}^{-1}_{VE_{ve,o,a}} \tag{3} $$

While the welding head is fixed in the robot's workspace and ${}^{W}\mathbf{T}_{DK}$ is thus constant for all volume elements, a unique transformation ${}^{Obj}\mathbf{T}_{VE_{ve,o,a}}$ results for each pose of the object coordinate system. Consequently, for a workspace analysis with respect to the optimal welding head position, all possible volume elements and the associated build orientation/welding direction must be analysed for each selected welding head position.

With a discretization of 0.05 m of a possible printable volume, 5 build orientations and 8 welding directions, which are determined in a 1 m^3 cube above the defined substrate plate, 370 440 analysis runs must be carried out:

$$n_1 = (\frac{l}{d} + 1)^3 oa = (\frac{1m}{0.05m} + 1)^3 \cdot 5 \cdot 8 = 370440 \tag{4}$$

The symmetry of the workspace of a 6-axis industrial robot is used to search exclusively in one plane (here $e_{x,W}$-$e_{z,W}$-plane, y-component $= const$). In addition to that, the calculated pose of the robot flange can be reused when the welding head is varied along its z-axis, though the new meaning in the object coordinate system must be taken into account. If you define this situation mathematically, you can determine all unique poses and calculate only these. This reduces the computational effort to a fraction of what would have been needed if each weld head height along this axis had been calculated individually. That axis is represented by the z axis of the L coordinate system (cf. Fig. 1). In our specific case we consider a 1.2 m long displacement span, the discretization used here must be identical to the previous one. Instead of 793 800 ($25 \cdot n_1$) analyses we only have to calculate the following number:

$$n_{25} = n_1 + (\frac{l}{d} + 1)^2 oa \frac{s}{d} = n_1 + (\frac{1m}{0.05m} + 1)^2 \cdot 5 \cdot 8 \cdot \frac{1.2m}{0.05m} = 793800 \tag{5}$$

To take advantage of this, we need to establish a relationship between all these individual setups. Since the L coordinate system is the same for all of them, we use this to gain further understanding.

Two different transformation sets for ${}^{L}\mathbf{T}_{Obj}(ve, a, o)$ can be established:

$${}^{L}\mathbf{T}_{Obj}(ve, a, o) = \begin{bmatrix} \mathbf{I} & \begin{matrix} 0 \\ 0 \\ {}^{L}z_{DK_O,{}^{L}O} \end{matrix} \\ \mathbf{0}^T & 1 \end{bmatrix} {}^{VE_{ve,o,a}}\mathbf{T}_{Obj} \tag{6}$$

$${}^{L}\mathbf{T}_{Obj} = \begin{bmatrix} \mathbf{I} & \begin{matrix} {}^{L}x_{Obj_O,{}^{L}O} \\ {}^{L}y_{Obj_O,{}^{L}O} \\ {}^{L}z_{Obj_O,{}^{L}O} \end{matrix} \\ \mathbf{0}^T & 1 \end{bmatrix} \begin{bmatrix} {}^{VE_{ve,o}}\mathbf{R}^{-1}_{VE_{ve,o,a}} & 0 \\ \mathbf{0}^T & 1 \end{bmatrix} \begin{bmatrix} {}^{VE_{ve}}\mathbf{R}^{-1}_{VE_{ve,o}} & 0 \\ \mathbf{0}^T & 1 \end{bmatrix} \tag{7}$$

Since the coordinate system L has the same orientation as the coordinate system DK, ${}^{VE_{ve,o}}\mathbf{R}^{-1}_{VE_{ve,o,a}}$ and ${}^{VE_{ve}}\mathbf{R}^{-1}_{VE_{ve,o}}$ can be used here equally for rotation. By case-specific transformation of this formula, all unique poses can now be determined, calculated, analysed and later transferred for a specific welding head height.

For a volume element (position is defined in the object coordinate system: ${}^{Obj}\mathbf{r}_{VE_{ve_O},Obj_O}$) with defined build orientation, welding direction and welding head height, the information of the point ${}^{L}\mathbf{r}_{Obj_O,{}^{L}O}$ can be assigned directly. Keeping a specific build orientation and weld direction, it is possible to assign the displacement calculated for this volume element to another volume element within a different welding head height. Therefore, many calculations would be redundant if we calculate each welding head height individually. Due to this fact, everything will first be combined and transferred to the L coordinate system, calculated there, and then transferred back individually to the object coordinate system for the desired printhead height. In 2.5D printing, for example, a volume element v^* is part of the printable volume if at least one element ve^* exists which is reachable by the robot and furthermore does not exceed the limits of the joint velocities for characteristic printing motions. To assign volume elements to the printable volume or further to represent and evaluate the quality, different tools shall be used. Reachability, manipulability and compliance with joint limits can be analysed for all joint angle configurations. The pure reachability of the necessary poses is essential for the printing process. In addition, joint velocities must be maintained. Therefore, in addition to the analysis of the general manipulability (based on the condition number to the jacobian matrix), special movements are analysed. Corresponding analysis methods are known in the field of robotics under inverse differential kinematics. This establishes a relationship between the velocities of the end effector and the joint velocities required for it via the inverse of the geometric Jacobian matrix (**J**). Since 6 axis industrial robots are predominantly used for MDAM, the geometric Jacobian matrix has full rank and can therefore be inverted. For a first analysis, the determinant of the geometric Jacobian matrix is calculated for each of the poses. If the determinant becomes very small, then this results in comparatively high joint velocities at the specific pose. High velocities can be expected especially in areas of singularity. Thus, this analysis is accompanied by a singularity analysis at the same time. Figure 2 shows this analysis for three manufacturing types using the Kuka kr6.

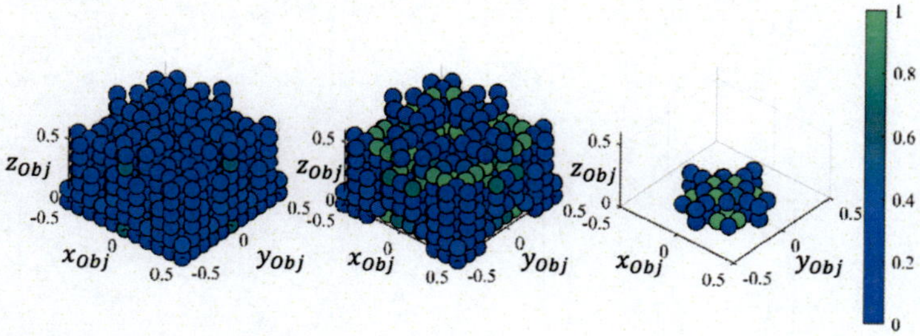

Fig. 2. Plot of printable volume and associated average $det(\mathbf{J})$ per volume element (left: 2.5D printing, center: 5D printing, and right: WAAM using MDAM (6D))

3 Results

Based on the WAAM parameter set, the position of the welding head relative to the robot should be found where the largest printable volume can be produced. The requirements for such a printable volume are again influenced by the selected parameter set. After initial results are available for the described parameter set, all parameters shall be systematically varied. Parameter variation and the associated different quality criteria for a printable volume are manifold and, moreover, do not make sense in every theoretical combination. However, it is possible to combine parameter sets according to the application.

MDAM using WAAM with its eccentric wire feeding restricts the printable volume due to the necessary accessibility of all volume elements in several orientations compared to 5D as well as standard 2.5D printing. While *2.5D printing* must reach at least one configuration along one (initial) build direction to build the part, *5D printing* must reach at least one orientation for each build direction. Furthermore, in *6D printing*, for each volume element min. 60% (resulting from the characteristics of zigzag infills (cf. Schmitz et al. [2]) as well as a possible grinding (cf. Schmitz et al. [2])) of all orientations around the vertical axis of the welding head must be achievable in order to fill a volume in a meaningful way, including all orientation changes of approx. 180°.

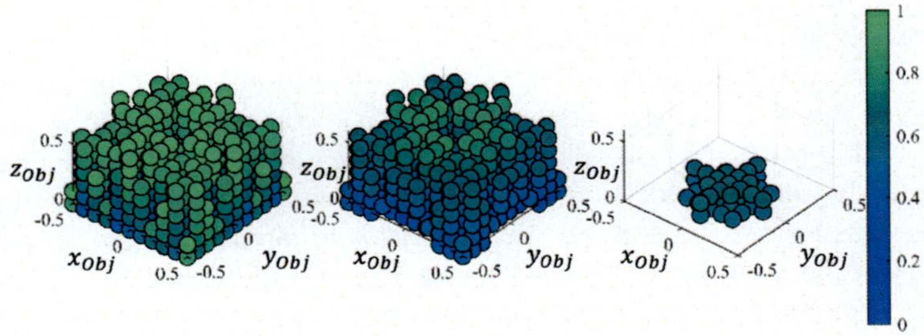

Fig. 3. Average proximity to the joint angle limits of the robot (left: 2.5D printing, center: 5D printing, and right: WAAM using MDAM (6D))

While for 2.5D printing and 5D printing all velocities in the welding plane were calculated for all usable combinations of build orientations and welding directions, for 6D printing only a single direction was calculated for the translational velocity. In addition, for 6D printing, a rotation of the printing platform around a center of rotation of a 180° turn was considered [2]. The resulting joint velocities were set relative to the respective joint velocity limits. The colored representation indicates the respective mean value of all calculated values for the volume elements. Volume elements whose joint velocities require exceeding the limit values were removed from the printable volume. It can be summarized that compared to 2.5D printing, 5D printing can generate more volumes that maintain a greater distance from the joint angle limits. This fact can be exploited in process preparation depending on the part. However, this requires adapted slicing. Within the small printable volume of 6D printing, it is possible to keep distance from the joint angle limits (see Fig. 3) and the joint velocity limits (see Fig. 4). 2.5D and 5D printing are always far from the joint velocity

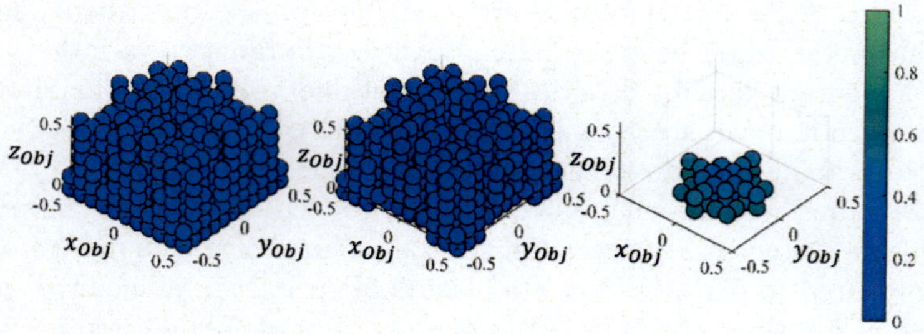

Fig. 4. Average proximity to the joint velocity limits of the robot (left: 2.5D printing, center: 5D printing, and right: WAAM using MDAM (6D))

limits. The character of pure translational motions along the path does not pose major challenges to the robot.

If the specific *shape and size of a component* are known, the quality of the printable volume can be actively influenced if the components are fixed at different locations on the substrate plate. As already shown in Fig. 4, the printable volumes are subject to a symmetrical structure. Components should therefore be positioned in the center of the robot flange. As the component volume increases, the quality of the robot manipulability in the edge areas decreases. This effect is consequently amplified for asymmetric components of the same volume. The influences investigated here relate to a fixed welding head pose. However, as Fig. 5 shows, the welding head pose has a relevant influence on the possible shape and volume of the printable components.

Furthermore, it should be made clear that *the robot kinematics and dynamics* in particular affect the printable volume. The type of robot, kinematic differences in the link lengths and dynamic properties of the drives, lead to printable volumes with different volumes and shapes. The analysation shows, that while serial 6-axis industrial robots from Kuka with different payloads and kinematics obviously differ in printable volume, the characteristic cube shape of the Omnifire from Heidelberger Druck [4] stands out. The Paragrip enables the printable volume with the lowest volume.

A key task of this developed analysis tool is the *optimal positioning of the welding head in the working space of the robot*. At this point, the effect on the volume and the quality of the printable volume is to be visualized and clarified on the basis of different welding head poses. For this purpose, characteristic poses and the corresponding printable volumes are analysed in Fig. 5. The printable volumes for three welding head poses in the 6D scenario with the Kuka kr6 and a fixed wire feed direction are shown. If only the position of the welding head in the working space of the robot is varied, this has an influence on the possible shape and volume of the printable volume. Figure 5 shows printable volumes with similar volumes but different shapes. The optimal welding head position with respect to the size of the printable volume (${}^{W}\mathbf{r}_{DK_O,W_O} = \left[0.7\ 0\ 0.6\right]$) calculated here enables the production of a volume of about 30 dm^3. Although this position is generally optimal, it may be that a specific part geometry could benefit from a different position, due to the differences in the shape of the printable volumes mentioned above.

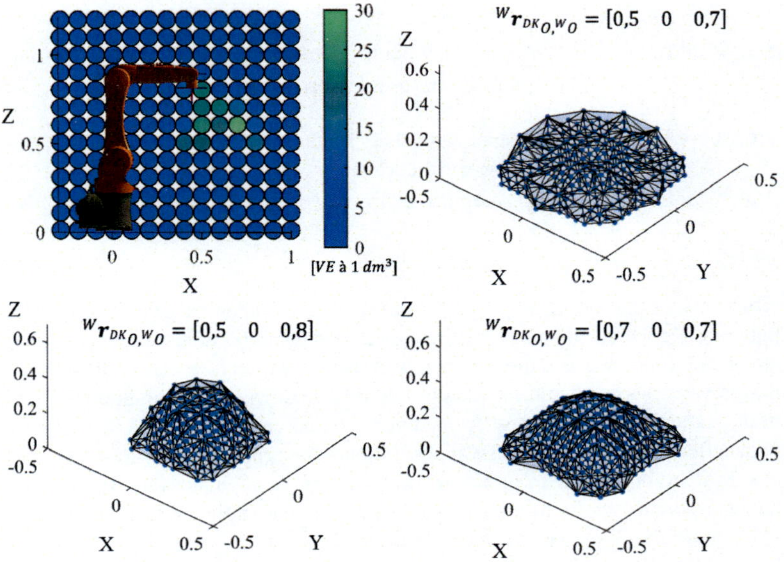

Fig. 5. Positioning of the welding head in the workspace of the Kuka kr6 and exemplary printable volumes

References

1. Corves, B., et al.: Robotik 4.0. In: Frenz, W. (ed.) Handbuch Industrie 4.0: Recht, Technik, Gesellschaft, pp. 569–589. Springer, Heidelberg (2020). https://doi.org/10.1007/978-3-662-58474-3_29
2. Schmitz, M., Wiartalla, J., Gelfgren, M., Mann, S., Corves, B., Hüsing, M.: A robot-centered path-planning algorithm for multidirectional additive manufacturing for WAAM processes and pure object manipulation. Appl. Sci. **11** (2021) https://doi.org/10.3390/app11135759
3. Parmar, K., et al.: Development of a multidirectional wire arc additive manufacturing (WAAM) process with pure object manipulation: process introduction and first prototypes. J. Manuf. Mater. Process. (2021). https://doi.org/10.3390/jmmp5040134
4. Niggemann, H.: Entwicklung des kinematischen Systems einer Digitaldruckmaschine zur Dekoration beliebig gekrümmter Oberflächen. Bewegungstechnik (2016). https://doi.org/10.51202/9783181022863-3
5. Riedel, M.: Flexible Bauteilhandhabung auf Basis einer rekonfigurierbaren parallelkinematischen Struktur. Aachen (2014)

An Artificial Potential Field Algorithm for Path Planning of Redundant Manipulators Based on Navigation Functions

Iman Soodmand[1(✉)], Maeruan Kebbach[1], Sven Herrmann[1], Rainer Bader[1], and Christoph Woernle[2]

[1] Department of Orthopedics, Rostock University Medical Center, 18057 Rostock, Germany
Iman.soodmand@med.uni-rostock.de
[2] Chair of Technical Mechanics/Dynamics, University of Rostock, 18051 Rostock, Germany

Abstract. In this work, we propose a novel artificial potential field off-line path planning algorithm for robot manipulators. The potential field is defined using navigation functions, and the parameters of the navigation function are defined as design variables of an optimization problem through which the minimum length of the path of control points on the manipulator links are achieved and the planned path is free of local minima. The applicability of the proposed algorithm is evaluated by applying it on a 5 degrees-of-freedom robot manipulator representing the human upper extremity to plan the so-called arm abduction and flexion motions in the presence of known workspace obstacles.

Keywords: Path planning · Artificial potential field · Navigation function · Manipulator · Human upper extremity

1 Introduction

Recently, artificial potential field (APF) algorithms have gained attention due to their mathematical simplicity and efficiency for global path planning of kinematically redundant robot manipulators. Using APF algorithms, one potential field pulls the robot toward its goal position by applying artificial attractive forces, while several potential fields repel the robot from workspace space obstacles by applying artificial repulsive forces on the robot. Gradient of the superposition potential of these fields guides the robot from high to low potential field [1]. However, the significant problem of this method is trapping in local minima where the sum of attractive and repulsive forces is equal to zero, and the goal is not reached.

Several methods are proposed to solve the issue of the local minima [2]. According to the aim of the planning, each method has different advantages and disadvantages. For example, using the local minima escape methods, the robot first moves toward the goal and when it traps in a local minimum it tries to escape. This strategy does not require prior information about the environment. However, it would generate a long path. The local minima removal methods on the other hand, e.g. navigation function (NF), are a

O. Altuzarra and A. Kecskéthy (Eds.): ARK 2022, SPAR 24, pp. 470–477, 2022.
https://doi.org/10.1007/978-3-031-08140-8_51

particular class of APFs that have the advantage of generating more optimal and smoother paths but on the cost of requiring global information about the environment prior to the path planning [3, 4].

The present study proposes a novel APF algorithm for off-line optimal path planning of redundant manipulators based on NFs defined on the robot configuration space. The final aim is to apply the proposed algorithm on a human upper extremity to plan human-like motions. In the future, it is planned to integrate this method into the synthesizing algorithms [5] to predict upper extremity goal-oriented motions in the presence of obstacles. Therefore, the planner should have a vision and information about the position of the workspace obstacles to avoid them before reaching them, similar to the way that the human central nervous system controls the upper extremity. For this aim, navigation function method has the advantage of being a global planner. In addition, similar to the other APF methods, it is based on artificial forces that can be integrated into the equations of motion of the (bio)mechanical systems.

2 Navigation Function for Path Planning Algorithm

In contrast to classical APF method [1], NF designs a global potential field in such a way that it includes information about all potential fields in a single function [3, 4]. For the sake of simplicity, here the equations are first formulated for a point-like robot, moving in a plane and then expanded for robot manipulators in 3D space. The position of the point-like robot (or later a control point on a robot manipulator) is $\mathbf{r}$, and the goal position which the robot searches for is $\mathbf{r}_\mathrm{g}$. The APF based on the NF is then defined as

$$\varphi(k, \mathbf{r}) = \frac{\left\|\mathbf{r} - \mathbf{r}_\mathrm{g}\right\|^2}{\left[\left\|\mathbf{r} - \mathbf{r}_\mathrm{g}\right\|^{2k} + \beta(\mathbf{r})\right]^{1/k}} \tag{1}$$

For the point-like robot, an exemplary 2D potential field $\varphi(k, \mathbf{r})$ in the x-y motion plane is generated (Fig. 1). The admissible workspace is a circle in the motion plane with center $\mathbf{r}_0$ and radius ρ_0. There are $n = 3$ circular obstacles in the workspace with centers $\mathbf{r}_\mathrm{j}$ and radii ρ_j. The function $\beta(\mathbf{r})$ includes the effect of the workspace boundary and the obstacles on the APF φ by

$$\beta(\mathbf{r}) = \prod_{j=0}^{\mathrm{n}} \beta_j(\mathbf{r}) \quad \text{with} \quad \beta_j(\mathbf{r}) = \begin{cases} -\|\mathbf{r} - \mathbf{r}_0\|^2 + \boldsymbol{\rho}_0^2 & \text{for } j = 0 \\ \left\|\mathbf{r} - \mathbf{r}_\mathrm{j}\right\|^2 - \boldsymbol{\rho}_j^2 & \text{for } j = 1, \ldots, n \end{cases} \tag{2}$$

The functions $\beta_j(\mathbf{r})$ are negative for forbidden robot positions $\mathbf{r}$, either outside the feasible workspace or inside the obstacles. The factor k in Eq. (1) directly influences the shape of the potential field $\varphi(k, \mathbf{r})$ as exemplarily shown in Fig. 1.

The factor k is defined by trial and error to be in the range of 2 and 10 [4]. When it has smaller values, the robot does not get close enough to the obstacles and generates a longer path. When it has larger values, the robot may overlook the obstacles and just pass over and collide with them. For example, for $k = 4$ obstacle 1 is overlooked (Fig. 1c) and

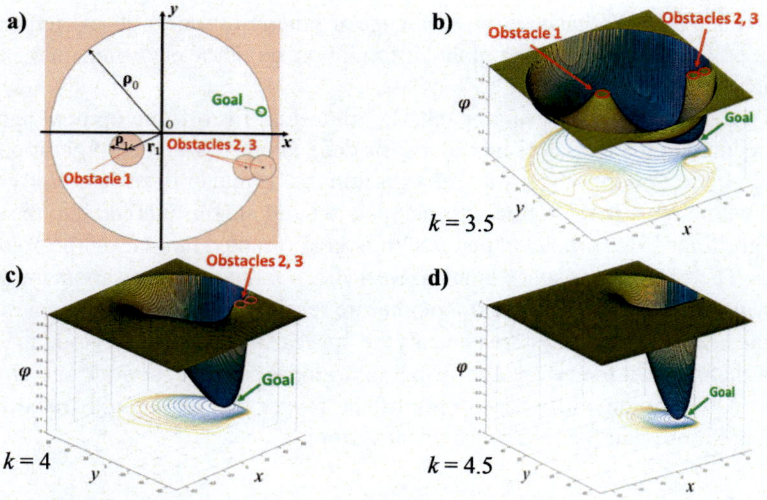

Fig. 1. Effect of the factor k on the shape of the potential field $\varphi(k, \mathbf{r})$ for a point-like robot. **a)** Workspace with 3 circular obstacles. **b–d)** Potential field for 3 different values of k

for $k = 4.5$ all obstacles are overlooked by the robot (Fig. 1d). Therefore, this factor has to be chosen precisely to generate an optimal path, free of collision and local minima.

The potential force acting on the robot to guide it toward its goal position is the negative gradient of the NF

$$\mathbf{f}(k, \mathbf{r}) = -\nabla\varphi(k, \mathbf{r}) \tag{3}$$

with

$$\nabla\varphi(k, \mathbf{r}) = \frac{2(\mathbf{r} - \mathbf{r}_g) - \|\mathbf{r} - \mathbf{r}_g\|^2\left[2k(\mathbf{r} - \mathbf{r}_g)\|\mathbf{r} - \mathbf{r}_g\|^{2(k-1)} + \nabla\beta\right]\left[\|\mathbf{r} - \mathbf{r}_g\|^{2k} + \beta\right]^{-1}\frac{1}{k}}{\left[\|\mathbf{r} - \mathbf{r}_g\|^{2k} + \beta\right]^{1/k}}, \tag{4}$$

$$\nabla\beta(\mathbf{r}) = \sum_{j=0}^{n}\left(\nabla\beta_j(\mathbf{r})\prod_{p=0,p\neq j}^{\mathrm{n}}\beta_p(\mathbf{r})\right), \nabla\beta_j(\mathbf{r}) = \begin{cases} -2(\mathbf{r} - \mathbf{r}_j) & \text{for} \quad j = 0 \\ 2(\mathbf{r} - \mathbf{r}_j) & \text{for} \quad j = 1, \ldots, n. \end{cases} \tag{5}$$

If k is chosen correctly, φ is a NF that generates an optimum (shortest) path without local minima for point robots.

For multi-segment robot manipulators, control points are defined at the end of each manipulator link (Fig. 2), and each control point is treated as the point in the introductory example. Thus, it has its own goal position $\mathbf{r}_{g,i}$ and its own potential field $\varphi_i(k_i, \mathbf{r}_i)$ with an individual factor k_i. According to Eq. (3) the negative gradient of each field generates a navigator force $\mathbf{f}_i(\mathbf{r}_i)$ on the corresponding control point. These navigator force vectors are transformed into the artificial joint torques $\boldsymbol{\tau} = [\tau_1 \ldots \tau_N]^{\mathrm{T}}$ by means of the transposed Jacobian matrix $\boldsymbol{J}_i^{\mathrm{T}}$ corresponding to the location of the control point

$\mathbf{r}_i$ on the robot links. With the number m of control points being equal to the number of the manipulator links it holds

$$\boldsymbol{\tau}(k_i, \mathbf{r}_i) = \sum_{i=1}^{m} \boldsymbol{J}_i^{\mathrm{T}} \mathbf{f}_i(k_i, \mathbf{r}_i). \tag{6}$$

The navigator joint torque vectors $\boldsymbol{\tau}$ are used within a gradient descent algorithm to guide the robot towards its final configuration in M iteration steps [6],

$$\boldsymbol{\theta}^{s+1} = \boldsymbol{\theta}^{s} + \alpha \frac{\boldsymbol{\tau}(\boldsymbol{\theta}^{s})}{\|\boldsymbol{\tau}(\boldsymbol{\theta}^{s})\|}, s = 1, \ldots, M. \tag{7}$$

The factor α determines the step size which can be constant or variable, and $\boldsymbol{\theta}^{s} = \left[\theta_1^s \ldots \theta_N^s\right]^{\mathrm{T}}$ is the vector of the manipulator joint angles in the s-th iteration step. The corresponding control point positions $\mathbf{r}_i^s(\boldsymbol{\theta}^s)$ are obtained by forward kinematic calculation. Using Eq. (7), the joint angles for all solving steps between initial and final configuration are calculated. It has to be noted that the sequence of waypoints $\boldsymbol{\theta}^s$ is time-independent. Defining a motion trajectory along the waypoints is a subsequent step that is not treated here.

Using the aforementioned approach, the manipulator is considered as a set of control points whose positions are constrained by the kinematic chain. However, other points on the robot links may collide with the obstacles. Thus, repulsive forces based on FIRAS function [1] are applied from any workspace obstacle to floating points on the manipulator links which are closest to the obstacles (Fig. 2).

The iteration of the path planning algorithm terminates if the distance between the control point on the robot end-effector and its goal position falls below a user-defined tolerance.

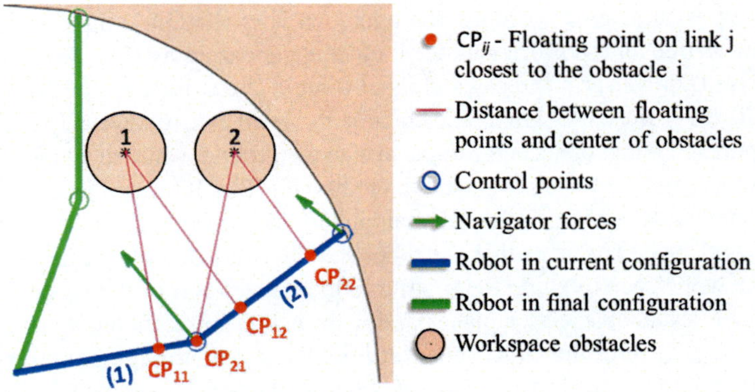

Fig. 2. Closest points (CP) on manipulator links to the workspace obstacles

3 Determination of Optimal Factors *k*

As illustrated in Fig. 1, the planner is highly sensitive to the factor k. In order to find the optimum factor k_i for the potential field of each of the m control point an optimization problem is defined in which k_i are the design variables. A weighted sum of the length of the paths between initial and goal positions for all control points is defined as the cost function,

$$\min_{k_1 \dots k_m} L = \sum_{j=1}^{m} w_j L_j(k_j, \mathbf{r}_j) \quad \text{with} \quad L_j = \sum_{u=1}^{M-1} \|\mathbf{r}_j^{u+1} - \mathbf{r}_j^{u}\| \tag{8}$$

Here M is the number of the waypoints through which the control points move and is equal to the number of iterations. Throughout the optimization process, if for a set of factors $k_1 \dots k_m$, three conditions as follow are not satisfied, a penalty is considered for this set of k factors to help the solver converge to the optimum values: Conditions are (1) The planner exceeds the maximum number of solving steps (defined by the user) or (2) the distance between the control point of the end-effector at the terminal solving step and its goal position exceeding a prescribed tolerance, or (3) any point on the robot links collides with an obstacle. In these cases the path length L is multiplied by an infinite value as a penalty. This optimization problem is solved using the genetic algorithm in MATLAB optimization toolbox (v8.4, 2014b, The MathWorks Inc., Natick, MA, USA), whereby parallel computing is used to increase the efficiency of the solver.

4 Results and Discussion

Two scenarios of path planning using the proposed algorithm are simulated and discussed in this section. The kinematics of the system is described by the joint coordinates, i.e. minimal coordinates. Inputs to the algorithm are the joint angles of the robot at the initial and final configurations, radius and position of the center of the obstacles modeled as spheres, the maximum number of iterations for the algorithm, and tolerance value to reach the end-effector goal position. The approach is applied on a three-link, $N = 3$, five degrees-of-freedom (dofs), right-side human upper extremity (Fig. 3a), comprising upper arm (red), lower arm (green) and hand (blue). The workspace of the robot is a sphere of 70 cm radius. This kinematic chain is modeled as a robot manipulator in 3D space. A spherical joint connects the upper arm to the ground, while revolute joints with parallel axes are modeled as elbow and wrist joints (Fig. 3a). If the end point of the hand link is considered as the end-effector, the manipulator has two redundant dofs. In addition to the real workspace obstacles that the robot should avoid, some virtual obstacles are added to the workspace to guide the robot into the correct part of the space. To define the position of virtual obstacles, a plane passing through 3 points including center of the spherical joint and tip of the end-effector at initial and goal configurations is defined. Normal vector of this plane is used to locate two virtual obstacles symmetrically on both sides of this plane.

4.1 Example 1 - Upper Extremity Abduction for Hand Raising

The task in this motion scenario is to reach from an initial configuration to the end-effector goal position (Fig. 3b) while avoiding a real obstacle in the workspace (O_1). As the manipulator has two redundant dofs, its configuration is not completely defined by the end-effector position. To guide the kinematic chain within the redundant dofs, two virtual obstacles are introduced ($VO_{1,2}$).

Since robot manipulators are usually free to move in the free space, virtual obstacles are not required. However, in the case of the human upper extremity, if these two virtual obstacles are not added to the potential field, the potential of the $-Y$ half space is lower than the $+Y$ half space (Fig. 3d) and the robot is guided into the $-Y$ half space. It is known that for a right-side human upper extremity the correct direction for abduction motion takes place in the $+Y$ half space.

This planner is sensitive to the initial configuration. If instead of zero angles for all dofs at the initial step (Fig. 3a), small initial angles toward the desired final configuration are applied (Fig. 4), the planner generates a different path in the $+Y$ and $-X$ octant of space instead of $+Y$ and $+X$ octant (compare Figs. 3b and 4).

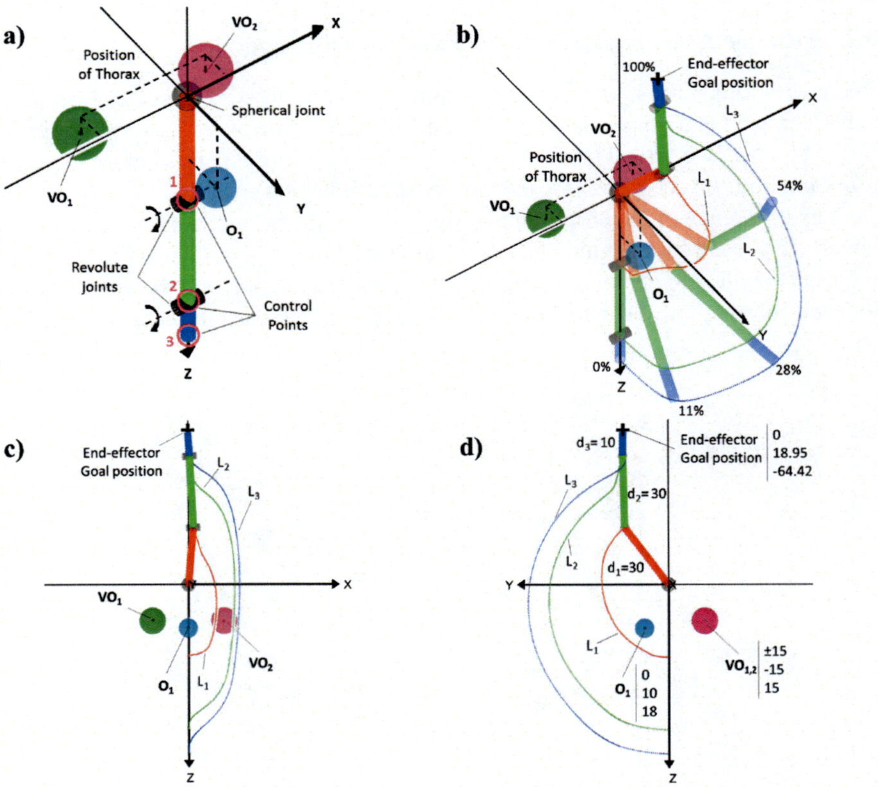

Fig. 3. Planned path for an upper extremity abduction motion for hand raising. **a)** Initial configuration **b)** planned path 3D view **c)** X-Z plane view **d)** Y-Z plane view. O, VO, L, and d stand for obstacle, virtual obstacles, path length, and link length, respectively.

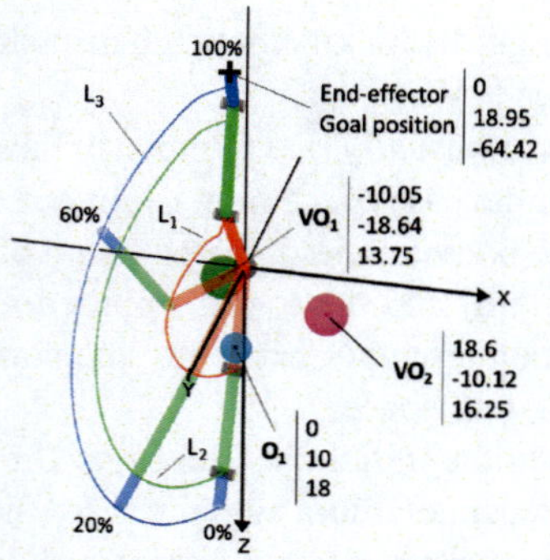

Fig. 4. Path planning dependency on the initial configuration. O and VO stand for obstacle and virtual obstacles, respectively. L refers to the path length.

4.2 Example 2 - Upper Extremity Flexion for Grasping

The goal in this motion scenario is that through an arm flexion, the end-effector as so-called human hand, grasps an object in the 3D space while avoiding two real obstacles ($O_{1,2}$, Fig. 5). Two virtual obstacles ($VO_{1,2}$) are placed as described before to guide the robot into the correct space (+X half space for arm flexion).

The difference of the proposed algorithm in this study compared to other studies in which NF is used for path planning of robot manipulators [4] is that artificial navigator forces are applied to all control points on the manipulator, not only the end-effector and for each control point a suitable factor k (in Eq. (1)) is designed through an optimization

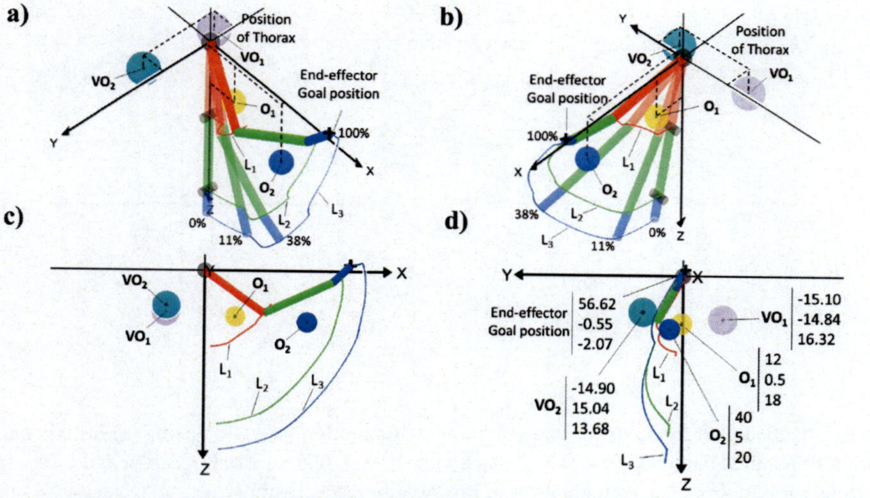

Fig. 5. Planned path for an upper extremity flexion motion for grasping. **a)** 3D view 1 **b)** 3D view 2 **c)** X-Z plane view **d)** Y-Z plane view. O and VO stand for obstacle and virtual obstacles, respectively. L refers to the path length.

according to Eq. (8). In this manner, both position of the end-effector and the posture of the manipulator are controlled at the same time.

If the identical values of k are used for all control points to define their potential fields, sum of the generated path lengths for all control points ΣL is 13.91% larger than the case that different values of k are used for each point (Fig. 6).

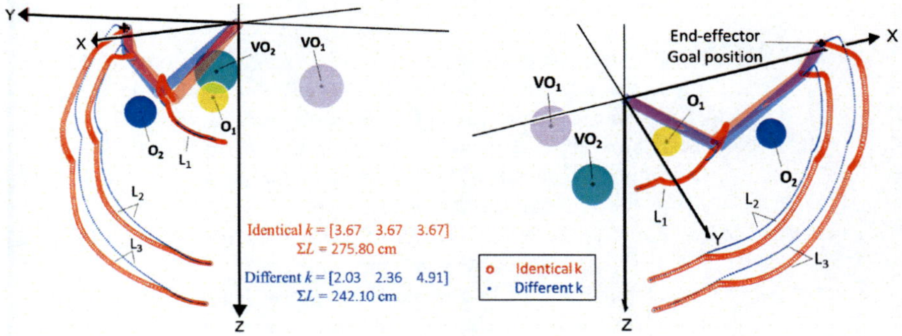

Fig. 6. Generated path using different or identical values of *k* for control points

Our proposed method has some limitations that have to be addressed in the refined versions of the algorithm. As explained in the upper extremity abduction example, for some motion scenarios especially where the end-effector has to move a large distance in the 3D space similar to the hand raising, virtual obstacles are required due to the redundant dofs to guide the robot into the correct space and the location of these obstacles are important. Another possible solution to address this problem is to use a set of spheres forming an approximate shape of the human thorax. Another limitation of this algorithm is that the goal positions of the control points $\mathbf{r}_{g,i}$ are required to form the NFs φ_i. This means that for each new goal position of the end-effector, an inverse kinematic problem has to be solved to find the goal position of the control points. The computation time required for solving the optimization problem to find the factors k_i for the potential field of control points should also be taken into account.

References

1. Khatib, O.: Real-time obstacle avoidance for manipulators and mobile robots. Int. J. Robot. Res. **5**, 396–404 (1986)
2. Zhang, T., Zhu, Y., Song, J.: Real-time motion planning for mobile robots by means of artificial potential field method in unknown environment. Ind. Robot. **37**, 384–400 (2010)
3. Rimon, E., Koditschek, D.E.: Exact robot navigation using artificial potential functions. IEEE Robot. Autom. **8**(1), 501–518 (1992)
4. Chan, C.C., Tsai, C.C.: Collision-free path planning based on new navigation function for an industrial robotic manipulator in human-robot coexistence environments. J. Chin. Inst. Eng. **43**(6), 508–518 (2020)
5. Khatib, O., Demircan, E., De Sapio, V., Sentis, L., Besier, T., Delp, S.: Robotics-based synthesis of human motion. J. Physiol. **103**(3–5), 211–219 (2009)
6. Spong, M.W., Hutchinson, S., Vidyasagar, M.: Robot Modeling and Control, pp. 232–233. Wiley, Chichester (2020)

Reactive Locomotion of a Hexapod for Navigation Across Irregular Ground

Joana Coelho[1(✉)], Bruno Dias[2], Gil Lopes[3], Fernando Ribeiro[4], and Paulo Flores[1]

[1] CMEMS-UMinho, Department of Mechanical Engineering, University of Minho, Campus de Azurém, 4800-058 Guimarães, Portugal
id8667@alunos.uminho.pt, pflores@dem.uminho.pt

[2] Center Algoritmi, Department of Informatics, University of Minho, Campus de Gualtar, 4710-057 Braga, Portugal
bruno.dias@di.uminho.pt

[3] University of Maia, Avenida Carlos de Oliveira Campos, 4475-690 Maia, Portugal
alopes@ismai.pt

[4] Center Algoritmi, Department of Industrial Electronics, University of Minho, Campus de Azurém, 4800-058 Guimarães, Portugal
fernando@dei.uminho.pt

Abstract. In controlled environments, the hexapod limbs actuation can be controlled as a closed system. However, the increase of the terrain complexity implies an adaptation of their trajectory based on the robot interactions with the environment. Thus, the implementation of terrain data to the legs actuation potentially improves the hexapod quasi-static stability in these scenarios. This paper presents an adaptive control system based on the limbs reactive behavior for navigation across complex environments. Through force sensors placed on the foot-tips, the model detects the foot-ground interactions and adjusts the limbs trajectory accordingly. Furthermore, to ensure that the robot posture remains stable throughout locomotion, an impedance control is implemented in each limb. The proposed control architecture was tested in an irregular ground through dynamic simulations with five different configurations. Through result analysis, an optimized model was achieved which reduces the oscillations of the torso and slippage of the feet when walking across obstacles.

1 Introduction

The six limbs of hexapods ensure them a higher static stability in comparison to other legged robots [4]. However, with the increase of the environment complexity, this ability is not sufficient to maintain an efficient locomotion and adequate torso posture. In fact, the system requires more information both about the environment and the robot internal state to adjust the limbs actuation accordingly. Based on the most recent publications about the control of hexapods, the majority still focuses on the generation of adaptive locomotion for indoor

O. Altuzarra and A. Kecskéméthy (Eds.): ARK 2022, SPAR 24, pp. 478–485, 2022.
https://doi.org/10.1007/978-3-031-08140-8_52

environments with uneven ground and proposes control strategies which resort to kinematic or dynamic models of the robot [2]. Thus, the generation of an adaptive control for unstructured environments remains an open question. In this case, the perception of the foot-ground interaction is important to plan the actuation of its legs according to the terrain topology. Thereby, Xia et al. [6] and Liu et al. [5] resorted to force sensors placed on the feet to evaluate the normal contact forces and adjust the limbs trajectory. Both studies aim at detecting a variation of the contact forces to switch the actuation of the legs. Along with this reactive behavior, it is important to ensure that the posture of the body remains unaltered with the irregularities of the ground. Thus, to avoid high attitude and height fluctuations of the torso, some research projects implemented virtual elastic elements to stabilize the feet positions with the normal contact forces [1,3]. On the other hand, a ground topology identification method and a virtual suspension model to the hexapod and foot-force compensation model was presented by Liu et al. [5], which requires a more accurate model of the robot interactions.

This work deals with a new approach for a proprioceptive adaptive design. For this purpose, a set of force sensors is implemented to achieve a reactive control of the limbs, changing their trajectory when an early contact force is detected. The advantage of this model is not requiring previous knowledge of the ground irregularity, as well as not resorting to a large amount of data from exteroceptive sensors to adjust locomotion. Furthermore, the posture of the hexapod is controlled using impedance models on each foot. The proposed control strategy was built as a remote Python API, and was tested and verified in CoppeliaSim. The results of the computational simulations performed on an irregular ground were compared against the results when a non-adaptive control method was used and could confirm the applicability of the proposed solution.

The remaining of this article is organized as follows. Section 2 describes the kinematic design of the hexapod selected for this study and Sect. 3 presents the proposed control architecture. Section 4 discusses the computational simulations and the obtained results. Finally, Sect. 5 presents some concluding remarks and future developments.

2 Model Description

This section presents the hexapod kinematic model, in order to describe its inverse kinematics. This analysis is important for the legs trajectory generation, in order to obtain the joints relative angular positions based on the imposed trajectory of the feet. The robot adopted is presented in Fig. 1(a). It contains three revolute joints per limb and an overall of 18 degrees of freedom. The radius of its hexagonal-shaped body is 0.0799 m. Each limb has three segments which in this work are denominated as coxa (l_1), femur (l_2) and tibia (l_3), and their dimensions are respectively, 0.050, 0.072, and 0.116 m. Its overall mass has a value of 1.347 kg.

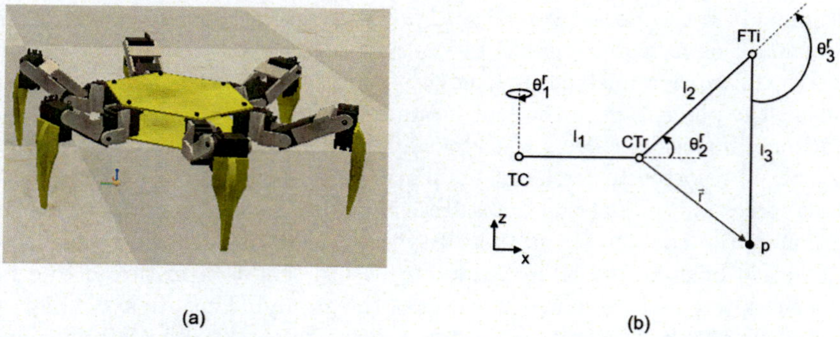

Fig. 1. Hexapod design: (a) portrays the model adopted for this study, and (b) presents the limbs kinematic design, with the relative angular position of each joint.

Figure 1(b) portrays the legs kinematic design of the hexapod. The adopted nomenclature for the joints is as follows: Thorax-Coxa (TC), Coxa-Trochanterofemur (CTr) and Femur-Tibia (FTi). Based on this representation, the value of θ_1^r is obtained through the following expression,

$$\theta_1^r = \arctan_2\left(p_y^{TC}, p_x^{TC}\right) \tag{1}$$

where p^{TC} is the relative position of the foot using the TC joint as reference. Since the controller uses the torso as reference to generate the feet coordinates, this variable is expressed as,

$$p^{TC} = (T_{TC}^{Torso})^{-1} p^{Torso} \tag{2}$$

where T_{TC}^{Torso} is the transformation matrix between the torso and the TC joint, and p^{Torso} is the foot coordinate with respect to the torso reference. With this transformation, the following relative angular positions can be obtained with respect to the first quadrant. Considering r, which is the norm vector between the foot and the CTr joint, the values for θ_2^r and θ_3^r can be obtained by,

$$\theta_2^r = \arccos\left(\frac{-l_3^2 + l_2^2 + r^2}{2 l_2 r}\right) - \arcsin\left(\frac{CTr_z^{Torso} - p_z^{Torso}}{r}\right) \tag{3}$$

$$\theta_3^r = -\pi + \arccos\left(\frac{-r^2 + l_2^2 + l_3^2}{2 l_2 l_3}\right). \tag{4}$$

3 Control Strategy

The hexapod locomotion control has the two hierarchical control layers presented in Fig. 2. The high-level control is responsible for the trajectory planning, while the low-level adjusts the motor commands according to the ground topology.

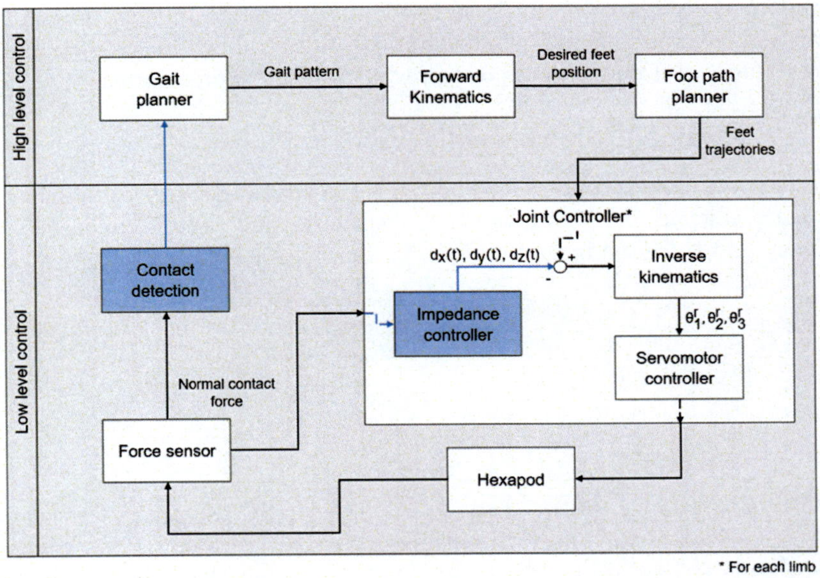

Fig. 2. Hierarchical control architecture.

Based on the gait planner, which provides data about the torso desired motion, the forward kinematics obtains the desired foot positions. Consequently, the footpath planner generates the adequate foothold trajectories according to the gait phase. During the swing phase, the limb searches for a new foothold position by executing a Bézier curve. Its intermediate coordinates are calculated using the desired leg stroke and step height. As for the stance phase, since the foot remains in contact with the ground, its actuation is as follows [4],

$$p(t) = \begin{bmatrix} p_x^0 \\ p_y^0 + \frac{S_y}{T}t \\ p_z^0 \end{bmatrix} \tag{5}$$

where p^0 is the initial foot position, S_y is the leg stroke along the Y-axis and T is the gait phase period. Using the desired foot position, the joints angular positions are obtained using the inverse kinematics. The reactive behavior comes with the contact detection module. Similarly, to the model presented by Xia et al. [6], the robot locomotion in an irregular ground is adjusted according to the early detection of foot-forces. For this purpose, the system reads the values from the force sensors placed on each foot-tip to detect irregularities in the normal contact forces. Thus, in the swing phase, if the normal force is higher than a threshold value then the trajectory of the limb stops and switches to the stance phase. Moreover, in an unstructured environment, the variable foot-force distribution and the different characteristics of the ground can cause deviations in the legs actuation [3]. Consequently, this implies an adjustment of the legs stiffness to respond to the changes of the system. Similarly, to the models presented by Irawan and Nonami [3] and Bjelonic et al. [1], this strategy adds a virtual mass-spring-damper system to each foot-tip, which can be expressed as,

$$-F_z(t) = M_{virt}\ddot{z}(t) + D_{virt}\dot{z}(t) + K_{virt}z(t) \tag{6}$$

where F_z is the normal contact force, M_{virt} is the virtual mass, D_{virt} is the virtual damping coefficient and K_{virt} is the virtual stiffness coefficient. The characteristic frequency and the damping ratio of the system [5] are, respectively,

$$\omega_n = \sqrt{\frac{K_{virt}}{M_{virt}}} \tag{7}$$

$$\xi = \frac{D_{virt}}{2\omega_n M_{virt}}. \tag{8}$$

4 Results and Discussion

The designed control strategy was implemented using the dynamic simulator CoppeliaSim. For the computational simulation setup, the Bullet physics library version 2.78, was selected, with a time-step of 50 ms. Since the control system is designed as a remote Python API, a Robot Operating System (ROS) framework establishes the communication between the software and the controller. To ensure real-time data acquisition, its rate is set 20 Hz.

Figure 3 presents the tested scenario. It includes six 0.1 × 0.1 m blocks with 0.02 m of height. The goal is to walk across the ground and adjust the legs trajectories, ensuring the body posture remains quasi-statically stable. To study the hexapod behavior, five different configurations for the control system, presented in Table 1 were selected. The first one consists of a non-adaptive model, without impedance controllers on the limbs and the next following cases considered values of ξ which could increase the system response, as suggested by Liu et al. [5]. In all simulations, the hexapod adopted a tripod gait, with swing and stance period of 1 s, a step height of 0.06 m and a stroke of 0.015 m. During the transition between gait phases, all six limbs remain in contact with the ground

Fig. 3. Simulation scenario where the potential obstacles are visible.

Table 1. Parameters of the five control configurations.

Case	Adaptive control	ω_n (s^{-1})	ξ
1	No	–	–
2	Yes	300	0.85
3	Yes	240	0.85
4	Yes	300	0.7
5	Yes	240	0.7

for a short time instant. Consequently, this event is not taken into consideration in this study. Thus, it is assumed that only three legs remain in contact with the ground, and the value of M_{virt} was set to one third of the system total mass, i.e., 0.4491 kg kg. The force threshold for the contact detection was set to 1.0 N.

The analysis of the roll and pitch angles variation of the body, which are portrayed in Fig. 4(a) and 4(b), leads to the conclusion that the first configuration has a higher perturbation of the hexapod posture, due to the lack of reflexive and impedance control. Additionally, the parameters selected for case 5 increased fluctuation of the posture parameters. This could be influenced by the value of ω_n, which may not be high enough to reduce resonance according to the implemented ξ. Thereby, this configuration is not considered an adequate solution to the tested scenario. Along with the body posture, the variation of the foothold height, and the consequent transition between the gait phases, also influences the torso oscillation in the Z-axis, which can decrease the system stability. From Fig. 4(c), when the hexapod has no control adaptation, its height has a higher variation, which is caused by the lack of response to the ground disturbances. Similarly, to the previous analysis, the fourth case presents the lowest variation in the Z-axis, which emphasizes the stability provided by this configuration.

To observe the influence of the reactive behavior, Fig. 5 compares the trajectory of one of the mid limbs throughout the simulation for the case 4 and the non-adaptive system. As it can be observed in the Y-axis, the limb walks across one of the steps between the position 0.15 and 0.30 m. Since the first method cannot absorb any impact caused by the normal contact forces, it continues to send the desired angular positions for the swing trajectory. As a consequence, the feet slippage and their position has a higher oscillation when surmounting this obstacle. On the contrary, the foot motion in the fourth experiment remains the same in this condition.

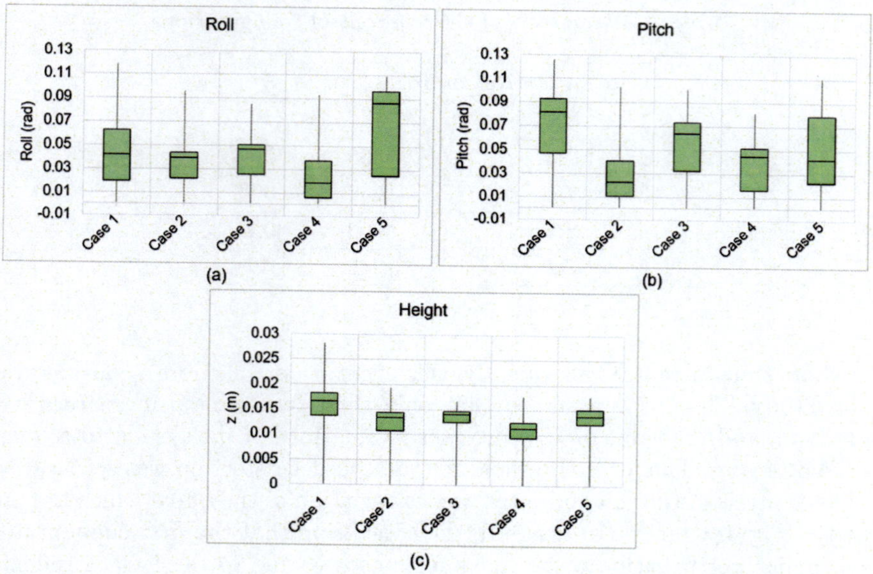

Fig. 4. Results obtained from the computational simulations: (a) roll angle, (b) pitch angle, and (c) torso height.

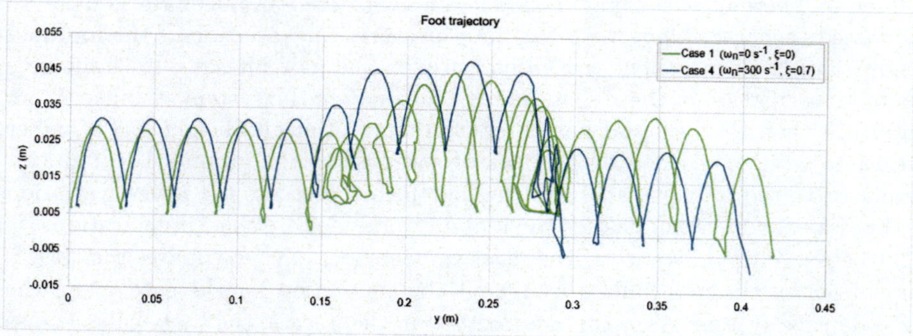

Fig. 5. Influence of the control on the limbs trajectory.

5 Conclusions

This paper focuses on the navigation of a hexapod in unstructured ground and discusses the proposed control architecture that relies on proprioceptive information and is inspired by the reactive behavior of insects to detect terrain obstacles and adjusted their gait. Moreover, to avoid the body posture oscillations caused by the foot-force distribution variation, each foot-tip has an impedance controller to decrease the contact force impact. The designed architecture was tested in five different configurations. In comparison to the non-adaptive controller, the configuration with a ω_n of $300\,\text{s}^{-1}$ and ξ of 0.7 provided the optimal response to the system disturbances, having the smaller fluctuations in terms of the roll, pitch and height of the torso. By analysis of the trajectories, it was also concluded that, without the reactive behavior, the feet are forced to try to reach a position which is not achievable. This increases foot slippage when walking across the step and, consequently, also increases the deviations of the feet positions. On the contrary, the trajectory of the limbs for the fourth model remains unchanged throughout the simulation, being clear that there is a transition of the limb trajectory when traversing the block. Considering the obtained results, the future stage of this work must test the proposed control strategy in more unstructured ground, to evaluate the torso posture control when the hexapod navigates across different slopes. Besides that, it is important to study the locomotion efficiency, by efficiently tuning the impedance control parameters to avoid body oscillation and feet slippage in different types of terrain.

Acknowledgements. The first author received funding through a doctoral scholarship from the Portuguese Foundation for Science and Technology (FCT) (Grant No. SFRH/BD/145818/2019), with funds from the Portuguese Ministry of Science, Technology and Higher Education and the European Social Fund through the Programa Operacional Regional Norte. This work has been supported by FCT within the R&D Units Project Scope: UIDB/00319/2020, UIDB/04436/2020 and UIDP/04436/2020.

References

1. Bjelonic, M., Kottege, N., Beckerle, P.: Proprioceptive control of an over-actuated hexapod robot in unstructured terrain. In: 2016 IEEE/RSJ International Conference on Intelligent Robots and Systems (IROS), pp. 2042–2049 (2016). https://doi.org/10.1109/IROS.2016.7759321
2. Coelho, J., Ribeiro, F., Dias, B., Lopes, G., Flores, P.: Trends in the control of hexapod robots: a survey. Robotics **10**(3), 100 (2021). https://doi.org/10.3390/robotics10030100
3. Irawan, A., Nonami, K.: Optimal impedance control based on body inertia for a hydraulically driven hexapod robot walking on uneven and extremely soft terrain. J. Field Robot. **28**(5), 690–713 (2011). https://doi.org/10.1002/rob.20404
4. Liu, Y., Fan, X., Ding, L., Wang, J., Liu, T., Gao, H.: Fault-tolerant tripod gait planning and verification of a hexapod robot. Appl. Sci. **10**(8), 2959 (2020). https://doi.org/10.3390/app10082959
5. Liu, Y., Wang, C., Zhang, H., Zhao, J.: Research on the posture control method of hexapod robot for rugged terrain. Appl. Sci. **10**(19), 6725 (2020). https://doi.org/10.3390/app10196725
6. Xia, H., Zhang, X., Zhang, H.: A new foot trajectory planning method for legged robots and its application in hexapod robots. Appl. Sci. **11**(19), 9217 (2021). https://doi.org/10.3390/app11199217

Trajectory Tracking by Fuzzy-Based Super-Twist Sliding Mode Control of a Parallel PnP Robot

Guanglei Wu[1], Chuangchuang Cui[1], and Huiping Shen[2](✉)

[1] School of Mechanical Engineering, Dalian University of Technology, Dalian 116024, China
gwu@dlut.edu.cn

[2] Research Center for Advanced Mechanism Theory, Changzhou University, Changzhou 213016, China
shp65@126.com

Abstract. This paper presents trajectory tracking control of a fast parallel robot for pick-and-place operations. Aiming at high tracking accuracy of the robot end-effector, fuzzy-based super-twist sliding mode control is designed and is evaluated by observing the joint dynamics, in comparison with the classical computed torque control and among others. The experimental results and comparison show the effectiveness of the developed control scheme, for fast parallel pick-and-place robots. The main contribution of this work is reflected in the integrated fuzzy algorithm and second-order sliding mode control for the model-based control design, with acceptable computational burden and trajectory tracking precision.

1 Introduction

High-speed parallel robots with Schönflies (i.e., SCARA) motion dedicate to fast pick-and-place (PnP) applications serving in many industrial sectors [1–4]. Such a robot usually work with fast movements, introducing the decreased position accuracy in turn due to the end-effector EE vibrations caused by frequent switchings of the active joint motions, thus, to ensure the EE positioning accuracy and dynamic stability, control design is of utmost importance. The focus of this work is to design the controller for high robot EE accuracy.

In the motion control of robot, highly advanced control algorithms are desired to be adopted for the controller deign for the expected dynamic performance. The advantages of fuzzy sliding mode variable structure control (FSMVSC) have been verified in terms of trajectory tracking accuracy [5], compared to computed-torque control (CTC) [6,7] and SMVSC [8,9]. On the other hand, due to the existence of

The reported work was in part supported by Natural Science Foundation of Liaoning Province (Grant No. 20180520028) and National Natural Science Foundation of China (Grant No. 51975062).

O. Altuzarra and A. Kecskémthy (Eds.): ARK 2022, SPAR 24, pp. 486–493, 2022.
https://doi.org/10.1007/978-3-031-08140-8_53

dynamic modeling errors and discontinuity of the switching function in the first-order sliding model control (SMC), frequent switching occurs to the system state on both sides of the SMC sliding surface, causing the frequent buffetings of the control variables in turn. Besides, the joint buffetings will lead to the induced vibrations of the overall robot. In light of this, two-order SMC (super-twist SMC) is applicable to enhance the dynamic control, as it can effectively overcome the variable buffeting problem by reducing the frequency of the switching function. Moreover, fuzzy algorithm [10,11] can be integrated into the SMVSC to handle the buffeting problem of the control variables, for the enhanced control stability. This work will deal with the control design to improve the trajectory tracking precision and the main contribution lies in the integrated fuzzy algorithm and second-order sliding mode control for the model-based control design.

This paper deals with controller design of a fast parallel robot for pick-and-place operations, aiming at high trajectory tracking precision. Fuzzy-based super-twist sliding mode control is designed, in comparison with the classical computed torque control and first-order sliding mode variable structure control. The experimental results and comparison show the effectiveness of the developed controller in terms of the trajectory tracking precision.

2 Parallel PnP Robot Under Study

Figure 1 depicts the CAD model of the parallel robot under study, with a non-fully symmetrical architecture and a screw-pair based end-effector. This robot admits a rectangular workspace for efficient use of shop-floor space and realizes the full-circle EE rotation for enhanced rotational capability [12], allowing more robots to be deployed side-by-side to improve the efficiency of the production lines [13].

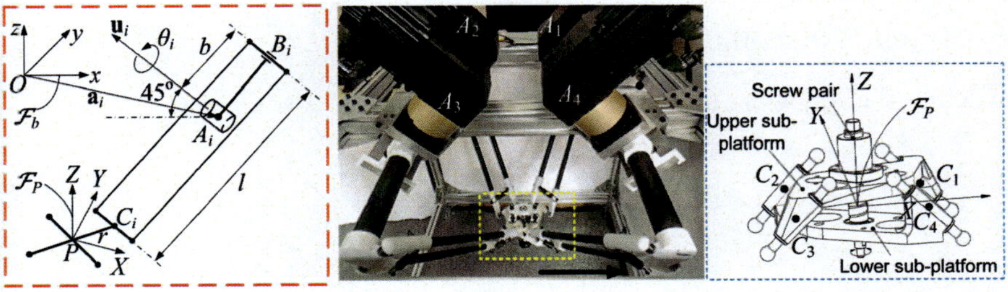

Fig. 1. Kinematic structure and parameters of the robot [5].

The reference coordinate frame $\mathcal{F}_b$ is built with the origin located at the geometric center of the base frame. Let $\mathbf{i}$, $\mathbf{j}$ and $\mathbf{k}$ stand for the unit vectors of x-, y- and z-axis, respectively, the axis of rotation of the ith actuated joint is denoted by $\mathbf{u}_i = R_y(\alpha_i)\mathbf{k}$, $-\alpha_{1(4)} = \alpha_{2(3)} = \pi/4$, and the Cartesian coordinates of points A_i, B_i and C_i, $i = 1, \ldots, 4$, are expressed in the reference frame as[1]

[1] sgn$(\cdot)$ stands for the sign function of $(\cdot)$, and mod( ) stands for the modulo operation.

$$\mathbf{a}_1 = -\mathbf{a}_3 = \left[\text{sgn}(\cos\beta_i)a_x \; \text{sgn}(\sin\beta_i)a_y \; 0\right]^T \tag{1a}$$

$$\mathbf{b}_i = b\left[\cos\alpha_i \cos\theta_i \; \sin\theta_i \; -\sin\alpha_i \cos\theta_i\right]^T \tag{1b}$$

$$\mathbf{c}_i = \left[r\cos\beta_i \; r\sin\beta_i \; \text{mod}(i,\, 2)h\phi/(2\pi)\right]^T + \mathbf{p} \tag{1c}$$

where $\mathbf{p} = \left[x \; y \; z\right]^T$ and ϕ represent EE position and orientation, respectively, and $\beta_i = (2i-1)\pi/4$. The end-effector pose can be represented by $\boldsymbol{\chi} = \left[\mathbf{p}^T \; \phi\right]^T$. The robot kinematics and dynamics have been well documented [12,13], which will be skipped. The dynamic model can be derived with the principle of virtual work as below

$$\boldsymbol{\tau} = \mathbf{M}(\boldsymbol{\theta})\ddot{\boldsymbol{\theta}} + \mathbf{C}(\boldsymbol{\theta},\, \dot{\boldsymbol{\theta}})\dot{\boldsymbol{\theta}} + \mathbf{G}(\boldsymbol{\theta}) \tag{2}$$

of which the details have been presented in the previous work [14]. Moreover, the geometric parameters of the robot in the unit of mm are designed as

$$a_x = 287, \; a_y = 132, \; b = 296, \; l = 600, \; r = 100, \; h = 10 \tag{3}$$

3 Model-Based Control Design

Based upon Eq. (2), and taking the frictions and time-varying disturbances into consideration, the robot dynamic model is rewritten as [15]

$$\boldsymbol{\tau} = \mathbf{M}(\boldsymbol{\theta})\ddot{\boldsymbol{\theta}} + \mathbf{C}(\boldsymbol{\theta},\, \dot{\boldsymbol{\theta}})\dot{\boldsymbol{\theta}} + \mathbf{G}(\boldsymbol{\theta}) + \mathbf{f}(\boldsymbol{\theta},\, t) \tag{4}$$

where $\mathbf{f}(\boldsymbol{\theta},\, t)$ characterizes the unmodeled errors. To suppress the joint buffetings caused by the frequent switching, integrated two-order SMC and fuzzy algorithm can be considered for improved dynamic control.

3.1 Super-Twist Sliding Mode Control

Let define the following state vector

$$\begin{cases} \mathbf{x}_1 = \boldsymbol{\theta} \\ \mathbf{x}_2 = \dot{\boldsymbol{\theta}} = \frac{\mathrm{d}}{\mathrm{d}t}\boldsymbol{\theta} \end{cases} \tag{5}$$

where $\boldsymbol{\theta}$ represents the vector of joint variables. For the design of sliding surface in the field of industrial robots, the following sliding surface is applicable

$$\mathbf{s} = \dot{\mathbf{e}} + \boldsymbol{\Lambda}\mathbf{e} \tag{6}$$

where $\mathbf{e} = \mathbf{x}_1 - \mathbf{x}_d$, $\mathbf{x}_d$ being the expected value for $\mathbf{x}_1$, and $\boldsymbol{\Lambda} = \text{diag}\left[\lambda_1 \; \lambda_2 \; \lambda_3 \; \lambda_4\right]$, $\lambda_i > 0$. Moreover, $\dot{\mathbf{e}} = \dot{\mathbf{x}}_1 - \dot{\mathbf{x}}_d$. Differentiating Eq. (6) w.r.t. time leads to the derivative sliding surface below

$$\dot{\mathbf{s}} = \ddot{\mathbf{e}} + \boldsymbol{\Lambda}\dot{\mathbf{e}} = \dot{\mathbf{x}}_2 - \ddot{\mathbf{x}}_d + \boldsymbol{\Lambda}(\mathbf{x}_2 - \dot{\mathbf{x}}_d) \tag{7}$$

and in accordance with Eq. (4), one obtains

$$\begin{cases} \dot{\mathbf{x}}_1 = \mathbf{x}_2 \\ \dot{\mathbf{x}}_2 = \mathbf{M}^{-1}(\mathbf{x}_1)\left[\mathbf{u} - \mathbf{C}(\mathbf{x}_1,\, \mathbf{x}_2)\mathbf{x}_2 - \mathbf{G}(\mathbf{x}_1) - \mathbf{f}(\mathbf{x}_1,\, t)\right] \end{cases} \tag{8}$$

where $\mathbf{u} = \boldsymbol{\tau}$ is the vector of control variables.

For the second step, the design of control law should meet $\mathbf{s} = \dot{\mathbf{s}} = \mathbf{0}$, thus, Eq. (7) can be written as

$$\dot{\mathbf{s}} = \mathbf{M}^{-1}(\mathbf{x}_1)\left[\mathbf{u} - \mathbf{C}(\mathbf{x}_1,\, \mathbf{x}_2)\mathbf{x}_2 - \mathbf{G}(\mathbf{x}_1) - \mathbf{f}(\mathbf{x}_1,\, t)\right] - \ddot{\mathbf{x}}_d + \boldsymbol{\Lambda}(\mathbf{x}_2 - \dot{\mathbf{x}}_d) \tag{9}$$

Based on the design principle of sliding mode control, the control variables $\mathbf{u}$ can be designed as

$$\mathbf{u} = \mathbf{u}_{eq} + \mathbf{u}_{sw} \tag{10}$$

where $\mathbf{u}_{eq}$ is the equivalent control law for the fundamental dynamics, which is derived from Eq. (2) as

$$\mathbf{u}_{eq} = \mathbf{M}(\mathbf{x}_1)\left(\ddot{\mathbf{x}}_d - \boldsymbol{\Lambda}\dot{\mathbf{e}}\right) + \mathbf{C}(\mathbf{x}_1,\, \mathbf{x}_2)\mathbf{x}_2 + \mathbf{G}(\mathbf{x}_1) \tag{11}$$

and $\mathbf{u}_{sw}$ is the switching function for the compensation of unmodeled errors $\mathbf{f}(\mathbf{x}_1,\, t)$. Compared to the traditional first-order SMC that may introduce discontinuity to the switching function, due to the sgn(·) function for the determination of the control variable signs, leading to the frequent chattering of the control variables, second-order sliding mode controller can be adopted to suppress the vibrations, for which can be set to

$$\begin{cases} \mathbf{u}_{sw} = -\mathbf{K}_1\mathbf{f}_{s1} + \boldsymbol{\nu} \\ \dot{\boldsymbol{\nu}} = -\mathbf{K}_2\mathbf{f}_{s2} \end{cases} \tag{12}$$

with

$$\mathbf{K}_1 = \mathrm{diag}\left[k_{11}\; k_{12}\; k_{13}\; k_{14}\right] \tag{13a}$$

$$\mathbf{K}_2 = \mathrm{diag}\left[k_{21}\; k_{22}\; k_{23}\; k_{24}\right] \tag{13b}$$

$$\mathbf{f}_{s1} = \left[f_{11}\; f_{12}\; f_{13}\; f_{14}\right]^T,\; f_{1i} = \sqrt{|s_i|}\mathrm{sgn}(s_i) \tag{13c}$$

$$\mathbf{f}_{s2} = \left[f_{21}\; f_{22}\; f_{23}\; f_{24}\right]^T,\; f_{2i} = \mathrm{sgn}(s_i) \tag{13d}$$

where k_{1i} and k_{2i} are the control variables. It is seen from Eq. (12) that the switching function is composed of two parts. The term $\sqrt{|s_i|}\mathrm{sgn}(s_i)$ in the first part is to ensure the continuity to reduce the chattering due to the discontinuous switch, and $\mathrm{sgn}(s_i)\int 1 dt$ in the second part is for error compensation in the switch. Moreover,

$$f_{2i} = \begin{cases} 1,\ \mathrm{sgn}(s_i) > 0 \\ -1,\ \mathrm{sgn}(s_i) \le 0 \end{cases},\; i = 1,\, ...,\, 4 \tag{14}$$

3.2 Fuzzy Control Design

Fuzzy algorithm can be integrated to make the control variables defined in the SMC adaptive to the external disturbances, by means of real-time adjustment, in order to improve the precision of trajectory tracking. In general, fuzzy control aims to keep the system state variables in the sliding surface and at the equilibrium position. Besides, it requires the reduced switching frequency of the control variables on both sides of the sliding surface, to decrease the frequent system chattering [16]. Thus, the two following control laws are designed:

- **Control Law of Switching Gains.** When the system is far away from the steady state, larger gains k_{1i} can make the system reach to the sliding surface quickly to reduce the responding time; conversely, when the system is close to the sliding surface, it requires smaller gains to buffer the switching functions.
- **Control Law of Sliding Surface Slope.** When the systems is far away from the sliding surface, the slope of the sliding surface can be increased to make the sliding surface close to the system state, in order to ensure the state vectors reaching to the sliding surface quickly; when the system state vectors are close to the sliding surface, thus, the current state is acceptable without changes.

In order to avoid numerous control variables, a common fuzzy controller of each tuning law is applied to the four joints. To ease the controller design, the input variables $(s_i, \dot{s}_i)$ and $(e_i, \dot{e}_i)$ are defined by identical fuzzy subsets, namely,

$$\{s_i, \dot{s}_i, e_i, \dot{e}_i\} \in \{NB\ NM\ NS\ Z\ PS\ PM\ PB\} \tag{15}$$

while the output variables are defined by the following fuzzy subsets

$$k_{1i} \in \{NB\ NM\ NS\ Z\ PS\ PM\ PB\} \tag{16a}$$

$$k_{2i} \in \{NH\ NVB\ NB\ NM\ NS\ NVS\ Z\ PVS\ PS\ PM\ PB\ PVB\ PH\} \tag{16b}$$

where the fuzzy subsets are defined in the fuzzy control language [10].

3.3 Integrated Controller Design

By integrating the second-order SMC and fuzzy algorithms, the control scheme of the robot, namely, fuzzy-based super-twist sliding mode control, is structured as depicted in Fig. 2. The function $\mathbf{f}(\boldsymbol{\chi}, \boldsymbol{\theta})$ is used to solve the robot kinematics between the motion parameters of the robot end-effector (input variables) $(\boldsymbol{\chi}, \dot{\boldsymbol{\chi}}, \ddot{\boldsymbol{\chi}})$ and the expected joint variables $(\mathbf{x}_d, \dot{\mathbf{x}}_d, \ddot{\mathbf{x}}_d)$. The dynamic Eq. (4) is written in the block "Robot Dynamics".

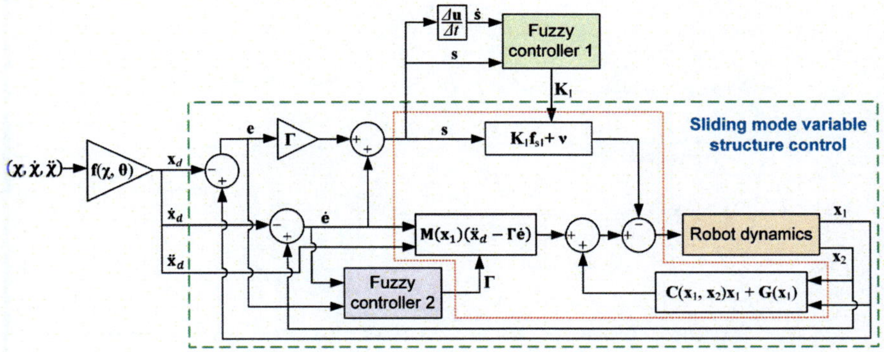

Fig. 2. The overall FSMVSC control scheme of the robot.

4 Experimental Evaluation of the Model-Based Control

In the dynamic control, the control variables, namely, the gains k_{1i} and k_{2i} in Eq. (12), are set to 20 and 1000, respectively, after multiple tunings by tracking different trajectories. Moreover, the bandwidth of all the fuzzy subsets equals to 0.1. Experimental implementation of robot dynamics is carried out along an industrial PnP testing trajectory with working frequency 2 Hz.

The experimental results of joint dynamics are illustrated in Figs. 3, along with the comparative study with the CTC and FSMVSC results, from which it is seen that the joints have significant shaking when the robot end-effector accelerates/decelerates during the pick and place periods, while, the dynamics profiles are much smoother when the robot moves with constant velocities except the acceleration profiles. Figure 3(a) show that the differences of the joint angular displacements and rates of the fuzzy based SMCs are much smaller than those of the CTC, particularly, the designed super-twist SMC, which means higher accuracy of trajectory tracking. Thus, the predicted position and orientation errors are shown in Fig. 3(b), showing that the tracking errors by the fuzzy-based super-twist SMC are the smallest.

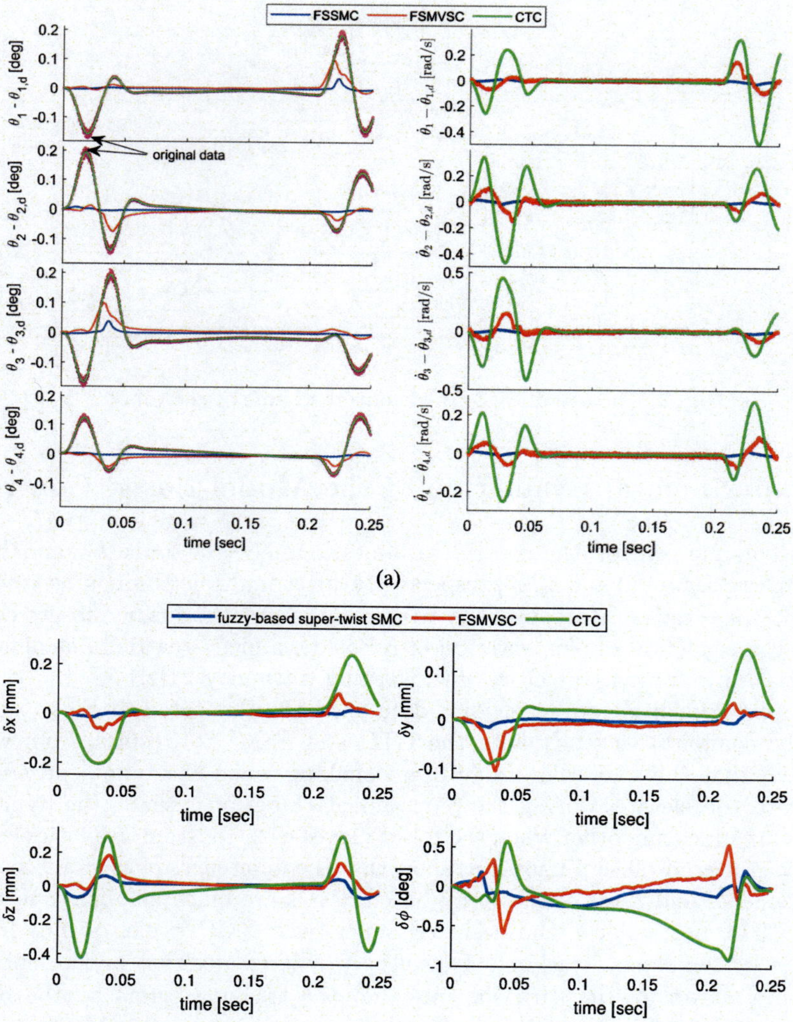

Fig. 3. Experimental motion errors after filtering with different control schemes: (a) joint displacement and rate errors; (b) predicted pose errors of robot end-effector.

5 Conclusions

This paper presents the model-based trajectory tracking control of a fast parallel PnP robot. Aiming to improve the tracking accuracy of the robot end-effector, an integrated fuzzy and two-order sliding mode controller, i.e., fuzzy-based super-twist sliding mode control, is designed. This two-order controller integrated with fuzzy subsets can effectively ensure the continuous switchings to reduce the gain switchings and chattering, together with error compensation in the switch. The main contribution of this work lies in the integrated fuzzy algorithm and second-order sliding mode control for the model-based control design, with acceptable computational burden and trajectory tracking precision.

References

1. Pierrot, F., Nabat, V., Company, O., Krut, S., Poignet, P.: Optimal design of a 4-DOF parallel manipulator: from academia to industry. IEEE Trans. Robot. **25**(2), 213–224 (2009)
2. Altuzarra, O., Şandru, B., Pinto, Ch., Petuya, V.: A symmetric parallel Schönflies-motion manipulator for pick-and-place operations. Robotica **29**, 853–862 (2011)
3. Belzile, B., Eskandary, P.K., Angeles, J.: Workspace determination and feedback control of a pick-and-place parallel robot: analysis and experiments. IEEE Robot. Autom. Lett. **5**(1), 40–47 (2020)
4. Gosselin, C., Isaksson, M., Marlow, K., Laliberté, T.: Workspace and sensitivity analysis of a novel nonredundant parallel SCARA robot featuring infinite tool rotation. IEEE Robot. Autom. Lett. **1**(2), 776–783 (2016)
5. Wu, G., Zhang, X., Zhu, L., Lin, Z., Liu, J.: Fuzzy sliding mode variable structure control of a high-speed parallel PnP robot. Mech. Mach. Theor. **162**, 104349 (2021)
6. Vivas, A., Poignet, P.: Predictive functional control of a parallel robot. Control. Eng. Pract. **13**(7), 863–874 (2005)
7. Liu, K., Xu, J., Ge, Z., Wang, Y., Zhao, D., Lu, Y.: Robust control of 3-DOF parallel robot driven by PMAs based on nominal stiffness model. Adv. Robot. **31**(10), 531–543 (2017)
8. Begon, P., Pierrot, F., Dauchez, P.: Fuzzy sliding mode control of a fast parallel robot. In: Proceedings of the 1995 IEEE International Conference on Robotics and Automation, vol. 1, pp. 1178–1183. IEEE (1995)
9. Fei, J., Ding, H.: Adaptive sliding mode control of dynamic system using RBF neural network. Nonlinear Dyn. **70**(2), 1563–1573 (2012)
10. Zadeh, L.A.: Fuzzy algorithms. Inf. Control **12**(2), 94–102 (1968)
11. Kayacan, E., Kayacan, E., Ramon, H., Kaynak, O., Saeys, W.: Towards Agrobots: trajectory control of an autonomous tractor using type-2 fuzzy logic controllers. IEEE/ASME Trans. Mechatron. **20**(1), 287–298 (2015)
12. Krut, S., Company, O., Nabat, V., Pierrot, F.: Heli4: a parallel robot for scara motions with a very compact traveling plate and a symmetrical design. In: 2006 IEEE/RSJ International Conference on Intelligent Robots and Systems, Beijing, China, 9–15 October 2006, pp. 1656–1661 (2006)
13. Wu, G., Bai, S., Hjørnet, P.: Architecture optimization of a parallel Schönflies-motion robot for pick-and-place applications in a predefined workspace. Mech. Mach. Theor. **106**, 148–165 (2016)
14. Lin, Z., Cui, C., Wu, G.: Dynamic modeling and torque feedforward based optimal fuzzy PD control of a high-speed parallel manipulator. J. Robot. Control **2**(6), 527–538 (2021)
15. Baruh, H.: Applied Dynamics. CRC Press, Boca Raton (2015)
16. Qiu, Z., Han, J., Liu, J.: Experiments on fuzzy sliding mode variable structure control for vibration suppression of a rotating flexible beam. J. Vib. Control **21**(2), 343–358 (2015)

Author Index

O. Altuzarra and A. Kecskémthy (Eds.): ARK 2022, SPAR 24, pp. 495–497, 2022.
https://doi.org/10.1007/978-3-031-08140-8